INTEGRAL TRANSFORMS and THEIR APPLICATIONS

INTEGRAL TRANSFORMS and THEIR APPLICATIONS

Lokenath Debnath
Professor and Chair of Mathematics and
Professor of Mechanical and Aerospace Engineering
University of Central Florida
Orlando, Florida

CRC Press
Boca Raton   New York   London   Tokyo

QA
432
D36
1995

Library of Congress Cataloging-in-Publication Data

Debnath, Lokenath.
Integral transforms and their applications / Lokenath Debnath.
  p. cm.
Includes bibliographical references and index.
ISBN 0-8493-9458-9 (alk. paper)
  1. Integral transforms. I. Title.
QA432.D36 1995
515.723--dc20
                                                                                95-10552
                                                                                      CIP

This book contains information obtained from authentic and highly regarded sources. Reprinted material is quoted with permission, and sources are indicated. A wide variety of references are listed. Reasonable efforts have been made to publish reliable data and information, but the authors and the publisher cannot assume responsibility for the validity of all materials or for the consequences of their use.

Neither this book nor any part may be reproduced or transmitted in any form or by any means, electronic or mechanical, including photocopying, microfilming, and recording, or by any information storage or retrieval system, without prior permission in writing from the publisher.

CRC Press, Inc.'s consent does not extend to copying for general distribution, for promotion, for creating new works, or for resale. Specific permission must be obtained in writing from CRC Press for such copying.

Direct all inquiries to CRC Press, Inc., 2000 Corporate Blvd., N.W., Boca Raton, Florida 33431.

© 1995 by CRC Press, Inc.

No claim to original U.S. Government works
International Standard Book Number 0-8493-9458-9
Library of Congress Card Number 95-10552
Printed in the United States of America 1 2 3 4 5 6 7 8 9 0
Printed on acid-free paper

TO MY WIFE

SADHANA

with gratitude and admiration

Contents

Preface

1. Integral Transforms — 1

    1.1 Brief Historical Introduction — 1
    1.2 Basic Concepts and Definitions — 3

2. Fourier Transforms — 5

    2.1 Introduction — 5
    2.2 The Fourier Integral Formulas — 5
    2.3 Definition of the Fourier Transform and Examples — 7
    2.4 Basic Properties of Fourier Transforms — 14
    2.5 Applications of Fourier Transforms to Ordinary Differential Equations — 25
    2.6 Solutions of Integral Equations — 29
    2.7 Solutions of Partial Differential Equations — 32
    2.8 Fourier Cosine and Sine Transforms with Examples — 50
    2.9 Properties of Fourier Cosine and Sine Transforms — 52
    2.10 Applications of Fourier Cosine and Sine Transforms to Partial Differential Equations — 54
    2.11 Evaluation of Definite Integrals — 58
    2.12 Applications of Fourier Transforms in Mathematical Statistics — 60
    2.13 Multiple Fourier Transforms and Their Applications — 64
    2.14 Exercises — 72

3. Laplace Transforms — 83

    3.1 Introduction — 83
    3.2 Definition of the Laplace Transform and Examples — 83
    3.3 Existence Conditions for the Laplace Transform — 87
    3.4 Basic Properties of Laplace Transforms — 88

	3.5	The Convolution Theorem and Properties of Convolution	92
	3.6	Differentiation and Integration of Laplace Transforms	96
	3.7	The Inverse Laplace Transform and Examples	98
	3.8	Tauberian Theorems and Watson's Lemma	109
	3.9	Laplace Transforms of Fractional Integrals and Fractional Derivatives	113
	3.10	Exercises	117

4. Applications of Laplace Transforms — 123

	4.1	Introduction	123
	4.2	Solutions of Ordinary Differential Equations	123
	4.3	Partial Differential Equations, and Initial and Boundary Value Problems	147
	4.4	Solutions of Integral Equations	159
	4.5	Solutions of Boundary Value Problems	163
	4.6	Evaluation of Definite Integrals	166
	4.7	Solutions of Difference and Differential-Difference Equations	168
	4.8	Applications of the Joint Laplace and Fourier Transform	173
	4.9	Summation of Infinite Series	183
	4.10	Exercises	185

5. Hankel Transforms — 193

	5.1	Introduction	193
	5.2	The Hankel Transform and Examples	193
	5.3	Operational Properties of the Hankel Transform	195
	5.4	Applications of Hankel Transforms to Partial Differential Equations	198
	5.5	Exercises	205

6. Mellin Transforms — 211

	6.1	Introduction	211
	6.2	Definition of the Mellin Transform and Examples	211
	6.3	Basic Operational Properties	214
	6.4	Applications of Mellin Transforms	218
	6.5	Mellin Transforms of the Weyl Fractional Integral and the Weyl Fractional Derivative	222
	6.6	Application of Mellin Transforms to Summation of Series	227
	6.7	Generalized Mellin Transforms	229
	6.8	Exercises	232

7. Hilbert and Stieltjes Transforms — 237

- 7.1 Introduction — 237
- 7.2 Definition of the Hilbert Transform and Examples — 237
- 7.3 Basic Properties of Hilbert Transforms — 239
- 7.4 Hilbert Transforms in the Complex Plane — 242
- 7.5 Applications of Hilbert Transforms — 243
- 7.6 Asymptotic Expansions of One-Sided Hilbert Transforms — 248
- 7.7 Definition of the Stieltjes Transform and Examples — 250
- 7.8 Basic Operational Properties of Stieltjes Transforms — 252
- 7.9 Inversion Theorems for Stieltjes Transforms — 254
- 7.10 Applications of Stieltjes Transforms — 257
- 7.11 The Generalized Stieltjes Transform — 259
- 7.12 Basic Properties of the Generalized Stieltjes Transform — 260
- 7.13 Exercises — 261

8. Finite Fourier Cosine and Sine Transforms — 265

- 8.1 Introduction — 265
- 8.2 Definitions of the Finite Fourier Sine and Cosine Transforms and Examples — 265
- 8.3 Basic Properties of Finite Fourier Sine and Cosine Transforms — 267
- 8.4 Applications of Finite Fourier Sine and Cosine Transforms — 272
- 8.5 Multiple Finite Fourier Transforms and Their Applications — 277
- 8.6 Exercises — 280

9. Finite Laplace Transforms — 283

- 9.1 Introduction — 283
- 9.2 Definition of the Finite Laplace Transform and Examples — 283
- 9.3 Basic Operational Properties of the Finite Laplace Transform — 288
- 9.4 Applications of Finite Laplace Transforms — 290
- 9.5 Tauberian Theorems — 294
- 9.6 Exercises — 294

10. Z Transforms — 295

- 10.1 Introduction — 295
- 10.2 Dynamic Linear Systems and Impulse Response — 295
- 10.3 Definition of the Z Transform and Examples — 298
- 10.4 Basic Operational Properties — 301

10.5	The Inverse Z Transform and Examples	306
10.6	Applications of Z Transforms to Finite Difference Equations	308
10.7	Summation of Infinite Series	311
10.8	Exercises	313

11. Finite Hankel Transforms 317

11.1	Introduction	317
11.2	Definition of the Finite Hankel Transform and Examples	317
11.3	Basic Operational Properties	319
11.4	Applications of Finite Hankel Transforms	319
11.5	Exercises	323

12. Legendre Transforms 325

12.1	Introduction	325
12.2	Definition of the Legendre Transform and Examples	325
12.3	Basic Operational Properties of Legendre Transforms	328
12.4	Applications of Legendre Transforms to Boundary Value Problems	333
12.5	Exercises	335

13. Jacobi and Gegenbauer Transforms 337

13.1	Introduction	337
13.2	Definition of the Jacobi Transform and Examples	337
13.3	Basic Operational Properties	339
13.4	Applications of Jacobi Transforms to the Generalized Heat Conduction Problem	340
13.5	The Gegenbauer Transform and its Basic Operational Properties	341
13.6	Application of the Gegenbauer Transform	344

14. Laguerre Transforms 345

14.1	Introduction	345
14.2	Definition of the Laguerre Transform and Examples	345
14.3	Basic Operational Properties	348
14.4	Applications of Laguerre Transforms	352
14.5	Exercises	354

15. Hermite Transforms — 355

    15.1 Introduction — 355
    15.2 Definition of the Hermite Transform and Examples — 355
    15.3 Basic Operational Properties — 358
    15.4 Exercises — 365

Appendix A   Some Special Functions and Their Properties — 367

    A-1 Gamma, Beta, and Error Functions — 367
    A-2 Bessel and Airy Functions — 372
    A-3 Legendre and Associated Legendre Functions — 377
    A-4 Jacobi and Gegenbauer Polynomials — 379
    A-5 Laguerre and Associated Laguerre Functions — 383
    A-6 Hermite and Weber-Hermite Functions — 385

Appendix B   Tables of Integral Transforms — 387

    Table B-1 Fourier Transforms — 387
    Table B-2 Fourier Cosine Transforms — 391
    Table B-3 Fourier Sine Transforms — 393
    Table B-4 Laplace Transforms — 395
    Table B-5 Hankel Transforms — 400
    Table B-6 Mellin Transforms — 403
    Table B-7 Hilbert Transforms — 406
    Table B-8 Stieltjes Transforms — 409
    Table B-9 Finite Fourier Cosine Transforms — 413
    Table B-10 Finite Fourier Sine Transforms — 415
    Table B-11 Finite Laplace Transforms — 417
    Table B-12 Z Transforms — 420
    Table B-13 Finite Hankel Transforms — 422

Answers and Hints to Selected Exercises — 423

Bibliography — 441

Index — 449

Preface

Historically, the concept of an integral transform originated from the celebrated Fourier integral formula. The importance of integral transforms is that they provide powerful operational methods for solving initial value problems and initial-boundary value problems for linear differential and integral equations. In fact, one of the main impulses for the development of the operational calculus of integral transforms was the study of differential and integral equations arising in applied mathematics, mathematical physics, and engineering science; it was in this setting that integral transforms arose and achieved their early successes. With ever greater demand for mathematical methods to provide both theory and applications for science and engineering, the utility and interest of integral transforms seems more clearly established than ever. In spite of the fact that integral transforms have many mathematical and physical applications, their use is still predominant in advanced study and research. Keeping these features in mind, our main goal in this book is to provide a systematic exposition of the basic properties of various integral transforms and their applications to the solution of boundary and initial value problems in applied mathematics, mathematical physics, and engineering. In addition, the operational calculus of integral transforms is applied to integral equations, difference equations, fractional integrals and fractional derivatives, summation of infinite series, evaluation of definite integrals, and problems of probability and statistics.

There appear to be many books available for students studying integral transforms with applications. Some are excellent but too advanced for the beginner. Some are too elementary or have limited scope. Some are out of print. While teaching transform methods, operational mathematics, and/or mathematical physics with applications, the author has had difficulty choosing textbooks to accompany the lectures. This book, which was developed as a result of many years of experience teaching advanced undergraduates and first-year graduate students in mathematics, physics, and engineering, is an attempt to meet that need. It is based essentially on a set of mimeographed lecture notes developed for courses given by the author at the University of Central Florida, East Carolina University, and the University of Calcutta.

This book is designed as an introduction to the theory and applications of integral transforms to problems in linear differential equations, and to boundary and initial value problems in partial differential equations. It is appropriate for a one-semester course. There are two basic prerequisites for the course: a standard calculus sequence and ordinary differential equations. The book assumes only a limited knowledge of complex variables and contour integration, partial differential equations, and continuum mechanics. Many new examples of applications dealing with problems in applied mathematics, physics, chemistry, biology, and engineering are included. It is *not* essential for the reader to know everything about these topics, but limited knowledge of at least some of them would be useful. Besides, the book is intended to serve as a reference work for those seriously interested in advanced study and research in the subject, whether for its own sake or for its applications to other fields of applied mathematics, mathematical physics, and engineering.

The first chapter gives a brief historical introduction and the basic ideas of integral transforms. The second chapter deals with the theory and applications of Fourier transforms, and of Fourier cosine and sine transforms. Important examples of applications of interest in applied mathematics, physics, statistics, and engineering are included. The theory and applications of Laplace transforms are discussed in Chapters 3 and 4 in considerable detail. The fifth chapter is concerned with the operational calculus of Hankel transforms with applications. Chapter 6 gives a detailed treatment of Mellin transforms and its various applications. Included are Mellin transforms of the Weyl fractional integral, Weyl fractional derivatives, and generalized Mellin transforms. Hilbert and Stieltjes transforms and their applications are discussed in Chapter 7.

Chapter 8 provides a short introduction to finite Fourier cosine and sine transforms and their basic operational properties. Applications of these transforms are also presented. The finite Laplace transform and its applications to boundary value problems are included in Chapter 9. Chapter 10 deals with a detailed theory and applications of Z transforms.

Chapter 12 is devoted to the operational calculus of Legendre transforms and their applications to boundary value problems in potential theory. Jacobi and Gegenbauer transforms and their applications are included in Chapter 13. Chapter 14 deals with the theory and applications of Laguerre transforms. The final chapter is concerned with the Hermite transform and its basic operational properties including the Convolution Theorem. Most of the material of these chapters has been developed since the early sixties and appears here in book form for the first time.

The book includes two important appendices. The first one deals with several special functions and their basic properties. The second appendix includes *thirteen* short tables of integral transforms. Many standard texts and reference books and a set of selected classic and recent research papers are included in the Bibliography that will be very useful for the reader interested in learning more about the subject.

The book contains 750 worked examples, applications, and exercises which include some that have been chosen from many standard books as well as recent papers. It is hoped that they will serve as helpful self-tests for understanding of the theory and mastery of the transform methods. These examples of applications and exercises were chosen from the areas of differential and difference equations, electric circuits and networks, vibration and wave propagation, heat conduction in solids, quantum mechanics, fractional calculus and fractional differential equations, dynamical systems, signal processing, integral equations, physical chemistry, mathematical biology, probability and statistics, and solid and fluid mechanics. This varied number of examples and exercises should provide something of interest for everyone. The exercises truly complement the text and range from the elementary to the challenging. Answers and hints to many selected exercises are provided at the end of the book.

This is a *text* and a *reference* book designed for use by the student and the reader of mathematics, science, and engineering. A serious attempt has been made to present almost all the standard material, and some new material as well. Those interested in more advanced rigorous treatment of the topics covered may consult standard books and treatises by Churchill, Doetsch, Sneddon, Titchmarsh, and Widder listed in the Bibliography. Many ideas, results, theorems, methods, problems, and exercises presented in this book are either motivated by or borrowed from the works cited in the Bibliography. The author wishes to acknowledge his gratitude to the authors of these works.

This book is designed as a new source for both classical and modern topics dealing with integral transforms and their applications for the future development of this useful subject. Its main features are:

1. A systematic mathematical treatment of the theory and method of integral transforms that gives the reader a clear understanding of the subject and its varied applications.
2. A detailed and clear explanation of every concept and method which is introduced, accompanied by carefully selected worked examples, with special emphasis being given to those topics in which students experience difficulty.
3. A wide variety of diverse examples of applications carefully selected from areas of applied mathematics, mathematical physics, and engineering science to provide motivation, and to illustrate how operational methods can be applied effectively to solve them.
4. A broad coverage of the essential standard material on integral transforms and their applications together with some new material that is *not* usually covered in familiar texts or reference books.
5. Most of the recent developments in the subject since the early sixties appear here in book form for the first time.
6. A wide spectrum of exercises has been carefully selected and included at the end of each chapter so that the reader may further develop both manipulative skills in the applications of integral transforms and a deeper insight into the subject.
7. Two appendices have been included in order to make the book self-contained.
8. Answers and hints to selected exercises are provided at the end of the book for additional help to students.
9. An updated bibliography is included to stimulate new interest in future study and research.

In preparing the book, the author has been encouraged by and has benefited from the helpful comments/criticism of a number of graduate students and faculty of several universities in the United States, Canada, and India. The author expresses his grateful thanks to all these individuals for their interest in the book. My special thanks to Jackie Callahan and Ronee Trantham who typed the manuscript and cheerfully put up with constant changes and revisions. In spite of the best efforts of everyone involved, some typographical errors doubtlessly remain. I do hope that these are both few and obvious, and will cause minimal confusion. The author also wishes to thank his friends and colleagues including Drs. Sudipto Roy Choudhury and Carroll A. Webber for their interest and help during the preparation of the book. Finally, the author wishes to express his special thanks to Dr. Wayne Yuhasz, Executive Editor, and the staff of CRC Press for their help and cooperation. I am also deeply indebted to my wife, Sadhana, for all her understanding and tolerance while the book was being written.

Lokenath Debnath
University of Central Florida

Author

Lokenath Debnath is Professor and Chair of the Department of Mathematics, and Professor of Mechanical and Aerospace Engineering at the University of Central Florida in Orlando.

Professor Debnath received his M.Sc. and Ph.D. degrees in pure mathematics from the University of Calcutta, and obtained his D.I.C. and Ph.D. degrees in applied mathematics from the Imperial College of Science and Technology, University of London. He was a Senior Research Fellow at the Department of Applied Mathematics and Theoretical Physics at the University of Cambridge and has had several visiting appointments at the University of Oxford, Florida State University, University of Maryland, and the University of Calcutta. He was Acting Chair of the Department of Statistics at the University of Central Florida and has served at East Carolina University as Professor of Mathematics and Professor of Physics.

Among many other honors and awards, he has received a Senior Fulbright Fellowship and an NSF Scientist award to visit India for lectures and research. He was a University Grants Commission Research Professor at the University of Calcutta and was elected President of the Calcutta Mathematical Society for a period of three years. He has served as a speaker of the SIAM Visiting Lecturer Program. He also has served as Organizer of several professional meetings and conferences at regional, national, and international levels; and as Director of three NSF-CBMS research conferences at the University of Central Florida and East Carolina University.

Dr. Debnath is author or co-author of seven graduate level books and research monographs, including *Introduction to Hilbert Spaces with Applications, Nonlinear Water Waves, Continuum Mechanics* published by Academic Press and *Partial Differential Equations for Scientists and Engineers* published by North Holland. He has also edited four research monographs including *Nonlinear Waves* published by Cambridge University Press. He is an author or co-author of over 250 research papers in pure and applied mathematics, including applied partial differential equations, integral transforms and special functions, solid and fluid mechanics, linear and nonlinear waves, solitons, magnetohydrodynamics, unsteady boundary layers, dynamics of oceans, and stability theory.

Professor Debnath is a member of many scientific organizations at both national and international levels. He has been a member of the editorial boards of several refereed journals, and he currently serves on the Editorial Board of the *Bulletin of the Calcutta Mathematical Society, Integral Transforms and Special Functions,* and *International Journal of Mathematics and Statistics.* He is the current and founding Managing Editor of the *International Journal of Mathematics and Mathematical Sciences.*

Dr. Debnath has delivered ten invited lectures at national and international meetings, has presented over 75 research papers at national and international conferences, and given over 150 seminar and colloquium lectures at universities and institutes in the United States and abroad.

Chapter 1

Integral Transforms

1.1 Brief Historical Introduction

The operational calculus may be considered as essentially a discovery of the first quarter of the nineteenth century. However, it was G.W. Leibnitz (1646-1716) who first introduced the idea of a symbolic method in calculus. Subsequently, both J.L. Lagrange (1736-1813) and P.S. Laplace (1749-1827) made considerable contributions to symbolic methods. Similar methods were employed by S.D. Poisson (1781-1840) for the solutions of the equation of heat conduction in three dimensions and the wave equation.

A major step forward in the development of a first systematic operational calculus was taken by both J. Fourier (1768-1830) and A.L. Cauchy (1789-1857). It is generally believed that the concept of an integral transform originated from the famous Fourier Integral Theorem as stated in Fourier's monumental treatise entitled *La Théorie Analytique de la Chaleur* that provided the basis for the modern mathematical theory of heat conduction. It was the work of Cauchy that contained the exponential form of the Fourier Integral Theorem as

$$f(x) = \frac{1}{2\pi} \int_{-\infty}^{\infty} \int_{-\infty}^{\infty} e^{ik(x-y)} f(y) \, dy \, dk . \tag{1.1.1}$$

Cauchy's work also contained the following formula for functions of the operator D:

$$\phi(D) f(x) = \frac{1}{2\pi} \int_{-\infty}^{\infty} \int_{-\infty}^{\infty} \phi(ik) e^{ik(x-y)} f(y) \, dy \, dk . \tag{1.1.2}$$

This is essentially the modern form of the operational calculus. His famous treatise entitled *Memoire sur l'emploi des equations symboliques* provided a fairly rigorous description of symbolic methods. The deep significance of the Fourier Integral Theorem was recognized by mathematicians and mathematical physicists of the nineteenth and twentieth centuries. Indeed, this theorem is regarded as one of the most fundamental results of modern mathematical analysis and has widespread physical and engineering applications.

It was Oliver Heaviside (1850-1925) who recognized the power and success of operational calculus and first used the operational method as a powerful and effective tool for the solutions of telegraph equation and the second order hyperbolic partial differential equations with constant coefficients. In his two papers entitled "On Operational Methods in Physical Mathematics," Parts I and II, published in *The Proceedings of the Royal Society*, London, in 1892 and 1893, Heaviside developed operational methods. His 1899 book on *Electromagnetic Theory* also contained the use and application of the operational methods to the analysis of electrical circuits or networks. Heaviside replaced the differential operator $D \equiv \dfrac{d}{dt}$ by p and treated the latter as an element of the ordinary laws of algebra. The development of his operational methods paid little attention to questions of mathematical rigor. The widespread use of Heaviside's methods prior to its vindication by the theory of the Fourier or Laplace transforms created a lot of controversy. This was similar to the controversy put forward against the widespread use of the delta function as one of the most useful mathematical devices in Dirac's logical formulation of quantum mechanics during the 1920s. In fact, P.A.M. Dirac (1902-1984) said: "All electrical engineers are familiar with the idea of a pulse, and the δ-function is just a way of expressing a pulse mathematically." Dirac's study of Heaviside's operator calculus in electromagnetic theory, his training as an electrical engineer, and his deep knowledge of the modern theory of electrical pulses seemed to have a tremendous impact on his ingenious development of modern quantum mechanics.

Apparently, the ideas of operational methods in both Fourier's and Cauchy's work inspired Heaviside to develop his new but less rigorous operational mathematics. In spite of the striking success of Heaviside's calculus as one of the most useful mathematical methods, contemporary mathematicians hardly recognized Heaviside's work in his lifetime, primarily due to lack of mathematical rigor. In his lecture on Heaviside and the Operational Calculus at the Birth Centenary of Oliver Heaviside, J.L.B. Cooper (1952) revealed some of the controversial issues surrounding Heaviside's work, and declared: "As a mathematician he was gifted with manipulative skill and with a genius for finding convenient methods of calculation. He simplified Maxwell's theory enormously; according to Hertz, the four equations known as Maxwell's were first given by Heaviside. He is one of the founders of vector analysis...." Reviewing the history of Heaviside's calculus, Cooper gave a fairly complete account of early history of the subject along with mathematicians' varying opinions about Heaviside's contributions to operational calculus. According to Cooper, a widely publicized story that operational calculus was discovered by Heaviside remained controversial. In spite of the controversies, it is generally believed that Heaviside's real achievement was to develop operational calculus, which is one of the most useful mathematical devices in applied mathematics, mathematical physics, and engineering sciences. In this context Lord Rayleigh's following quotation seems to be most appropriate from a physical point of view: "In the mathematical investigation I have usually employed such methods as present themselves naturally to a physicist. The pure mathematician will complain, and (it must be confessed) sometimes with justice, of deficient rigor. But to this question there are two sides. For, however important it may be to maintain a uniformly high standard in pure mathematics, the physicist may occasionally do well to rest content with arguments which are fairly satisfactory and conclusive from his point of view. To his mind, exercised in a different order of ideas, the more severe procedure of the pure mathematician may appear not more but less demonstrative. And further, in many cases of difficulty to insist upon highest standard would mean the exclusion of the subject altogether in view of the space that would be required."

With the exception of a group of pure mathematicians, everyone has found Heaviside's work a remarkable achievement even though he did not provide a rigorous demonstration of his operational calculus. In defense of Heaviside, Richard P. Feynman's thought seems to be worth quoting. "However, the emphasis should be somewhat more on how to do the mathematics quickly and easily, and what formulas are true, rather than the mathematicians' interest in methods of rigorous proof." The development of operational calculus was somewhat similar to that of calculus. Mathematicians of the seventeenth century who invented the calculus did not provide a rigorous formulation of it. The rigorous formulation came only in the nineteenth century, even though in the transition the non-rigorous demonstration of the calculus that is still admired. It is well known that twentieth-century mathematicians have provided a rigorous foundation of the Heaviside operational calculus. So, by any standard, Heaviside deserves a lot of credit for his work.

The next phase of the development of operational calculus is characterized by the effort to provide justifications of the heuristic methods by rigorous proofs. In this phase, T.J. Bromwich (1875-1930) first successfully introduced the theory of complex functions to give formal justification of Heaviside's calculus. In addition to his many contributions to this subject, he gave the formal derivation of the Heaviside Expansion Theorem and the correct interpretation of Heaviside's operational results. After Bromwich's work, notable contributions to rigorous formulation of operational calculus were made by J.R. Carson, B. van der Pol, G. Doetsch, and others.

In concluding our discussion on the historical development of operational calculus, we should add a note of caution against the controversial evaluation of Heaviside's work. From an applied mathematical point of view, Heaviside's operational calculus was an important achievement. In support of this statement, an assessment of Heaviside's work made by E.T. Whittaker in Heaviside's obituary is recorded below: "Looking back..., we should place the operational calculus with Poincaré's discovery of automorphic functions and Ricci's discovery of the tensor calculus as the three most important mathematical advances of the last quarter of the nineteenth century." Although Heaviside paid little attention to questions of mathematical rigor, he recognized that operational calculus is one of the most effective and useful mathematical methods in applied mathematical sciences. This has led naturally to rigorous mathematical analysis of integral transforms. Indeed, the Fourier or Laplace transform methods based on the rigorous foundation are essentially equivalent to the modern operational calculus.

1.2 Basic Concepts and Definitions

The *integral transform* of a function $f(x)$ defined in $a \leq x \leq b$ is denoted by $\mathcal{I}\{f(x)\} = F(k)$, and defined by

$$\mathcal{I}\{f(x)\} = F(k) = \int_a^b K(x,k)f(x)\,dx, \tag{1.2.1}$$

where $K(x,k)$, given function of two variables x and k, is called the *kernel* of the transform. The operator $\mathcal{I}$ is usually called an *integral transform operator* or simply an *integral transformation*. The transform function $F(k)$ is often referred to as the *image* of the given object function $f(x)$, and k is called the *transform variable*.

Similarly, the integral transform of a function of several variables is defined by

$$\mathcal{I}\{f(\mathbf{x})\} = F(\mathbf{\kappa}) = \int_S K(\mathbf{x},\mathbf{\kappa})f(\mathbf{x})\,d\mathbf{x}, \tag{1.2.2}$$

where $\mathbf{x} = (x_1, x_2, \ldots, x_n)$, $\mathbf{\kappa} = (k_1, k_2, \ldots, k_n)$ and $S \subset R^n$.

A mathematical theory of transformations of this type can be developed by using the properties of Banach spaces. From a mathematical point of view, such a program would be of great interest, but it may *not* be useful for practical applications. Our goal here is to study integral transforms as operational methods with emphasis on applications.

The idea of the integral transform operator is somewhat similar to that of the well-known linear differential operator, $D \equiv \dfrac{d}{dx}$, which acts on a function $f(x)$ to produce another function $f'(x)$, that is,

$$Df(x) = f'(x). \tag{1.2.3}$$

Usually, $f'(x)$ is called the *derivative* or the image of $f(x)$ under the linear transformation D.

There are a number of important integral transforms including the Fourier, Laplace, Hankel, and Mellin transforms. They are defined by choosing different kernels $K(x,k)$ and different values for a and b involved in (1.2.1). Obviously, $\mathcal{I}$ is a linear operator since it satisfies the property of *linearity*:

$$\mathcal{I}\{\alpha f(x) + \beta g(x)\} = \int_a^b \{\alpha f(x) + \beta g(x)\}K(x,k)\,dx$$

$$= \alpha \mathcal{I}\{f(x)\} + \beta \mathcal{I}\{g(x)\}, \tag{1.2.4}$$

where α and β are arbitrary constants. In order to obtain $f(x)$ from a given $F(k) = \mathcal{I}\{f(x)\}$, we introduce the inverse operator $\mathcal{I}^{-1}$ such that

$$\mathcal{I}^{-1}\{F(k)\} = f(x). \tag{1.2.5}$$

Accordingly $\mathcal{I}^{-1}\mathcal{I} = \mathcal{I}\mathcal{I}^{-1} = \mathbf{1}$ which is the identity operator. It can be proved that $\mathcal{I}^{-1}$ is also a linear operator.

Finally, it can also be proved that the integral transform is unique. In other words, if $\mathcal{I}\{f(x)\} = \mathcal{I}\{g(x)\}$, then $f(x) = g(x)$ under suitable conditions. This is known as the *uniqueness theorem*.

Chapter 2

Fourier Transforms

2.1 Introduction

Many linear boundary value and initial value problems in applied mathematics, mathematical physics, and engineering science can be effectively solved by the use of the Fourier transform, the Fourier cosine transform, or the Fourier sine transform. These transforms are very useful for solving differential or integral equations for the following reasons. First, these equations are replaced by simple algebraic equations, which enables us to find the solution of the transform function. The solution of the given equation is then obtained in the original variables by inverting the transform solution. Second, the Fourier transform of the elementary source term is used for determination of the fundamental solution that illustrates the basic ideas behind the construction and implementation of Green's functions. Third, the transform solution combined with the convolution theorem provides an elegant representation of the solution for the boundary value and initial value problems.

We begin this chapter with a formal derivation of the Fourier integral formulas. These results are then used to define the Fourier, Fourier cosine, and Fourier sine transforms. This is followed by a detailed discussion of the basic operational properties of these transforms with examples. Special attention is given to convolution and its main properties. Sections 2.5 and 2.6 deal with applications of the Fourier transform to the solution of ordinary differential equations and integral equations. In Section 2.7, a wide variety of partial differential equations are solved by the use of the Fourier transform method. The technique that is developed in this and other sections can be applied with little or no modification to different kinds of initial and boundary value problems that are encountered in applications. The Fourier cosine and sine transforms are introduced in Section 2.8. The properties and applications of these transforms are discussed in Sections 2.9 and 2.10. This is followed by evaluation of definite integrals with the aid of Fourier transforms. Section 2.12 is devoted to applications of Fourier transforms in mathematical statistics. The multiple Fourier transforms and their applications are discussed in Section 2.13.

2.2 The Fourier Integral Formulas

A function $f(x)$ is said to satisfy *Dirichlet's conditions* in the interval $-a < x < a$, if

(i) $f(x)$ has only a finite number of finite discontinuities in $-a < x < a$ and has no infinite discontinuities.

(ii) $f(x)$ has only a finite number of maxima and minima in $-a < x < a$.

From the theory of Fourier series we know that if $f(x)$ satisfies the Dirichlet conditions in $-a < x < a$, it can be represented as the complex Fourier series

$$f(x) = \sum_{n=-\infty}^{\infty} a_n \exp(in\pi x/a), \qquad (2.2.1)$$

where the coefficients are

$$a_n = \frac{1}{2a} \int_{-a}^{a} f(\xi) \exp(-in\pi \xi/a) d\xi. \qquad (2.2.2)$$

This representation is evidently periodic of period $2a$ in the interval. However, the right hand side of (2.2.1) cannot represent $f(x)$ *outside* the interval $-a < x < a$ unless $f(x)$ is periodic of period $2a$. Thus problems on finite intervals lead to Fourier series, and problems on the whole line $-\infty < x < \infty$ lead to the Fourier integrals. We now attempt to find an integral representation of a non-periodic function $f(x)$ in $(-\infty, \infty)$ by letting $a \to \infty$. As the interval grows $(a \to \infty)$ the values $k_n = \frac{n\pi}{a}$ become closer together, and form a dense set. If we write $\delta k = (k_{n+1} - k_n) = \frac{\pi}{a}$ and substitute coefficients a_n into (2.2.1), we obtain

$$f(x) = \frac{1}{2\pi} \sum_{n=-\infty}^{\infty} (\delta k) \left[\int_{-a}^{a} f(\xi) \exp(-i\xi k_n) d\xi \right] \exp(ixk_n). \qquad (2.2.3)$$

In the limit as $a \to \infty$, k_n becomes a continuous variable k and δk becomes dk. Consequently, the sum can be replaced by the integral in the limit and (2.2.3) reduces to the result

$$f(x) = \frac{1}{2\pi} \int_{-\infty}^{\infty} \left[\int_{-\infty}^{\infty} f(\xi) e^{-ik\xi} d\xi \right] e^{ikx} dk. \qquad (2.2.4)$$

This is known as the *Fourier integral formula*. Although the above arguments do not constitute a rigorous proof of (2.2.4), the formula is correct and valid for functions that are piecewise continuously differentiable in every finite interval and is absolutely integrable on the whole real line.

A function $f(x)$ is said to be *absolutely integrable* on $(-\infty, \infty)$ if

$$\int_{-\infty}^{\infty} |f(x)| dx < \infty \qquad (2.2.5)$$

exists.

It can be shown that the formula (2.2.4) is valid under more general conditions. The result is contained in the following theorem:

Theorem 2.2.1 If $f(x)$ satisfies Dirichlet's conditions in $(-\infty,\infty)$, and is absolutely integrable on $(-\infty,\infty)$, then the Fourier integral (2.2.4) converges to the function $\frac{1}{2}[f(x+0)+f(x-0)]$ at a finite discontinuity at x. In other words,

$$\frac{1}{2}[f(x+0)+f(x-0)] = \frac{1}{2\pi}\int_{-\infty}^{\infty} e^{ikx}\left[\int_{-\infty}^{\infty} f(\xi)e^{-ik\xi}d\xi\right]dk. \qquad (2.2.6)$$

This is usually called the *Fourier integral theorem*.

If the function $f(x)$ is continuous at point x, then $f(x+0) = f(x-0) = f(x)$, then (2.2.6) reduces to (2.2.4).

The Fourier integral theorem was originally stated in Fourier's famous treatise entitled *La Théorie Analytique da la Chaleur* (1822), and its deep significance was recognized by mathematicians and mathematical physicists. Indeed, this theorem is one of the most monumental results of modern mathematical analysis and has widespread physical and engineering applications.

We express the exponential factor $\exp[ik(x-\xi)]$ in (2.2.4) in terms of trigonometric functions and use the even and odd nature of the cosine and the sine functions respectively as functions of k so that (2.2.4) can be written as

$$f(x) = \frac{1}{\pi}\int_0^{\infty} dk \int_{-\infty}^{\infty} f(\xi)\cos k(x-\xi)d\xi. \qquad (2.2.7)$$

This is another version of the *Fourier integral formula*. In many physical problems, the function $f(x)$ vanishes very rapidly as $|x|\to\infty$, which ensures the existence of the repeated integrals as expressed.

We now assume that $f(x)$ is an even function and expand the cosine function in (2.2.7) to obtain

$$f(x) = f(-x) = \frac{2}{\pi}\int_0^{\infty}\cos kx\, dk \int_0^{\infty} f(\xi)\cos k\xi\, d\xi. \qquad (2.2.8)$$

This is called the *Fourier cosine integral formula*.

Similarly, for an odd function $f(x)$, we obtain the *Fourier sine integral formula*

$$f(x) = -f(-x) = \frac{2}{\pi}\int_0^{\infty}\sin kx\, dk\int_0^{\infty} f(\xi)\sin k\xi\, d\xi. \qquad (2.2.9)$$

These integral formulas were discovered independently by Cauchy in his work on the propagation of waves on the surface of water.

2.3 Definition of the Fourier Transform and Examples

We use the Fourier integral formula (2.2.4) to give a formal definition of the Fourier transform.

Definition 2.3.1 The Fourier transform of $f(x)$ is denoted by $\mathscr{F}\{f(x)\} = F(k)$, and defined by the integral

$$\mathscr{F}\{f(x)\} = F(k) = \frac{1}{\sqrt{2\pi}} \int_{-\infty}^{\infty} e^{-ikx} f(x) dx, \qquad (2.3.1)$$

where $\mathscr{F}$ is called the *Fourier transform operator* or the *Fourier transformation* and the factor $\frac{1}{\sqrt{2\pi}}$ is obtained by splitting the factor $\frac{1}{2\pi}$ involved in (2.2.4). This is often called the *complex Fourier transform*. A sufficient condition for $f(x)$ to have a Fourier transform is that $f(x)$ is absolutely integrable on $(-\infty, \infty)$.

The *inverse Fourier transform*, denoted by $\mathscr{F}^{-1}\{F(k)\} = f(x)$, is defined by

$$\mathscr{F}^{-1}\{F(k)\} = f(x) = \frac{1}{\sqrt{2\pi}} \int_{-\infty}^{\infty} e^{ikx} F(k) dk, \qquad (2.3.2)$$

where $\mathscr{F}^{-1}$ is called the *inverse Fourier transform operator*. Clearly, both $\mathscr{F}$ and $\mathscr{F}^{-1}$ are linear integral operators. In applied mathematics, x usually represents a space variable and $k\left(=\frac{2\pi}{\lambda}\right)$ is a wavenumber variable where λ is the wavelength. However, in electrical engineering, x is replaced by the time variable t and k is replaced by the frequency variable $\omega(=2\pi\nu)$ where ν is the frequency in cycles per second. The function $F(\omega) = \mathscr{F}\{f(t)\}$ is called the *spectrum* of the *time signal function* $f(t)$. In electrical engineering literature, the Fourier transform pairs are defined slightly differently by

$$\mathscr{F}\{f(t)\} = F(\nu) = \int_{-\infty}^{\infty} f(t) e^{-2\pi i \nu t} dt, \qquad (2.3.3)$$

and

$$\mathscr{F}^{-1}\{F(\nu)\} = f(t) = \int_{-\infty}^{\infty} F(\nu) e^{2\pi i \nu t} d\nu = \frac{1}{2\pi} \int_{-\infty}^{\infty} F(\omega) e^{i\omega t} d\omega, \qquad (2.3.4)$$

where $\omega = 2\pi\nu$ is called the *angular frequency*. The Fourier integral formula implies that any function of time $f(t)$ which has a Fourier transform can be equally specified by its spectrum. Physically, the signal $f(t)$ is represented as a spectral superposition of simple harmonic oscillations of frequency ν and complex amplitude $F(\nu)$.

Before we give examples of Fourier transforms, it seems pertinent to make some important comments. The above conditions for the existence of the Fourier transform of a function are sufficient rather than necessary conditions. Moreover, they are too strong for many practical applications. With these existence conditions, many simple functions such as a constant function, $\sin \sigma t$, and $t^n H(t)$ do not have Fourier transforms, even though they arise frequently in applications. In other words, the Fourier transform has been defined for a very restricted class of functions. However, attempts have been made (see Lighthill (1958), Jones (1982)) to generalize the definition of the Fourier transform for a more general class of functions in order to include the above and other functions. No attempt will be made in this book to include the formal treatment of the theory of Fourier transforms of generalized functions; our objective here is to simply show that there is a sense, useful in practical applications, in which the above stated functions do have Fourier transforms, and to indicate heuristic methods of finding them.

Example 2.3.1 $F(k) = \mathscr{F}\{\exp(-ax^2)\} = \dfrac{1}{\sqrt{2a}} \exp\left(-\dfrac{k^2}{4a}\right).$ (2.3.5)

$$F(k) = \frac{1}{\sqrt{2\pi}} \int_{-\infty}^{\infty} e^{-ikx-ax^2} dx$$

$$= \frac{1}{\sqrt{2\pi}} \int_{-\infty}^{\infty} \exp\left[-a\left(x+\frac{ik}{2a}\right)^2 - \frac{k^2}{4a}\right] dx$$

$$= \frac{1}{\sqrt{2\pi}} \exp(-k^2/4a) \int_{-\infty}^{\infty} e^{-ay^2} dy = \frac{1}{\sqrt{2a}} \exp\left(-\frac{k^2}{4a}\right),$$

in which the change of variable $y = x + \dfrac{ik}{2a}$ is used. The above result is correct, but the change of variable can be justified by the method of complex analysis because $(ik/2a)$ is complex. If $a = \dfrac{1}{2}$

$$\mathscr{F}\{e^{-x^2/2}\} = e^{-k^2/2}.$$ (2.3.6)

This shows $\mathscr{F}\{f(x)\} = f(k)$. Such a function is said to be *self-reciprocal* under the Fourier transformation.

Example 2.3.2 $\mathscr{F}\{\exp(-a|x|)\} = \sqrt{\dfrac{2}{\pi}} \cdot \dfrac{a}{(a^2+k^2)}.$ (2.3.7)

$$\mathscr{F}\{e^{-a|x|}\} = \frac{1}{\sqrt{2\pi}} \int_{-\infty}^{\infty} e^{-a|x|-ikx} dx$$

$$= \frac{1}{\sqrt{2\pi}} \left[\int_{0}^{\infty} e^{-(a+ik)x} dx + \int_{-\infty}^{0} e^{(a-ik)x} dx\right]$$

$$= \frac{1}{\sqrt{2\pi}} \left[\frac{1}{a+ik} + \frac{1}{a-ik}\right] = \sqrt{\frac{2}{\pi}} \frac{a}{(a^2+k^2)}.$$

Example 2.3.3 Find the Fourier transform of

$$f(x) = \left(1 - \frac{|x|}{a}\right) H\left(1 - \frac{|x|}{a}\right)$$

where $H(x)$ is the Heaviside unit step function defined by

$$H(x) = \begin{cases} 1, & x > 0 \\ 0, & x < 0 \end{cases}. \tag{2.3.8}$$

Or more generally,

$$H(x-a) = \begin{cases} 1, & x > a \\ 0, & x < a \end{cases}, \tag{2.3.9}$$

where a is a fixed real number. So the Heaviside function $H(x-a)$ has a finite discontinuity at $x = a$.

$$\mathcal{F}\{f(x)\} = \frac{1}{\sqrt{2\pi}} \int_{-a}^{a} e^{-ikx}\left(1-\frac{|x|}{a}\right)dx = \frac{2}{\sqrt{2\pi}} \int_{0}^{a}\left(1-\frac{x}{a}\right)\cos kx\, dx$$

$$= \frac{2a}{\sqrt{2\pi}} \int_{0}^{1} (1-x)\cos(akx)\, dx = \frac{2a}{\sqrt{2\pi}} \int_{0}^{1} (1-x)\frac{d}{dx}\left(\frac{\sin akx}{ak}\right) dx$$

$$= \frac{2a}{\sqrt{2\pi}} \int_{0}^{1} \frac{\sin(akx)}{ak}\, dx = \frac{a}{\sqrt{2\pi}} \int_{0}^{1} \frac{d}{dx}\left[\frac{\sin^2\left(\frac{akx}{2}\right)}{\left(\frac{ak}{2}\right)^2}\right] dx$$

$$= \frac{a}{\sqrt{2\pi}} \frac{\sin^2\left(\frac{ak}{2}\right)}{\left(\frac{ak}{2}\right)^2}. \tag{2.3.10}$$

Example 2.3.4 Find the Fourier transform of the gate function $f_a(x)$ used in electrical engineering where

$$f_a(x) = H(a-|x|) = \begin{cases} 1, & |x| < a \\ 0, & |x| > a \end{cases} \tag{2.3.11}$$

$$F_a(k) = \mathcal{F}\{f_a(x)\} = \frac{1}{\sqrt{2\pi}} \int_{-\infty}^{\infty} e^{-ikx} f_a(x)\, dx$$

$$= \frac{1}{\sqrt{2\pi}} \int_{-a}^{a} e^{-ikx} dx = \sqrt{\frac{2}{\pi}}\left(\frac{\sin ak}{k}\right). \tag{2.3.12}$$

The graphs of $f_a(x)$ and $F_a(k)$ are given in Figure 2.1.

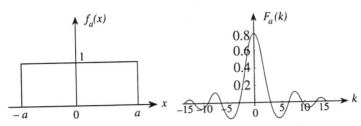

Figure 2.1 Graphs of $f_a(x)$ and $F_a(k)$.

A continuous function $f(x)$ defined in $(-\infty, \infty)$ such that its Fourier transform $F(k)$ vanishes identically outside a finite interval $-a < k < a$ is called a *band limited function*.

The Fourier transform of the gate function given in Example 2.3.4 is a simple example of a band limited function.

In 1920, Dirac introduced the delta function $\delta(x)$ having the following properties

$$\delta(x) = 0, \qquad x \neq 0,$$

$$\int_{-\infty}^{\infty} \delta(x)\, dx = 1.$$

(2.3.13)

These properties cannot be satisfied by any ordinary functions in classical mathematics. Hence the delta function is not a function in the classical sense. However, it can be treated as a function in the generalized sense, and in fact, $\delta(x)$ is called a *generalized function* or *distribution*. The concept of the delta function is clear and simple in modern mathematics. It is very useful in physics and engineering. Physically, the delta function represents a point mass, that is, a particle of unit mass located at the origin. In this context, it may be called a *mass-density* function. This leads to the result for a point particle that can be considered as the limit of a sequence of continuous distributions which become more and more concentrated. Even though $\delta(x)$ is not a function in the classical sense, it can be approximated by a sequence of ordinary functions. As an example, we consider the sequence

$$\delta_n(x) = \sqrt{\frac{n}{\pi}} \exp(-nx^2), \quad n = 1, 2, 3, \ldots. \qquad (2.3.14)$$

Clearly, $\delta_n(x) \to 0$ as $n \to \infty$ for any $x \neq 0$ and $\delta_n(0) \to \infty$ as $n \to \infty$ as shown in Figure 2.2. Also, for all $n = 1, 2, 3, \ldots$,

$$\int_{-\infty}^{\infty} \delta_n(x)\, dx = 1$$

and

$$\lim_{n \to \infty} \int_{-\infty}^{\infty} \delta_n(x)\, dx = \int_{-\infty}^{\infty} \delta(x)\, dx = 1$$

as expected. So the delta function can be considered as the limit of a sequence of ordinary functions, and we write

$$\delta(x) = \lim_{n \to \infty} \sqrt{\frac{n}{\pi}} \exp(-nx^2). \qquad (2.3.15)$$

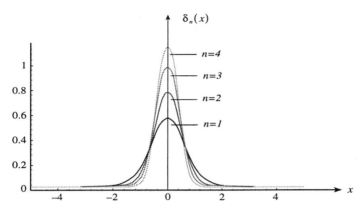

Figure 2.2 The sequence of delta functions.

Sometimes, the delta function $\delta(x)$ is defined by its fundamental property

$$\int_{-\infty}^{\infty} f(x)\delta(x-a)\,dx = f(a), \qquad (2.3.16)$$

where $f(x)$ is continuous in any interval containing the point $x = a$. Clearly,

$$\int_{-\infty}^{\infty} f(a)\delta(x-a)\,dx = f(a)\int_{-\infty}^{\infty} \delta(x-a)\,dx = f(a). \qquad (2.3.17)$$

Thus (2.3.16) and (2.3.17) lead to the result

$$f(x)\delta(x-a) = f(a)\delta(x-a). \qquad (2.3.18)$$

The following results are also true

$$x\,\delta(x) = 0 \qquad (2.3.19)$$
$$\delta(x-a) = \delta(a-x). \qquad (2.3.20)$$

Result (2.3.20) shows that $\delta(x)$ is an even function.

Clearly, the result

$$\int_{-\infty}^{x} \delta(y)\,dy = \begin{cases} 1, & x > 0 \\ 0, & x < 0 \end{cases} = H(x)$$

shows that

$$\frac{d}{dx} H(x) = \delta(x). \qquad (2.3.21)$$

The Fourier transform of the Dirac delta function is

$$\mathcal{F}\{\delta(x)\} = \frac{1}{\sqrt{2\pi}} \int_{-\infty}^{\infty} e^{-ikx} \delta(x) dx = \frac{1}{\sqrt{2\pi}}. \tag{2.3.22}$$

Hence

$$\delta(x) = \mathcal{F}^{-1}\left\{\frac{1}{\sqrt{2\pi}}\right\} = \frac{1}{2\pi} \int_{-\infty}^{\infty} e^{ikx} dk. \tag{2.3.23}$$

This is an integral representation of the delta function extensively used in quantum mechanics. Also, (2.3.23) can be rewritten as

$$\delta(k) = \frac{1}{2\pi} \int_{-\infty}^{\infty} e^{ikx} dx. \tag{2.3.24}$$

We now define another function $sgn(x)$, read as "sign of x" by

$$sgn(x) = \begin{cases} 1, & x > 0 \\ -1, & x < 0 \end{cases}. \tag{2.3.25}$$

Clearly,

$$H(x) = \frac{1}{2}[1 + sgn(x)]. \tag{2.3.26}$$

Or

$$sgn(x) = 2H(x) - 1, \tag{2.3.27}$$

$$\frac{d}{dx}[sgn(x)] = 2H'(x) = 2\delta(x). \tag{2.3.28}$$

Example 2.3.5 $\mathcal{F}\{e^{-ax}H(x)\} = \frac{1}{\sqrt{2\pi}(ik+a)}, \quad a > 0. \tag{2.3.29}$

We have, by definition,

$$\mathcal{F}\{e^{-ax}H(x)\} = \frac{1}{\sqrt{2\pi}} \int_0^{\infty} \exp\{-x(ik+a)\} dx = \frac{1}{\sqrt{2\pi}(ik+a)}.$$

Example 2.3.6 By considering the function (see Figure 2.3)
$$f_a(x) = e^{-ax}H(x) - e^{ax}H(-x), \quad a > 0, \tag{2.3.30}$$
find the Fourier transform of $sgn(x)$.

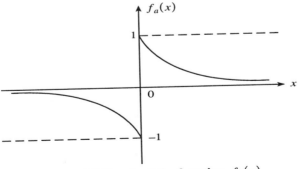

Figure 2.3 Graph of the function $f_a(x)$.

We have, by definition,
$$\mathscr{F}\{f_a(x)\} = -\frac{1}{\sqrt{2\pi}}\int_{-\infty}^{0}\exp\{(a-ik)x\}dx + \frac{1}{\sqrt{2\pi}}\int_{0}^{\infty}\exp\{-(a+ik)x\}dx$$
$$= \frac{1}{\sqrt{2\pi}}\left[\frac{1}{a+ik} - \frac{1}{a-ik}\right] = \sqrt{\frac{2}{\pi}} \cdot \frac{(-ik)}{a^2+k^2}.$$

In the limit as $a \to 0$, $f_a(x) \to \operatorname{sgn}(x)$ and then
$$\mathscr{F}\{\operatorname{sgn}(x)\} = \sqrt{\frac{2}{\pi}} \cdot \frac{1}{ik}.$$

Or
$$\mathscr{F}\left\{\sqrt{\frac{\pi}{2}}\, i\operatorname{sgn}(x)\right\} = \frac{1}{k}.$$

2.4 Basic Properties of Fourier Transforms

Theorem 2.4.1 If $\mathscr{F}\{f(x)\} = F(k)$, then

(a) (Shifting) $\quad\mathscr{F}\{f(x-a)\} = e^{-ika}\mathscr{F}\{f(x)\}, \quad\quad (2.4.1)$

(b) (Scaling) $\quad\mathscr{F}\{f(ax)\} = \frac{1}{|a|}\mathscr{F}\{f(x)\}, \quad\quad (2.4.2)$

(c) (Conjugate) $\quad\mathscr{F}\{\overline{f(-x)}\} = \overline{\mathscr{F}\{f(x)\}}, \quad\quad (2.4.3)$

(d) (Translation) $\quad\mathscr{F}\{e^{iax}f(x)\} = F(k-a), \quad\quad (2.4.4)$

(e) (Duality) $\quad\mathscr{F}\{F(x)\} = f(-k). \quad\quad (2.4.5)$

Proof. (a) We obtain, from the definition,
$$\mathscr{F}\{f(x-a)\} = \frac{1}{\sqrt{2\pi}}\int_{-\infty}^{\infty}e^{-ikx}f(x-a)dx$$
$$= \frac{1}{\sqrt{2\pi}}\int_{-\infty}^{\infty}e^{-ik(\xi+a)}f(\xi)\,d\xi \quad (x-a=\xi)$$
$$= e^{-ika}\mathscr{F}\{f(x)\}.$$

The proofs of results (b)-(d) follow easily from the definition of the Fourier transform. We give a proof of the duality property (e).

We have, by definition,
$$f(x) = \frac{1}{\sqrt{2\pi}}\int_{-\infty}^{\infty}e^{ikx}F(k)\,dk = \mathscr{F}^{-1}\{F(k)\}.$$

Interchanging x and k, and then replacing k by $-k$, we obtain

$$f(-k) = \frac{1}{\sqrt{2\pi}} \int_{-\infty}^{\infty} e^{-ikx} F(x)\, dx = \mathcal{F}\{F(x)\}.$$

Theorem 2.4.2 If $f(x)$ is piecewise continuously differentiable and absolutely integrable, then
(i) $F(k)$ is bounded for $-\infty < k < \infty$.
(ii) $F(k)$ is continuous for $-\infty < k < \infty$.

Proof. It follows from the definition that

$$|F(k)| \leq \frac{1}{\sqrt{2\pi}} \int_{-\infty}^{\infty} |e^{-ikx}| |f(x)|\, dx = \frac{1}{\sqrt{2\pi}} \int_{-\infty}^{\infty} |f(x)|\, dx = \frac{c}{\sqrt{2\pi}},$$

where $c = \int_{-\infty}^{\infty} |f(x)|\, dx = \text{constant}$. This proves result (i).

To prove (ii), we have

$$|F(k+h) - F(k)| \leq \frac{1}{\sqrt{2\pi}} \int_{-\infty}^{\infty} |e^{-ihx} - 1| |f(x)|\, dx \leq \sqrt{\frac{2}{\pi}} \int_{-\infty}^{\infty} |f(x)|\, dx.$$

Since $\lim_{h \to 0} |e^{-ihx} - 1| = 0$ for all $x \in R$, we obtain

$$\lim_{h \to 0} |F(k+h) - F(k)| \leq \lim_{h \to 0} \frac{1}{\sqrt{2\pi}} \int_{-\infty}^{\infty} |e^{-ihx} - 1| |f(x)|\, dx = 0.$$

This shows that $F(k)$ is continuous.

Theorem 2.4.3 (*Riemann-Lebesgue Lemma*). If $F(k) = \mathcal{F}\{f(x)\}$, then

$$\lim_{|k| \to \infty} |F(k)| = 0. \tag{2.4.6}$$

Proof. Since $e^{-ikx} = -e^{-ikx - i\pi}$, we have

$$F(k) = -\frac{1}{\sqrt{2\pi}} \int_{-\infty}^{\infty} e^{-ik\left(x + \frac{\pi}{k}\right)} f(x)\, dx = -\frac{1}{\sqrt{2\pi}} \int_{-\infty}^{\infty} e^{-ikx} f\left(x - \frac{\pi}{k}\right) dx.$$

Hence

$$F(k) = \frac{1}{2}\left\{ \frac{1}{\sqrt{2\pi}} \left[\int_{-\infty}^{\infty} e^{-ikx} f(x)\, dx - \int_{-\infty}^{\infty} e^{-ikx} f\left(x - \frac{\pi}{k}\right) dx \right] \right\}$$

$$= \frac{1}{2} \frac{1}{\sqrt{2\pi}} \int_{-\infty}^{\infty} e^{-ikx} \left[f(x) - f\left(x - \frac{\pi}{k}\right) \right] dx.$$

Therefore

$$|F(k)| \leq \frac{1}{2\sqrt{2\pi}} \int_{-\infty}^{\infty} \left| f(x) - f\left(x - \frac{\pi}{k}\right) \right| dx.$$

Thus, we obtain

$$\lim_{|k|\to\infty}|F(k)| \leq \frac{1}{2\sqrt{2\pi}} \lim_{|k|\to\infty}\int_{-\infty}^{\infty}\left|f(x)-f\left(x-\frac{\pi}{k}\right)\right|dx = 0.$$

Theorem 2.4.4 If $f(x)$ is continuously differentiable and $f(x) \to 0$ as $|x| \to \infty$, then

$$\mathcal{F}\{f'(x)\} = (ik)\mathcal{F}\{f(x)\} = ik\ F(k). \tag{2.4.7}$$

Proof. We have, by definition,

$$\mathcal{F}\{f'(x)\} = \frac{1}{\sqrt{2\pi}}\int_{-\infty}^{\infty} e^{-ikx} f'(x)\,dx$$

which is, by integrating by parts,

$$= \frac{1}{\sqrt{2\pi}}\left[f(x)e^{-ikx}\right]_{-\infty}^{\infty} + \frac{ik}{\sqrt{2\pi}}\int_{-\infty}^{\infty} e^{-ikx} f(x)\,dx$$

$$= (ik)F(k).$$

If $f(x)$ is continuously n-times differentiable and $f^{(k)}(x) \to 0$ as $|x| \to \infty$ for $k = 1,2,\ldots,(n-1)$, then

$$\mathcal{F}\{f^{(n)}(x)\} = (ik)^n \mathcal{F}\{f(x)\} = (ik)^n F(k). \tag{2.4.8}$$

A repeated application of Theorem 2.4.4 to higher derivatives gives the result.

The operational results similar to those of (2.4.7) and (2.4.8) hold for partial derivatives of a function of two or more independent variables. For example, if $u(x,t)$ is a function of space variable x and time variable t, then

$$\mathcal{F}\left\{\frac{\partial u}{\partial x}\right\} = ik\, U(k,t), \quad \mathcal{F}\left\{\frac{\partial^2 u}{\partial x^2}\right\} = -k^2\, U(k,t)$$

$$\mathcal{F}\left\{\frac{\partial u}{\partial t}\right\} = \frac{dU}{dt}, \quad \mathcal{F}\left\{\frac{\partial^2 u}{\partial t^2}\right\} = \frac{d^2 U}{dt^2}.$$

Definition. The *convolution* of two integrable functions $f(x)$ and $g(x)$, denoted by $f(x) * g(x)$, is defined by

$$f(x) * g(x) = \frac{1}{\sqrt{2\pi}}\int_{-\infty}^{\infty} f(x-\xi)g(\xi)\,d\xi. \tag{2.4.9}$$

Theorem 2.4.5 (*Convolution Theorem*). If $\mathcal{F}\{f(x)\} = F(k)$ and $\mathcal{F}\{g(x)\} = G(k)$, then

$$\mathcal{F}\{f(x) * g(x)\} = F(k)G(k), \tag{2.4.10}$$

or

$$f(x) * g(x) = \mathcal{F}^{-1}\{F(k)G(k)\}. \tag{2.4.11}$$

or equivalently

$$\int_{-\infty}^{\infty} f(x-\xi)g(\xi)d\xi = \int_{-\infty}^{\infty} e^{ikx} F(k)G(k)dk. \qquad (2.4.12)$$

Proof. We have, by the definition of the Fourier transform,

$$\mathscr{F}\{f(x)*g(x)\} = \frac{1}{2\pi}\int_{-\infty}^{\infty} e^{-ikx} dx \int_{-\infty}^{\infty} f(x-\xi)g(\xi)d\xi$$

$$= \frac{1}{2\pi}\int_{-\infty}^{\infty} e^{-ik\xi} g(\xi)d\xi \int_{-\infty}^{\infty} e^{-ik(x-\xi)} f(x-\xi)dx$$

$$= \frac{1}{2\pi}\int_{-\infty}^{\infty} e^{-ik\xi} g(\xi)d\xi \int_{-\infty}^{\infty} e^{-ik\eta} f(\eta)d\eta$$

$$= G(k)F(k).$$

This completes the proof.

The convolution has the following algebraic properties:

$$f*g = g*f \quad \text{(Commutative)}, \qquad (2.4.13)$$
$$f*(g*h) = (f*g)*h \quad \text{(Associative)}, \qquad (2.4.14)$$
$$f*(g+h) = f*g + f*h \quad \text{(Distributive)}, \qquad (2.4.15)$$
$$f*\sqrt{2\pi}\,\delta = f = \sqrt{2\pi}\,\delta*f \quad \text{(Identity)}. \qquad (2.4.16)$$

In view of the commutative property of the convolution, (2.4.12) can be written as

$$\int_{-\infty}^{\infty} f(\xi)g(x-\xi)d\xi = \int_{-\infty}^{\infty} e^{ikx} F(k)G(k)dk. \qquad (2.4.17)$$

This is valid for all real x, and hence putting $x=0$ gives

$$\int_{-\infty}^{\infty} f(\xi)g(-\xi)d\xi = \int_{-\infty}^{\infty} f(x)g(-x)dx = \int_{-\infty}^{\infty} F(k)G(k)dk. \qquad (2.4.18)$$

We substitute $g(x) = \overline{f(-x)}$ to obtain

$$G(k) = \mathscr{F}\{g(x)\} = \mathscr{F}\{\overline{f(-x)}\} = \overline{\mathscr{F}\{f(x)\}} = \overline{F(k)}.$$

Evidently (2.4.18) becomes

$$\int_{-\infty}^{\infty} f(x)\overline{f(x)}\,dx = \int_{-\infty}^{\infty} F(k)\overline{F(k)}\,dk \qquad (2.4.19)$$

or

$$\int_{-\infty}^{\infty} |f(x)|^2\,dx = \int_{-\infty}^{\infty} |F(k)|^2\,dk. \qquad (2.4.20)$$

This is well known as *Parseval's relation*.

For square integrable functions $f(x)$ and $g(x)$, the *inner product* (f,g) is defined by

$$(f,g) = \int_{-\infty}^{\infty} f(x)\overline{g(x)}\,dx \qquad (2.4.21)$$

so the *norm* $\|f\|$ is defined by

$$\|f\|^2 = (f,f) = \int_{-\infty}^{\infty} f(x)\overline{f(x)}\,dx = \int_{-\infty}^{\infty} |f(x)|^2\,dx. \qquad (2.4.22)$$

In terms of the norm, the Parseval relation takes the form

$$\|f\| = \|F\| = \|\mathcal{F}f\|. \qquad (2.4.23)$$

This means that the Fourier transform action is *unitary*. Physically, the quantity $\|f\|$ is a measure of energy and $\|F\|$ represents the *power spectrum* of f.

Theorem 2.4.6 (*General Parseval's Relation*). If $\mathcal{F}\{f(x)\} = F(k)$ and $\mathcal{F}\{g(x)\} = G(k)$ then

$$\int_{-\infty}^{\infty} f(x)\overline{g(x)}\,dx = \int_{-\infty}^{\infty} F(k)\overline{G(k)}\,dk. \qquad (2.4.24)$$

Proof. We proceed formally to obtain

$$\int_{-\infty}^{\infty} F(k)\overline{G(k)}\,dk = \int_{-\infty}^{\infty} dk \cdot \frac{1}{2\pi} \int_{-\infty}^{\infty} e^{-iky} f(y)\,dy \overline{\int_{-\infty}^{\infty} e^{-ikx} g(x)\,dx}$$

$$= \frac{1}{2\pi} \int_{-\infty}^{\infty} f(y)\,dy \int_{-\infty}^{\infty} \overline{g(x)}\,dx \int_{-\infty}^{\infty} e^{ik(x-y)}\,dk$$

$$= \int_{-\infty}^{\infty} \overline{g(x)}\,dx \int_{-\infty}^{\infty} \delta(x-y) f(y)\,dy = \int_{-\infty}^{\infty} f(x)\overline{g(x)}\,dx.$$

In particular, when $g(x) = f(x)$, the above result agrees with (2.4.19).

We now use an indirect method to obtain the Fourier transform of $sgn(x)$, that is,

$$\mathcal{F}\{sgn(x)\} = \sqrt{\frac{2}{\pi}} \frac{1}{ik}. \qquad (2.4.25)$$

From (2.3.28), we find

$$\mathcal{F}\left\{\frac{d}{dx} sgn(x)\right\} = \mathcal{F}\{2H'(x)\} = 2\mathcal{F}\{\delta(x)\} = \sqrt{\frac{2}{\pi}},$$

which is, by (2.4.7),

$$ik\,\mathcal{F}\{sgn(x)\} = \sqrt{\frac{2}{\pi}}.$$

or

$$\mathcal{F}\{sgn(x)\} = \sqrt{\frac{2}{\pi}} \cdot \frac{1}{ik}.$$

The Fourier transform of $H(x)$ follows from (2.3.24) and (2.4.25):

$$\mathscr{F}\{H(x)\} = \frac{1}{2}\mathscr{F}\{1+sgn(x)\} = \frac{1}{2}\big[\mathscr{F}\{1\} + \mathscr{F}\{sgn(x)\}\big]$$

$$= \sqrt{\frac{\pi}{2}}\left[\delta(k) + \frac{1}{i\pi k}\right]. \qquad (2.4.26)$$

Example 2.4.1 (*Synthesis and Resolution of a Signal; Physical Interpretation of Convolution*). In electrical engineering problems, a time-dependent electric, optical or electromagnetic *pulse* is usually called a *signal*. Such a signal can be considered as a superposition of plane waves of all real frequencies so that it can be represented by the inverse Fourier transform

$$f(t) = \mathscr{F}^{-1}\{F(\omega)\} = \frac{1}{2\pi}\int_{-\infty}^{\infty} F(\omega) e^{i\omega t}\, d\omega, \qquad (2.4.27)$$

where $F(\omega) = \mathscr{F}\{f(t)\}$, the factor $(1/2\pi)$ is introduced because the angular frequency ω is related to linear frequency v by $\omega = 2\pi v$, and negative frequencies are introduced for mathematical convenience so that we can avoid dealing with the cosine and sine functions separately. Clearly, $F(\omega)$ can be represented by the Fourier transform of the signal $f(t)$ as

$$F(\omega) = \int_{-\infty}^{\infty} f(t) e^{-i\omega t}\, dt. \qquad (2.4.28)$$

This represents the *resolution* of the signal into its angular frequency components, and (2.4.27) gives a *synthesis* of the signal from its individual components.

Suppose a signal $f(t)$ is a continuous function of time t defined in $-\infty < t < \infty$ and its Fourier transform $F(\omega)$ vanishes outside a finite frequency band, that is, $F(\omega) = 0$ for $|\omega| > a$. Such a signal $f(t)$ is called *bandlimited*.

In particular, if

$$F(\omega) = \begin{cases} 1, & |\omega| < a \\ 0, & |\omega| > a \end{cases} \qquad (2.4.29)$$

then $F(\omega)$ is called a *gate function* and is denoted by $F_a(\omega)$, and the associated signal is denoted by $f_a(t)$ so that

$$f_a(t) = \frac{1}{2\pi}\int_{-\infty}^{\infty} F(\omega) e^{i\omega t}\, d\omega = \frac{1}{2\pi}\int_{-a}^{a} e^{i\omega t}\, d\omega = \frac{\sin at}{\pi t}. \qquad (2.4.30)$$

Both $F(\omega)$ and $f_a(t)$ are shown in Figure 2.4.

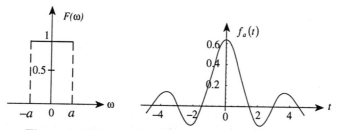

Figure 2.4 The gate function and its Fourier transform.

Consider the limit as $a \to \infty$ of the Fourier integral for $-\infty < \omega < \infty$

$$1 = \lim_{a \to \infty} \int_{-\infty}^{\infty} e^{-i\omega t} f_a(t)\,dt = \lim_{a \to \infty} \int_{-\infty}^{\infty} e^{-i\omega t} \frac{\sin at}{\pi t}\,dt$$

$$= \int_{-\infty}^{\infty} e^{-i\omega t}\left[\lim_{a \to \infty} \frac{\sin at}{\pi t}\right] dt = \int_{-\infty}^{\infty} e^{-i\omega t} \delta(t)\,dt.$$

Clearly, the delta function $\delta(t)$ can be thought as the limit of the sequence of functions $f_a(t)$. More precisely,

$$\delta(t) = \lim_{a \to \infty}\left(\frac{\sin at}{\pi t}\right). \tag{2.4.31}$$

We next consider the bandlimited signal

$$f_a(t) = \frac{1}{2\pi} \int_{-a}^{a} F(\omega) e^{i\omega t}\,d\omega = \frac{1}{2\pi} \int_{-\infty}^{\infty} F(\omega) F_a(\omega) e^{i\omega t}\,d\omega,$$

which is, by the Convolution Theorem 2.4.5,

$$= \int_{-\infty}^{\infty} f(\tau) f_a(t-\tau)\,d\tau = \int_{-\infty}^{\infty} \frac{\sin a(t-\tau)}{\pi(t-\tau)} f(\tau)\,d\tau. \tag{2.4.32}$$

Integral (2.4.32) represents the *sampling integral representation* of the bandlimited signal $f_a(t)$.

Consider a simple electrical device such as an amplifier with an input signal $f(t)$, and an output signal $g(t)$. For an input of a single frequency ω, $f(t) = e^{i\omega t}$. The amplifier will change the amplitude and may also change the phase so that the output can be expressed in terms of the input, the amplitude and the phase modifying function $\Phi(\omega)$ as

$$g(t) = \Phi(\omega) f(t), \tag{2.4.33}$$

where $\Phi(\omega)$ is usually known as the *transfer function* and is, in general, a complex function of the real variable ω. This function is generally independent of the presence or absence of any other frequency components. Thus the total output may be found by integrating over the entire input as modified by the amplifier

$$g(t) = \frac{1}{2\pi} \int_{-\infty}^{\infty} \Phi(\omega) F(\omega) e^{i\omega t} d\omega. \qquad (2.4.34)$$

Thus the total output signal can readily be calculated from any given input signal $f(t)$. On the other hand, the transfer function $\Phi(\omega)$ is obviously characteristic of the amplifier device and can in general be obtained as the Fourier transform of some function $\phi(t)$ so that

$$\Phi(\omega) = \int_{-\infty}^{\infty} \phi(t) e^{-i\omega t} dt. \qquad (2.4.35)$$

The Convolution Theorem 2.4.5 allows us to rewrite (2.4.34) as

$$g(t) = \mathcal{F}^{-1}\{\Phi(\omega) F(\omega)\} = f(t) * \phi(t) = \int_{-\infty}^{\infty} f(\tau) \phi(t-\tau) d\tau. \qquad (2.4.36)$$

Physically, this result represents an output signal $g(t)$ as the integral superposition of an input signal $f(t)$ modified by $\phi(t-\tau)$. Linear translation invariant systems, such as sensors and filters, are modeled by the convolution equations $g(t) = f(t) * \phi(t)$, where $\phi(t)$ is the system impulse response function. In fact (2.4.36) is the most general mathematical representation of an output (effect) function in terms of an input (cause) function modified by the amplifier where t is the time variable. Assuming the principle of causality, that is, every effect has a cause, we must require $\tau < t$. The principle of causality is imposed by requiring

$$\phi(t-\tau) = 0 \quad \text{when } \tau > t. \qquad (2.4.37)$$

Consequently, (2.4.36) gives

$$g(t) = \int_{-\infty}^{t} f(\tau) \phi(t-\tau) d\tau. \qquad (2.4.38)$$

In order to determine the significance of $\phi(t)$, we use an impulse function $f(\tau) = \delta(\tau)$ so that (2.4.38) becomes

$$g(t) = \int_{-\infty}^{t} \delta(\tau) \phi(t-\tau) d\tau = \phi(t) H(t). \qquad (2.4.39)$$

This recognizes $\phi(t)$ as the output corresponding to a unit impulse at $t = 0$, and the Fourier transform of $\phi(t)$ is

$$\Phi(\omega) = \mathcal{F}\{\phi(t)\} = \int_{0}^{\infty} \phi(t) e^{-i\omega t} dt, \qquad (2.4.40)$$

with $\phi(t) = 0$ for $t < 0$.

Example 2.4.2 (*The Series Sampling Expansion of a Bandlimited Signal*). Consider a bandlimited signal $f_a(t)$ with Fourier transform $F(\omega) = 0$ for $|\omega| > a$. We write the Fourier series expansion of $F(\omega)$ on the interval

$-a < \omega < a$ in terms of the orthogonal set of functions $\left\{\exp\left(-\dfrac{in\pi\omega}{a}\right)\right\}$ in the form

$$F(\omega) = \sum_{n=-\infty}^{\infty} a_n \exp\left(-\dfrac{in\pi}{a}\omega\right), \qquad (2.4.41)$$

where the Fourier coefficients a_n are given by

$$a_n = \dfrac{1}{2a}\int_{-a}^{a} F(\omega)\exp\left(\dfrac{in\pi}{a}\omega\right)d\omega = \dfrac{1}{2a}f_a\left(\dfrac{n\pi}{a}\right). \qquad (2.4.42)$$

Thus the Fourier series expansion (2.4.41) becomes

$$F(\omega) = \dfrac{1}{2a}\sum_{n=-\infty}^{\infty} f_a\left(\dfrac{n\pi}{a}\right)\exp\left(-\dfrac{in\pi}{a}\omega\right). \qquad (2.4.43)$$

The signal function $f_a(t)$ is obtained by multiplying (2.4.43) by $e^{i\omega t}$ and integrating over $(-a, a)$ so that

$$f_a(t) = \int_{-a}^{a} F(\omega) e^{i\omega t} d\omega$$

$$= \dfrac{1}{2a}\int_{-a}^{a} e^{i\omega t} d\omega \left[\sum_{n=-\infty}^{\infty} f_a\left(\dfrac{n\pi}{a}\right)\exp\left(-\dfrac{in\pi}{a}\omega\right)\right]$$

$$= \dfrac{1}{2a}\sum_{n=-\infty}^{\infty} f_a\left(\dfrac{n\pi}{a}\right)\int_{-a}^{a}\exp\left[i\omega\left(t - \dfrac{n\pi}{a}\right)\right]d\omega$$

$$= \sum_{n=-\infty}^{\infty} f_a\left(\dfrac{n\pi}{a}\right)\dfrac{\sin a\left(t - \dfrac{n\pi}{a}\right)}{a\left(t - \dfrac{n\pi}{a}\right)}. \qquad (2.4.44)$$

This sampling expansion expresses the continuous bandlimited signal $f_a(t)$ in terms of the infinite set of discrete samples $\left\{f_a\left(\dfrac{n\pi}{a}\right)\right\}$. In practice, a discrete set of samples is useful in the sense that most systems receive discrete samples $\{f(t_n)\}$ as an input. Result (2.4.44) can be obtained from the convolution theorem by using discrete input samples

$$\sum_{n=-\infty}^{\infty} \dfrac{\pi}{a} f_a\left(\dfrac{n\pi}{a}\right)\delta\left(t - \dfrac{n\pi}{a}\right) = f(t). \qquad (2.4.45)$$

Hence the sampling expansion (2.4.32) gives the bandlimited signal

$$f_a(t) = \int_{-\infty}^{\infty} \frac{\sin a(t-\tau)}{\pi(t-\tau)} \left[\sum_{n=-\infty}^{\infty} \frac{\pi}{a} f_a\left(\frac{n\pi}{a}\right) \delta\left(\tau - \frac{n\pi}{a}\right) \right] d\tau$$

$$= \sum_{n=-\infty}^{\infty} f_a\left(\frac{n\pi}{a}\right) \int_{-\infty}^{\infty} \frac{\sin a(t-\tau)}{a(t-\tau)} \delta\left(\tau - \frac{n\pi}{a}\right) d\tau$$

$$= \sum_{n=-\infty}^{\infty} f_a\left(\frac{n\pi}{a}\right) \frac{\sin a\left(t - \frac{n\pi}{a}\right)}{a\left(t - \frac{n\pi}{a}\right)}. \tag{2.4.46}$$

Example 2.4.3 (*Poisson's Summation Formula*). We consider a continuous function $f(x)$ on the real line $(-\infty < x < \infty)$ which vanishes for large $|x|$. Then, for each x, the series

$$\sum_{n=-\infty}^{\infty} f(x + 2na) = g(x) \tag{2.4.47}$$

converges absolutely to a continuous function $g(x)$. We assume that this convergence is uniform for $-a \leq x \leq a$. It is easy to check that $g(x)$ is periodic with period $2a$. The complex Fourier series for $g(x)$ is

$$g(x) = \sum_{m=-\infty}^{\infty} c_m \exp(im\pi x/a), \tag{2.4.48}$$

where the coefficients for $m = 0, \pm 1, \pm 2, \ldots$ are

$$c_m = \frac{1}{2a} \int_{-a}^{a} g(x) \exp(-im\pi x/a) \, dx. \tag{2.4.49}$$

We replace $g(x)$ by the limit of the sum

$$g(x) = \lim_{N \to \infty} \sum_{n=-N}^{N} f(x + 2na), \tag{2.4.50}$$

so that (2.4.49) reduces to

$$c_m = \frac{1}{2a} \lim_{N \to \infty} \sum_{n=-N}^{N} \int_{-a}^{a} f(x + 2na) \exp(-im\pi x/a) \, dx$$

$$= \frac{1}{2a} \lim_{N \to \infty} \sum_{n=-N}^{N} \int_{(2n-1)a}^{(2n+1)a} f(y) \exp(-im\pi y/a) \, dy$$

$$= \frac{1}{2a} \lim_{N \to \infty} \int_{-(2N+1)a}^{(2N+1)a} f(x) \exp(-im\pi x/a) \, dx$$

$$= \frac{\sqrt{2\pi}}{2a} F\left(\frac{m\pi}{a}\right), \tag{2.4.51}$$

where $F\left(\dfrac{m\pi}{a}\right)$ is the Fourier transform of $f(x)$.

Evidently,

$$\sum_{n=-\infty}^{\infty} f(x+2na) = g(x) = \sum_{m=-\infty}^{\infty} \sqrt{\dfrac{\pi}{2}\dfrac{1}{a}} F\left(\dfrac{m\pi}{a}\right) \exp(im\pi x/a). \quad (2.4.52)$$

We let $x = 0$ in (2.4.52) to obtain the *Poisson summation formula*

$$\sum_{n=-\infty}^{\infty} f(2na) = \sum_{m=-\infty}^{\infty} \sqrt{\dfrac{\pi}{2}\dfrac{1}{a}} F\left(\dfrac{m\pi}{a}\right). \quad (2.4.53)$$

When $2a = 1$, this formula becomes

$$\sum_{n=-\infty}^{\infty} f(n) = \sqrt{2\pi} \sum_{n=-\infty}^{\infty} F(2n\pi). \quad (2.4.54)$$

We apply these formulas to prove the following series

(a) $\quad \displaystyle\sum_{n=-\infty}^{\infty} \dfrac{1}{(n^2+b^2)} = \dfrac{\pi}{b}\coth(\pi b), \quad (2.4.55)$

(b) $\quad \displaystyle\sum_{n=-\infty}^{\infty} \exp(-\pi n^2 t) = \dfrac{1}{\sqrt{t}} \sum_{n=-\infty}^{\infty} \exp\left(-\dfrac{\pi n^2}{t}\right). \quad (2.4.56)$

To show (a), we write $f(x) = (x^2+b^2)^{-1}$ so that $F(k) = \sqrt{\dfrac{\pi}{2}\dfrac{1}{b}} \exp(-b|k|)$. We now use (2.4.54) to derive

$$\sum_{n=-\infty}^{\infty} \dfrac{1}{(n^2+b^2)} = \dfrac{\pi}{b} \sum_{n=-\infty}^{\infty} \exp(-2n\pi b)$$

$$= \dfrac{\pi}{b}\left[\sum_{n=1}^{\infty} \exp(-2n\pi b) + \sum_{n=1}^{\infty} \exp(2n\pi b)\right]$$

which is, by writing $r = \exp(-2\pi b)$,

$$= \dfrac{\pi}{b}\left[\sum_{n=1}^{\infty} r^n + \sum_{n=1}^{\infty}\left(\dfrac{1}{r}\right)^n\right] = \dfrac{\pi}{b}\left(\dfrac{r}{1-r} + \dfrac{1}{1-r}\right)$$

$$= \dfrac{\pi}{b}\left(\dfrac{1+r}{1-r}\right) = \dfrac{\pi}{b}\coth(\pi b).$$

For (b), we assume $f(x) = \exp(-\pi t x^2)$ so that $F(k) = \dfrac{1}{\sqrt{2\pi t}} \exp\left(-\dfrac{k^2}{4\pi t}\right)$. Thus, the Poisson formula (2.4.54) gives

$$\sum_{n=-\infty}^{\infty} \exp(-\pi t n^2) = \dfrac{1}{\sqrt{t}} \sum_{n=-\infty}^{\infty} \exp(-\pi n^2/t).$$

The *Jacobi theta function* $\theta(t)$ is defined by

$$\theta(t) = \sum_{n=-\infty}^{\infty} \exp(-\pi t n^2), \qquad (2.4.57)$$

so that (2.4.56) gives the functional equation for the theta function

$$\sqrt{t}\,\theta(t) = \theta\left(\frac{1}{t}\right). \qquad (2.4.58)$$

2.5 Applications of Fourier Transforms to Ordinary Differential Equations

We consider the nth order linear ordinary differential equation with constant coefficients

$$Ly(x) = f(x), \qquad (2.5.1)$$

where L is the nth order differential operator given by

$$L \equiv a_n D^n + a_{n-1} D^{n-1} + \cdots + a_1 D + a_0, \qquad (2.5.2)$$

where $a_n, a_{n-1}, \ldots, a_1, a_0$ are constants, $D \equiv \dfrac{d}{dx}$ and $f(x)$ is a given function.

Application of the Fourier transform to both sides of (2.5.1) gives

$$\left[a_n(ik)^n + a_{n-1}(ik)^{n-1} + \cdots + a_1(ik) + a_0\right]Y(k) = F(k),$$

where $\mathscr{F}\{y(x)\} = Y(k)$ and $\mathscr{F}\{f(x)\} = F(k)$.

Or equivalently

$$P(ik)Y(k) = F(k),$$

where

$$P(z) = \sum_{r=0}^{n} a_r z^r.$$

Thus

$$Y(k) = \frac{F(k)}{P(ik)} = F(k)Q(k), \qquad (2.5.3)$$

where $Q(k) = \dfrac{1}{P(ik)}$.

Applying the Convolution Theorem 2.4.5 to (2.5.3) gives the formal solution

$$y(x) = \frac{1}{\sqrt{2\pi}} \int_{-\infty}^{\infty} f(\xi) q(x-\xi)\, d\xi, \qquad (2.5.4)$$

provided $q(x) = \mathscr{F}^{-1}\{Q(k)\}$ is known explicitly.

In order to give a physical interpretation of the solution (2.5.4), we consider the differential equation with a suddenly applied impulse function $f(x) = \delta(x)$ so that

$$L\{G(x)\} = \delta(x). \qquad (2.5.5)$$

The solution of this equation can be written from the inversion of (2.5.3) in the form

$$G(x) = \mathscr{F}^{-1}\left\{\frac{1}{\sqrt{2\pi}}Q(k)\right\} = \frac{1}{\sqrt{2\pi}}q(x). \tag{2.5.6}$$

Thus the solution (2.5.4) takes the form

$$y(x) = \int_{-\infty}^{\infty} f(\xi)G(x-\xi)d\xi. \tag{2.5.7}$$

Clearly, $G(x)$ behaves like a *Green's function*, that is, it is the response to a *unit impulse*. In any physical system, $f(x)$ usually represents the *input function*, while $y(x)$ is referred to as the *output* obtained by the superposition principle. The Fourier transform of $\{\sqrt{2\pi}\,G(x)\} = q(x)$ is called the *admittance*. In order to find the response to a given input, we determine the Fourier transform of the input function, multiply the result by the admittance, and then apply the inverse Fourier transform to the product so obtained.

We illustrate these ideas by solving a simple problem in the electrical circuit theory.

Example 2.5.1 (*Electric Current in a Simple Circuit*). The current $I(t)$ in a simple circuit containing the resistance R and inductance L satisfies the equation

$$L\frac{dI}{dt} + RI = E(t), \tag{2.5.8}$$

where $E(t)$ is the applied electromagnetic force.

With $E(t) = E_0 \exp(-a|t|)$, we use the Fourier transform with respect to time t to obtain

$$(ikL + R)\hat{I}(k) = E_0\sqrt{\frac{2}{\pi}}\frac{a}{(a^2 + k^2)}.$$

Or

$$\hat{I}(k) = \frac{aE_0}{iL}\sqrt{\frac{2}{\pi}}\frac{1}{\left(k - \frac{Ri}{L}\right)(k^2 + a^2)},$$

where $\mathscr{F}\{I(t)\} = \hat{I}(k)$. The inverse Fourier transform gives

$$I(t) = \frac{aE_0}{i\pi L}\int_{-\infty}^{\infty}\frac{\exp(ikt)\,dk}{\left(k - \frac{Ri}{L}\right)(k^2 + a^2)}. \tag{2.5.9}$$

This integral can be evaluated by the Cauchy Residue Theorem. For $t > 0$

$$I(t) = \frac{aE_0}{i\pi L} \cdot 2\pi i \left[\text{Residue at } k = \frac{Ri}{L} + \text{Residue at } k = ia\right]$$

$$= \frac{2aE_0}{L}\left[\frac{e^{-\frac{R}{L}t}}{\left(a^2 - \frac{R^2}{L^2}\right)} - \frac{e^{-at}}{2a\left(a - \frac{R}{L}\right)}\right]$$

$$= E_0\left[\frac{e^{-at}}{R-aL} - \frac{2aLe^{-\frac{R}{L}t}}{R^2 - a^2L^2}\right]. \qquad (2.5.10)$$

Similarly, for $t < 0$, the residue theorem gives

$$I(t) = -\frac{aE_0}{i\pi L} \cdot 2\pi i\left[\text{Residue at } k = -ia\right]$$

$$= -\frac{2aE_0}{L}\left[\frac{-Le^{at}}{(aL+R)2a}\right] = \frac{E_0 e^{at}}{(aL+R)}. \qquad (2.5.11)$$

At $t = 0$, the current is continuous and therefore

$$I(0) = \lim_{t \to 0} I(t) = \frac{E_0}{R+aL}.$$

If $E(t) = \delta(t)$, then $\hat{E}(k) = \frac{1}{\sqrt{2\pi}}$ and the solution is obtained by using the inverse Fourier transform

$$I(t) = \frac{1}{2\pi iL}\int_{-\infty}^{\infty}\frac{e^{ikt}}{k - \frac{iR}{L}}dk$$

which is, by the theorem of residues,

$$= \frac{1}{L}[\text{Residue at } k = iR/L]$$

$$= \frac{1}{L}\exp\left(-\frac{Rt}{L}\right). \qquad (2.5.12)$$

Thus the current tends to zero as $t \to \infty$ as expected.

Example 2.5.2 Find the solution of the ordinary differential equation

$$-\frac{d^2u}{dx^2} + a^2u = f(x), \quad -\infty < x < \infty \qquad (2.5.13)$$

by the Fourier transform method.

Application of the Fourier transform to (2.5.13) gives

$$U(k) = \frac{F(k)}{k^2 + a^2}.$$

This can readily be inverted by the Convolution Theorem 2.4.5 to obtain

$$u(x) = \frac{1}{\sqrt{2\pi}}\int_{-\infty}^{\infty}f(\xi)g(x-\xi)d\xi, \qquad (2.5.14)$$

where $g(x) = \mathscr{F}^{-1}\left\{\dfrac{1}{k^2+a^2}\right\} = \dfrac{1}{a}\sqrt{\dfrac{\pi}{2}}\exp(-a|x|)$ by Example 2.3.2.
Thus the final solution is

$$u(x) = \frac{1}{2a}\int_{-\infty}^{\infty} f(\xi)e^{-a|x-\xi|}\,d\xi. \tag{2.5.15}$$

Example 2.5.3 (*The Bernoulli-Euler Beam Equation*). We consider the vertical deflection $u(x)$ of an infinite beam on an elastic foundation under the action of a prescribed vertical load $W(x)$. The deflection $u(x)$ satisfies the ordinary differential equation

$$EI\frac{d^4u}{dx^4} + \kappa u = W(x), \quad -\infty < x < \infty \tag{2.5.16}$$

where EI is the flexural rigidity and κ is the foundation modulus of the beam. We find the solution assuming that $W(x)$ has a compact support and u, u', u'', u''' all tend to zero as $|x| \to \infty$.

We first rewrite (2.5.16) as

$$\frac{d^4u}{dx^4} + a^4 u = w(x) \tag{2.5.17}$$

where $a^4 = \kappa/EI$ and $w(x) = W(x)/EI$. Using the Fourier transform to (2.5.17) gives

$$U(k) = \frac{W(k)}{k^4 + a^4}.$$

The inverse Fourier transform gives the solution

$$u(x) = \frac{1}{\sqrt{2\pi}}\int_{-\infty}^{\infty}\frac{W(k)}{k^4+a^4}e^{ikx}\,dk$$

$$= \frac{1}{2\pi}\int_{-\infty}^{\infty}\frac{e^{ikx}}{k^4+a^4}dk\int_{-\infty}^{\infty} w(\xi)e^{-ik\xi}\,d\xi$$

$$= \int_{-\infty}^{\infty} w(\xi)G(\xi,x)\,d\xi, \tag{2.5.18}$$

where

$$G(\xi,x) = \frac{1}{2\pi}\int_{-\infty}^{\infty}\frac{e^{ik(x-\xi)}}{k^4+a^4}dk = \frac{1}{\pi}\int_{0}^{\infty}\frac{\cos k(x-\xi)\,dk}{k^4+a^4}. \tag{2.5.19}$$

This integral can be evaluated by the theorem of residues, or by using the table of Fourier integrals. We simply state the result

$$G(\xi,x) = \frac{1}{2a^3}\exp\left(-\frac{a}{\sqrt{2}}|x-\xi|\right)\sin\left[\frac{a(x-\xi)}{\sqrt{2}} + \frac{\pi}{4}\right]. \tag{2.5.20}$$

In particular, we find the explicit solution due to a concentrated load of unit strength acting at some point x_0, that is, $w(x) = \delta(x-x_0)$. Then the solution for this case becomes

$$u(x) = \int_{-\infty}^{\infty} \delta(\xi - x_0) G(x,\xi) d\xi = G(x, x_0). \qquad (2.5.21)$$

Thus the kernel $G(x,\xi)$ involved in the solution (2.5.18) has the physical significance of being the deflection, as a function of x, due to a unit point load acting at ξ. Thus the deflection due to a point load of strength $w(\xi)d\xi$ at ξ is $w(\xi)d\xi \cdot G(x,\xi)$, and hence (2.5.18) represents the superposition of all such incremental deflections.

The reader is referred to a more general dynamic problem of an infinite Bernoulli-Euler beam with damping and elastic foundation that has been solved by Stadler and Shreeves (1970), and also by Sheehan and Debnath (1972). These authors used the Fourier-Laplace transform method to determine the steady state and the transient solutions of the beam problem.

2.6   Solutions of Integral Equations

The method of Fourier transforms can be used to solve simple integral equations of the convolution type. We illustrate the method by examples.

We first solve the Fredholm integral equation with convolution kernel in the form

$$\int_{-\infty}^{\infty} f(t) g(x-t) dt + \lambda f(x) = u(x), \qquad (2.6.1)$$

where $g(x)$ and $u(x)$ are given functions and λ is a known parameter.

Application of the Fourier transform to (2.6.1) gives

$$\sqrt{2\pi}\, F(k) G(k) + \lambda\, F(k) = U(k).$$

Or

$$F(k) = \frac{U(k)}{\sqrt{2\pi}\, G(k) + \lambda}. \qquad (2.6.2)$$

The inverse Fourier transform leads to a formal solution

$$f(x) = \frac{1}{\sqrt{2\pi}} \int_{-\infty}^{\infty} \frac{U(k) e^{ikx}\, dk}{\sqrt{2\pi}\, G(k) + \lambda}. \qquad (2.6.3)$$

In particular, if $g(x) = \dfrac{1}{x}$ so that

$$G(k) = i \sqrt{\frac{\pi}{2}}\, \operatorname{sgn} k,$$

then the solution becomes

$$f(x) = \frac{1}{\sqrt{2\pi}} \int_{-\infty}^{\infty} \frac{U(k) e^{ikx}\, dk}{\lambda + i\pi \operatorname{sgn} k}. \qquad (2.6.4)$$

If $\lambda = 1$ and $g(x) = \frac{1}{2}\left(\frac{x}{|x|}\right)$ so that $G(k) = \frac{1}{\sqrt{2\pi}}\frac{1}{(ik)}$, the solution (2.6.3) reduces to the form

$$f(x) = \frac{1}{\sqrt{2\pi}} \int_{-\infty}^{\infty} (ik) \frac{U(k) e^{ikx} dk}{(1+ik)}$$

$$= \frac{1}{\sqrt{2\pi}} \int_{-\infty}^{\infty} \mathcal{F}\{u'(x)\} \, \mathcal{F}\{\sqrt{2\pi}\, e^{-x}\} e^{ikx}\, dk$$

$$= u'(x) * \sqrt{2\pi}\, e^{-x}$$

$$= \int_{-\infty}^{\infty} u'(\xi)\exp(\xi - x)\, d\xi. \tag{2.6.5}$$

Example 2.6.1 Find the solution of the integral equation

$$\int_{-\infty}^{\infty} f(x-\xi) f(\xi)\, d\xi = \frac{1}{x^2 + a^2}. \tag{2.6.6}$$

Application of the Fourier transform gives

$$\sqrt{2\pi}\, F(k) F(k) = \sqrt{\frac{\pi}{2}}\, \frac{e^{-a|k|}}{a}.$$

Or

$$F(k) = \frac{1}{\sqrt{2a}} \exp\left\{-\frac{1}{2} a|k|\right\}. \tag{2.6.7}$$

The inverse Fourier transform gives the solution

$$f(x) = \frac{1}{\sqrt{2\pi}}\frac{1}{\sqrt{2a}} \int_{-\infty}^{\infty} \exp\left(ikx - \frac{1}{2} a|k|\right) dk$$

$$= \frac{1}{2\sqrt{\pi a}}\left[\int_{0}^{\infty} \exp\left\{-k\left(\frac{a}{2}+ix\right)\right\} dk + \int_{0}^{\infty} \exp\left\{-k\left(\frac{a}{2}-ix\right)\right\} dk\right]$$

$$= \frac{1}{2\sqrt{\pi a}}\left[\frac{4a}{(4x^2 + a^2)}\right] = \sqrt{\frac{a}{\pi}} \cdot \frac{2}{(4x^2 + a^2)}.$$

Example 2.6.2 Solve the integral equation

$$\int_{-\infty}^{\infty} \frac{f(t)\, dt}{(x-t)^2 + a^2} = \frac{1}{(x^2 + b^2)}. \tag{2.6.8}$$

Taking the Fourier transform, we obtain

$$\sqrt{2\pi}\, F(k)\, \mathcal{F}\left\{\frac{1}{x^2 + a^2}\right\} = \sqrt{\frac{\pi}{2}}\, \frac{e^{-b|k|}}{b}$$

or

$$\sqrt{2\pi}\, F(k) \sqrt{\frac{\pi}{2}} \cdot \frac{e^{-a|k|}}{a} = \sqrt{\frac{\pi}{2}}\, \frac{e^{-b|k|}}{b}.$$

Integral Transforms and Their Applications 31

Thus
$$F(k) = \frac{1}{\sqrt{2\pi}}\left(\frac{a}{b}\right)\exp\{-|k|(b-a)\}. \tag{2.6.9}$$

The inverse Fourier transform leads to the solution

$$f(x) = \frac{a}{2\pi b}\int_{-\infty}^{\infty}\exp\left[ikx - |k|(b-a)\right]dk$$

$$= \frac{a}{2\pi b}\left[\int_0^{\infty}\exp\left[-k\{(b-a)+ix\}\right]dk + \int_0^{\infty}\exp\left[-k\{(b-a)-ix\}\right]dk\right]$$

$$= \frac{a}{2\pi b}\left[\frac{1}{(b-a)+ix} + \frac{1}{(b-a)-ix}\right]$$

$$= \left(\frac{a}{\pi b}\right)\frac{(b-a)}{(b-a)^2 + x^2}. \tag{2.6.10}$$

Example 2.6.3 Solve the integral equation
$$f(t) + 4\int_{-\infty}^{\infty}e^{-a|x-t|}f(t)dt = g(x). \tag{2.6.11}$$

Application of the Fourier transform gives
$$F(k) + 4\sqrt{2\pi}\,F(k)\cdot\frac{2a}{\sqrt{2\pi}(a^2+k^2)} = G(k)$$

$$F(k) = \frac{(a^2+k^2)}{a^2+k^2+8a}G(k). \tag{2.6.12}$$

The inverse Fourier transform gives
$$f(x) = \frac{1}{\sqrt{2\pi}}\int_{-\infty}^{\infty}\frac{(a^2+k^2)G(k)}{a^2+k^2+8a}e^{ikx}\,dk. \tag{2.6.13}$$

In particular, if $a=1$ and $g(x) = e^{-|x|}$ so that $G(k) = \sqrt{\frac{2}{\pi}}\frac{1}{1+k^2}$, then the solution (2.6.13) becomes
$$f(x) = \frac{1}{\pi}\int_{-\infty}^{\infty}\frac{e^{ikx}}{k^2+3^2}\,dk. \tag{2.6.14}$$

For $x > 0$, we use a semicircular closed contour in the lower half of the complex plane to evaluate (2.6.14). It turns out that
$$f(x) = \frac{1}{3}e^{-3x}. \tag{2.6.15}$$

Similarly, for $x < 0$, a semicircular closed contour in the upper half of the complex plane is used to evaluate (2.6.14) so that
$$f(x) = \frac{1}{3}e^{3x}, \quad x < 0. \tag{2.6.16}$$

Thus the final solution is
$$f(x) = \frac{1}{3}\exp(-3|x|). \qquad (2.6.17)$$

2.7 Solutions of Partial Differential Equations

In this section we illustrate how the Fourier transform method can be used to obtain the solution of boundary value and initial value problems for linear partial differential equations of different kinds.

Example 2.7.1 (*Dirichlet's Problem in the Half-Plane*). We consider the solution of the Laplace equation in the half-plane
$$u_{xx} + u_{yy} = 0, \quad -\infty < x < \infty, \quad y \geq 0, \qquad (2.7.1)$$
with the boundary conditions
$$u(x,0) = f(x), \quad -\infty < x < \infty, \qquad (2.7.2)$$
$$u(x,y) \to 0 \quad \text{as } |x| \to \infty, \ y \to \infty. \qquad (2.7.3)$$

We introduce the Fourier transform with respect to x
$$U(k,y) = \frac{1}{\sqrt{2\pi}} \int_{-\infty}^{\infty} e^{-ikx} u(x,y) dx \qquad (2.7.4)$$
so that (2.7.1)-(2.7.3) become
$$\frac{d^2 U}{dy^2} - k^2 U = 0 \qquad (2.7.5)$$
$$U(k,0) = F(k)$$
$$U(k,y) \to 0 \quad \text{as } y \to \infty. \qquad (2.7.6\text{ab})$$

Thus the solution of this transformed system is
$$U(k,y) = F(k) e^{-|k|y}. \qquad (2.7.7)$$

Application of the Convolution Theorem 2.4.5 gives the solution
$$u(x,y) = \frac{1}{\sqrt{2\pi}} \int_{-\infty}^{\infty} f(\xi) g(x-\xi) d\xi, \qquad (2.7.8)$$
where
$$g(x) = \mathcal{F}^{-1}\{e^{-|k|y}\} = \sqrt{\frac{2}{\pi}} \frac{y}{(x^2 + y^2)}. \qquad (2.7.9)$$

Consequently, the solution (2.7.8) becomes
$$u(x,y) = \frac{y}{\pi} \int_{-\infty}^{\infty} \frac{f(\xi) d\xi}{(x-\xi)^2 + y^2}, \quad y > 0. \qquad (2.7.10)$$

This is the well known *Poisson integral formula* in the half-plane. It is noted that
$$\lim_{y \to 0^+} u(x,y) = \int_{-\infty}^{\infty} f(\xi) \left[\lim_{y \to 0^+} \frac{y}{\pi} \cdot \frac{1}{(x-\xi)^2 + y^2} \right] d\xi = \int_{-\infty}^{\infty} f(\xi) \delta(x-\xi) d\xi, \qquad (2.7.11)$$
where Cauchy's definition of the delta function is used, that is,

$$\delta(x-\xi) = \lim_{y \to 0^+} \frac{y}{\pi} \cdot \frac{1}{(x-\xi)^2 + y^2}. \qquad (2.7.12)$$

This may be recognized as a solution of the Laplace equation for a dipole source at $(x, y) = (\xi, 0)$.

In particular, when

$$f(x) = T_0 H(a - |x|) \qquad (2.7.13)$$

the solution (2.7.10) reduces to

$$u(x, y) = \frac{y T_0}{\pi} \int_{-a}^{a} \frac{d\xi}{(\xi - x)^2 + y^2}$$

$$= \frac{T_0}{\pi} \left[\tan^{-1}\left(\frac{x+a}{y}\right) - \tan^{-1}\left(\frac{x-a}{y}\right) \right]$$

$$= \frac{T_0}{\pi} \tan^{-1}\left(\frac{2ay}{x^2 + y^2 - a^2}\right). \qquad (2.7.14)$$

The curves in the upper half-plane for which the steady state temperature is constant are known as *isothermal curves*. In this case, these curves represent a family of circular arcs

$$x^2 + y^2 - \alpha y = a^2 \qquad (2.7.15)$$

with centers on the y-axis and the fixed end points on the x-axis at $x = \pm a$.

Another special case deals with

$$f(x) = \delta(x). \qquad (2.7.16)$$

The solution for this case follows from (2.7.10) and is

$$u(x, t) = \frac{y}{\pi} \int_{-\infty}^{\infty} \frac{\delta(\xi) d\xi}{(x-\xi)^2 + y^2} = \frac{y}{\pi} \frac{1}{(x^2 + y^2)}. \qquad (2.7.17)$$

Further, we can readily deduce the solution of the *Neumann problem* in the half-plane from the solution of the Dirichlet problem.

Example 2.7.2 (*Neumann's Problem in the Half-Plane*). Find a solution of the Laplace equation

$$u_{xx} + u_{yy} = 0, \quad -\infty < x < \infty, \quad y > 0, \qquad (2.7.18)$$

with the boundary condition

$$u_y(x, 0) = f(x), \quad -\infty < x < \infty. \qquad (2.7.19)$$

This condition specifies the normal derivative on the boundary, and physically, it describes the fluid flow or heat flux at the boundary.

We define a new function $v(x, y) = u_y(x, y)$ so that

$$u(x, y) = \int^{y} v(x, \eta) \, d\eta, \qquad (2.7.20)$$

where an arbitrary constant can be added to the right hand side. Clearly, the function v satisfies the Laplace equation

$$\frac{\partial^2 v}{\partial x^2}+\frac{\partial^2 v}{\partial y^2}=\frac{\partial^2 u_y}{\partial x^2}+\frac{\partial^2 u_y}{\partial y^2}=\frac{\partial}{\partial y}\left(u_{xx}+u_{yy}\right)=0,$$

with the boundary condition
$$v(x,0)=u_y(x,0)=f(x) \text{ for } -\infty<x<\infty.$$

Thus $v(x,y)$ satisfies the Laplace equation with the Dirichlet condition on the boundary. Obviously, the solution is given by (2.7.10), that is,

$$v(x,y)=\frac{y}{\pi}\int_{-\infty}^{\infty}\frac{f(\xi)d\xi}{(x-\xi)^2+y^2}. \tag{2.7.21}$$

Then the solution $u(x,y)$ can be obtained from (2.7.20) in the form

$$u(x,y)=\int_{-\infty}^{y}v(x,\eta)d\eta=\frac{1}{\pi}\int_{-\infty}^{y}\eta d\eta\int_{-\infty}^{\infty}\frac{f(\xi)d\xi}{(x-\xi)^2+\eta^2}$$

$$=\frac{1}{\pi}\int_{-\infty}^{\infty}f(\xi)d\xi\int_{-\infty}^{y}\frac{\eta d\eta}{(x-\xi)^2+\eta^2}, \quad y>0$$

$$=\frac{1}{2\pi}\int_{-\infty}^{\infty}f(\xi)\log\left[(x-\xi)^2+y^2\right]d\xi, \tag{2.7.22}$$

where an arbitrary constant can be added to this solution. In other words, the solution of any Neumann problem is uniquely determined up to an arbitrary constant.

Example 2.7.3 (*The Cauchy Problem for the Diffusion Equation*). We consider the initial value problem for a one-dimensional diffusion equation with no sources or sinks

$$u_t=\kappa u_{xx}, \quad -\infty<x<\infty, \quad t>0, \tag{2.7.23}$$

where κ is a diffusivity constant with the initial condition

$$u(x,0)=f(x), \quad -\infty<x<\infty. \tag{2.7.24}$$

We solve this problem using the Fourier transform in the space variable x defined by (2.7.4). Application of this transform to (2.7.23)-(2.7.24) gives

$$U_t=-\kappa k^2 U, \quad t>0, \tag{2.7.25}$$
$$U(k,0)=F(k). \tag{2.7.26}$$

The solution of the transformed system is

$$U(k,t)=F(k)e^{-\kappa k^2 t}. \tag{2.7.27}$$

The inverse Fourier transform gives the solution

$$u(x,t)=\frac{1}{\sqrt{2\pi}}\int_{-\infty}^{\infty}F(k)\exp\left[(ikx-\kappa k^2 t)\right]dk$$

which is, by the Convolution Theorem 2.4.5,

$$=\frac{1}{\sqrt{2\pi}}\int_{-\infty}^{\infty}f(\xi)g(x-\xi)\,d\xi, \tag{2.7.28}$$

where

$$g(x) = \mathscr{F}^{-1}\{e^{-\kappa k^2 t}\} = \frac{1}{\sqrt{2\kappa t}} \exp\left(-\frac{x^2}{4\kappa t}\right), \text{ by (2.3.5)}.$$

Thus solution (2.7.28) becomes

$$u(x,t) = \frac{1}{\sqrt{4\pi\kappa t}} \int_{-\infty}^{\infty} f(\xi) \exp\left[-\frac{(x-\xi)^2}{4\kappa t}\right] d\xi. \tag{2.7.29}$$

The integrand involved in the solution consists of the initial value $f(x)$ and Green's function (or *elementary solution*) $G(x-\xi,t)$ of the diffusion equation for the infinite interval:

$$G(x-\xi,t) = \frac{1}{\sqrt{4\pi\kappa t}} \exp\left[-\frac{(x-\xi)^2}{4\kappa t}\right]. \tag{2.7.30}$$

So, in terms of $G(x-\xi,t)$, solution (2.7.29) can be written as

$$u(x,t) = \int_{-\infty}^{\infty} f(\xi)\, G(x-\xi,t)\, d\xi \tag{2.7.31}$$

so that, in the limit as $t \to 0+$, this formally becomes

$$u(x,0) = f(x) = \int_{-\infty}^{\infty} f(\xi) \lim_{t \to 0+} G(x-\xi,t)\, d\xi.$$

The limit of $G(x-\xi,t)$ represents the Dirac delta function

$$\delta(x-\xi) = \lim_{t \to 0+} \frac{1}{2\sqrt{\pi\kappa t}} \exp\left[-\frac{(x-\xi)^2}{4\kappa t}\right]. \tag{2.7.32}$$

It is important to point out that the integrand in (2.7.31) consists of the initial temperature distribution $f(x)$ and Green's function $G(x-\xi,t)$ which represents the temperature response along the rod at time t due to an initial unit impulse of heat at $x = \xi$. The physical meaning of the solution (2.7.31) is that the initial temperature distribution $f(x)$ is decomposed into a spectrum of impulses of magnitude $f(\xi)$ at each point $x = \xi$ to form the resulting temperature $f(\xi)\, G(x-\xi,t)$. Thus the resulting temperature is integrated to find solution (2.7.31). This is called the *principle of superposition*.

We make the change of variable

$$\frac{\xi - x}{2\sqrt{\kappa t}} = \zeta, \quad d\zeta = \frac{d\xi}{2\sqrt{\kappa t}}$$

to express solution (2.7.29) in the form

$$u(x,t) = \frac{1}{\sqrt{\pi}} \int_{-\infty}^{\infty} f\left(x + 2\sqrt{\kappa t}\,\zeta\right) \exp\left(-\zeta^2\right) d\zeta. \tag{2.7.33}$$

The integral solution (2.7.33) or (2.7.29) is called the *Poisson integral representation* of the temperature distribution. This integral is convergent for all time $t > 0$, and the integrals obtained from (2.7.33) by differentiation under the integral sign with respect to x and t are uniformly convergent in the neighborhood of the point (x,t). Hence the solution $u(x,t)$ and its derivatives of all orders exist for $t > 0$.

Finally, we consider a special case where
$$f(x) = T_0 H(x) \tag{2.7.34}$$
where T_0 is a constant. In this case, solution (2.7.29) becomes
$$u(x,t) = \frac{T_0}{2\sqrt{\pi\kappa t}} \int_0^\infty \exp\left[-\frac{(x-\xi)^2}{4\kappa t}\right] d\xi. \tag{2.7.35}$$

Introducing the change of variable $\eta = \dfrac{\xi - x}{2\sqrt{\kappa t}}$, we can express solution (2.7.35) in the form
$$u(x,t) = \frac{T_0}{\sqrt{\pi}} \int_{-x/2\sqrt{\kappa t}}^{\infty} e^{-\eta^2} d\eta = \frac{T_0}{2} \operatorname{erfc}\left(-\frac{x}{2\sqrt{\kappa t}}\right)$$
$$= \frac{T_0}{2}\left[1 + \operatorname{erf}\left(\frac{x}{2\sqrt{\kappa t}}\right)\right]. \tag{2.7.36}$$

Example 2.7.4 (*The Cauchy Problem for the Wave Equation*). Obtain the d'Alembert solution of the initial value problem for the wave equation
$$u_{tt} = c^2 u_{xx}, \quad -\infty < x < \infty, \quad t > 0, \tag{2.7.37}$$
with the initial data
$$u(x,0) = f(x), \quad u_t(x,0) = g(x), \quad -\infty < x < \infty. \tag{2.7.38ab}$$
Application of the Fourier transform $\mathscr{F}\{u(x,t)\} = U(k,t)$ to this system gives
$$\frac{d^2 U}{dt^2} + c^2 k^2 U = 0$$
$$U(k,0) = F(k), \quad \left(\frac{dU}{dt}\right)_{t=0} = G(k).$$

The solution of the transformed system is
$$U(k,t) = A\, e^{ickt} + B\, e^{-ickt},$$
where A and B are constants to be determined from the transformed data $A + B = F(k)$ and $A - B = \dfrac{1}{ikc} G(k)$. Solving for A and B, we obtain the solution
$$U(k,t) = \frac{1}{2} F(k)\left(e^{ickt} + e^{-ickt}\right) + \frac{G(k)}{2ick}\left(e^{ickt} - e^{-ickt}\right). \tag{2.7.39}$$

Thus the inverse Fourier transform of (2.7.39) yields the solution

$$u(x,t) = \frac{1}{2}\left[\frac{1}{\sqrt{2\pi}} \int_{-\infty}^{\infty} F(k)\left\{e^{ik(x+ct)} + e^{ik(x-ct)}\right\} dk\right]$$
$$+ \frac{1}{2c}\left[\frac{1}{\sqrt{2\pi}} \int_{-\infty}^{\infty} \frac{G(k)}{ik}\left\{e^{ik(x+ct)} - e^{ik(x-ct)}\right\} dk\right]. \quad (2.7.40)$$

We use the following results

$$f(x) = \mathscr{F}^{-1}\{F(k)\} = \frac{1}{\sqrt{2\pi}} \int_{-\infty}^{\infty} e^{ikx} F(k) dk,$$

$$g(x) = \mathscr{F}^{-1}\{G(k)\} = \frac{1}{\sqrt{2\pi}} \int_{-\infty}^{\infty} e^{ikx} G(k) dk,$$

to obtain the solution as

$$u(x,t) = \frac{1}{2}[f(x-ct) + f(x+ct)] + \frac{1}{2c}\frac{1}{\sqrt{2\pi}} \int_{-\infty}^{\infty} G(k) dk \int_{x-ct}^{x+ct} e^{ik\xi} d\xi$$

$$= \frac{1}{2}[f(x-ct) + f(x+ct)] + \frac{1}{2c} \int_{x-ct}^{x+ct} d\xi \left[\frac{1}{\sqrt{2\pi}} \int_{-\infty}^{\infty} e^{ik\xi} G(k) dk\right]$$

$$= \frac{1}{2}[f(x-ct) + f(x+ct)] + \frac{1}{2c} \int_{x-ct}^{x+ct} g(\xi) d\xi. \quad (2.7.41)$$

This is the well known *d'Alembert's solution* of the wave equation.

The method and the form of the solution reveal several important features of the wave equation. First, the method of solution essentially proves the existence of the d'Alembert solution and the solution is unique provided $f(x)$ is twice continuously differentiable and $g(x)$ is continuously differentiable. Second, the terms involving $f(x \pm ct)$ in (2.7.41) show that disturbances are propagated along the characteristics with velocity c. Both terms combined together suggest that the value of the solution at position x and at time t depends only on the initial values of $f(x)$ at $x-ct$ and $x+ct$ and the values of $g(x)$ between these two points. The interval $(x-ct, x+ct)$ is called the *domain of dependence* of the variable (x,t). Finally, the solution depends continuously on the initial data, that is, the problem is well posed. In other words, a small change in either $f(x)$ or $g(x)$ results in a correspondingly small change in the solution $u(x,t)$.

In particular if $f(x) = \exp(-x^2)$ and $g(x) \equiv 0$, the time development of solution (2.7.41) with $c=1$ is shown in Figure 2.5. In this case, the solution becomes

$$u(x,t) = \frac{1}{2}\left[e^{-(x-t)^2} + e^{-(x+t)^2}\right]. \quad (2.7.42)$$

As shown in Figure 2.5, the initial form $f(x) = \exp(-x^2)$ is found to split into two waves propagating in opposite direction with unit velocity.

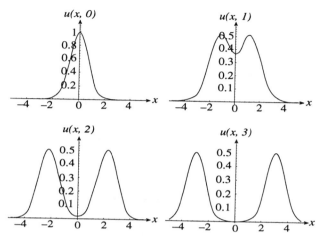

Figure 2.5 The time development of solution (2.7.42).

Example 2.7.5 (*The Schrödinger Equation in Quantum Mechanics*). The time-dependent Schrödinger equation of a particle of mass m is

$$i\hbar \psi_t = \left[V(\mathbf{x}) - \frac{\hbar^2}{2m} \nabla^2 \right] \psi = H\psi, \qquad (2.7.43)$$

where $h = 2\pi\hbar$ is the Planck constant, $\psi(\mathbf{x},t)$ is the wave function, $V(\mathbf{x})$ is the potential, $\nabla^2 = \frac{\partial^2}{\partial x^2} + \frac{\partial^2}{\partial y^2} + \frac{\partial^2}{\partial z^2}$ is the three-dimensional Laplacian, and H is the Hamiltonian.

If $V(\mathbf{x}) = \text{constant} = V$, we can seek a plane wave solution of the form

$$\psi(\mathbf{x},t) = A \exp[i(\boldsymbol{\kappa} \cdot \mathbf{x} - \omega t)], \qquad (2.7.44)$$

where A is a constant amplitude, $\boldsymbol{\kappa} = (k, l, m)$ is the wavenumber vector, and ω is the frequency.

Substituting this solution into (2.7.43), we conclude that this solution is possible provided the following relation is satisfied:

$$i\hbar(-i\omega) = V - \frac{\hbar^2}{2m}(i\kappa)^2, \quad \kappa^2 = k^2 + l^2 + m^2.$$

Or

$$\hbar \omega = V + \frac{\hbar^2 \kappa^2}{2m}. \qquad (2.7.45)$$

This is called the *dispersion relation* and shows that the sum of the potential energy V and the kinematic energy $\dfrac{(\hbar k)^2}{2m}$ is equal to the total energy $\hbar\omega$. Further, the kinetic energy

$$K.E. = \frac{1}{2m}(\hbar\kappa)^2 = \frac{p^2}{2m}, \qquad (2.7.46)$$

where $p = \hbar\kappa$ is the momentum of the particle. In the one-dimensional case, the group velocity is

$$C_g = \frac{\partial\omega}{\partial k} = \frac{\hbar k}{m} = \frac{p}{m} = \frac{mv}{v} = v. \qquad (2.7.47)$$

This shows that the group velocity is equal to the classical particle velocity v.

We now use the Fourier transform method to solve the one-dimensional Schrödinger equation for a free particle ($V \equiv 0$), that is,

$$i\hbar\,\psi_t = -\frac{\hbar^2}{2m}\psi_{xx}, \quad -\infty < x < \infty,\ t > 0, \qquad (2.7.48)$$

$$\psi(x,0) = \psi_0(x), \quad -\infty < x < \infty, \qquad (2.7.49)$$

$$\psi(x,t) \to 0 \quad \text{as } |x| \to \infty. \qquad (2.7.50)$$

Application of the Fourier transform to (2.7.48)-(2.7.50) gives

$$\Psi_t = -\frac{i\hbar k^2}{2m}\Psi, \qquad (2.7.51)$$

$$\Psi(k,0) = \Psi_0(k). \qquad (2.7.52)$$

The solution of this transformed system is

$$\Psi(k,t) = \Psi_0(k)\exp(-i\alpha k^2 t), \quad \alpha = \frac{\hbar}{2m}. \qquad (2.7.53)$$

The inverse Fourier transform gives the formal solution

$$\psi(x,t) = \frac{1}{\sqrt{2\pi}}\int_{-\infty}^{\infty}\Psi_0(k)\exp\{ik(x - \alpha kt)\}\,dk$$

$$= \frac{1}{2\pi}\int_{-\infty}^{\infty}e^{-iky}\psi(y,0)\,dy\int_{-\infty}^{\infty}\exp\{ik(x - \alpha kt)\}\,dk$$

$$= \frac{1}{2\pi}\int_{-\infty}^{\infty}\psi(y,0)\,dy\int_{-\infty}^{\infty}\exp\{ik(x - y - \alpha kt)\}\,dk, \qquad (2.7.54)$$

We rewrite the integrand of the second integral

$$\exp[ik(x - y - \alpha kt)]$$

$$= \exp\left[-i\alpha t\left\{k^2 - 2k\cdot\frac{x-y}{2\alpha t} + \left(\frac{x-y}{2\alpha t}\right)^2 - \left(\frac{x-y}{2\alpha t}\right)^2\right\}\right]$$

$$= \exp\left[-i\alpha t\left\{k - \frac{x-y}{2\alpha t}\right\}^2\right]\exp\left[\frac{i(x-y)^2}{4\alpha t}\right]$$

$$= \exp\left[\frac{i(x-y)^2}{4\alpha t}\right] \exp(-i\alpha t \xi^2), \quad \xi = k - \frac{x-y}{2\alpha t}.$$

Using this result in (2.7.54), we obtain

$$\psi(x,t) = \frac{1}{2\pi} \int_{-\infty}^{\infty} \exp\left[\frac{i(x-y)^2}{4\alpha t}\right] \psi(y,0) dy \int_{-\infty}^{\infty} \exp(-i\alpha t \xi^2) d\xi$$

$$= \frac{1}{2\pi} \sqrt{\frac{\pi}{2\alpha t}} (1-i) \int_{-\infty}^{\infty} \exp\left[\frac{i(x-y)^2}{4\alpha t}\right] \psi(y,0) dy$$

$$= \frac{(1-i)}{2\sqrt{2\alpha\pi t}} \int_{-\infty}^{\infty} \exp\left[\frac{i(x-y)^2}{4\alpha t}\right] \psi(y,0) dy. \tag{2.7.55}$$

This is the integral solution of the problem.

Example 2.7.6 (*Slowing Down of Neutrons*). We consider the problem of slowing down neutrons in an infinite medium with a source of neutrons governed by

$$u_t = u_{xx} + \delta(x)\delta(t), \quad -\infty < x < \infty, \quad t > 0, \tag{2.7.56}$$

$$u(x,0) = \delta(x), \quad -\infty < x < \infty, \tag{2.7.57}$$

$$u(x,t) \to 0 \quad \text{as } |x| \to \infty \text{ for } t > 0, \tag{2.7.58}$$

where $u(x,t)$ represents the number of neutrons per unit volume per unit time which reach the age t and $\delta(x)\delta(t)$ is the source function.

Application of the Fourier transform method gives

$$\frac{dU}{dt} + k^2 U = \frac{1}{\sqrt{2\pi}} \delta(t),$$

$$U(k,0) = \frac{1}{\sqrt{2\pi}}.$$

The solution of this transformed system is

$$U(k,t) = \frac{1}{\sqrt{2\pi}} e^{-k^2 t}. \tag{2.7.59}$$

The inverse Fourier transform gives the solution

$$u(x,t) = \frac{1}{2\pi} \int_{-\infty}^{\infty} e^{ikx - k^2 t} dk = \frac{1}{\sqrt{2\pi}} \mathcal{F}^{-1}\{e^{-k^2 t}\} = \frac{1}{2\sqrt{\pi}} \exp\left(-\frac{x^2}{4t}\right). \tag{2.7.60}$$

Example 2.7.7 (*One-Dimensional Wave Equation*). Obtain the solution of the one-dimensional wave equation

$$u_{tt} = c^2 u_{xx}, \quad -\infty < x < \infty, \quad t > 0, \tag{2.7.61}$$

$$u(x,0) = 0, \quad u_t(x,0) = \delta(x), \quad -\infty < x < \infty. \tag{2.7.62ab}$$

Making reference to Example 2.7.4, we find $f(x) \equiv 0$ and $g(x) = \delta(x)$ so that $F(k) = 0$ and $G(k) = \frac{1}{\sqrt{2\pi}}$. The solution for $U(k,t)$ is given by

$$U(k,t) = \frac{1}{2c\sqrt{2\pi}} \left[\frac{e^{ickt}}{ik} - \frac{e^{-ickt}}{ik} \right].$$

Thus the inverse Fourier transform gives

$$u(x,t) = \frac{1}{2c\sqrt{2\pi}} \mathscr{F}^{-1}\left\{ \frac{e^{ickt}}{ik} - \frac{e^{-ickt}}{ik} \right\}$$

$$= \frac{1}{2c\sqrt{2\pi}} \left[\sqrt{\frac{\pi}{2}} \{sgn(x+ct) - sgn(x-ct)\} \right]$$

$$= \frac{1}{4c}\left[sgn(x+ct) - sgn(x-ct)\right]$$

$$= \begin{cases} \frac{1-1}{4c} = 0, & |x| > ct > 0 \\ \frac{1+1}{4c} = \frac{1}{2c}, & |x| < ct \end{cases}.$$

In other words, the solution can be written as

$$u(x,t) = \frac{1}{2c} H(c^2 t^2 - x^2).$$

Example 2.7.8 (*Linearized Shallow Water Equations in a Rotating Ocean*). The horizontal equations of motion of a uniformly rotating inviscid homogeneous ocean of constant depth h are

$$u_t - fv = -g\,\eta_x, \qquad (2.7.63)$$
$$v_t + fu = 0, \qquad (2.7.64)$$
$$\eta_t + hu_x = 0, \qquad (2.7.65)$$

where $f = 2\Omega \sin\theta$ is the Coriolis parameter, which is constant in the present problem, g is the acceleration due to gravity, $\eta(x,t)$ is the free surface elevation, $u(x,t)$ and $v(x,t)$ are the velocity fields. The wave motion is generated by the prescribed free surface elevation at $t = 0$ so that the initial conditions are

$$u(x,0) = 0 = v(x,0), \quad \eta(x,0) = \eta_0\, H(a - |x|), \qquad (2.7.66\text{abc})$$

and the velocity field and elevation function vanish at infinity.

We apply the Fourier transform with respect to x defined by

$$\mathscr{F}\{f(x,t)\} = F(k,t) = \frac{1}{\sqrt{2\pi}} \int_{-\infty}^{\infty} e^{-ikx} f(x,t)\, dx \qquad (2.7.67)$$

to the system (2.7.63)-(2.7.65) so that the system becomes

$$\frac{dU}{dt} - fV = -gikE$$

$$\frac{dV}{dt} + fU = 0$$

$$\frac{dE}{dt} = -hikU$$

$$U(k,0) = 0 = V(k,0), \quad E(k,0) = 2\eta_0 \left(\frac{\sin ak}{k}\right), \quad (2.7.68abc)$$

where $E(k,t) = \mathcal{F}\{\eta(x,t)\}$.

Elimination of U and V from the transformed system gives a single equation for $E(k,t)$ as

$$\frac{d^3E}{dt^3} + \omega^2 \frac{dE}{dt} = 0, \quad (2.7.69)$$

where $\omega^2 = f^2 + c^2k^2$ and $c^2 = gh$. The general solution of (2.7.69) is

$$E(k,t) = A + B\cos\omega t + C\sin\omega t, \quad (2.7.70)$$

where A, B, and C are arbitrary constants to be determined from (2.7.68c) and

$$\left(\frac{d^2E}{dt^2}\right)_{t=0} = -c^2k^2 \; E(k,0) = -c^2k^2 \cdot 2\eta_0 \frac{\sin ak}{k},$$

which gives

$$B = 2\eta_0 \frac{\sin ak}{k} \cdot \left(\frac{c^2k^2}{\omega^2}\right).$$

Also $\left(\frac{dE}{dt}\right)_{t=0} = 0$ gives $C \equiv 0$ and (2.7.68c) implies $A + B = 2\eta_0 \frac{\sin ak}{k}$.

Consequently, the solution (2.7.70) becomes

$$E(k,t) = 2\eta_0 \left(\frac{\sin ak}{k}\right) \frac{f^2 + c^2k^2 \cos\omega t}{(f^2 + c^2k^2)}. \quad (2.7.71)$$

Similarly

$$U(k,t) = \frac{2\eta_0 \sin ak}{ih} \cdot \frac{c^2 \sin\omega t}{\sqrt{c^2k^2 + f^2}}, \quad (2.7.72)$$

$$V(k,t) = \frac{1}{f}\left(\frac{dU}{dt} + gik\, E\right). \quad (2.7.73)$$

The inverse Fourier transform gives the formal solution for $\eta(x,t)$

$$\eta(x,t) = \frac{2\eta_0}{\sqrt{2\pi}} \int_{-\infty}^{\infty} \frac{\sin ak}{k} \cdot \frac{f^2 + c^2k^2 \cos\omega t}{(f^2 + c^2k^2)} e^{ikx} \, dk. \quad (2.7.74)$$

Similar integral expressions for $u(x,t)$ and $v(x,t)$ can be obtained.

Example 2.7.9 (*Sound Waves Induced by a Spherical Body*). We consider propagation of sound waves in an unbounded fluid medium generated by an impulsive radial acceleration of a sphere of radius a. Such waves are assumed to be spherically symmetric and the associated velocity potential on the pressure field $p(r,t)$ satisfies the wave equation

$$\frac{\partial^2 p}{\partial t^2} = c^2 \left[\frac{1}{r^2}\frac{\partial}{\partial r}\left(r^2 \frac{\partial p}{\partial r}\right)\right], \quad (2.7.75)$$

where c is the speed of sound. The boundary condition required for the problem is

$$\frac{1}{\rho_0}\left(\frac{\partial p}{\partial r}\right) = -a_0\,\delta(t) \qquad \text{at } r = a, \tag{2.7.76}$$

where ρ_0 is the mean density of the fluid and a_0 is a constant.

Application of the Fourier transform of $p(r,t)$ with respect to time t gives

$$\frac{1}{r^2}\frac{d}{dr}\left(r^2\frac{dP}{dr}\right) = -k^2 P(r,\omega), \tag{2.7.77}$$

$$\frac{dP}{dr} = -\frac{a_0 \rho_0}{\sqrt{2\pi}}, \text{ at } r = a, \tag{2.7.78}$$

where $\mathcal{F}\{p(r,t)\} = P(r,\omega)$ and $k^2 = \frac{\omega^2}{c^2}$.

The general solution of (2.7.77)-(2.7.78) is

$$P(r,\omega) = \frac{A}{r}e^{ikr} + \frac{B}{r}e^{-ikr}, \tag{2.7.79}$$

where A and B are arbitrary constants.

The inverse Fourier transform gives the solution

$$p(r,t) = \frac{1}{\sqrt{2\pi}}\int_{-\infty}^{\infty}\left[\frac{A}{r}e^{i(\omega t + kr)} + \frac{B}{r}e^{i(\omega t - kr)}\right]d\omega. \tag{2.7.80}$$

The first term of the integrand represents incoming spherical waves generated at infinity and the second term corresponds to outgoing spherical waves due to the impulsive radial acceleration of the sphere. Since there is no disturbance at infinity, we impose the *Sommerfeld radiation condition* at infinity to eliminate the incoming waves so that $A = 0$, and B is calculated using (2.7.78). Thus the inverse Fourier transform gives the formal solution

$$p(r,t) = \left(\frac{a_0\rho_0 a^2}{2\pi r}\right)\int_{-\infty}^{\infty}\frac{\exp\left[i\omega\left\{t - \frac{r-a}{c}\right\}\right]d\omega}{\left(1 + \frac{i\omega a}{c}\right)}. \tag{2.7.81}$$

We next choose a closed contour with a semicircle in the upper half plane and the real ω-axis. Using the Cauchy theory of residues, we calculate the residue contribution from the pole at $\omega = ic/a$. Finally, it turns out that the final solution is

$$u(r,t) = \frac{\rho_0 a_0 ca}{r}\exp\left[-\frac{c}{a}\left(t - \frac{r-a}{c}\right)\right]H\left(t - \frac{r-a}{c}\right). \tag{2.7.82}$$

Example 2.7.10 (*The Linearized Korteweg–de Vries Equation*). The linearized KdV equation for the free surface elevation $\eta(x,t)$ in an inviscid water of constant depth h is

$$\eta_t + c\eta_x + \frac{ch^2}{6}\eta_{xxx} = 0, \quad -\infty < x < \infty, \quad t > 0, \tag{2.7.83}$$

where $c = \sqrt{gh}$ is the shallow water speed.

Solve equation (2.7.83) with the initial condition
$$\eta(x,0) = f(x), \quad -\infty < x < \infty. \tag{2.7.84}$$
Application of the Fourier transform $\mathscr{F}\{\eta(x,t)\} = E(k,t)$ to the KdV system gives the solution for $E(k,t)$ in the form
$$E(k,t) = F(k)\exp\left[ikct\left(\frac{k^2h^2}{6} - 1\right)\right].$$
The inverse transform gives
$$\eta(x,t) = \frac{1}{\sqrt{2\pi}} \int_{-\infty}^{\infty} F(k)\exp\left[ik\left\{(x-ct) + \left(\frac{cth^2}{6}\right)k^2\right\}\right] dk. \tag{2.7.85}$$
In particular, if $f(x) = \delta(x)$, then (2.7.85) reduces to the Airy integral
$$\eta(x,t) = \frac{1}{\pi} \int_0^{\infty} \cos\left[k(x-ct) + \left(\frac{cth^2}{6}\right)k^3\right] dk \tag{2.7.86}$$
which is, in terms of the Airy function,
$$= \left(\frac{cth^2}{2}\right)^{-\frac{1}{3}} Ai\left[\left(\frac{cth^2}{2}\right)^{-\frac{1}{3}}(x-ct)\right], \tag{2.7.87}$$
where the Airy function $Ai(z)$ is defined by
$$Ai(z) = \frac{1}{\pi} \int_0^{\infty} \cos\left(kz + \frac{1}{3}k^3\right) dk. \tag{2.7.88}$$

Example 2.7.11 (*Biharmonic Equation in Fluid Mechanics*). Usually, the biharmonic equation arises in fluid mechanics and in elasticity. The equation can readily be solved by using the Fourier transform method. We first derive a biharmonic equation from the *Navier-Stokes equations* of motion in a viscous fluid which is given by
$$\frac{\partial \mathbf{u}}{\partial t} + (\mathbf{u} \cdot \nabla)\mathbf{u} = \mathbf{F} - \frac{1}{\rho}\nabla p + \nu\nabla^2\mathbf{u}, \tag{2.7.89}$$
where $\mathbf{u} = (u, v, w)$ is the velocity field, $\mathbf{F}$ is the external force per unit mass of the fluid, p is the pressure, ρ is the density and ν is the kinematic viscosity of the fluid.

The conservation of mass of an incompressible fluid is described by the *continuity equation*
$$\text{div } \mathbf{u} = 0. \tag{2.7.90}$$
In terms of some representative length scale L and velocity scale U, it is convenient to introduce the nondimensional flow variables
$$\mathbf{x}' = \frac{\mathbf{x}}{L}, \quad t' = \frac{Ut}{L}, \quad \mathbf{u}' = \frac{\mathbf{u}}{U}, \quad p' = \frac{p}{\rho U^2}. \tag{2.7.91}$$
In terms of these nondimensional variables, equation (2.7.89) without the external force can be written, dropping the primes, as

$$\frac{\partial \mathbf{u}}{\partial t} + (\mathbf{u} \cdot \nabla)\mathbf{u} = -\nabla p + \frac{1}{R}\nabla^2 \mathbf{u}, \qquad (2.7.92)$$

where $R = UL/v$ is called the *Reynolds number*. Physically, it measures the ratio of inertial forces of the order U^2/L to viscous forces of the order vU/L^2, and it has special dynamical significance. This is one of the most fundamental nondimensional parameters for the specification of the dynamical state of viscous flow fields.

In the absence of the external force, $\mathbf{F} = \mathbf{0}$, it is preferable to write the Navier-Stokes equations (2.7.89) in the form (since $\mathbf{u} \times \boldsymbol{\omega} = \frac{1}{2}\nabla u^2 - \mathbf{u} \cdot \nabla \mathbf{u}$)

$$\frac{\partial \mathbf{u}}{\partial t} - \mathbf{u} \times \boldsymbol{\omega} = -\nabla\left(\frac{p}{\rho} + \frac{1}{2}u^2\right) - v\nabla^2 \mathbf{u}, \qquad (2.7.93)$$

where $\boldsymbol{\omega} = \text{curl } \mathbf{u}$ is the *vorticity vector* and $u^2 = \mathbf{u} \cdot \mathbf{u}$.

We can eliminate the pressure p from (2.7.93) by taking the curl of (2.7.93), giving

$$\frac{\partial \boldsymbol{\omega}}{\partial t} - \text{curl }(\mathbf{u} \times \boldsymbol{\omega}) = v\nabla^2 \boldsymbol{\omega} \qquad (2.7.94)$$

which becomes, by div $\mathbf{u} = 0$ and div $\boldsymbol{\omega} = 0$,

$$\frac{\partial \boldsymbol{\omega}}{\partial t} = (\boldsymbol{\omega} \cdot \nabla)\mathbf{u} - (\mathbf{u} \cdot \nabla)\boldsymbol{\omega} + v\nabla^2 \boldsymbol{\omega}. \qquad (2.7.95)$$

This is universally known as the *vorticity transport equation*. The left hand-side represents the rate of change of vorticity. The first two terms on the right-hand side represent the rate of change of vorticity due to stretching and twisting of vortex lines. The last term describes the diffusion of vorticity by molecular viscosity.

In case of two-dimensional flow, $(\boldsymbol{\omega} \cdot \nabla)\mathbf{u} = 0$, equation (2.7.95) becomes

$$\frac{D\boldsymbol{\omega}}{dt} = \frac{\partial \boldsymbol{\omega}}{\partial t} + (\mathbf{u} \cdot \nabla)\boldsymbol{\omega} = v\nabla^2 \boldsymbol{\omega}, \qquad (2.7.96)$$

where $\mathbf{u} = (u, v, 0)$ and $\boldsymbol{\omega} = (0, 0, \zeta)$, and $\zeta = v_x - u_y$. Equation (2.7.96) shows that only convection and conduction occur. In terms of the stream function $\psi(x, y)$ where

$$u = \psi_y, \quad v = -\psi_x, \quad \omega = -\nabla^2 \psi, \qquad (2.7.97)$$

which satisfy (2.7.90) identically, equation (2.7.96) assumes the form

$$\frac{\partial}{\partial t}(\nabla^2 \psi) + \left(\frac{\partial \psi}{\partial y}\frac{\partial}{\partial x} - \frac{\partial \psi}{\partial x}\frac{\partial}{\partial y}\right)\nabla^2 \psi = v\nabla^4 \psi. \qquad (2.7.98)$$

In case of slow motion (velocity is small) or in case of a very viscous fluid (v very large), the Reynolds number R is very small. For a steady flow in such cases of an incompressible viscous fluid, $\frac{\partial}{\partial t} \equiv 0$, while $(\mathbf{u} \cdot \nabla)\boldsymbol{\omega}$ is negligible in comparison with the viscous term. Consequently, (2.7.98) reduces to the standard *biharmonic equation*

$$\nabla^4 \psi = 0. \tag{2.7.99}$$

Or, more explicitly,

$$\nabla^2(\nabla^2)\psi \equiv \psi_{xxxx} + 2\psi_{xxyy} + \psi_{yyyy} = 0. \tag{2.7.100}$$

We solve this equation in a semi-infinite viscous fluid bounded by an infinite horizontal plate at $y = 0$, and the fluid is introduced normally with a prescribed velocity through a strip $-a < x < a$ of the plate. Thus the required boundary conditions are

$$u \equiv \frac{\partial \psi}{\partial y} = 0, \quad v \equiv \frac{\partial \psi}{\partial x} = H(a - |x|) f(x) \quad \text{on } y = 0, \quad (2.7.101\text{ab})$$

where $f(x)$ is a given function of x.

Furthermore, the fluid is assumed to be at rest at large distances from the plate, that is,

$$(\psi_x, \psi_y) \to (0,0) \quad \text{as } y \to \infty \qquad \text{for } -\infty < x < \infty. \tag{2.7.102}$$

To solve the biharmonic equation (2.7.100) with the boundary conditions (2.7.101ab) and (2.7.102), we introduce the Fourier transform with respect to x

$$\Psi(k,y) = \frac{1}{\sqrt{2\pi}} \int_{-\infty}^{\infty} e^{-ikx} \psi(x,y) dx. \tag{2.7.103}$$

Thus the Fourier transformed problem is

$$\left(\frac{d^2}{dy^2} - k^2\right)^2 \Psi(k,y) = 0, \tag{2.7.104}$$

$$\frac{d\Psi}{dy} = 0, \quad (ik)\Psi = F(k), \quad y = 0, \tag{2.7.105ab}$$

where

$$F(k) = \frac{1}{\sqrt{2\pi}} \int_{-a}^{a} e^{-ikx} f(x) dx. \tag{2.7.106}$$

In view of the Fourier transform of (2.7.102), the bounded solution of (2.7.104) is

$$\Psi(k,y) = (A + B|k|y) \exp(-|k|y), \tag{2.7.107}$$

where A and B can be determined from (2.7.105ab) so that $A = B = (ik)^{-1} F(k)$. Consequently, the solution (2.7.107) becomes

$$\Psi(k,y) = (ik)^{-1}(1 + |k|y) F(k) \exp(-|k|y). \tag{2.7.108}$$

The inverse Fourier transform gives the formal solution

$$\psi(x,y) = \frac{1}{\sqrt{2\pi}} \int_{-\infty}^{\infty} F(k) G(k) \exp(ikx) dk, \tag{2.7.109}$$

where

$$G(k) = (ik)^{-1}(1 + |k|y) \exp(-|k|y)$$

so that

Integral Transforms and Their Applications 47

$$g(x) = \mathscr{F}_s^{-1}\{G(k)\} = \mathscr{F}_s^{-1}\{(ik)^{-1}\exp(-|k|y)\} + y\,\mathscr{F}_s^{-1}\{(ik)^{-1}|k|\exp(-|k|y)\}$$
$$= \mathscr{F}_s^{-1}\{k^{-1}\exp(-ky)\} + y\,\mathscr{F}_s^{-1}\{e^{-ky}\}$$

which is, by (2.8.7) and (2.8.8),

$$= \sqrt{\frac{2}{\pi}}\tan^{-1}\left(\frac{x}{y}\right) + \sqrt{\frac{2}{\pi}}\frac{xy}{(x^2+y^2)}. \tag{2.7.110}$$

Using the Convolution Theorem 2.4.5 in (2.7.109) gives the final solution

$$\psi(x,y) = \frac{1}{\pi}\int_{-\infty}^{\infty} f(x-\xi)\left[\tan^{-1}\left(\frac{\xi}{y}\right) + \frac{y\xi}{\xi^2+y^2}\right]d\xi. \tag{2.7.111}$$

In particular, if $f(x) = \delta(x)$, then solution (2.7.111) becomes

$$\psi(x,y) = \frac{1}{\pi}\left[\tan^{-1}\left(\frac{x}{y}\right) + \frac{xy}{x^2+y^2}\right]. \tag{2.7.112}$$

The velocity fields u and v can be determined from (2.7.112).

Example 2.7.12 (*Biharmonic Equation in Elasticity*). We derive the biharmonic equation in elasticity from the two-dimensional equilibrium equations and the compatibility condition. In two-dimensional elastic medium, the strain components e_{xx}, e_{xy}, e_{yy} in terms of the displacement functions $(u, v, 0)$ are

$$e_{xx} = \frac{\partial u}{\partial x}, \quad e_{yy} = \frac{\partial v}{\partial y}, \quad e_{xy} = \frac{1}{2}\left(\frac{\partial u}{\partial y} + \frac{\partial v}{\partial x}\right). \tag{2.7.113}$$

Differentiating these results gives the compatibility condition

$$\frac{\partial^2 e_{xx}}{\partial y^2} + \frac{\partial^2 e_{yy}}{\partial x^2} = 2\frac{\partial^2 e_{xy}}{\partial x \partial y}. \tag{2.7.114}$$

In terms of the Poisson ratio v and Young's modulus E of the elastic material, the strain component in the z direction is expressed in terms of stress components

$$E e_{zz} = \sigma_{zz} - v(\sigma_{xx} + \sigma_{yy}). \tag{2.7.115}$$

In the case of plane strain, $e_{zz} = 0$, so that

$$\sigma_{zz} = v(\sigma_{xx} + \sigma_{yy}). \tag{2.7.116}$$

Substituting this result in other stress-strain relations, we obtain the strain components e_{xx}, e_{xy}, e_{yy} that are related to stress components $\sigma_{xx}, \sigma_{xy}, \sigma_{yy}$ by

$$E e_{xx} = \sigma_{xx} - v(\sigma_{yy} + \sigma_{zz}) = (1-v^2)\sigma_{xx} - v(1+v)\sigma_{yy}, \tag{2.7.117}$$
$$E e_{yy} = \sigma_{yy} - v(\sigma_{xx} + \sigma_{zz}) = (1-v^2)\sigma_{yy} - v(1+v)\sigma_{xx}, \tag{2.7.118}$$
$$E e_{xy} = (1+v)\sigma_{xy}. \tag{2.7.119}$$

Putting (2.7.117)-(2.7.119) into (2.7.114) gives

$$\frac{\partial^2}{\partial y^2}\left[\sigma_{xx} - v(\sigma_{xx}+\sigma_{yy})\right] + \frac{\partial^2}{\partial x^2}\left[\sigma_{yy} - v(\sigma_{xx}+\sigma_{yy})\right] = 2\frac{\partial^2 \sigma_{xy}}{\partial x \partial y}. \tag{2.7.120}$$

The basic differential equations for the stress components $\sigma_{xx}, \sigma_{yy}, \sigma_{xy}$ in the medium under the action of body forces X and Y are

$$\frac{\partial \sigma_{xx}}{\partial x} + \frac{\partial \sigma_{xy}}{\partial y} + \rho X = \rho \frac{\partial^2 u}{\partial t^2}, \qquad (2.7.121)$$

$$\frac{\partial \sigma_{xy}}{\partial x} + \frac{\partial \sigma_{yy}}{\partial y} + \rho Y = \rho \frac{\partial^2 v}{\partial t^2}, \qquad (2.7.122)$$

where ρ is the mass density of the elastic material.

The equilibrium equations follow from (2.7.121)-(2.7.122) in the absence of the body forces $(X = Y = 0)$ as

$$\frac{\partial}{\partial x} \sigma_{xx} + \frac{\partial}{\partial y} \sigma_{xy} = 0, \qquad (2.7.123)$$

$$\frac{\partial}{\partial x} \sigma_{xy} + \frac{\partial}{\partial y} \sigma_{yy} = 0. \qquad (2.7.124)$$

It is obvious that the expressions

$$\sigma_{xx} = \frac{\partial^2 \chi}{\partial y^2}, \quad \sigma_{xy} = -\frac{\partial^2 \chi}{\partial x \partial y}, \quad \sigma_{yy} = \frac{\partial^2 \chi}{\partial x^2} \qquad (2.7.125)$$

satisfy the equilibrium equations for any arbitrary function $\chi(x,y)$. Substituting from equations (2.7.125) into the compatibility condition (2.7.120), we see that χ must satisfy the biharmonic equation

$$\frac{\partial^4 \chi}{\partial x^4} + 2 \frac{\partial^4 \chi}{\partial x^2 \partial y^2} + \frac{\partial^4 \chi}{\partial y^4} = 0, \qquad (2.7.126)$$

which may be written symbolically as

$$\nabla^4 \chi = 0. \qquad (2.7.127)$$

The function χ was first introduced by Airy in 1862 and is known as the *Airy stress function*.

We determine the stress distribution in a semi-infinite elastic medium bounded by an infinite plane at $x = 0$ due to an external pressure to its surface. The x-axis is normal to this plane and assumed positive in the direction into the medium. We assume that the external surface pressure p varies along the surface so that the boundary conditions are

$$\sigma_{xx} = -p(y), \quad \sigma_{xy} = 0 \quad \text{on } x = 0 \quad \text{for all } y \text{ in } (-\infty, \infty). \quad (2.7.128)$$

We derive solutions so that stress components σ_{xx}, σ_{yy}, and σ_{xy} all vanish as $x \to \infty$.

In order to solve the biharmonic equation (2.7.127), we introduce the Fourier transform $\tilde{\chi}(x,k)$ of the Airy stress function with respect to y so that (2.7.127)-(2.7.128) reduce to

$$\left(\frac{d^2}{dx^2} - k^2 \right)^2 \tilde{\chi} = 0, \qquad (2.7.129)$$

$$k^2 \tilde{\chi}(0,k) = \tilde{p}(k), \quad (ik)\left(\frac{d\tilde{\chi}}{dx}\right)_{x=0} = 0, \qquad (2.7.130)$$

where $\tilde{p}(k) = \mathcal{F}\{p(y)\}$. The bounded solution of the transformed problem is

$$\tilde{\chi}(x,k) = (A + Bx)\exp(-|k|x), \qquad (2.7.131)$$

where A and B are constants of integration to be determined from (2.7.130). It turns out that $A = \tilde{p}(k)/k^2$ and $B = \tilde{p}(k)/|k|$ and hence the solution becomes

$$\tilde{\chi}(x,k) = \frac{\tilde{p}(k)}{k^2}\{1 + |k|x\}\exp(-|k|x). \qquad (2.7.132)$$

The inverse Fourier transform yields the formal solution

$$\chi(x,y) = \frac{1}{\sqrt{2\pi}} \int_{-\infty}^{\infty} \frac{\tilde{p}(k)}{k^2} (1 + |k|x)\exp(iky - |k|x)\,dk. \qquad (2.7.133)$$

The stress components are obtained from (2.7.125) in the form

$$\sigma_{xx}(x,y) = -\frac{1}{\sqrt{2\pi}} \int_{-\infty}^{\infty} k^2\, \tilde{\chi}(x,k)\exp(iky)\,dk, \qquad (2.7.134)$$

$$\sigma_{xy}(x,y) = -\frac{1}{\sqrt{2\pi}} \int_{-\infty}^{\infty} (ik)\left(\frac{d\tilde{\chi}}{dx}\right)\exp(iky)\,dk, \qquad (2.7.135)$$

$$\sigma_{yy}(x,y) = \frac{1}{\sqrt{2\pi}} \int_{-\infty}^{\infty} \frac{d^2\tilde{\chi}}{dx^2}\exp(iky)\,dk, \qquad (2.7.136)$$

where $\tilde{\chi}(x,k)$ are given by (2.7.132). In particular, if $p(y) = P\delta(y)$ so that $\tilde{p}(k) = P(2\pi)^{-\frac{1}{2}}$. Consequently, from (2.7.133)-(2.7.136) we obtain

$$\chi(x,y) = \frac{P}{2\pi} \int_{-\infty}^{\infty} k^{-2}(1 + |k|x)\exp(iky - |k|x)\,dk$$

$$= \frac{P}{\pi} \int_{0}^{\infty} k^{-2}(1 + kx)\cos ky\,\exp(-kx)\,dk. \qquad (2.7.137)$$

$$\sigma_{xx} = -\frac{P}{\pi} \int_{0}^{\infty} (1 + kx)e^{-kx}\cos ky\,dk = -\frac{2Px^3}{\pi(x^2 + y^2)^2}. \qquad (2.7.138)$$

$$\sigma_{xy} = -\frac{Px}{\pi} \int_{0}^{\infty} k\sin ky\,\exp(-kx)\,dk = -\frac{2Px^2 y}{\pi(x^2 + y^2)^2}. \qquad (2.7.139)$$

$$\sigma_{yy} = -\frac{P}{\pi} \int_{0}^{\infty} (1 - kx)\exp(-kx)\cos ky\,dk = -\frac{2Pxy^2}{\pi(x^2 + y^2)^2}. \qquad (2.7.140)$$

Another physically realistic pressure distribution is

$$p(y) = P\,H(|a| - y), \qquad (2.7.141)$$

where P is a constant, so that

$$\tilde{p}(k) = \sqrt{\frac{2}{\pi}} \frac{P}{k} \sin ak. \qquad (2.7.142)$$

Substituting this value for $\tilde{p}(k)$ into (2.7.133)-(2.7.136), we obtain the integral expression for $\chi, \sigma_{xx}, \sigma_{xy}$, and σ_{yy}.

It is noted here that if a point force of magnitude P_0 acts at the origin located on the boundary, then we put $P = (P_0/2a)$ in (2.7.142) and find

$$\tilde{p}(k) = \lim_{a \to 0} \sqrt{\frac{2}{\pi}} \frac{P_0}{2} \left(\frac{\sin ak}{ak} \right) = \frac{P_0}{\sqrt{2\pi}}. \qquad (2.7.143)$$

Thus the stress components can also be written in this case.

2.8 Fourier Cosine and Sine Transforms with Examples

The Fourier cosine integral formula (2.2.8) leads to the *Fourier cosine transform* and its inverse defined by

$$\mathscr{F}_c \{f(x)\} = F_c(k) = \sqrt{\frac{2}{\pi}} \int_0^\infty \cos kx \, f(x) \, dx, \qquad (2.8.1)$$

$$\mathscr{F}_c^{-1} \{F_c(k)\} = f(x) = \sqrt{\frac{2}{\pi}} \int_0^\infty \cos kx \, F_c(k) \, dk, \qquad (2.8.2)$$

where $\mathscr{F}_c$ is the Fourier cosine transform operator and $\mathscr{F}_c^{-1}$ is its inverse operator.

Similarly, the Fourier sine integral formula (2.2.9) leads to the *Fourier sine transform* and its inverse defined by

$$\mathscr{F}_s \{f(x)\} = F_s(k) = \sqrt{\frac{2}{\pi}} \int_0^\infty \sin kx \, f(x) \, dx, \qquad (2.8.3)$$

$$\mathscr{F}_s^{-1} \{F_s(k)\} = f(x) = \sqrt{\frac{2}{\pi}} \int_0^\infty \sin kx \, F_s(k) \, dk, \qquad (2.8.4)$$

where $\mathscr{F}_s$ is the Fourier sine transform operator and $\mathscr{F}_s^{-1}$ is its inverse.

Example 2.8.1 Show that

(a) $\mathscr{F}_c \{e^{-ax}\} = \sqrt{\dfrac{2}{\pi}} \dfrac{a}{(a^2 + k^2)}, \quad (a > 0).$ \hfill (2.8.5)

(b) $\mathscr{F}_s \{e^{-ax}\} = \sqrt{\dfrac{2}{\pi}} \dfrac{k}{(a^2 + k^2)}, \quad (a > 0).$ \hfill (2.8.6)

We have

$$\mathcal{F}_c\{e^{-ax}\} = \sqrt{\frac{2}{\pi}} \int_0^\infty e^{-ax} \cos kx \, dx$$

$$= \frac{1}{2}\sqrt{\frac{2}{\pi}} \int_0^\infty \left[e^{-(a-ik)x} + e^{-(a+ik)x}\right] dx$$

$$\mathcal{F}_c\{e^{-ax}\} = \frac{1}{2}\sqrt{\frac{2}{\pi}}\left[\frac{1}{a-ik} + \frac{1}{a+ik}\right] = \sqrt{\frac{2}{\pi}}\frac{a}{(a^2+k^2)}.$$

The proof of the other result is similar and hence left to the reader.

Example 2.8.2 Show that
$$\mathcal{F}_s^{-1}\left\{\frac{1}{k}\exp(-sk)\right\} = \sqrt{\frac{2}{\pi}}\tan^{-1}\left(\frac{x}{s}\right). \qquad (2.8.7)$$

We have the standard definite integral
$$\sqrt{\frac{\pi}{2}}\,\mathcal{F}_s^{-1}\{\exp(-sk)\} = \int_0^\infty \exp(-sk)\sin kx \, dk = \frac{x}{s^2+x^2}. \qquad (2.8.8)$$

Integrating both sides with respect to s from s to ∞ gives
$$\int_0^\infty \frac{e^{-sk}}{k}\sin kx \, dk = \int_s^\infty \frac{x\,ds}{x^2+s^2} = \left[\tan^{-1}\frac{s}{x}\right]_s^\infty$$

$$= \frac{\pi}{2} - \tan^{-1}\left(\frac{s}{x}\right) = \tan^{-1}\left(\frac{x}{s}\right). \qquad (2.8.9)$$

Thus
$$\mathcal{F}_s^{-1}\left\{\frac{1}{k}\exp(-sk)\right\} = \sqrt{\frac{2}{\pi}}\int_0^\infty \frac{1}{k}\exp(-sk)\sin xk \, dk$$

$$= \sqrt{\frac{2}{\pi}}\tan^{-1}\left(\frac{x}{s}\right).$$

Example 2.8.3 Show that
$$\mathcal{F}_s\{erfc(ax)\} = \sqrt{\frac{2}{\pi}}\frac{1}{k}\left[1 - \exp\left(-\frac{k^2}{4a^2}\right)\right]. \qquad (2.8.10)$$

We have
$$\mathcal{F}_s\{erfc(ax)\} = \sqrt{\frac{2}{\pi}}\int_0^\infty erfc(ax)\sin kx \, dx$$

$$= \frac{2\sqrt{2}}{\pi}\int_0^\infty \sin kx \, dx \int_{ax}^\infty e^{-t^2} dt.$$

Interchanging the order of integration, we obtain

$$\mathcal{F}_s\{erf(ax)\} = \frac{2\sqrt{2}}{\pi} \int_0^\infty \exp(-t^2) dt \int_0^{t/a} \sin kx \, dx$$

$$= \frac{2\sqrt{2}}{\pi k} \int_0^\infty \exp(-t^2) \left\{1 - \cos\left(\frac{kt}{a}\right)\right\} dt$$

$$= \frac{2\sqrt{2}}{\pi k} \left[\frac{\sqrt{\pi}}{2} - \frac{\sqrt{\pi}}{2} \exp\left(-\frac{k^2}{4a^2}\right)\right].$$

Thus

$$\mathcal{F}_s\{erfc(ax)\} = \sqrt{\frac{2}{\pi}} \frac{1}{k}\left[1 - \exp\left(-\frac{k^2}{4a^2}\right)\right].$$

2.9 Properties of Fourier Cosine and Sine Transforms

Theorem 2.9.1 If $\mathcal{F}_c\{f(x)\} = F_c(k)$ and $\mathcal{F}_s\{f(x)\} = F_s(k)$, then

$$\mathcal{F}_c\{f(ax)\} = \frac{1}{a} F_c\left(\frac{k}{a}\right), \quad a > 0. \tag{2.9.1}$$

$$\mathcal{F}_s\{f(ax)\} = \frac{1}{a} F_s\left(\frac{k}{a}\right), \quad a > 0. \tag{2.9.2}$$

Under appropriate conditions, the following properties also hold:

$$\mathcal{F}_c\{f'(x)\} = k F_s(k) - \sqrt{\frac{2}{\pi}} f(0), \tag{2.9.3}$$

$$\mathcal{F}_c\{f''(x)\} = -k^2 F_c(k) - \sqrt{\frac{2}{\pi}} f'(0), \tag{2.9.4}$$

$$\mathcal{F}_s\{f'(x)\} = -k F_c(k), \tag{2.9.5}$$

$$\mathcal{F}_s\{f''(x)\} = -k^2 F_s(k) + \sqrt{\frac{2}{\pi}} k f(0). \tag{2.9.6}$$

These results can be generalized for the cosine and sine transforms of higher order derivatives of a function. They are left as exercises.

Theorem 2.9.2 (*Convolution Theorem for the Fourier Cosine Transform*). If $\mathcal{F}_c\{f(x)\} = F_c(k)$ and $\mathcal{F}_c\{g(k)\} = G_c(k)$, then

$$\mathcal{F}_c^{-1}\{F_c(k) G_c(k)\} = \frac{1}{\sqrt{2\pi}} \int_0^\infty f(\xi)\left[g(x+\xi) + g(|x-\xi|)\right] d\xi. \tag{2.9.7}$$

Or equivalently

$$\int_0^\infty F_c(k) G_c(k) \cos kx \, dk = \frac{1}{2} \int_0^\infty f(\xi)\left[g(x+\xi) + g(|x-\xi|)\right] d\xi. \tag{2.9.8}$$

Proof. Using the definition of the inverse Fourier cosine transform, we have

$$\mathscr{F}_c^{-1}\{F_c(k)G_c(k)\}$$

$$= \sqrt{\frac{2}{\pi}} \int_0^\infty F_c(k)\, G_c(k) \cos kx\, dk$$

$$= \left(\frac{2}{\pi}\right) \int_0^\infty G_c(k) \cos kx\, dk \int_0^\infty f(\xi) \cos k\xi\, d\xi.$$

Hence

$$\mathscr{F}_c^{-1}\{F_c(k)G_c(k)\}$$

$$= \left(\frac{2}{\pi}\right) \int_0^\infty f(\xi)\,d\xi \int_0^\infty \cos kx \cos k\xi\, G_c(k)\, dk$$

$$= \frac{1}{2}\sqrt{\frac{2}{\pi}} \int_0^\infty f(\xi)\,d\xi \sqrt{\frac{2}{\pi}} \int_0^\infty \left[\cos k(x+\xi) + \cos k(|x-\xi|)\right] G_c(k)\, dk$$

$$= \frac{1}{\sqrt{2\pi}} \int_0^\infty f(\xi)\left[g(x+\xi) + g(|x-\xi|)\right] d\xi,$$

in which the definition of the inverse Fourier cosine transform is used. This proves (2.9.7).

It also follows from the proof of Theorem 2.9.2 that

$$\int_0^\infty F_c(k)G_c(k)\cos kx\, dk = \frac{1}{2}\int_0^\infty f(\xi)\left[g(x+\xi) + g(|x-\xi|)\right] d\xi.$$

This proves result (2.9.8).

Putting $x = 0$ in (2.9.8), we obtain

$$\int_0^\infty F_c(k)G_c(k)\, dk = \int_0^\infty f(\xi)g(\xi)\,d\xi = \int_0^\infty f(x)g(x)\,dx.$$

Substituting $g(x) = \overline{f(x)}$ gives, since $G_c(k) = \overline{F_c(k)}$,

$$\int_0^\infty |F_c(k)|^2\, dk = \int_0^\infty |f(x)|^2\, dx. \qquad (2.9.9)$$

This is the *Parseval relation* for the Fourier cosine transform.

Similarly, we obtain

$$\int_0^\infty F_s(k)G_s(k)\cos kx\, dk$$

$$= \sqrt{\frac{2}{\pi}} \int_0^\infty G_s(k)\cos kx\, dk \int_0^\infty f(\xi)\sin k\xi\, d\xi$$

which is, by interchanging the order of integration,

$$= \sqrt{\frac{2}{\pi}} \int_0^\infty f(\xi)\, d\xi \int_0^\infty G_s(k) \sin k\xi \cos kx\, dk$$

$$= \frac{1}{2} \int_0^\infty f(\xi)\, d\xi\, \sqrt{\frac{2}{\pi}} \int_0^\infty G_s(k) \left[\sin k(\xi+x) + \sin k(\xi-x)\right] dk$$

$$= \frac{1}{2} \int_0^\infty f(\xi) \left[g(\xi+x) + g(\xi-x)\right] d\xi,$$

in which the inverse Fourier sine transform is used. Thus we find

$$\int_0^\infty F_s(k) G_s(k) \cos kx\, dk = \frac{1}{2} \int_0^\infty f(\xi) \left[g(\xi+x) + g(\xi-x)\right] d\xi. \qquad (2.9.10)$$

Or equivalently

$$\mathscr{F}_c^{-1}\{F_s(k) G_s(k)\} = \frac{1}{\sqrt{2\pi}} \int_0^\infty f(\xi) \left[g(\xi+x) + g(\xi-x)\right] d\xi. \qquad (2.9.11)$$

Result (2.9.10) or (2.9.11) is also called the *Convolution Theorem* of the Fourier cosine transform.

Putting $x = 0$ in (2.9.10) gives

$$\int_0^\infty F_s(k) G_s(k)\, dk = \int_0^\infty f(\xi) g(\xi)\, d\xi = \int_0^\infty f(x) g(x)\, dx.$$

Replacing $g(x)$ by $\overline{f(x)}$ gives the *Parseval relation* for the Fourier sine transform

$$\int_0^\infty |F_s(k)|^2\, dk = \int_0^\infty |f(x)|^2\, dx. \qquad (2.9.12)$$

2.10 Applications of Fourier Cosine and Sine Transforms to Partial Differential Equations

Example 2.10.1 (*One-Dimensional Diffusion Equation on a Half Line*). Consider the initial-boundary value problem for the one-dimensional diffusion equation in $0 < x < \infty$ with no sources or sinks:

$$\frac{\partial u}{\partial t} = \kappa \frac{\partial^2 u}{\partial x^2}, \quad 0 < x < \infty, \quad t > 0, \qquad (2.10.1)$$

where κ is a constant, with the initial condition

$$u(x, 0) = 0, \quad 0 < x < \infty, \qquad (2.10.2)$$

and the boundary conditions

(a) $u(0, t) = f(t), \quad t \geq 0, \quad u(x, t) \to 0 \quad \text{as } x \to \infty, \qquad (2.10.3)$

or

(b) $u_x(0, t) = f(t), \quad t \geq 0, \quad u(x, t) \to 0 \quad \text{as } x \to \infty. \qquad (2.10.4)$

This problem with the boundary conditions (2.10.3) is solved by using the Fourier sine transform

$$U_s(k,t) = \sqrt{\frac{2}{\pi}} \int_0^\infty \sin kx \; u(x,t) \, dx.$$

Application of the Fourier sine transform gives

$$\frac{dU_s}{dt} = -\kappa \, k^2 \, U_s(k,t) + \sqrt{\frac{2}{\pi}} \, \kappa \, k \, f(t), \quad (2.10.5)$$

$$U_s(k,0) = 0. \quad (2.10.6)$$

The bounded solution of this differential system with $U_s(k,0) = 0$ is

$$U_s(k,t) = \sqrt{\frac{2}{\pi}} \, \kappa k \int_0^t f(\tau) \exp\left[-\kappa(t-\tau)k^2\right] d\tau. \quad (2.10.7)$$

The inverse transform gives the solution

$$u(x,t) = \sqrt{\frac{2}{\pi}} \, \kappa \int_0^t f(\tau) \, \mathscr{F}_s^{-1}\left\{k \exp\left[-\kappa(t-\tau)k^2\right]\right\} d\tau$$

$$= \frac{x}{\sqrt{4\pi\kappa}} \int_0^t f(\tau) \exp\left[-\frac{x^2}{4\kappa(t-\tau)}\right] \frac{d\tau}{(t-\tau)^{3/2}} \quad (2.10.8)$$

in which $\mathscr{F}_s^{-1}\left\{k \exp(-t\kappa k^2)\right\} = \frac{x}{2\sqrt{2}} \cdot \frac{\exp(-x^2/4\kappa t)}{(\kappa t)^{3/2}}$ is used.

In particular, $f(t) = T_0 = $ constant, (2.10.7) reduces to

$$U_s(k,t) = \sqrt{\frac{2}{\pi}} \frac{T_0}{k} \left[1 - \exp(-\kappa t k^2)\right]. \quad (2.10.9)$$

Inversion gives the solution

$$u(x,t) = \left(\frac{2T_0}{\pi}\right) \int_0^\infty \frac{\sin kx}{k} \left[1 - \exp(-\kappa t k^2)\right] dk. \quad (2.10.10)$$

Making use of the integral

$$\int_0^\infty e^{-k^2 a^2} \frac{\sin kx}{k} \, dk = \frac{\pi}{2} \, \text{erf}\left(\frac{x}{2a}\right), \quad (2.10.11)$$

the solution becomes

$$u(x,t) = \frac{2T_0}{\pi} \left[\frac{\pi}{2} - \frac{\pi}{2} \, \text{erf}\left(\frac{x}{2\sqrt{\kappa t}}\right)\right]$$

$$= T_0 \, \text{erfc}\left(\frac{x}{2\sqrt{\kappa t}}\right), \quad (2.10.12)$$

where the *error function*, $\text{erf}(x)$ is defined by

$$\text{erf}(x) = \frac{2}{\sqrt{\pi}} \int_0^x e^{-\alpha^2} d\alpha, \quad (2.10.13)$$

so that

$$erf(0) = 0, \quad erf(\infty) = \frac{2}{\sqrt{\pi}} \int_0^\infty e^{-\alpha^2} d\alpha = 1, \text{ and } erf(-x) = -erf(x),$$

and the *complementary error function*, $erfc(x)$ is defined by

$$erfc(x) = 1 - erf(x) = \frac{2}{\sqrt{\pi}} \int_x^\infty e^{-\alpha^2} d\alpha, \qquad (2.10.14)$$

so that

$$erfc(x) = 1 - erf(x), \quad erfc(0) = 1, \quad erfc(\infty) = 0,$$

and

$$erfc(-x) = 1 - erf(-x) = 1 + erf(x) = 2 - erfc(x).$$

Equation (2.10.1) with boundary condition (2.10.4) is solved by the Fourier cosine transform

$$U_c(k,t) = \sqrt{\frac{2}{\pi}} \int_0^\infty \cos kx \, u(x,t) \, dx.$$

Application of this transform to (2.10.1) gives

$$\frac{dU_c}{dt} + \kappa k^2 U_c = -\sqrt{\frac{2}{\pi}} \kappa f(t). \qquad (2.10.15)$$

The solution of (2.10.15) with $U_c(k,0) = 0$ is

$$U_c(k,t) = -\sqrt{\frac{2}{\pi}} \kappa \int_0^t f(\tau) \exp\left[-k^2 \kappa (t-\tau)\right] d\tau. \qquad (2.10.16)$$

Since

$$\mathscr{F}_c^{-1}\left\{\exp(-t\kappa k^2)\right\} = \frac{1}{\sqrt{2\kappa t}} \exp\left(-\frac{x^2}{4\kappa t}\right), \qquad (2.10.17)$$

the inverse Fourier cosine transform gives the final form of the solution

$$u(x,t) = -\sqrt{\frac{\kappa}{\pi}} \int_0^t \frac{f(\tau)}{\sqrt{t-\tau}} \exp\left[-\frac{x^2}{4\kappa(t-\tau)}\right] d\tau. \qquad (2.10.18)$$

Example 2.10.2 (*The Laplace Equation in the Quarter Plane*). Solve the Laplace equation

$$u_{xx} + u_{yy} = 0, \quad 0 < x, \, y < \infty, \qquad (2.10.19)$$

with the boundary conditions

$$u(0,y) = a, \qquad u(x,0) = 0, \qquad (2.10.20a)$$

$$\nabla u \to 0 \quad \text{as } r = \sqrt{x^2 + y^2} \to \infty, \qquad (2.10.20b)$$

where a is a constant.

We apply the Fourier sine transform with respect to x to find

$$\frac{d^2 U_s}{dy^2} - k^2 U_s + \sqrt{\frac{2}{\pi}} ka = 0.$$

The solution of this inhomogeneous equation is

$$U_s(k,y) = A e^{-ky} + \sqrt{\frac{2}{\pi}} \cdot \frac{a}{k},$$

where A is a constant to be determined from $U_s(k,0) = 0$. Consequently,
$$U_s(k,y) = \frac{a}{k}\sqrt{\frac{2}{\pi}}\left(1 - e^{-ky}\right). \tag{2.10.21}$$

The inverse transformation gives the formal solution
$$u(x,y) = \frac{2a}{\pi}\int_0^\infty \frac{1}{k}\left(1 - e^{-ky}\right)\sin kx\, dk.$$

Or
$$u(x,y) = \frac{2a}{\pi}\left[\int_0^\infty \frac{\sin kx}{k}\, dk - \int_0^\infty \frac{1}{k}e^{-ky}\sin kx\, dk\right]$$

$$= a - \frac{2a}{\pi}\left(\frac{\pi}{2} - \tan^{-1}\frac{y}{x}\right) = \frac{2a}{\pi}\tan^{-1}\left(\frac{x}{y}\right), \tag{2.10.22}$$

in which (2.8.9) is used.

Example 2.10.3 *(The Laplace Equation in a Semi-Infinite Strip with the Dirichlet Data).* Solve the Laplace equation
$$u_{xx} + u_{yy} = 0, \quad 0 < x < \infty, \quad 0 < y < b, \tag{2.10.23}$$

with the boundary conditions
$$u(0,y) = 0, \quad u(x,y) \to 0 \quad \text{as} \quad x \to \infty \quad \text{for } 0 < y < b \tag{2.10.24}$$
$$u(x,b) = 0, \quad u(x,0) = f(x) \quad \text{for } 0 < x < \infty. \tag{2.10.25}$$

In view of the Dirichlet data, the Fourier sine transform with respect to x can be used to solve this problem. Applying the Fourier sine transform to (2.10.23)-(2.10.25) gives
$$\frac{d^2 U_s}{dy^2} - k^2 U_s = 0, \tag{2.10.26}$$
$$U_s(k,b) = 0, \quad U_s(k,0) = F_s(k). \tag{2.10.27}$$

The solution of (2.10.26) with (2.10.27) is
$$U_s(k,y) = F_s(k)\frac{\sinh[k(b-y)]}{\sinh kb}. \tag{2.10.28}$$

The inverse Fourier sine transform gives the formal solution
$$u(x,y) = \sqrt{\frac{2}{\pi}}\int_0^\infty F_s(k)\frac{\sinh[k(b-y)]}{\sinh kb}\sin kx\, dk$$

$$= \frac{2}{\pi}\int_0^\infty\left[\int_0^\infty f(l)\sin kl\, dl\right]\frac{\sinh[k(b-y)]}{\sinh kb}\sin kx\, dk. \tag{2.10.29}$$

In the limit as $kb \to \infty$, $\dfrac{\sinh[k(b-y)]}{\sinh kb} \sim \exp(-ky)$, hence the above problem reduces to the corresponding problem in the quarter plane, $0 < x < \infty$, $0 < y < \infty$. Thus solution (2.10.29) becomes

$$u(x,y) = \frac{2}{\pi}\int_0^\infty f(l)dl \int_0^\infty \sin kl \sin kx \, \exp(-ky)\,dk$$

$$= \frac{1}{\pi}\int_0^\infty f(l)dl \int_0^\infty \{\cos k(x-l) - \cos k(x+l)\}\exp(-ky)\,dk$$

$$= \frac{1}{\pi}\int_0^\infty f(l)\left[\frac{y}{(x-l)^2+y^2} - \frac{y}{(x+l)^2+y^2}\right]dl. \tag{2.10.30}$$

This is the exact integral solution of the problem. If $f(x)$ is an odd function of x, then solution (2.10.30) reduces to the solution (2.7.10) of the same problem in the half plane.

2.11 Evaluation of Definite Integrals

The Fourier transform can be employed to evaluate certain definite integrals. Although the method of evaluation may not be very rigorous, it is quite simple and straightforward. The method can be illustrated by means of examples.

Example 2.11.1 Evaluate the integral

$$I(a,b) = \int_{-\infty}^\infty \frac{dx}{(x^2+a^2)(x^2+b^2)}, \quad a>0, b>0. \tag{2.11.1}$$

If we write $f(x) = e^{-a|x|}$ and $g(x) = e^{-b|x|}$ then $F(k) = \sqrt{\frac{2}{\pi}}\frac{a}{(k^2+a^2)}$,

$G(k) = \sqrt{\frac{2}{\pi}}\frac{b}{(k^2+b^2)}$. The Convolution Theorem 2.4.5 gives (2.4.18), that is,

$$\int_{-\infty}^\infty F(k)G(k)\,dk = \int_{-\infty}^\infty f(x)g(-x)\,dx.$$

Or equivalently

$$\int_{-\infty}^\infty \frac{dk}{(k^2+a^2)(k^2+b^2)} = \frac{\pi}{2ab}\int_{-\infty}^\infty e^{-|x|(a+b)}dx = \frac{\pi}{ab}\int_0^\infty e^{-(a+b)x}dx = \frac{\pi}{ab(a+b)}. \tag{2.11.2}$$

This is the desired result.
Further

$$\int_0^\infty \frac{dx}{(x^2+a^2)(x^2+b^2)} = \frac{\pi}{2ab(a+b)}. \tag{2.11.3}$$

Example 2.11.2 Show that

$$\int_0^\infty \frac{x^{-p}dx}{(a^2+x^2)} = \frac{\pi}{2}a^{-(p+1)}\sec\left(\frac{\pi p}{2}\right). \tag{2.11.4}$$

Proof. We write

$$f(x) = e^{-ax} \quad \text{so that} \quad F_c(k) = \sqrt{\frac{2}{\pi}} \frac{a}{(a^2 + k^2)}.$$

$$g(x) = x^{p-1} \quad \text{so that} \quad G_c(k) = \sqrt{\frac{2}{\pi}} k^{-p} \Gamma(p) \cos\left(\frac{\pi p}{2}\right).$$

Using Parseval's result for the Fourier cosine transform gives

$$\int_0^\infty F_c(k) G_c(k) dk = \int_0^\infty f(x) g(x) dx.$$

Or

$$\frac{2a}{\pi} \cos\left(\frac{\pi p}{2}\right) \Gamma(p) \int_0^\infty \frac{k^{-p} dk}{k^2 + a^2} = \int_0^\infty x^{p-1} e^{-ax} dx$$

$$= \frac{1}{a^p} \int_0^\infty e^{-t} t^{p-1} dt, \quad (ax = t)$$

$$= \frac{\Gamma(p)}{a^p}.$$

Thus

$$\int_0^\infty \frac{k^{-p} dk}{a^2 + k^2} = \frac{\pi}{2a^{p+1}} \sec\left(\frac{\pi p}{2}\right).$$

Example 2.11.3 If $a > 0$, $b > 0$, show that

$$\int_0^\infty \frac{x^2 \, dx}{(a^2 + x^2)(b^2 + x^2)} = \frac{\pi}{2(a+b)}. \tag{2.11.5}$$

Proof. We consider

$$\mathcal{F}_s\{e^{-ax}\} = \sqrt{\frac{2}{\pi}} \frac{k}{k^2 + a^2} = F_s(k)$$

$$\mathcal{F}_s\{e^{-bx}\} = \sqrt{\frac{2}{\pi}} \frac{k}{k^2 + b^2} = G_s(k).$$

Then the Convolution Theorem for the Fourier cosine transform gives

$$\int_0^\infty F_s(k) G_s(k) \cos kx \, dk = \frac{1}{2} \int_0^\infty g(\xi) [f(\xi + x) + f(\xi - x)] \, d\xi.$$

Putting $x = 0$ gives

$$\int_0^\infty F_s(k) G_s(k) \, dk = \int_0^\infty g(\xi) f(\xi) \, d\xi$$

or

$$\int_0^\infty \frac{k^2 \, dk}{(k^2 + a^2)(k^2 + b^2)} = \frac{\pi}{2} \int_0^\infty e^{-(a+b)\xi} d\xi = \frac{\pi}{2(a+b)}.$$

Example 2.11.4 Show that

$$\int_0^\infty \frac{x^2 dx}{\left(x^2+a^2\right)^4} = \frac{\pi}{(2a)^5}, \quad a>0. \tag{2.11.6}$$

We write $f(x) = \dfrac{1}{2(x^2+a^2)}$ so that $f'(x) = -\dfrac{x}{(x^2+a^2)^2}$, and

$$\mathcal{F}\{f(x)\} = F(k) = \sqrt{\frac{\pi}{2}}\left(\frac{1}{2a}\right)\exp(-a|k|).$$

Making reference to the Parseval's relation (2.4.19), we obtain

$$\int_{-\infty}^{\infty}|f'(x)|^2 dx = \int_{-\infty}^{\infty}|\mathcal{F}\{f'(x)\}|^2 dk = \int_{-\infty}^{\infty}|(ik)\mathcal{F}\{f(x)\}|^2 dk.$$

Thus

$$\int_{-\infty}^{\infty}\frac{x^2}{\left(x^2+a^2\right)^4}dx = \frac{\pi}{2}\int_{-\infty}^{\infty}k^2 \cdot \frac{1}{(2a)^2}\exp(-2a|k|)dk$$

$$= \frac{\pi}{(2a)^2}\int_0^\infty k^2 \exp(-2ak)dk = \frac{2\pi}{(2a)^5}.$$

This gives the desired result.

2.12 Applications of Fourier Transforms in Mathematical Statistics

In probability theory and mathematical statistics, the characteristic function of a random variable is defined by the Fourier transform or by the Fourier-Stieltjes transform of the distribution function of a random variable. Many important results in probability theory and mathematical statistics can be obtained, and their proofs can be simplified with rigor by using the methods of characteristic functions. Thus the Fourier transforms play an important role in probability theory and statistics.

Definition 2.12.1 (*Distribution Function*). The *distribution* function $F(x)$ of a random variable X is defined as the probability, that is, $F(x) = P(X < x)$ for every real number x.

It is immediately evident from this definition that the distribution function satisfies the following properties:
(i) $F(x)$ is a non-decreasing function, that is, $F(x_1) \leq F(x_2)$ if $x_1 < x_2$.
(ii) $F(x)$ is continuous only from the left at a point x, that is, $F(x-0) = F(x)$, but $F(x+0) \neq F(x)$.
(iii) $F(-\infty) = 0$ and $F(+\infty) = 1$.

If X is a continuous variable and if there exists a non-negative function $f(x)$ such that for every real x the following relation holds:

$$F(x) = \int_{-\infty}^{x} f(x)dx, \qquad (2.12.1)$$

where $F(x)$ is the distribution function of the random variable X, then the function $f(x)$ is called the *probability density* or simply the *density function* of the random variable X.

It is immediately obvious that every density function $f(x)$ satisfies the following properties:

(i) $F(+\infty) = \int_{-\infty}^{\infty} f(x)dx = 1.$ \qquad (2.12.2a)

(ii) For every real a and b where $a < b$,

$$P(a \leq X \leq b) = F(b) - F(a) = \int_{a}^{b} f(x)dx. \qquad (2.12.2b)$$

(iii) If $f(x)$ is continuous at some point x, then $F'(x) = f(x)$.

It is noted that every real function $f(x)$ which is non-negative, and integrable over the whole real line and satisfies (2.12.2), is the probability density function of a continuous random variable X. On the other hand, the function $F(x)$ defined by (2.12.1) satisfies all properties of a distribution function.

Definition 2.12.2 (*Characteristic Function*). If X is a continuous random variable with the density function $f(x)$, then the *characteristic function*, $\phi(t)$ of the random variable X or of the distribution function $F(x)$ is defined by the formula

$$\phi(t) = E(\exp(itX)) = \int_{-\infty}^{\infty} f(x)\exp(itx)dx, \qquad (2.12.3)$$

where $E[g(X)]$ is called the *expected value* of the random variable $g(X)$.

In problems of mathematical statistics, it is convenient to define the Fourier transform of $f(x)$ and its inverse in a slightly different way by

$$\mathscr{F}\{f(x)\} = \phi(t) = \int_{-\infty}^{\infty} \exp(itx)f(x)dx, \qquad (2.12.4)$$

$$\mathscr{F}^{-1}\{\phi(t)\} = f(x) = \frac{1}{2\pi}\int_{-\infty}^{\infty} \exp(-itx)\phi(t)dt. \qquad (2.12.5)$$

Evidently, the characteristic function of $F(x)$ is the Fourier transform of the density function $f(x)$. The Fourier transform of the distribution function follows from the fact that

$$\mathcal{F}\{F'(x)\} = \mathcal{F}\{f(x)\} = \phi(t),$$

or

$$\mathcal{F}\{F(x)\} = it^{-1}\phi(t). \qquad (2.12.6)$$

The *composition of two distribution functions* $F_1(x)$ and $F_2(x)$ is defined by

$$F(x) = F_1(x) * F_2(x) = \int_{-\infty}^{\infty} F_1(x-y) F_2'(y) dy. \qquad (2.12.7)$$

Thus the Fourier transform of (2.12.7) gives

$$it^{-1}\phi(t) = \mathcal{F}\left\{\int_{-\infty}^{\infty} F_1(x-y) F_2'(y) dy\right\}$$

$$= \mathcal{F}\{F_1(x)\} \mathcal{F}\{f_2(x)\} = it^{-1}\phi_1(t)\phi_2(t),$$

whence an important result follows:

$$\phi(t) = \phi_1(t)\phi_2(t), \qquad (2.12.8)$$

where $\phi_1(t)$ and $\phi_2(t)$ are the characteristic functions of the distribution functions $F_1(x)$ and $F_2(x)$ respectively.

The *nth moment* of a random variable X is defined by

$$m_n = E[X^n] = \int_{-\infty}^{\infty} x^n f(x) dx, \quad n = 1, 2, 3, \ldots. \qquad (2.12.9)$$

provided this integral exists. The first moment m_1 (or simply m) is called the *expectation* of X and has the form

$$m = E(X) = \int_{-\infty}^{\infty} x f(x) dx. \qquad (2.12.10)$$

Thus the moment of any order n is calculated by evaluating the integral (2.12.9). However, the evaluation of the integral is, in general, a difficult task. This difficulty can be resolved with the help of the characteristic function defined by (2.12.4). Differentiating (2.12.4) n times and putting $t = 0$ gives a fairly simple formula

$$m_n = \int_{-\infty}^{\infty} x^n f(x) dx = (-i)^n \phi^{(n)}(0), \qquad (2.12.11)$$

where $n = 1, 2, 3, \ldots$.

When $n = 1$, the expectation of a random variable X becomes

$$m_1 = E(X) = \int_{-\infty}^{\infty} x f(x) dx = (-i)\phi'(0). \qquad (2.12.12)$$

Thus the simple formula (2.12.11) involving the derivatives of the characteristic function provides for the existence and the computation of the moment of any arbitrary order.

Similarly, the variance σ^2 of a random variable is given in terms of the characteristic function as

$$\sigma^2 = \int_{-\infty}^{\infty}(x-m)^2 f(x)dx = m_2 - m_1^2$$

$$= \{\phi'(0)\}^2 - \phi''(0). \qquad (2.12.13)$$

Example 2.12.1 Find the moments of the normal distribution defined by the density function

$$f(x) = \frac{1}{\sigma\sqrt{2\pi}} \exp\left\{-\frac{(x-m)^2}{2\sigma^2}\right\}. \qquad (2.12.14)$$

The characteristic function of the normal distribution is the Fourier transform of $f(x)$, which is

$$\phi(t) = \frac{1}{\sigma\sqrt{2\pi}} \int_{-\infty}^{\infty} e^{itx} \exp\left[-\frac{(x-m)^2}{2\sigma^2}\right]dx.$$

We substitute $x - m = y$ and use Example 2.3.1 to obtain

$$\phi(t) = \frac{\exp(itm)}{\sigma\sqrt{2\pi}} \int_{-\infty}^{\infty} e^{ity} \exp\left(-\frac{y^2}{2\sigma^2}\right)dy = \exp\left(itm - \frac{1}{2}t^2\sigma^2\right). \qquad (2.12.15)$$

Thus

$$m_1 = (-i)\phi'(0) = m,$$

$$m_2 = -\phi''(0) = (m^2 + \sigma^2),$$

$$m_3 = m(m^2 + 3\sigma^2).$$

Finally, the variance of the normal distribution is

$$m_2 - m_1^2 = \sigma^2. \qquad (2.12.16)$$

The above discussion reveals that characteristic functions are very useful for investigation of certain problems in mathematical statistics. We close this section by discussing more properties of characteristic functions.

Theorem 2.12.1 (*Addition Theorem*). The characteristic function of the sum of a finite number of independent random variables is equal to the product of their characteristic functions.

Proof. Suppose $X_1, X_2, \ldots, X_n$ are n independent random variables and $Z = X_1 + X_2 + \cdots + X_n$. Further, suppose $\phi_1(t), \phi_2(t), \ldots, \phi_n(t)$, and $\phi(t)$ are the characteristic functions of $X_1, X_2, \ldots, X_n$ and Z respectively.

Then we have

$$\phi(t) = E[\exp(itZ)] = E\left[\exp\{it(X_1 + X_2 + \cdots + X_n)\}\right],$$

which is, by the independence of the random variables,

$$= E(e^{itX_1}) E(e^{itX_2}) \cdots E(e^{itX_n})$$

$$= \phi_1(t) \phi_2(t) \cdots \phi_n(t). \qquad (2.12.17)$$

This proves the *addition theorem*.

Example 2.12.2 Find the expected value and the standard deviation of the sum of n independent normal random variables.

Suppose $X_1, X_2, \ldots, X_n$ are n independent random variables with the normal distributions $N(m_r, \sigma_r)$, where $r = 1, 2, \ldots, n$. The respective characteristic functions of these distributions are

$$\phi_r(t) = \exp\left[itm_r - \frac{1}{2}t^2\sigma_r^2\right], \quad r = 1, 2, 3, \ldots, n. \quad (2.12.18)$$

Because of the independence of $X_1, X_2, \ldots, X_n$, the random variable $Z = X_1 + X_2 + \cdots + X_n$ has the characteristic function

$$\phi(t) = \phi_1(t)\phi_2(t)\cdots\phi_n(t)$$
$$= \exp\left[it(m_1 + m_2 + \cdots + m_n) - \frac{1}{2}(\sigma_1^2 + \sigma_2^2 + \cdots + \sigma_n^2)t^2\right]. \quad (2.12.19)$$

This represents the characteristic function of the normal distribution $N\left(m_1 + m_2 + \cdots + m_n, \sqrt{\sigma_1^2 + \sigma_2^2 + \cdots + \sigma_n^2}\right)$. Thus the expected value of Z is $(m_1 + m_2 + \cdots + m_n)$ and its standard deviation is $(\sigma_1^2 + \sigma_2^2 + \cdots + \sigma_n^2)^{\frac{1}{2}}$.

Finally, we state the fundamental *Central Limit Theorem* without proof.

Theorem 2.12.2 (*The Lévy-Cramér Theorem*). Suppose $\{X_n\}$ is a sequence of random variables, $F_n(x)$ and $\phi_n(t)$ are respectively the distribution and characteristic functions of X_n. Then the sequence $\{F_n(x)\}$ is convergent to a distribution function $F(x)$ if and only if the sequence $\{\phi_n(t)\}$ is convergent at every point t on the real line to a function $\phi(t)$ continuous in some neighborhood of the origin. The limit function $\phi(t)$ is then the characteristic function of the limit distribution function $F(x)$, and the convergence $\phi_n(t) \to \phi(t)$ is uniform in every finite interval on the t-axis.

All these ideas developed in this section can be generalized for the multidimensional distribution functions by the use of multiple Fourier transforms. We refer interested readers to Lukacs (1960).

2.13 Multiple Fourier Transforms and Their Applications

Definition 2.13.1 Under the assumptions on $f(\mathbf{x})$ similar to those made for the one dimensional case, the *multiple Fourier transform* of $f(\mathbf{x})$, where $\mathbf{x} = (x_1, x_2, \ldots, x_n)$ is the n-dimensional vector, defined by

$$\mathscr{F}\{f(\mathbf{x})\} = F(\boldsymbol{\kappa}) = \frac{1}{(2\pi)^{n/2}} \int_{-\infty}^{\infty}\cdots\int_{-\infty}^{\infty} \exp\{-i(\boldsymbol{\kappa}\cdot\mathbf{x})\} f(\mathbf{x})\,d\mathbf{x}, \quad (2.13.1)$$

where $\boldsymbol{\kappa} = (k_1, k_2, \ldots, k_n)$ is the n-dimensional transform vector and $\boldsymbol{\kappa}\cdot\mathbf{x} = (k_1 x_1 + k_2 x_2 + \cdots + k_n x_n)$.

The inverse Fourier transform is similarly defined by

$$\mathcal{F}^{-1}\{F(\boldsymbol{\kappa})\} = f(\mathbf{x}) = \frac{1}{(2\pi)^{n/2}} \int_{-\infty}^{\infty} \cdots \int_{-\infty}^{\infty} \exp\{i(\boldsymbol{\kappa}\cdot\mathbf{x})\} F(\boldsymbol{\kappa}) d\boldsymbol{\kappa}. \quad (2.13.2)$$

In particular, the *double Fourier transform* is defined by

$$\mathcal{F}\{f(x,y)\} = F(k,\ell) = \frac{1}{2\pi} \int_{-\infty}^{\infty} \int_{-\infty}^{\infty} \exp\{-i(\boldsymbol{\kappa}\cdot\mathbf{r})\} f(x,y) dx dy, \quad (2.13.3)$$

where $\mathbf{r} = (x, y)$ and $\boldsymbol{\kappa} = (k, \ell)$.

The inverse Fourier transform is given by

$$\mathcal{F}^{-1}\{F(k,\ell)\} = f(x,y) = \frac{1}{2\pi} \int_{-\infty}^{\infty} \int_{-\infty}^{\infty} \exp\{i(\boldsymbol{\kappa}\cdot\mathbf{r})\} F(k,\ell) dk\, d\ell. \quad (2.13.4)$$

Similarly, the three-dimensional Fourier transform and its inverse are defined by the integrals

$$\mathcal{F}\{f(x,y,z)\} = F(k,\ell,m)$$

$$= \frac{1}{(2\pi)^{3/2}} \int_{-\infty}^{\infty}\int_{-\infty}^{\infty}\int_{-\infty}^{\infty} \exp\{-i(\boldsymbol{\kappa}\cdot\mathbf{r})\} f(x,y,z) dx\, dy\, dz, \quad (2.13.5)$$

$$\mathcal{F}^{-1}\{F(k,\ell,m)\} = f(x,y,z)$$

$$= \frac{1}{(2\pi)^{3/2}} \int_{-\infty}^{\infty}\int_{-\infty}^{\infty}\int_{-\infty}^{\infty} \exp\{i(\boldsymbol{\kappa}\cdot\mathbf{r})\} F(k,\ell,m) dk\, d\ell\, dm. \quad (2.13.6)$$

The operational properties of these multiple Fourier transforms are similar to those of the one-dimensional case. In particular, results (2.4.7) and (2.4.8) relating the Fourier transforms of derivatives to the Fourier transforms of given functions are valid for the higher dimensional case as well. In higher dimensions, they are applied to the transforms of partial derivatives of $f(\mathbf{x})$ under the assumptions that f and its partial derivatives vanish at infinity.

We illustrate the multiple Fourier transform method by the following examples of applications:

Example 2.13.1 (*The Dirichlet Problem for the Three-Dimensional Laplace Equation in the Half-Space*). The boundary value problem for $u(x,y,z)$ satisfies the following equation and boundary conditions:

$$\nabla^2 u \equiv u_{xx} + u_{yy} + u_{zz} = 0, \quad -\infty < x, y < \infty, \quad z > 0, \quad (2.13.7)$$

$$u(x, y, 0) = f(x, y), \quad -\infty < x, y < \infty \quad (2.13.8)$$

$$u(x, y, z) \to 0 \quad \text{as } r = \sqrt{x^2 + y^2 + z^2} \to \infty. \quad (2.13.9)$$

We use the double Fourier transform defined by (2.13.3) to the system (2.13.7)-(2.13.9) which reduces to

$$\frac{d^2 U}{dz^2} - \kappa^2 U = 0 \quad \text{for } z > 0,$$

$$U(k, \ell, 0) = F(k, \ell).$$

Thus the solution of this transformed problem is

$$U(k,\ell,z) = F(k,\ell)\exp(-|\kappa|z) = F(k,\ell)G(k,\ell), \qquad (2.13.10)$$

where $\kappa = (k,\ell)$ and $G(k,\ell) = \exp(-|\kappa|z)$ so that

$$g(x,y) = \mathscr{F}^{-1}\{\exp(-|\kappa|z)\} = \frac{z}{(x^2+y^2+z^2)^{3/2}}. \qquad (2.13.11)$$

Applying the Convolution Theorem to (2.13.10), we obtain the formal solution

$$u(x,y,z) = \frac{1}{2\pi}\int_{-\infty}^{\infty}\int_{-\infty}^{\infty}f(\xi,\eta)g(x-\xi,y-\eta,z)d\xi d\eta$$

$$= \frac{z}{2\pi}\int_{-\infty}^{\infty}\int_{-\infty}^{\infty}\frac{f(\xi,\eta)d\xi d\eta}{\left[(x-\xi)^2+(y-\eta)^2+z^2\right]^{3/2}}. \qquad (2.13.12)$$

Example 2.13.2 (*The Two-Dimensional Diffusion Equation*). We solve the two-dimensional diffusion equation

$$u_t = K\nabla^2 u, \quad -\infty < x,y < \infty, \quad t > 0, \qquad (2.13.13)$$

with the boundary conditions

$$u(x,y,0) = f(x,y) \quad -\infty < x,y < \infty, \qquad (2.13.14)$$

$$u(x,y,t) \to 0 \quad \text{as } r = \sqrt{x^2+y^2} \to \infty, \qquad (2.13.15)$$

where K is the diffusivity constant.

The double Fourier transform of $u(x,y,t)$ defined by (2.13.3) is used to reduce the system (2.13.13)-(2.13.14) into the form

$$\frac{dU}{dt} = -\kappa^2 KU, \quad t > 0,$$

$$U(k,\ell,0) = F(k,\ell).$$

The solution of this system is

$$U(k,\ell,t) = F(k,\ell)\exp(-tK\kappa^2) = F(k,\ell)G(k,\ell), \qquad (2.13.16)$$

where

$$G(k,\ell) = \exp(-K\kappa^2 t),$$

so that

$$g(x,y) = \mathscr{F}^{-1}\{\exp(-tK\kappa^2)\} = \frac{1}{2Kt}\exp\left[-\frac{x^2+y^2}{4Kt}\right]. \qquad (2.13.17)$$

Finally, the Convolution Theorem gives the formal solution

$$u(x,y,t) = \frac{1}{4\pi Kt}\int_{-\infty}^{\infty}\int_{-\infty}^{\infty}f(\xi,\eta)\exp\left[-\frac{(x-\xi)^2+(y-\eta)^2}{4Kt}\right]d\xi d\eta. \qquad (2.13.18)$$

Or equivalently

$$u(x,y,t) = \frac{1}{4\pi Kt}\int_{-\infty}^{\infty}\int_{-\infty}^{\infty}f(\mathbf{r}')\exp\left\{-\frac{|\mathbf{r}-\mathbf{r}'|^2}{4Kt}\right\}d\mathbf{r}' \qquad (2.13.19)$$

where $\mathbf{r}' = (\xi,\eta)$.

We make the change of variable $(\mathbf{r}' - \mathbf{r}) = \sqrt{4Kt}\,\mathbf{R}$ to reduce (2.13.19) in the form

$$u(x,y,t) = \frac{1}{\pi\sqrt{4Kt}} \int_{-\infty}^{\infty}\int_{-\infty}^{\infty} f(\mathbf{r} + \sqrt{4Kt}\,\mathbf{R})\exp(-R^2)\,d\mathbf{R}. \quad (2.13.20)$$

Similarly, the formal solution of the initial value problem for the three-dimensional diffusion equation

$$u_t = K(u_{xx} + u_{yy} + u_{zz}), \quad -\infty < x, y, z < \infty, \quad t > 0 \quad (2.13.21)$$

$$u(x,y,z,0) = f(x,y,z), \quad -\infty < x, y, z < \infty \quad (2.13.22)$$

is given by

$$u(x,y,z,t) = \frac{1}{(4\pi Kt)^{3/2}} \int\int\int_{-\infty}^{\infty} f(\xi,\eta,\zeta)\exp\left(-\frac{r^2}{4Kt}\right) d\xi\,d\eta\,d\zeta, \quad (2.13.23)$$

where

$$r^2 = (x-\xi)^2 + (y-\eta)^2 + (z-\zeta)^2.$$

Or equivalently

$$u(x,y,z,t) = \frac{1}{(4\pi Kt)^{3/2}} \int\int\int_{-\infty}^{\infty} f(\mathbf{r}')\exp\left\{-\frac{|\mathbf{r}-\mathbf{r}'|^2}{4Kt}\right\} d\xi\,d\eta\,d\zeta, \quad (2.13.24)$$

where $\mathbf{r} = (x,y,z)$ and $\mathbf{r}' = (\xi,\eta,\zeta)$.

Making the change of variable $\mathbf{r}' - \mathbf{r} = \sqrt{4tK}\,\mathbf{R}$, solution (2.13.24) reduces to

$$u(x,y,z,t) = \frac{1}{\pi^{3/2} 4Kt} \int\int\int_{-\infty}^{\infty} f(\mathbf{r} + \sqrt{4Kt}\,\mathbf{R})\exp(-R^2)\,d\mathbf{R}. \quad (2.13.25)$$

This is known as the *Fourier solution*.

Example 2.13.3 (*The Cauchy Problem for the Two–Dimensional Wave Equation*). The initial value problem for the wave equation in two dimensions is governed by

$$u_{tt} = c^2(u_{xx} + u_{yy}), \quad -\infty < x, y < \infty, \quad t > 0, \quad (2.13.26)$$

with the initial data

$$u(x,y,0) = 0, \quad u_t(x,y,0) = f(x,y), \quad -\infty < x, y < \infty, \quad (2.13.27\text{ab})$$

where c is a constant. We assume that u and its first partial derivatives vanish at infinity.

We apply the two-dimensional Fourier transform defined by (2.13.3) to the system (2.13.26)-(2.13.27ab), which becomes

$$\frac{d^2 U}{dt^2} + c^2 \kappa^2 U = 0, \quad \kappa^2 = k^2 + \ell^2,$$

$$U(k,\ell,0) = 0, \quad \left(\frac{dU}{dt}\right)_{t=0} = F(k,\ell).$$

The solution of this transformed system is

$$U(k,\ell,t) = F(k,\ell)\,\frac{\sin(c\kappa t)}{c\kappa}. \quad (2.13.28)$$

The inverse Fourier transform gives the formal solution

$$u(x,y,t) = \frac{1}{2\pi c} \int\!\!\int_{-\infty}^{\infty} \exp(i\boldsymbol{\kappa}\cdot\mathbf{r})\frac{\sin(c\kappa t)}{\kappa} F(\boldsymbol{\kappa})\,d\boldsymbol{\kappa} \qquad (2.13.29)$$

$$= \frac{1}{4i\pi c}\int\!\!\int_{-\infty}^{\infty}\frac{F(\boldsymbol{\kappa})}{\kappa}\left[\exp\left\{i\kappa\left(\frac{\boldsymbol{\kappa}\cdot\mathbf{r}}{\kappa}+ct\right)\right\} - \exp\left\{i\kappa\left(\frac{\boldsymbol{\kappa}\cdot\mathbf{r}}{\kappa}-ct\right)\right\}\right]d\boldsymbol{\kappa}. \qquad (2.13.30)$$

The form of this solution reveals an interesting feature of the wave equation. The exponential terms $\exp\left\{i\kappa\left(ct\pm\frac{\boldsymbol{\kappa}\cdot\mathbf{r}}{\kappa}\right)\right\}$ involved in the integral solution (2.13.30) represent plane wave solutions of the wave equation (2.13.26). Thus the solutions remain constant on the planes $\boldsymbol{\kappa}\cdot\mathbf{r} = $ constant that move parallel to themselves with velocity c. Evidently, solution (2.13.30) represents a superposition of the plane wave solutions traveling in all possible directions.

Similarly, the solution of the Cauchy problem for the three-dimensional wave equation

$$u_{tt} = c^2(u_{xx}+u_{yy}+u_{zz}), \quad -\infty < x,y,z < \infty, \quad t>0, \qquad (2.13.31)$$

$$u(x,y,z,0)=0, \quad u_t(x,y,z,0)=f(x,y,z), \quad -\infty < x,y,z < \infty \qquad (2.13.32ab)$$

is given by

$$u(\mathbf{r},t) = \frac{1}{2ic(2\pi)^{3/2}}\int\!\!\int\!\!\int_{-\infty}^{\infty}\frac{F(\boldsymbol{\kappa})}{\kappa}\left[\exp\left\{i\kappa\left(\frac{\boldsymbol{\kappa}\cdot\mathbf{r}}{\kappa}+ct\right)\right\}\right.$$

$$\left. - \exp\left\{i\kappa\left(\frac{\boldsymbol{\kappa}\cdot\mathbf{r}}{\kappa}-ct\right)\right\}\right]d\boldsymbol{\kappa}, \qquad (2.13.33)$$

where $\mathbf{r}=(x,y,z)$ and $\boldsymbol{\kappa}=(k,\ell,m)$.

In particular, when $f(x,y,z) = \delta(x)\delta(y)\delta(z)$ so that $F(\boldsymbol{\kappa}) = (2\pi)^{-3/2}$, solution (2.13.33) becomes

$$u(\mathbf{r},t) = \frac{1}{(2\pi)^3}\int\!\!\int\!\!\int_{-\infty}^{\infty}\frac{\sin c\kappa t}{c\kappa}\exp(i(\boldsymbol{\kappa}\cdot\mathbf{r}))\,d\boldsymbol{\kappa}. \qquad (2.13.34)$$

In terms of the spherical polar coordinates (κ,θ,ϕ) where the polar axis (the z-axis) is taken along the $\mathbf{r}$ direction with $\boldsymbol{\kappa}\cdot\mathbf{r}=\kappa r\cos\theta$, we write (2.13.34) in the form

$$u(r,t) = \frac{1}{(2\pi)^3}\int_0^{2\pi}d\phi\int_0^{\pi}d\theta\int_0^{\infty}\exp(i\kappa r\cos\theta)\frac{\sin c\kappa t}{c\kappa}\cdot\kappa^2\sin\theta\,d\kappa$$

$$= \frac{1}{2\pi^2 cr}\int_0^{\infty}\sin(c\kappa t)\sin(\kappa r)\,d\kappa$$

$$= \frac{1}{8\pi^2 cr}\int_{-\infty}^{\infty}\left[e^{i\kappa(ct-r)} - e^{i\kappa(ct+r)}\right]d\kappa.$$

Or

$$u(r,t) = \frac{1}{4\pi cr}\left[\delta(ct-r) - \delta(ct+r)\right]. \tag{2.13.35}$$

For $t > 0$, $ct + r > 0$ so that $\delta(ct+r) = 0$ and hence

$$u(\mathbf{r},t) = \frac{1}{4\pi cr}\delta(ct-r). \tag{2.13.36}$$

Example 2.13.4 (*The Three-Dimensional Poisson Equation*). The solution of the Poisson equation

$$-\nabla^2 u = f(\mathbf{r}), \tag{2.13.37}$$

where $\mathbf{r} = (x, y, z)$ is given by

$$u(\mathbf{r}) = \int\!\!\!\int\!\!\!\int_{-\infty}^{\infty} G(\mathbf{r},\boldsymbol{\xi}) f(\boldsymbol{\xi}) d\boldsymbol{\xi}, \tag{2.13.38}$$

where the Green's function $G(\mathbf{r},\boldsymbol{\xi})$ of the operator, $-\nabla^2$, is

$$G(\mathbf{r},\boldsymbol{\xi}) = \frac{1}{4\pi}\frac{1}{|\mathbf{r}-\boldsymbol{\xi}|}. \tag{2.13.39}$$

To obtain the fundamental solution, we need to solve the equation

$$-\nabla^2 G(\mathbf{r},\boldsymbol{\xi}) = \delta(x-\xi)\delta(y-\eta)\delta(z-\zeta), \quad \mathbf{r} \neq \boldsymbol{\xi}. \tag{2.13.40}$$

Application of the three-dimensional Fourier transform defined by (2.13.5) to (2.13.40) gives

$$\kappa^2 \hat{G}(\boldsymbol{\kappa},\boldsymbol{\xi}) = \frac{1}{(2\pi)^{3/2}}\exp(-i\boldsymbol{\kappa}\cdot\boldsymbol{\xi}), \tag{2.13.41}$$

where $\hat{G}(\boldsymbol{\kappa},\boldsymbol{\xi}) = \mathscr{F}\{G(\mathbf{r},\boldsymbol{\xi})\}$ and $\boldsymbol{\kappa} = (k,\ell,m)$.

The inverse Fourier transform gives the formal solution

$$G(\mathbf{r},\boldsymbol{\xi}) = \frac{1}{(2\pi)^3}\int\!\!\!\int\!\!\!\int_{-\infty}^{\infty}\exp\{i\boldsymbol{\kappa}\cdot(\mathbf{r}-\boldsymbol{\xi})\}\frac{d\boldsymbol{\kappa}}{\kappa^2}$$

$$= \frac{1}{(2\pi)^3}\int\!\!\!\int\!\!\!\int_{-\infty}^{\infty}\exp(i\boldsymbol{\kappa}\cdot\mathbf{x})\frac{d\boldsymbol{\kappa}}{\kappa^2}, \tag{2.13.42}$$

where $\mathbf{x} = |\mathbf{r}-\boldsymbol{\xi}|$.

We evaluate this integral using polar coordinates in the $\boldsymbol{\kappa}$-space with the axis along the x-axis. In terms of spherical polar coordinates (κ,θ,ϕ) so that $\boldsymbol{\kappa}\cdot\mathbf{x} = \kappa R\cos\theta$ where $R = |\mathbf{x}|$. Thus (2.13.42) becomes

$$G(\mathbf{r},\boldsymbol{\xi}) = \frac{1}{(2\pi)^3}\int_0^{2\pi}d\phi\int_0^{\pi}d\theta\int_0^{\infty}\exp(i\kappa R\cos\theta)\kappa^2\sin\theta\cdot\frac{d\kappa}{\kappa^2}$$

$$= \frac{1}{(2\pi)^2}\int_0^{\infty}2\frac{\sin(\kappa R)}{\kappa R}d\kappa = \frac{1}{4\pi R} = \frac{1}{4\pi|\mathbf{r}-\boldsymbol{\xi}|}, \tag{2.13.43}$$

provided $R > 0$.

In electrodynamics, the fundamental solution (2.13.43) has a well-known interpretation. Physically, it represents the potential at point $\mathbf{r}$ generated by the unit point charge distribution at point $\boldsymbol{\xi}$. This is what can be expected because $\delta(\mathbf{r}-\boldsymbol{\xi})$ is the charge density corresponding to a unit point charge at $\boldsymbol{\xi}$.

The solution of (2.13.37) is then given by

$$u(\mathbf{r}) = \int\int\int_{-\infty}^{\infty} G(\mathbf{r},\boldsymbol{\xi}) f(\boldsymbol{\xi}) d\boldsymbol{\xi} = \frac{1}{4\pi} \int\int\int_{-\infty}^{\infty} \frac{f(\boldsymbol{\xi}) d\boldsymbol{\xi}}{|\mathbf{r}-\boldsymbol{\xi}|}. \tag{2.13.44}$$

The integrand in (2.13.44) consists of the given charge distribution $f(\mathbf{r})$ at $\mathbf{r} = \boldsymbol{\xi}$ and Green's function $G(\mathbf{r},\boldsymbol{\xi})$. Physically, $G(\mathbf{r},\boldsymbol{\xi})$ f($\boldsymbol{\xi}$) represents the resulting potentials due to elementary point charges, and the total potential due to a given charge distribution $f(\mathbf{r})$ is then obtained by the integral superposition of the resulting potentials. This is called the *principle of superposition*.

Example 2.13.5 (*The Two-Dimensional Helmholtz Equation*). To find the fundamental solution of the two-dimensional Helmholtz equation

$$-\nabla^2 G + \alpha^2 G = \delta(x-\xi)\delta(y-\eta), \quad -\infty < x, y < \infty. \tag{2.13.45}$$

It is convenient to make the change of variables $x - \xi = x^*$, $y - \eta = y^*$. Consequently, (2.13.45) reduces to the form, dropping the asterisks,

$$G_{xx} + G_{yy} - \alpha^2 G = -\delta(x)\delta(y). \tag{2.13.46}$$

Application of the double Fourier transform $\hat{G}(\boldsymbol{\kappa}) = \mathscr{F}\{G(x,y)\}$ to (2.13.46) gives

$$\hat{G}(\boldsymbol{\kappa}) = \frac{1}{2\pi} \frac{1}{(\kappa^2 + \alpha^2)}, \tag{2.13.47}$$

where $\boldsymbol{\kappa} = (k,\ell)$ and $\kappa^2 = k^2 + \ell^2$.

The inverse Fourier transform yields the solution

$$G(x,y) = \frac{1}{4\pi^2} \int\int_{-\infty}^{\infty} \exp(i\boldsymbol{\kappa}\cdot\mathbf{x})(\kappa^2 + \alpha^2)^{-1} dk\, d\ell. \tag{2.13.48}$$

In terms of polar coordinates $(x,y) = r(\cos\theta,\sin\theta)$, $(k,\ell) = \rho(\cos\phi,\sin\phi)$, the integral solution (2.13.48) becomes

$$G(x,y) = \frac{1}{4\pi^2} \int_0^\infty \frac{\rho\, d\rho}{(\rho^2 + \alpha^2)} \int_0^{2\pi} \exp\{ir\rho\cos(\phi-\theta)\} d\phi,$$

which is, replacing the second integral by $2\pi J_0(r\rho)$,

$$= \frac{1}{2\pi} \int_0^\infty \frac{\rho J_0(r\rho) d\rho}{(\rho^2 + \alpha^2)}. \tag{2.13.49}$$

In terms of the original coordinates, the fundamental solution of (2.13.45) is given by

$$G(\mathbf{r},\boldsymbol{\xi}) = \frac{1}{2\pi} \int_0^\infty \frac{\rho J_0\left[\rho\sqrt{(x-\xi)^2 + (y-\eta)^2}\right] d\rho}{(\rho^2 + \alpha^2)}. \tag{2.13.50}$$

Accordingly, the solution of the inhomogeneous equation

$$(\nabla^2 - \alpha^2)u = -f(x,y) \tag{2.13.51}$$

is
$$u(x,y) = \int\int_{-\infty}^{\infty} G(\mathbf{r},\boldsymbol{\xi}) f(\boldsymbol{\xi}) d\boldsymbol{\xi}, \qquad (2.13.52)$$

where $G(\mathbf{r},\boldsymbol{\xi})$ is given by (2.13.50).

Since the integral solution (2.13.49) does not exist for $\alpha = 0$, Green's function for the two-dimensional Poisson equation (2.13.45) cannot be derived from (2.13.49). Instead, we differentiate (2.13.49) with respect to r to obtain

$$\frac{\partial G}{\partial r} = \frac{1}{2\pi} \int_0^\infty \frac{\rho^2 J_0'(r\rho) d\rho}{(\rho^2 + \alpha^2)}$$

which is, for $\alpha = 0$,

$$\frac{\partial G}{\partial r} = \frac{1}{2\pi} \int_0^\infty J_0'(r\rho) d\rho = -\frac{1}{2\pi r}.$$

Integrating this result gives

$$G(r,\theta) = -\frac{1}{2\pi} \log r.$$

In terms of the original coordinates, Green's function becomes

$$G(\mathbf{r},\boldsymbol{\xi}) = -\frac{1}{4\pi} \log\left[(x-\xi)^2 + (y-\eta)^2\right]. \qquad (2.13.53)$$

This is Green's function for the two-dimensional Poisson equation $\nabla^2 = -f(x,y)$. Thus the solution of the Poisson equation is

$$u(x,y) = \int\int_{-\infty}^{\infty} G(\mathbf{r},\boldsymbol{\xi}) f(\boldsymbol{\xi}) d\boldsymbol{\xi}. \qquad (2.13.54)$$

Example 2.13.6 (*Diffusion of Vorticity from a Vortex Sheet*). To solve the two-dimensional vorticity equation in the x, y plane given by

$$\zeta_t = \nu \nabla^2 \zeta \qquad (2.13.55)$$

with the initial condition

$$\zeta(x,y,0) = \zeta_0(x,y), \qquad (2.13.56)$$

where $\zeta = v_x - u_y$.

Application of the double Fourier transform defined by

$$\hat{\zeta}(k,\ell,t) = \frac{1}{2\pi} \int\int_{-\infty}^{\infty} \exp\left[-i(kx+\ell y)\right] \zeta(x,y,t) dx dy$$

to (2.13.55)-(2.13.56) gives

$$\frac{d\hat{\zeta}}{dt} = -\nu(k^2+\ell^2)\hat{\zeta},$$

$$\hat{\zeta}(k,\ell,0) = \hat{\zeta}_0(k,\ell).$$

Thus the solution of the transformed system is

$$\hat{\zeta}(k,\ell,t) = \hat{\zeta}_0(k,\ell) \exp\left[-\nu(k^2+\ell^2)t\right]. \qquad (2.13.57)$$

The inversion theorem for Fourier transform gives the formal solution

$$\zeta(x,y,t) = \frac{1}{2\pi} \int\!\!\!\int_{-\infty}^{\infty} \hat{\zeta}_0(k,\ell) \exp\left[i(\boldsymbol{\kappa}\cdot\mathbf{r}) - v\kappa^2 t\right] dk\, d\ell, \qquad (2.13.58)$$

where $\boldsymbol{\kappa} = (k,\ell)$ and $\kappa^2 = k^2 + \ell^2$.

In particular, if $\zeta_0(x,y) = V\delta(x)$ represents a vortex sheet of strength V per unit width in the plane $x = 0$, we find $\hat{\zeta}_0(k,\ell) = V\delta(\ell)$ and hence

$$\zeta(x,y,t) = \frac{V}{2\pi} \int_{-\infty}^{\infty} \exp\{ikx - vk^2 t\} dk$$

$$= \frac{V}{2\sqrt{\pi vt}} \exp\left(-\frac{x^2}{4vt}\right). \qquad (2.13.59)$$

Apart from a constant, the velocity field is given by

$$u(x,t) = 0, \quad v(x,t) = \frac{V}{\sqrt{\pi}} \operatorname{erf}\left(\frac{x}{2\sqrt{vt}}\right). \qquad (2.13.60)$$

2.14 Exercises

1. Find the Fourier transforms of each of the following functions:

   (a) $f(x) = \dfrac{1}{1+x^2}$,

   (b) $f(x) = \dfrac{x}{1+x^2}$,

   (c) $f(x) = \delta^{(n)}(x)$,

   (d) $f(x) = x \exp(-a|x|)$, $a > 0$;

   (e) $f(x) = e^x \exp(-e^x)$,

   (f) $f(x) = x \exp\left(-\dfrac{ax^2}{2}\right)$, $a > 0$;

   (g) $f(x) = x^2 \exp\left(-\dfrac{1}{2}x^2\right)$,

   (h) $f(x) = \begin{cases} 1-|x|, & |x| \leq 1 \\ 0, & |x| > 1 \end{cases}$,

   (i) $f(x) = \begin{cases} 1-x^2, & |x| \leq 1 \\ 0, & |x| > 1 \end{cases}$,

   (j) $f_n(x) = (-1)^n \exp\left(\dfrac{1}{2}x^2\right) \times \left(\dfrac{d}{dx}\right)^n \exp(-x^2)$.

2. Show that

   (a) $\mathcal{F}\{\delta(x-ct) + \delta(x+ct)\} = \sqrt{\dfrac{\pi}{2}} \cos(kct)$,

   (b) $\mathcal{F}\{H(ct-|x|)\} = \sqrt{\dfrac{\pi}{2}} \dfrac{\sin kct}{k}$.

3. Show that

   (a) $i\dfrac{d}{dk} F(k) = \mathcal{F}\{x f(x)\}$,

(b) $i^n \dfrac{d^n}{dk^n} F(k) = \mathcal{F}\{x^n f(x)\}$.

4. Use exercise 3(b) to find the Fourier transform of $f(x) = x^2 \exp(-ax^2)$.

5. Prove the following:

   (a) $\mathcal{F}\left\{(a^2 - x^2)^{-\frac{1}{2}} H(a - |x|)\right\} = \sqrt{\dfrac{\pi}{2}} J_0(ak)$, $a > 0$.

   (b) $\mathcal{F}\{P_n(x) H(1 - |x|)\} = (-i)^n \dfrac{1}{\sqrt{k}} J_{n+\frac{1}{2}}(k)$,

   where $P_n(x)$ is the Legendre polynomial of degree n.

6. If $f(x)$ has a finite discontinuity at a point $x = a$, show that

   $$\mathcal{F}\{f'(x)\} = (ik) F(k) - \dfrac{1}{\sqrt{2\pi}} \exp(-ika)[f]_a,$$

   where $[f]_a = f(a+0) - f(a-0)$.

   Generalize this result for $\mathcal{F}\{f^{(n)}(x)\}$.

7. Prove the following results for the convolution of the Fourier transform:

   (a) $\sqrt{2\pi}\, \delta(x) * f(x) = f(x)$, 　　(b) $\sqrt{2\pi}\, \delta'(x) * f(x) = f'(x)$.

   (c) $\dfrac{d}{dx}\{f(x) * g(x)\} = f'(x) * g(x) = f(x) * g'(x)$.

8. Use the Fourier transform to solve the following ordinary differential equations in $-\infty < x < \infty$:

   (a) $y''(x) - y(x) + 2f(x) = 0$, where $f(x) = 0$ when $x < -a$ and when $x > a$, and $y(x)$ and its derivatives vanish at $x = \pm\infty$.

   (b) $2 y''(x) + x y'(x) + y(x) = 0$,

   (c) $y''(x) + x y'(x) + y(x) = 0$,

   (d) $y''(x) + x y'(x) + xy(x) = 0$,

   (e) $\ddot{y}(t) + 2\alpha\, \dot{y}(t) + y(t) = f(t)$, $\dot{y}(t) = \dfrac{dy}{dt}$.

9. Solve the following integral equations for an unknown function $f(x)$:

   (a) $\displaystyle\int_{-\infty}^{\infty} \phi(x - t)\, f(t)\, dt = g(x)$.

   (b) $\displaystyle\int_{-\infty}^{\infty} \exp(-at^2)\, f(x - t)\, dt = \exp(-bx^2)$, $a, b > 0$.

   (c) $f(x) + \displaystyle\int_{-\infty}^{\infty} f(x - t)\, \exp(-at)\, dt = \dfrac{1}{(x^2 + b^2)}$.

   (d) $\displaystyle\int_{-\infty}^{\infty} \dfrac{f(t)\, dt}{(x - t)^2 + a^2} = \dfrac{\sqrt{2\pi}}{(x^2 + b^2)}$ 　for $b > a > 0$.

(e) $\dfrac{1}{\pi} \oint_{-\infty}^{\infty} \dfrac{f(t)dt}{x-t} = \phi(x),$

where the integral in (e) is treated as the Cauchy Principal value.

10. Solve the Cauchy problem for the Klein-Gordon equation
$$u_{tt} - c^2 u_{xx} + a^2 u = 0, \quad -\infty < x < \infty, \quad t > 0.$$
$$u(x, 0) = f(x), \quad \left(\dfrac{\partial u}{\partial t}\right)_{t=0} = g(x) \text{ for } -\infty < x < \infty.$$

11. Solve the telegraph equation
$$u_{tt} - c^2 u_{xx} + u_t - a u_x = 0, \quad -\infty < x < \infty, \quad t > 0.$$
$$u(x, 0) = f(x), \quad \left(\dfrac{\partial u}{\partial t}\right)_{t=0} = g(x) \quad \text{for } -\infty < x < \infty.$$

Show that the solution is unstable when $c^2 < a^2$. If $c^2 > a^2$, show that the bounded integral solution is
$$u(x, t) = \dfrac{1}{\sqrt{2\pi}} \int_{-\infty}^{\infty} A(k) \exp\left[-k^2(c^2 - a^2)t + ik(x + at)\right] dk$$

where $A(k)$ is given in terms of the transformed functions of the initial data. Hence deduce the asymptotic solution as $t \to \infty$ in the form
$$u(x, t) = A(0)\sqrt{\dfrac{\pi}{2(c^2 - a^2)t}} \exp\left[-\dfrac{(x + at)^2}{4(c^2 - a^2)t}\right].$$

12. Solve the equation
$$u_{tt} + u_{xxxx} = 0, \quad -\infty < x < \infty, \quad t > 0$$
$$u(x, 0) = f(x), \quad u_t(x, 0) = 0 \quad \text{for } -\infty < x < \infty.$$

13. Find the solution of the dissipative wave equation
$$u_{tt} - c^2 u_{xx} + \alpha u_t = 0, \quad -\infty < x < \infty, \quad t > 0,$$
$$u(x, 0) = f(x), \quad \left(\dfrac{\partial u}{\partial t}\right)_{t=0} = g(x) \quad \text{for } -\infty < x < \infty,$$

where $\alpha > 0$ is the dissipation parameter.

14. Obtain the Fourier cosine transforms of the following functions:
    (a) $f(x) = x \exp(-ax), \quad a > 0,$
    (b) $f(x) = e^{-ax} \cos x, \quad a > 0,$
    (c) $f(x) = \dfrac{1}{x},$
    (d) $K_0(ax),$

where $K_0(ax)$ is the *modified Bessel function*.

15. Find the Fourier sine transform of the following functions:
    (a) $f(x) = x \exp(-ax), \quad a > 0,$
    (b) $f(x) = \dfrac{1}{x} \exp(-ax), \quad a > 0,$
    (c) $f(x) = \dfrac{1}{x}.$

16. If $F_c(k) = \mathscr{F}_c\{\exp(-ax^2)\}$, show that $F_c(k)$ satisfies the equation
$$\frac{dF_c}{dk} + \left(\frac{k}{2a}\right)F_c = 0, \quad F_c(0) = 1.$$

17. Prove the following for the Fourier sine transform

(a) $\int_0^\infty F_s(k) G_c(k) \sin kx \, dk = \frac{1}{2} \int_0^\infty g(\xi)[f(\xi+x) - f(\xi-x)]d\xi,$

(b) $\int_0^\infty F_c(k) G_s(k) \sin kx \, dk = \frac{1}{2} \int_0^\infty f(\xi)[g(\xi+x) - g(\xi-x)]d\xi.$

18. Solve the integral equation
$$\int_0^\infty f(x) \sin kx \, dk = \begin{cases} 1-k, & 0 \le k < 1 \\ 0, & k > 1 \end{cases}.$$

19. Solve Example 2.10.1 with the boundary data
$u(0, t) = 0, \quad u(x, t) \to 0$ as $x \to \infty$, for $t > 0$.

20. Apply the Fourier cosine transform to find the solution $u(x, y)$ of the problem
$u_{xx} + u_{yy} = 0, \quad 0 < x < \infty, \; 0 < y < \infty,$
$u(x, 0) = H(a-x), \; a > x; \quad u_x(0, y) = 0, \quad 0 < x, y < \infty.$

21. Use the Fourier cosine (or sine) transform to solve the following integral equation:

(a) $\int_0^\infty f(x) \cos kx \, dx = \sqrt{\frac{\pi}{2k}},$ (b) $\int_0^\infty f(x) \sin kx \, dx = \frac{a}{a^2 + k^2},$

(c) $\left(\frac{2}{\pi}\right) \int_0^\infty f(x) \sin kx \, dx = J_0(ak),$ (d) $\int_0^\infty f(x) \cos kx \, dx = \frac{\sin ak}{k}.$

22. Solve the diffusion equation in the semi-infinite line
$u_t = \kappa u_{xx}, \quad 0 \le x < \infty, \; t > 0,$
with the boundary and initial data
$u(0, t) = 0$ for $t > 0$
$u(x, t) \to 0$ as $x \to \infty$ for $t > 0,$
$u(x, 0) = f(x)$ for $0 < x < \infty.$

23. Use the Parseval formula to evaluate the following integrals with $a > 0$ and $b > 0$:

(a) $\int_{-\infty}^\infty \frac{dx}{(x^2 + a^2)^2},$ (b) $\int_{-\infty}^\infty \frac{\sin ax}{x(x^2 + b^2)} dx$

(c) $\int_{-\infty}^\infty \frac{\sin^2 ax}{x^2} dx,$ (d) $\int_{-\infty}^\infty \frac{\exp(-bx^2) dx}{(x^2 + a^2)}.$

24. Show that
$$\int_0^\infty \frac{\sin ax \sin bx}{x^2} dx = \frac{\pi}{2} \min(a, b).$$

25. If $f(x) = \exp(-ax)$ and $g(x) = H(t-x)$, show that
$$\int_0^\infty \frac{\sin tx}{x(x^2+a^2)} dx = \frac{\pi}{2a^2}[1-\exp(-at)].$$

26. Use the Poisson summation formula to find the sum of each of the following series with non-zero a:

(a) $\sum_{n=0}^\infty \frac{1}{(a^2+n^2)}$, (b) $\sum_{n=1}^\infty \frac{\sin an}{n}$,

(c) $\sum_{n=1}^\infty \frac{\sin^2 an}{n^2}$, (d) $\sum_{n=-\infty}^\infty \frac{a}{n^2+a^2}$.

27. The Fokker-Planck equation (Reif, 1965) is used to describe the evolution of probability distribution functions $u(x, t)$ in nonequilibrium statistical mechanics and has the form
$$\frac{\partial u}{\partial t} = \frac{\partial}{\partial x}\left(\frac{\partial}{\partial x} + x\right)u.$$
The fundamental solution of this equation is defined by the equation
$$\left[\frac{\partial}{\partial t} - \frac{\partial}{\partial x}\left(\frac{\partial}{\partial x} + x\right)\right]G(x, \xi; t, \tau) = \delta(x-\xi)\,\delta(t-\tau).$$
Show that the fundamental solution is
$$G(x, \xi; t, \tau) = [2\pi\{1-\exp[-2(t-\tau)]\}]^{-\frac{1}{2}} \exp\left[-\frac{\{x-\xi \exp[-(t-\tau)]\}^2}{2[1-\exp\{-2(t-\tau)\}]}\right].$$
Hence derive
$$\lim_{t\to\infty} G(x, \xi; t, \tau) = \frac{1}{\sqrt{2\pi}} \exp\left(-\frac{1}{2}x^2\right).$$
With the initial condition $u(x, 0) = f(x)$, show that the function $u(x, t)$ tends to the normal distribution as $t \to \infty$, that is,
$$\lim_{t\to\infty} u(x, t) = \frac{1}{\sqrt{2\pi}} \exp\left(-\frac{1}{2}x^2\right) \int_{-\infty}^\infty f(\xi)\,d\xi.$$

28. The transverse vibration of an infinite elastic beam of mass m per unit length and the bending stiffness EI is governed by
$$u_{tt} + a^2 u_{xxxx} = 0, \quad a^2 = \frac{EI}{m}, \quad -\infty < x < \infty, \quad t > 0.$$
Solve this equation subject to the boundary and initial data
$$u(0, t) = 0 \quad \text{for all} \quad t > 0,$$
$$u(x, 0) = \phi(x), \quad u_t(x, 0) = \psi''(x) \quad \text{for} \quad 0 < x < \infty.$$

Show that the Fourier transform solution is
$$U(k,t) = \Phi(k)\cos(atk^2) - \Psi(k)\sin(atk^2).$$
Find the integral solution for $u(x,t)$.

29. Solve the Lamb (1904) problem in geophysics that satisfies the Helmholtz equation in an infinite elastic half-space
$$u_{xx} + u_{zz} + \frac{\omega^2}{c_2^2}u = 0, \quad -\infty < x < \infty, \quad z > 0,$$
where ω is the frequency and c_2 is the shear wave speed.

At the surface of the half-space $(z=0)$, the boundary condition relating the surface stress to the impulsive point load distribution is
$$\mu\frac{\partial u}{\partial z} = -P\delta(x) \text{ at } z=0,$$
where μ is one of the Lamé's constants, P is a constant and
$$u(x,z) \to 0 \text{ as } z \to \infty \text{ for } -\infty < x < \infty.$$
Show that the solution in terms of polar coordinates is
$$u(x,z) = \frac{P}{2i\mu} H_0^{(2)}\left(\frac{\omega r}{c_2}\right)$$
$$\sim \frac{P}{2i\mu}\left(\frac{2c_2}{\pi\omega r}\right)^{\frac{1}{2}} \exp\left(\frac{\pi i}{4} - \frac{i\omega r}{c_2}\right) \text{ for } \omega r \gg c_2.$$

30. Find the solution of the Cauchy-Poisson problem (Debnath, 1994, p 83) in an inviscid water of infinite depth which is governed by
$$\phi_{xx} + \phi_{zz} = 0, \quad -\infty < x < \infty, \quad -\infty < z \leq 0, \quad t > 0,$$
$$\left.\begin{array}{l}\phi_z - \eta_t = 0 \\ \phi_t + g\eta = 0\end{array}\right\} \text{ on } z=0, \quad t>0,$$
$$\phi_z \to 0 \text{ as } z \to -\infty.$$
$$\phi(x,0,0) = 0 \text{ and } \eta(x,0) = P\delta(x),$$
where $\phi = \phi(x,z,t)$ is the velocity potential, $\eta(x,t)$ is the free surface elevation, and P is a constant.

Derive the asymptotic solution for the free surface elevation in the limit as $t \to \infty$.

31. Obtain the solutions for the velocity potential $\phi(x,z,t)$ and the free surface elevation $\eta(x,t)$ involved in the two-dimensional surface waves in water of finite (or infinite) depth h. The governing equation, boundary, and free surface conditions and initial conditions (see Debnath 1994, p 92) are

$$\phi_{xx} + \phi_{zz} = 0, \quad -h \leq z \leq 0, \quad -\infty < x < \infty, \quad t > 0$$

$$\left.\begin{array}{l}\phi_t + g\eta = -\dfrac{P}{\rho} p(x)\exp(i\omega t) \\ \phi_z - \eta_t = 0\end{array}\right\} z = 0, \quad t > 0$$

$$\phi(x, z, 0) = 0 = \eta(x, 0) \text{ for all } x \text{ and } z.$$

32. Solve the steady-state surface wave problem (Debnath, 1994, p 47) on a running stream of infinite depth due to an external steady pressure applied to the free surface. The governing equation and the free surface conditions are

$$\phi_{xx} + \phi_{zz} = 0, \quad -\infty < x < \infty, \quad -\infty < z < 0, \quad t > 0,$$

$$\left.\begin{array}{l}\phi_x + U\phi_x + g\eta = -\dfrac{P}{\rho} \delta(x)\exp(\varepsilon t) \\ \eta_t + U\eta_x = \phi_z\end{array}\right\} z = 0, \quad (\varepsilon > 0),$$

$$\phi_z \to 0 \text{ as } z \to -\infty.$$

where U is the stream velocity, $\phi(x, z, t)$ is the velocity potential, and $\eta(x, t)$ is the free surface elevation.

33. Use the Fourier sine transform to solve the following initial and boundary value problem for the wave equation:

$$u_{tt} = c^2 u_{xx}, \quad 0 < x < \infty, \quad t > 0,$$

$$u(x, 0) = 0, \quad u_t(x, 0) = 0 \quad \text{for} \quad 0 < x < \infty,$$

$$u(0, t) = f(t) \text{ for } t > 0,$$

where $f(t)$ is a given function.

34. Solve the following initial and boundary value problem for the wave equation using the Fourier cosine transform:

$$u_{tt} = c^2 u_{xx}, \quad 0 < x < \infty, \quad t > 0,$$

$$u(0, t) = f(t) \text{ for } t > 0$$

$$u(x, 0) = 0, u_t(x, 0) = 0 \quad \text{for} \quad 0 < x < \infty,$$

where $f(t)$ is a known function.

35. Apply the Fourier transform to solve the initial value problem for the dissipative wave equation

$$u_{tt} = c^2 u_{xx} + \alpha u_{xxt}, \quad -\infty < x < \infty, \quad t > 0,$$

$$u(x, 0) = f(x), \; u_t(x, 0) = \alpha f''(x) \quad \text{for} \quad -\infty < x < \infty,$$

where α is a positive constant.

36. Use the Fourier sine transform to solve the initial and boundary value

problem for free vibrations of a semi-infinite string:
$$u_{tt} = c^2 u_{xx}, \quad 0 < x < \infty, \quad t > 0,$$
$$u(0, t) = 0, \quad t \geq 0,$$
$$u(x, 0) = f(x) \text{ and } u_t(x, 0) = g(x) \text{ for } 0 < x < \infty.$$

37. The static deflection $u(x, y)$ in a thin elastic disk in the form of a quadrant satisfies the boundary value problem
$$u_{xxxx} + 2 u_{xxyy} + u_{yyyy} = 0, \quad 0 < x < \infty, \quad 0 < y < \infty,$$
$$u(0, y) = u_{xx}(0, y) = 0 \quad \text{for} \quad 0 < y < \infty,$$
$$u(x, 0) = \frac{ax}{1+x^2}, \quad u_{yy}(x, 0) = 0 \quad \text{for} \quad 0 < x < \infty,$$
where a is a constant, and $u(x, y)$ and its derivatives vanish as $x \to \infty$ and $y \to \infty$.

Use the Fourier sine transform to show that
$$u(x, y) = \frac{a}{2} \int_0^\infty (2 + ky) \exp[-(1+y)k] \sin kx \, dx$$
$$= \frac{ax}{x^2 + (1+y)^2} + \frac{axy(1+y)}{\left[x^2 + (1+y)^2\right]^2}.$$

38. In exercise 37, replace the conditions on $y = 0$ by the conditions
$$u(x, 0) = 0, \quad u_{yy}(x, 0) = \frac{ax}{(1+x^2)^2} \quad \text{for} \quad 0 < x < \infty.$$
Show that the solution is
$$u(x, y) = -\frac{ax}{4} \int_0^\infty \exp[-(1+y)k] \sin kx \, dk$$
$$= -\frac{1}{4} \frac{axy}{\left[x^2 + (1+y)^2\right]}.$$

39. In exercise 37, solve the biharmonic equation in $0 < x < \infty$, $0 < y < b$ with the boundary conditions
$$u(0, y) = a \sin y, \quad u_{xx}(0, y) = 0 \quad \text{for} \quad 0 < y < b,$$
$$u(x, 0) = u_{yy}(x, 0) = u(x, b) = u_{yy}(x, b) = 0 \quad \text{for} \quad 0 < x < \infty,$$
and $u(x, y)$, $u_x(x, y)$ vanish as $x \to \infty$.

40. Use the Fourier transform to solve the boundary value problem
$$u_{xx} + u_{yy} = -x \exp(-x^2), \quad -\infty < x < \infty, \quad 0 < y < \infty,$$
$$u(x, 0) = 0, \text{ for } -\infty < x < \infty, u \text{ and its derivative vanish as } y \to \infty.$$

Show that
$$u(x, y) = \frac{1}{\sqrt{4\pi}} \int_0^\infty [1 - \exp(-ky)] \frac{\sin kx}{k} \exp\left(-\frac{k^2}{4}\right) dk.$$

41. Using the definition of the characteristic function for the discrete random variable X
$$\phi(t) = E[\exp(itX)] = \sum_r p_r \exp(itx_r)$$
where $p_r = P(X = x_r)$, show that the characteristic function of the binomial distribution
$$p_r = \binom{n}{r} p^r (1-p)^{n-r}$$
is
$$\phi(t) = [1 + p(e^{it} - 1)]^n$$
Find the moments.

42. Show that the characteristic function of the Poisson distribution defined by
$$p_r = P(X = r) = \frac{\lambda^r}{r!} e^{-\lambda}, \quad r = 0, 1, 2, \ldots$$
is
$$\phi(t) = \exp[\lambda(e^{it} - 1)].$$
Find the moments.

43. Find the characteristic function of
    (a) The Gamma distribution whose density function is
    $$f(x) = \frac{a^p}{\Gamma(p)} x^{p-1} e^{-ax} H(x),$$
    (b) The Beta distribution whose density function is
    $$f(x) = \begin{cases} \dfrac{x^{p-1}(1-x)^{q-1}}{B(p,q)} & \text{for } 0 < x < 1, \\ 0 & \text{for } x < 0 \text{ and } x > 1 \end{cases}$$
    (c) The Cauchy distribution whose density function is
    $$f(x) = \frac{1}{\pi} \frac{\lambda}{[\lambda^2 + (x-\mu)^2]},$$
    (d) The Laplace distribution whose density function is
    $$f(x) = \frac{1}{2\lambda} \exp\left(-\frac{|x-u|}{\lambda}\right), \quad \lambda > 0.$$

44. Find the density function of the random variable X whose characteristic function is $\phi(t) = (1 - |t|) H(1 - |t|)$.

45. Find the characteristic function of uniform distribution whose density function is
$$f(x) = \begin{cases} 0, & x < 0 \\ 1, & 0 \leq x \leq a \\ 0, & x > a \end{cases}$$

46. Solve the *initial value problem* (Debnath, 1994, p 115) for the two-dimensional surface waves at the free surface of a running stream of velocity U. The problem satisfies the equation, boundary, and initial conditions
$$\phi_{xx} + \phi_{zz} = 0, \quad -\infty < x < \infty, \quad -h \leq z \leq 0, \quad t > 0,$$
$$\left.\begin{array}{l}\phi_x + U\phi_x + g\eta = -\dfrac{P}{\rho}\delta(x)\exp(i\omega t) \\ \eta_t + U\eta_x - \phi_z = 0 \end{array}\right\} \text{ on } z = 0, \quad t > 0,$$
$$\phi(x, z, 0) = \eta(x, 0) = 0, \quad \text{for all } x \text{ and } z.$$

47. Apply the Fourier transform to solve the equation
$$u_{xxxx} + u_{yy} = 0, \quad -\infty < x < \infty, \quad y \geq 0;$$
satisfying the conditions
$$u(x, 0) = f(x), \quad u_y(x, 0) = 0 \quad \text{for} \quad -\infty < x < \infty,$$
$u(x, y)$ and its partial derivatives vanish as $|x| \to \infty$.

48. The transverse vibration of a thin membrane of great extent satisfies the wave equation
$$c^2(u_{xx} + u_{yy}) = u_{tt}, \quad -\infty < x, y < \infty, \quad t > 0,$$
with the initial and boundary conditions
$$u(x, y, t) \to 0 \quad \text{as} \quad |x| \to \infty, \quad |y| \to \infty \quad \text{for all } t \geq 0,$$
$$u(x, y, 0) = f(x, y), \quad u_t(x, y, 0) = 0 \quad \text{for all } x, y.$$
Apply the double Fourier transform method to solve this problem.

49. Solve the diffusion problem with a source $q(x, t)$
$$u_t = \kappa u_{xx} + q(x, t), \quad -\infty < x < \infty, \quad t > 0,$$
$$u(x, 0) = 0 \quad \text{for} \quad -\infty < x < \infty.$$
Show that the solution is
$$u(x, t) = \frac{1}{\sqrt{4\pi \kappa t}} \int_0^t (t-\tau)^{-\frac{1}{2}} d\tau \int_{-\infty}^{\infty} q(k, \tau) \exp\left[-\frac{(x-k)^2}{4\kappa(t-\tau)}\right] dk.$$

50. The function $u(x, t)$ satisfies the diffusion problem in a half-line
$$u_t = \kappa u_{xx} + q(x, t), \quad 0 \leq x < \infty, \quad t > 0,$$
$$u(x, 0) = 0, \quad u(0, t) = 0 \quad \text{for} \quad x \geq 0 \text{ and } t > 0.$$

Show that
$$u(x,t) = \sqrt{\frac{2}{\pi}} \int_0^t d\tau \int_0^\infty Q_s(k,\tau) \exp[-\kappa k^2(t-\tau)] \sin kx \, dk,$$
where $Q_s(k,t)$ is the Fourier sine transform of $q(x,t)$.

51. Apply the triple Fourier transform to solve the initial value problem
$$u_t = \kappa(u_{xx} + u_{yy} + u_{zz}), \quad -\infty < x, y, z < \infty, \quad t > 0,$$
$$u(\mathbf{x}, 0) = f(\mathbf{x}) \quad \text{for all } x, y, z,$$
where $\mathbf{x} = (x, y, z)$.

52. Use the double Fourier transform to solve the telegraph equation
$$u_{tt} + a u_t + bu = c^2 u_{xx}, \quad -\infty < x, t < \infty,$$
$$u(0,t) = f(t), \quad u_x(0,t) = g(t), \quad \text{for } -\infty < t < \infty,$$
where a, b, c are constants and $f(x)$ and $g(t)$ are arbitrary functions of t.

53. Determine the steady-state temperature distribution in a disk occupying the semi-infinite strip $0 < x < \infty$, $0 < y < 1$ if the edges $x = 0$ and $y = 0$ are insulated, and the edge $y = 1$ is kept at a constant temperature $T_0 H(a - x)$. Assuming that the disk loses heat due to its surroundings according to Newton's law with proportionality constant h, solve the boundary value problem
$$u_{xx} + u_{yy} - hu = 0, \quad 0 < x < \infty, \quad 0 < y < 1,$$
$$u(x, 1) = T_0 H(a - x), \quad \text{for } 0 < x < \infty,$$
$$u_x(0, y) = 0 = u_y(x, 0) \quad \text{for } 0 < x < \infty, \quad 0 < y < 1.$$

54. Use the double Fourier transform to solve the following equations:
   (a) $u_{xxxx} - u_{yy} + 2u = f(x, y),$
   (b) $u_{xx} + 2u_{yy} + 3u_x - 4u = f(x, y),$
where $f(x, y)$ is a given function.

55. Use the Fourier transform to solve the Rossby wave problem in an inviscid β-plane ocean bounded by walls at $y = 0$ and $y = 1$ where y and x represent vertical and horizontal directions. The fluid is initially at rest and then, at $t = 0+$, an arbitrary disturbance localized to the vicinity of $x = 0$ is applied to generate Rossby waves. This problem satisfies the Rossby wave equation
$$\frac{\partial}{\partial t}\left[\left(\nabla^2 - \kappa^2\right)\psi\right] + \beta \psi_x = 0, \quad -\infty < x < \infty, \quad 0 \leq y \leq 1, \quad t > 0,$$
with the boundary and initial conditions
$$\psi_x(x, y) = 0 \quad \text{for } 0 < x < \infty, \quad y = 0 \text{ and } y = 1,$$
$$\psi(x, y, t) = \psi_0(x, y) \text{ at } t = 0 \text{ for all } x \text{ and } y.$$

Chapter 3

Laplace Transforms

3.1 Introduction

In this chapter, we present the formal definition of the Laplace transform and calculate the Laplace transforms of some elementary functions directly from the definition. The existence conditions for the Laplace transform are stated in Section 3.3. The basic operational properties of the Laplace transforms including convolution and its properties, and the differentiation and integration of Laplace transforms are discussed in some detail. The inverse Laplace transform is introduced in Section 3.7, and four methods of evaluation of the inverse transform are developed with examples. The Heaviside Expansion Theorem and the Tauberian theorems for the Laplace transform are discussed. Finally, the Laplace transform of fractional integrals and fractional derivatives are included with examples.

3.2 Definition of the Laplace Transform and Examples

We start with the Fourier Integral Formula (2.2.4), which expresses the representation of a function $f_1(x)$ defined on $-\infty < x < \infty$ in the form

$$f_1(x) = \frac{1}{2\pi} \int_{-\infty}^{\infty} e^{ikx} dk \int_{-\infty}^{\infty} e^{-ikt} f_1(t) dt. \qquad (3.2.1)$$

We next set $f_1(x) \equiv 0$ in $-\infty < x < 0$ and write

$$f_1(x) = e^{-cx} f(x) H(x) = e^{-cx} f(x), \quad x > 0, \qquad (3.2.2)$$

where c is a positive fixed number, so that (3.2.1) becomes

$$f(x) = \frac{e^{cx}}{2\pi} \int_{-\infty}^{\infty} e^{ikx} dk \int_0^{\infty} \exp\{-t(c+ik)\} f(t) dt. \qquad (3.2.3)$$

With a change of variable, $c + ik = s$, $i\, dk = ds$ we rewrite (3.2.3) as

$$f(x) = \frac{e^{cx}}{2\pi i} \int_{c-i\infty}^{c+i\infty} \exp\{(s-c)x\} ds \int_0^{\infty} e^{-st} f(t) dt. \qquad (3.2.4)$$

Thus the *Laplace transform* of $f(t)$ is formally defined by

$$\mathcal{L}\{f(t)\} = \bar{f}(s) = \int_0^{\infty} e^{-st} f(t) dt, \quad \operatorname{Re} s > 0, \qquad (3.2.5)$$

where e^{-st} is the *kernel* of the transform and s is the *transform variable* which is a complex number. Under broad conditions on $f(t)$, its transform $\bar{f}(s)$ is analytic in s in the half-plane Re $s > a$.

Result (3.2.4) then gives the formal definition of the *inverse Laplace transform*

$$\mathscr{L}^{-1}\{\bar{f}(s)\} = f(t) = \frac{1}{2\pi i}\int_{c-i\infty}^{c+i\infty} e^{st}\bar{f}(s)\,ds, \qquad c > 0. \tag{3.2.6}$$

Obviously, $\mathscr{L}$ and $\mathscr{L}^{-1}$ are linear integral operators.

Using the definition (3.2.5), we can calculate the Laplace transforms of some simple and elementary functions.

Example 3.2.1 If $f(t) = 1$ for $t > 0$, then

$$\bar{f}(s) = \mathscr{L}\{1\} = \int_0^\infty e^{-st}\,dt = \frac{1}{s}. \tag{3.2.7}$$

Example 3.2.2 If $f(t) = e^{at}$, where a is a constant, then

$$\mathscr{L}\{e^{at}\} = \bar{f}(s) = \int_0^\infty e^{-(s-a)t}\,dt = \frac{1}{s-a}, \qquad s > a. \tag{3.2.8}$$

Example 3.2.3 If $f(t) = \sin at$, where a is a real constant, then

$$\mathscr{L}\{\sin at\} = \int_0^\infty e^{-st}\sin at\,dt = \frac{1}{2i}\int_0^\infty\left[e^{-t(s-ia)} - e^{-t(s+ia)}\right]dt$$

$$= \frac{1}{2i}\left[\frac{1}{s-ia} - \frac{1}{s+ia}\right] = \frac{a}{s^2+a^2}. \tag{3.2.9}$$

Similarly,

$$\mathscr{L}\{\cos at\} = \frac{s}{s^2+a^2}. \tag{3.2.10}$$

Example 3.2.4 If $f(t) = \sinh at$ or $\cosh at$, where a is a real constant, then

$$\mathscr{L}\{\sinh at\} = \int_0^\infty e^{-st}\sinh at\,dt = \frac{a}{s^2-a^2}, \tag{3.2.11}$$

$$\mathscr{L}\{\cosh at\} = \int_0^\infty e^{-st}\cosh at\,dt = \frac{s}{s^2-a^2}. \tag{3.2.12}$$

Example 3.2.5 If $f(t) = t^n$, where n is a positive integer, then

$$\bar{f}(s) = \mathscr{L}\{t^n\} = \frac{n!}{s^{n+1}}. \tag{3.2.13}$$

We recall (3.2.7) and formally differentiate it with respect to s. This gives

$$\int_0^\infty t\,e^{-st}\,dt = \frac{1}{s^2}, \tag{3.2.14}$$

which means that

$$\mathcal{L}\{t\} = \frac{1}{s^2}. \tag{3.2.15}$$

Differentiating (3.2.14) with respect to s gives

$$\mathcal{L}\{t^2\} = \int_0^\infty t^2 e^{-st}\, dt = \frac{2}{s^3}. \tag{3.2.16}$$

Similarly, differentiation of (3.2.7) n times yields

$$\mathcal{L}\{t^n\} = \int_0^\infty t^n e^{-st}\, dt = \frac{n!}{s^{n+1}}. \tag{3.2.17}$$

Example 3.2.6 If $a(>-1)$ is a real number, then

$$\mathcal{L}\{t^a\} = \frac{\Gamma(a+1)}{s^{a+1}}, \quad (s>0). \tag{3.2.18}$$

We have

$$\mathcal{L}\{t^a\} = \int_0^\infty t^a e^{-st}\, dt,$$

which is, by putting $st = x$,

$$= \frac{1}{s^{a+1}} \int_0^\infty x^a e^{-x}\, dx = \frac{\Gamma(a+1)}{s^{a+1}},$$

where $\Gamma(a)$ represents the *gamma function* defined by the integral

$$\Gamma(a) = \int_0^\infty x^{a-1} e^{-x}\, dx, \quad a>0. \tag{3.2.19}$$

It can be shown that the Gamma function satisfies the relation

$$\Gamma(a+1) = a\Gamma(a). \tag{3.2.20}$$

Obviously, result (3.2.18) is an extension of (3.2.17). The latter is a special case of the former when a is a positive integer.

In particular, when $a = -\frac{1}{2}$, result (3.2.18) gives

$$\mathcal{L}\left\{\frac{1}{\sqrt{t}}\right\} = \frac{\Gamma\left(\frac{1}{2}\right)}{\sqrt{s}} = \sqrt{\frac{\pi}{s}}, \quad \Gamma\left(\frac{1}{2}\right) = \sqrt{\pi}. \tag{3.2.21}$$

Similarly,

$$\mathcal{L}\{\sqrt{t}\} = \frac{\Gamma\left(\frac{3}{2}\right)}{s^{3/2}} = \frac{\sqrt{\pi}}{2} \frac{1}{s^{3/2}}, \tag{3.2.22}$$

where

$$\Gamma\left(\frac{3}{2}\right) = \Gamma\left(\frac{1}{2}+1\right) = \frac{1}{2}\Gamma\left(\frac{1}{2}\right) = \frac{\sqrt{\pi}}{2}.$$

Example 3.2.7 If $f(t) = \mathrm{erf}\left(\frac{a}{2\sqrt{t}}\right)$, then

$$\mathcal{L}\left\{erf\left(\frac{a}{2\sqrt{t}}\right)\right\} = \frac{1}{s}\left(1 - e^{-a\sqrt{s}}\right), \qquad (3.2.23)$$

where $erf(t)$ is the error function defined by (2.10.13).

To prove (3.2.23), we begin with the definition (3.2.5) so that

$$\mathcal{L}\left\{erf\left(\frac{a}{2\sqrt{t}}\right)\right\} = \int_0^\infty e^{-st}\left[\frac{2}{\sqrt{\pi}}\int_0^{a/2\sqrt{t}} e^{-x^2} dx\right] dt,$$

which is, by putting $x = \dfrac{a}{2\sqrt{t}}$ or $t = \dfrac{a^2}{4x^2}$ and interchanging the order of integration,

$$= \frac{2}{\sqrt{\pi}}\int_0^\infty e^{-x^2} dx \int_0^{a^2/4x^2} e^{-st} dt$$

$$= \frac{2}{\sqrt{\pi}}\int_0^\infty e^{-x^2} \frac{1}{s}\left\{1 - \exp\left(-\frac{a^2 s}{4x^2}\right)\right\} dx$$

$$= \frac{1}{s}\cdot\frac{2}{\sqrt{\pi}}\left[\int_0^\infty e^{-x^2} dx - \int_0^\infty \exp\left\{-\left(x^2 + \frac{sa^2}{4x^2}\right)\right\} dx\right],$$

where the integral

$$\int_0^\infty \exp\left\{-\left(x^2 + \frac{\alpha^2}{x^2}\right)\right\} dx = \frac{1}{2}\left[\int_0^\infty \left(1 - \frac{\alpha}{x^2}\right)\exp\left[-\left(x + \frac{\alpha}{x}\right)^2 + 2\alpha\right]\right.$$

$$\left. + \int_0^\infty \left(1 + \frac{\alpha}{x^2}\right)\exp\left[-\left(x - \frac{\alpha}{x}\right)^2 - 2\alpha\right]\right] dx$$

which is, by putting $y = \left(x \pm \dfrac{\alpha}{x}\right)$, $dy = \left(1 + \dfrac{\alpha}{x^2}\right) dx$, and observing that the first integral vanishes,

$$= \frac{1}{2}e^{-2\alpha}\int_{-\infty}^\infty e^{-y^2} dy = \frac{\sqrt{\pi}}{2}e^{-2\alpha}, \qquad \alpha = \frac{a\sqrt{s}}{2}.$$

Consequently,

$$\mathcal{L}\left\{erf\left(\frac{a}{2\sqrt{t}}\right)\right\} = \frac{1}{s}\cdot\frac{2}{\sqrt{\pi}}\left[\frac{\sqrt{\pi}}{2} - \frac{\sqrt{\pi}}{2}e^{-a\sqrt{s}}\right] = \frac{1}{s}\left[1 - e^{-a\sqrt{s}}\right].$$

We use (3.2.23) to find the Laplace transform of the complementary error function defined by (2.10.14) and obtain

$$\mathcal{L}\left\{erfc\left(\frac{a}{2\sqrt{t}}\right)\right\} = \frac{1}{s}e^{-a\sqrt{s}}. \qquad (3.2.24)$$

The proof of this result follows from $erfc(x) = 1 - erf(x)$ and $\mathcal{L}\{1\} = \dfrac{1}{s}$.

Example 3.2.8 If $f(t) = J_0(at)$ is a Bessel function of order zero, then
$$\mathscr{L}\{J_0(at)\} = \frac{1}{\sqrt{s^2 + a^2}}. \tag{3.2.25}$$

Using the series representation of $J_0(at)$, we obtain
$$\mathscr{L}\{J_0(at)\} = \mathscr{L}\left[1 - \frac{a^2 t^2}{2^2} + \frac{a^4 t^4}{2^2 \cdot 4^2} - \frac{a^6 t^6}{2^2 \cdot 4^2 \cdot 6^2} + \cdots\right]$$
$$= \frac{1}{s} - \frac{a^2}{2^2} \cdot \frac{2!}{s^3} + \frac{a^4}{2^2 \cdot 4^2} \cdot \frac{4!}{s^5} - \frac{a^6}{2^2 \cdot 4^2 \cdot 6^2} \cdot \frac{6!}{s^7} + \cdots$$
$$= \frac{1}{s}\left[1 - \frac{1}{2}\left(\frac{a^2}{s^2}\right) + \frac{1 \cdot 3}{2 \cdot 4}\left(\frac{a^4}{s^4}\right) - \frac{1 \cdot 3 \cdot 5}{2 \cdot 4 \cdot 6}\left(\frac{a^6}{s^6}\right) + \cdots\right]$$
$$= \frac{1}{s}\left[\left(1 + \frac{a^2}{s^2}\right)^{-\frac{1}{2}}\right] = \frac{1}{\sqrt{a^2 + s^2}}.$$

3.3 Existence Conditions for the Laplace Transform

A function $f(t)$ is said to be of order e^{at}, $(a > 0)$ as $t \to \infty$ if there exists a finite value T and a positive constant K such that
$$|f(t)| \leq K e^{at} \quad \text{for all } t > T, \tag{3.3.1}$$
and we write this symbolically as
$$f(t) = O(e^{at}) \quad \text{as } t \to \infty. \tag{3.3.2}$$
Or equivalently
$$\lim_{t \to \infty} e^{-at}|f(t)| = 0, \quad a > 0. \tag{3.3.3}$$

Such a function $f(t)$ is said to be of *exponential order* e^{at} as $t \to \infty$, and clearly, it does not grow faster than $K e^{at}$ as $t \to \infty$.

Theorem 3.3.1 If a function $f(t)$ is continuous or piecewise continuous in every finite interval $(0, T)$, and of exponential order e^{at}, then the Laplace transform of $f(t)$ exists for all s provided $\operatorname{Re} s > a$.

Proof. We have
$$|\bar{f}(s)| = \left|\int_0^\infty e^{-st} f(t) dt\right| \leq \int_0^\infty e^{-st}|f(t)| dt$$
$$\leq K \int_0^\infty e^{-t(s-a)} dt = \frac{K}{s - a}, \quad \text{for } \operatorname{Re} s > a. \tag{3.3.4}$$

Thus the proof is complete.

It is noted that the conditions as stated in Theorem 3.3.1 are sufficient rather than necessary conditions.

It also follows from (3.3.4) that $\lim_{s\to\infty}|\bar{f}(s)|=0$, that is, $\lim_{s\to\infty}\bar{f}(s)=0$. This result can be regarded as the limiting property of the Laplace transform. However, $\bar{f}(s) = s$ or s^2 is not the Laplace transform of any continuous (or piecewise continuous) function because $\bar{f}(s)$ does not tend to zero as $s \to \infty$.

Further, a function $f(t) = \exp(at^2)$, $a > 0$ cannot have a Laplace transform even though it is continuous but is *not* of the exponential order because
$$\lim_{t\to\infty} \exp(at^2 - st) = \infty.$$

3.4 Basic Properties of Laplace Transforms

Theorem 3.4.1 *(Heaviside's Shifting Theorem).* If $\mathcal{L}\{f(t)\} = \bar{f}(s)$, then
$$\mathcal{L}\{e^{-at}f(t)\} = \bar{f}(s+a), \tag{3.4.1}$$
where a is a real constant.

Proof. We have, by definition,
$$\mathcal{L}\{e^{-at}f(t)\} = \int_0^\infty e^{-(s+a)t}f(t)\,dt = \bar{f}(s+a).$$

Example 3.4.1 The following results readily follow from (3.4.1)
$$\mathcal{L}\{t^n e^{-at}\} = \frac{n!}{(s+a)^{n+1}}, \tag{3.4.2}$$
$$\mathcal{L}\{e^{-at}\sin bt\} = \frac{b}{(s+a)^2 + b^2}, \tag{3.4.3}$$
$$\mathcal{L}\{e^{-at}\cos bt\} = \frac{s+a}{(s+a)^2 + b^2}. \tag{3.4.4}$$

If $\mathcal{L}\{f(t)\} = \bar{f}(s)$, then
$$\mathcal{L}\{f(t-a)H(t-a)\} = e^{-as}\bar{f}(s), \quad a > 0, \tag{3.4.5}$$
where $H(t-a)$ is the Heaviside unit step function defined by (2.3.9).

It follows from the definition that
$$\mathcal{L}\{f(t-a)H(t-a)\} = \int_0^\infty e^{-st}f(t-a)H(t-a)\,dt$$
$$= \int_a^\infty e^{-st}f(t-a)\,dt,$$
which is, by putting $t - a = \tau$,
$$= e^{-sa}\int_0^\infty e^{-s\tau}f(\tau)\,d\tau = e^{-sa}\bar{f}(s).$$

In particular, if $f(t) = 1$, then

$$\mathcal{L}\{H(t-a)\} = \frac{1}{s}\exp(-sa). \tag{3.4.6}$$

Example 3.4.2 Show that the Laplace transform of the square wave function $f(t)$ defined by

$$f(t) = H(t) - 2H(t-a) + 2H(t-2a) - 2H(t-3a) + \cdots \tag{3.4.7}$$

is

$$\bar{f}(s) = \frac{1}{s}\tanh\left(\frac{as}{2}\right). \tag{3.4.8}$$

The graph of $f(t)$ is shown in Figure 3.1.

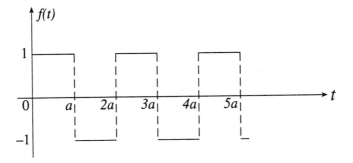

Figure 3.1 Square wave function.

$f(t) = H(t) - 2H(t-a) = 1 - 2 \cdot 0 = 1, \qquad 0 < t < a$

$f(t) = H(t) - 2H(t-a) + 2H(t-2a) = 1 - 2 \cdot 1 + 2 \cdot 0 = -1, \qquad 0 < a < t < 2a$

Thus

$$\bar{f}(s) = \frac{1}{s} - 2 \cdot \frac{e^{-as}}{s} + 2 \cdot \frac{e^{-2as}}{s} - 2 \cdot \frac{e^{-3as}}{s} + \cdots$$

$$= \frac{1}{s}\left[1 - 2r\left(1 - r + r^2 - \cdots\right)\right], \qquad \text{where } r = e^{-as}$$

$$= \frac{1}{s}\left[1 - \frac{2r}{1+r}\right] = \frac{1}{s}\left[1 - \frac{2e^{-as}}{1+e^{-as}}\right]$$

$$= \frac{1}{s}\left(\frac{1-e^{-as}}{1+e^{-as}}\right) = \frac{1}{s}\left(\frac{e^{\frac{sa}{2}} - e^{-\frac{as}{2}}}{e^{\frac{sa}{2}} + e^{-\frac{as}{2}}}\right) = \frac{1}{s}\tanh\left(\frac{as}{2}\right).$$

Example 3.4.3 (*The Laplace Transform of a Periodic Function*). If $f(t)$ is a periodic function of period a, and if $\mathcal{L}\{f(t)\}$ exists, show that

$$\mathcal{L}\{f(t)\} = [1 - \exp(-as)]^{-1} \int_0^a e^{-st} f(t) \, dt. \tag{3.4.9}$$

We have, by definition,

$$\mathcal{L}\{f(t)\} = \int_0^\infty e^{-st} f(t) \, dt = \int_0^a e^{-st} f(t) \, dt + \int_a^\infty e^{-st} f(t) \, dt.$$

Letting $t = \tau + a$ in the second integral gives

$$\bar{f}(s) = \int_0^a e^{-st} f(t) \, dt + \exp(-sa) \int_0^\infty e^{-s\tau} f(\tau + a) \, d\tau,$$

which is, due to $f(\tau + a) = f(\tau)$ and replacing the dummy variable τ by t in the second integral,

$$= \int_0^a e^{-st} f(t) \, dt + \exp(-sa) \int_0^\infty e^{-st} f(t) \, dt.$$

Finally, combining the second term with the left hand side, we obtain (3.4.9).

In particular, we calculate the Laplace transform of a rectified sine wave, that is, $f(t) = |\sin at|$. This is a periodic function with period $\dfrac{\pi}{a}$. We have

$$\int_0^{\frac{\pi}{a}} e^{-st} \sin at \, dt = \left[\frac{e^{-st}(-a\cos at - s\sin at)}{(s^2 + a^2)} \right]_0^{\frac{\pi}{a}} = \frac{a\left\{1 + \exp\left(-\dfrac{s\pi}{a}\right)\right\}}{(s^2 + a^2)}.$$

Clearly, (3.4.9) gives

$$\mathcal{L}\{f(t)\} = \frac{a}{(s^2 + a^2)} \cdot \frac{1 + \exp\left(-\dfrac{s\pi}{a}\right)}{1 - \exp\left(-\dfrac{s\pi}{a}\right)}$$

$$= \frac{a}{(s^2 + a^2)} \left[\frac{\exp\left(\dfrac{s\pi}{2a}\right) + \exp\left(-\dfrac{s\pi}{2a}\right)}{\exp\left(\dfrac{s\pi}{2a}\right) - \exp\left(-\dfrac{s\pi}{2a}\right)} \right]$$

$$= \frac{a}{(s^2 + a^2)} \coth\left(\frac{\pi s}{2a}\right).$$

Theorem 3.4.2 (*Laplace Transforms of Derivatives*). If $\mathcal{L}\{f(t)\} = \bar{f}(s)$, then

$$\mathcal{L}\{f'(t)\} = s\,\mathcal{L}\{f(t)\} - f(0) = s\bar{f}(s) - f(0), \tag{3.4.10}$$

$$\mathcal{L}\{f''(t)\} = s^2\,\mathcal{L}\{f(t)\} - s f(0) - f'(0) = s^2 \bar{f}(s) - s f(0) - f'(0), \tag{3.4.11}$$

More generally,

$$\mathcal{L}\{f^{(n)}(t)\} = s^n \bar{f}(s) - s^{n-1} f(0) - s^{n-2} f'(0) - \cdots - s f^{(n-2)}(0) - f^{(n-1)}(0), \tag{3.4.12}$$

where $f^{(r)}(0)$ is the value of $f^{(r)}(t)$ at $t = 0$.

Proof. We have, by definition,

$$\mathcal{L}\{f'(t)\} = \int_0^\infty e^{-st} f'(t) \, dt,$$

which is, by integrating by parts,

$$= \left[e^{-st} f(t) \right]_0^\infty + s \int_0^\infty e^{-st} f(t) \, dt$$

$$= s \bar{f}(s) - f(0),$$

in which we assumed $f(t) e^{-st} \to 0$ as $t \to \infty$.

Similarly,

$$\mathcal{L}\{f''(t)\} = s \mathcal{L}\{f'(t)\} - f'(0), \qquad \text{by (3.4.10)}$$

$$= s \left[s \bar{f}(s) - f(0) \right] - f'(0)$$

$$= s^2 \bar{f}(s) - s f(0) - f'(0),$$

where we have assumed $e^{-st} f'(t) \to 0$ as $t \to \infty$.

A similar procedure can be used to prove the general result (3.4.12).

It may be noted that similar results hold when the Laplace transform is applied to partial derivatives of a function of two or more independent variables. For example, if $u(x,t)$ is a function of two variables x and t, then

$$\mathcal{L}\left\{ \frac{\partial u}{\partial t} \right\} = s \bar{u}(x,s) - u(x,0), \tag{3.4.13}$$

$$\mathcal{L}\left\{ \frac{\partial^2 u}{\partial t^2} \right\} = s^2 \bar{u}(x,s) - s u(x,0) - \left[\frac{\partial u}{\partial t} \right]_{t=0}, \tag{3.4.14}$$

$$\mathcal{L}\left\{ \frac{\partial u}{\partial x} \right\} = \frac{d\bar{u}}{dx}, \quad \mathcal{L}\left\{ \frac{\partial^2 u}{\partial x^2} \right\} = \frac{d^2 \bar{u}}{dx^2}. \tag{3.4.15}$$

Results (3.4.10)-(3.4.12) imply that the Laplace transform reduces the operation of differentiation into algebraic operation. In view of this, the Laplace transform can be used effectively to solve ordinary or partial differential equations.

Example 3.4.4 Use (3.4.12) to find $\mathcal{L}\{t^n\}$.

Here $f(t) = t^n$, $f'(t) = n t^{n-1}, \ldots, f^{(n)}(t) = n!$ and $f(0) = f'(0) = \cdots = f^{(n-1)}(0) = 0$.

Thus

$$\mathcal{L}\{n!\} = s^n \mathcal{L}\{t^n\}.$$

Or

$$\mathcal{L}\{t^n\} = \frac{n!}{s^n} \mathcal{L}\{1\} = \frac{n!}{s^{n+1}}.$$

3.5 The Convolution Theorem and Properties of Convolution

Theorem 3.5.1 (Convolution Theorem). If $\mathcal{L}\{f(t)\} = \bar{f}(s)$ and $\mathcal{L}\{g(t)\} = \bar{g}(s)$, then

$$\mathcal{L}\{f(t) * g(t)\} = \mathcal{L}\{f(t)\} \mathcal{L}\{g(t)\} = \bar{f}(s)\bar{g}(s). \tag{3.5.1}$$

Or equivalently

$$\mathcal{L}^{-1}\{\bar{f}(s)\bar{g}(s)\} = f(t) * g(t), \tag{3.5.2}$$

where $f(t) * g(t)$ is called the *convolution* of $f(t)$ and $g(t)$ and is defined by the integral

$$f(t) * g(t) = \int_0^t f(t - \tau) g(\tau) d\tau. \tag{3.5.3}$$

The integral in (3.5.3) is often referred to as the *convolution integral* (or *Faltung*) and is denoted simply by $f * g$.

Proof. We have, by definition,

$$\mathcal{L}\{f(t) * g(t)\} = \int_0^\infty e^{-st} dt \int_0^t f(t - \tau) g(\tau) d\tau, \tag{3.5.4}$$

where the region of integration in the τ–t plane is as shown in Figure 3.2. The integration in (3.5.4) is first performed with respect to τ from $\tau = 0$ to $\tau = t$ of the vertical strip and then from $t = 0$ to ∞ by moving the vertical strip from $t = 0$ outwards to cover the whole region under the line $\tau = t$.

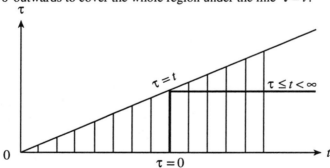

Figure 3.2 Region of integration.

We now change the order of integration so that we integrate first along the horizontal strip from $t = \tau$ to ∞ and then from $\tau = 0$ to ∞ by moving the horizontal strip vertically from $\tau = 0$ upwards. Evidently, (3.5.4) becomes

$$\mathcal{L}\{f(t) * g(t)\} = \int_0^\infty g(\tau) d\tau \int_{t=\tau}^\infty e^{-st} f(t - \tau) dt,$$

which is, by the change of variable $t - \tau = x$,

$$\mathcal{L}\{f(t) * g(t)\} = \int_0^\infty g(\tau) d\tau \int_0^\infty e^{-s(x+\tau)} f(x) dx$$

$$= \int_0^\infty e^{-s\tau} g(\tau) d\tau \int_0^\infty e^{-sx} f(x) dx$$

$$= \bar{g}(s)\bar{f}(s).$$

This completes the proof.

Second Proof. We have, by definition,

$$\bar{f}(s)\bar{g}(s) = \int_0^\infty e^{-s\sigma} f(\sigma) d\sigma \int_0^\infty e^{-s\mu} g(\mu) d\mu$$

$$= \int_0^\infty \int_0^\infty e^{-s(\sigma+\mu)} f(\sigma) g(\mu) d\sigma d\mu, \qquad (3.5.5)$$

where the double integral is taken over the entire first quadrant R of the $\sigma-\mu$ plane bounded by $\sigma = 0$ and $\mu = 0$ as shown in Figure 3.3(a).

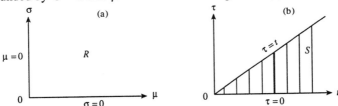

Figure 3.3 Regions of integration.

We make the change of variables $\mu = \tau$, $\sigma = t - \mu = t - \tau$ so that the axes $\sigma = 0$ and $\mu = 0$ transform into the lines $\tau = 0$ and $\tau = t$, respectively, in the $\tau - t$ plane. Consequently, (3.5.5) becomes

$$\bar{f}(s)\bar{g}(s) = \int_0^\infty e^{-st} dt \int_{\tau=0}^{\tau=t} f(t-\tau) g(\tau) d\tau$$

$$= \mathcal{L}\left\{\int_0^t f(t-\tau) g(\tau) d\tau\right\}$$

$$= \mathcal{L}\{f(t) * g(t)\}.$$

This proves the theorem.

Note: A more rigorous proof of the convolution theorem can be found in any standard treatise (see Doetsch, 1950) on Laplace transforms. The convolution operation has the following properties:

$$f(t) * \{g(t) * h(t)\} = \{f(t) * g(t)\} * h(t), \text{ (Associative)}, \qquad (3.5.6)$$
$$f(t) * g(t) = g(t) * f(t), \qquad \text{(Commutative)}, \qquad (3.5.7)$$
$$f(t) * \{g(t) + h(t)\} = f(t) * g(t) + f(t) * h(t), \text{ (Distributive)}, \qquad (3.5.8)$$

$$f(t)*\{ag(t)\} = \{af(t)\}*g(t) = a\{f(t)*g(t)\}, \text{ where } a \text{ is a constant}, \quad (3.5.9)$$

$$\mathcal{L}\{f_1*f_2*f_3*\cdots*f_n\} = \bar{f}_1(s)\bar{f}_2(s)\cdots\bar{f}_n(s). \quad (3.5.10)$$

$$\mathcal{L}\{f^{*n}\} = \{\bar{f}(s)\}^n, \quad (3.5.11)$$

where $f^{*n} = f*f*\cdots*f$ is sometimes called the *nth convolution*.

Remark: By virtue of (3.5.6) and (3.5.7), it is clear that the set of all Laplace transformable functions forms a commutative semigroup with respect to the operation *. The set of all Laplace transformable functions does not form a group because $f*g^{-1}$ does not, in general, have a Laplace transform.

We now prove the associative property. We have

$$f(t)*\{g(t)*h(t)\} = \int_0^t f(\tau) \int_0^{t-\tau} g(t-\sigma-\tau)h(\sigma)\,d\sigma\,d\tau \quad (3.5.12)$$

$$= \int_0^t h(\sigma) \int_0^{t-\sigma} g(t-\tau-\sigma)f(\tau)\,d\tau\,d\sigma$$

$$= h(t)*\{f(t)*g(t)\} = \{f(t)*g(t)\}*h(t), \quad (3.5.13)$$

where (3.5.13) is obtained from (3.5.12) by interchanging the order of integration combined with the fact that $0 \leq \sigma \leq t-\tau$ and $0 \leq \tau \leq t$ imply $0 \leq \tau \leq t-\sigma$ and $0 \leq \sigma \leq t$. Properties (3.5.10) and (3.5.11) follow immediately from the associative law of the convolution.

To prove (3.5.7), we recall the definition of the convolution and make a change of variable $t-\tau = t'$. This gives

$$f(t)*g(t) = \int_0^t f(t-\tau)g(\tau)\,d\tau = \int_0^t g(t-t')f(t')\,dt' = g(t)*f(t).$$

The proofs of (3.5.8)-(3.5.9) are very simple and hence may be omitted.

Example 3.5.1 Obtain the convolutions

(a) $t*e^{at}$, (b) $(\sin at * \sin at)$,

(c) $\dfrac{1}{\sqrt{\pi t}}*e^{at}$, (d) $1*\dfrac{a}{2}\dfrac{e^{-a^2/4t}}{\sqrt{\pi t^3}}$.

We have

(a) $t*e^{at} = \int_0^t \tau e^{a(t-\tau)}\,d\tau = e^{at}\int_0^t \tau e^{-a\tau}\,d\tau = \dfrac{1}{a^2}(e^{at} - at - 1).$

(b) $\sin at * \sin at = \int_0^t \sin a\tau \, \sin a(t-\tau)\,d\tau = \dfrac{1}{2a}(\sin at - at\cos at).$

(c) $\dfrac{1}{\sqrt{\pi t}}*e^{at} = \dfrac{1}{\sqrt{\pi}}\int_0^t \dfrac{1}{\sqrt{\tau}}e^{a(t-\tau)}\,d\tau,$

which is, by putting $\sqrt{a\tau} = x$,

Integral Transforms and Their Applications

$$\frac{1}{\sqrt{\pi t}} * e^{at} = \frac{2e^{at}}{\sqrt{\pi a}} \int_0^{\sqrt{at}} e^{-x^2} dx = \frac{e^{at}}{\sqrt{a}} \operatorname{erf}(\sqrt{at}).$$

(d) We have

$$1 * \frac{a}{2} \frac{e^{-a^2/4t}}{\sqrt{\pi t^3}} = \frac{a}{2\sqrt{\pi}} \int_0^t \frac{e^{-a^2/4\tau}}{\tau^{3/2}} d\tau,$$

which is, by letting $\dfrac{a}{2\sqrt{\tau}} = x$,

$$= \frac{2}{\sqrt{\pi}} \int_{\frac{a}{2\sqrt{t}}}^{\infty} e^{-x^2} dx = \operatorname{erfc}\left(\frac{a}{2\sqrt{t}}\right).$$

Example 3.5.2 Using the Convolution Theorem 3.5.1, prove that

$$B(m,n) = \frac{\Gamma(m)\Gamma(n)}{\Gamma(m+n)}, \qquad (3.5.14)$$

where $\Gamma(m)$ is the gamma function, and $B(m,n)$ is the beta function defined by

$$B(m,n) = \int_0^1 x^{m-1}(1-x)^{n-1} dx, \quad (m>0, n>0). \qquad (3.5.15)$$

To prove this, we consider

$$f(t) = t^{m-1} \quad (m>0) \quad \text{and} \quad g(t) = t^{n-1}, \quad (n>0).$$

Evidently, $\bar{f}(s) = \dfrac{\Gamma(m)}{s^m}$ and $\bar{g}(s) = \dfrac{\Gamma(n)}{s^n}$.

We have

$$f * g = \int_0^t \tau^{m-1}(t-\tau)^{n-1} d\tau = \mathcal{L}^{-1}\{\bar{f}(s)\bar{g}(s)\}$$

$$= \Gamma(m)\Gamma(n) \mathcal{L}^{-1}\{s^{-(m+n)}\}$$

$$= \frac{\Gamma(m)\Gamma(n)}{\Gamma(m+n)} t^{m+n-1}.$$

Letting $t = 1$, we derive the result

$$\int_0^1 \tau^{m-1}(1-\tau)^{n-1} d\tau = \frac{\Gamma(m)\Gamma(n)}{\Gamma(m+n)},$$

which proves the result (3.5.14).

3.6 Differentiation and Integration of Laplace Transforms

Theorem 3.6.1 If $f(t) = O(e^{at})$ as $t \to \infty$, then the Laplace integral

$$\int_0^\infty e^{-st} f(t)\, dt, \tag{3.6.1}$$

is uniformly convergent with respect to s provided $s \geq a_1$ where $a_1 > a$.

Proof. Since

$$\left| e^{-st} f(t) \right| \leq K e^{-t(s-a)} \leq K e^{-t(a_1-a)} \quad \text{for all } s \geq a_1$$

and $\int_0^\infty e^{-t(a_1-a)}\, dt$ exists for $a_1 > a$, by Weierstrass' test, the Laplace integral is uniformly convergent for all $s > a_1$ where $a_1 > a$. This completes the proof.

In view of the uniform convergence of (3.6.1), differentiation of (3.2.5) with respect to s within the integral sign is permissible. Hence

$$\frac{d}{ds}\bar{f}(s) = \frac{d}{ds}\int_0^\infty e^{-st} f(t)\, dt = \int_0^\infty \frac{\partial}{\partial s} e^{-st} f(t)\, dt$$

$$= -\int_0^\infty t f(t) e^{-st}\, dt = -\mathcal{L}\{t f(t)\}. \tag{3.6.2}$$

Similarly, we obtain

$$\frac{d^2}{ds^2}\bar{f}(s) = (-1)^2 \mathcal{L}\{t^2 f(t)\}, \tag{3.6.3}$$

$$\frac{d^3}{ds^3}\bar{f}(s) = (-1)^3 \mathcal{L}\{t^3 f(t)\}. \tag{3.6.4}$$

More generally,

$$\frac{d^n}{ds^n}\bar{f}(s) = (-1)^n \mathcal{L}\{t^n f(t)\}. \tag{3.6.5}$$

Result (3.6.5) can be stated in the following theorem:

Theorem 3.6.2 (Derivatives of the Laplace Transform). If $\mathcal{L}\{f(t)\} = \bar{f}(s)$, then

$$\mathcal{L}\{t^n f(t)\} = (-1)^n \frac{d^n}{ds^n}\bar{f}(s), \tag{3.6.6}$$

where $n = 0, 1, 2, 3, \ldots$.

Example 3.6.1 Show that

(a) $\mathcal{L}\{t^n e^{-at}\} = \dfrac{n!}{(s+a)^{n+1}}$,

(b) $\mathcal{L}\{t \cos at\} = \dfrac{s^2 - a^2}{(s^2 + a^2)^2}$,

(c) $\mathcal{L}\{t \sin at\} = \dfrac{2as}{(s^2 + a^2)^2}$,

(d) $\mathcal{L}\{t f'(t)\} = -\left\{ s \dfrac{d}{ds}\bar{f}(s) + \bar{f}(s) \right\}$.

(a) Application of Theorem 3.6.2 gives
$$\mathcal{L}\{t^n e^{-at}\} = (-1)^n \frac{d^n}{ds^n} \cdot \frac{1}{(s+a)} = (-1)^{2n} \frac{n!}{(s+a)^{n+1}}.$$

(b)
$$\mathcal{L}\{t \cos at\} = (-1)\frac{d}{ds}\left(\frac{s}{s^2+a^2}\right) = \frac{s^2-a^2}{(s^2+a^2)^2}.$$

Results (c) and (d) can be proved similarly.

Theorem 3.6.3 (Integral of the Laplace Transform). If $\mathcal{L}\{f(t)\} = \bar{f}(s)$, then

$$\mathcal{L}\left\{\frac{f(t)}{t}\right\} = \int_s^\infty \bar{f}(s)\,ds. \tag{3.6.7}$$

Proof. In view of the uniform convergence of (3.6.1), $\bar{f}(s)$ can be integrated with respect to s in (s,∞) so that

$$\int_s^\infty \bar{f}(s)\,ds = \int_s^\infty ds \int_0^\infty e^{-st} f(t)\,dt$$

$$= \int_0^\infty f(t)\,dt \int_s^\infty e^{-st}\,ds$$

$$= \int_0^\infty \frac{f(t)}{t} e^{-st}\,dt = \mathcal{L}\left\{\frac{f(t)}{t}\right\}.$$

This proves the theorem.

Example 3.6.2 Show that

(a) $\mathcal{L}\left\{\dfrac{\sin at}{t}\right\} = \tan^{-1}\left(\dfrac{a}{s}\right),$ (b) $\mathcal{L}\left\{\dfrac{e^{-a^2/4t}}{\sqrt{\pi t^3}}\right\} = \dfrac{2}{a}\exp(-a\sqrt{s}).$

(a) Using (3.6.7), we obtain

$$\mathcal{L}\left\{\frac{\sin at}{t}\right\} = a\int_s^\infty \frac{ds}{s^2+a^2} = \frac{\pi}{2} - \tan^{-1}\left(\frac{s}{a}\right) = \tan^{-1}\left(\frac{a}{s}\right).$$

(b) $\mathcal{L}\left\{\dfrac{1}{t}\cdot\dfrac{e^{-a^2/4t}}{\sqrt{\pi t}}\right\} = \int_s^\infty \bar{f}(s)\,ds = \int_s^\infty \dfrac{e^{-a\sqrt{s}}}{\sqrt{s}}\,ds$, by Table B-4 of Laplace transforms,

which is, by putting $a\sqrt{s} = x$,

$$= \frac{2}{a}\int_{a\sqrt{s}}^\infty e^{-x}\,dx = \frac{2}{a}\exp(-a\sqrt{s}).$$

Theorem 3.6.4 (The Laplace Transform of an Integral). If $\mathcal{L}\{f(t)\} = \bar{f}(s)$, then

$$\mathcal{L}\left\{\int_0^t f(\tau)\,d\tau\right\} = \frac{\bar{f}(s)}{s}. \tag{3.6.8}$$

Proof. We write
$$g(t) = \int_0^t f(\tau)\,d\tau$$
so that $g(0) = 0$ and $g'(t) = f(t)$. Then it follows from (3.4.10) that
$$\bar{f}(s) = \mathcal{L}\{f(t)\} = \mathcal{L}\{g'(t)\} = s\bar{g}(s) = s\mathcal{L}\left\{\int_0^t f(\tau)\,d\tau\right\}.$$
Dividing both sides by s, we obtain (3.6.8).

It is noted that the Laplace transform of an integral corresponds to the division of the transform of its integrand by s. Result (3.6.8) can be used for evaluation of the inverse Laplace transform.

Example 3.6.3 Use result (3.6.8) to find

(a) $\mathcal{L}\left\{\int_0^t \tau^n e^{-a\tau}\,d\tau\right\}$, (b) $\mathcal{L}\{Si(at)\} = \mathcal{L}\left\{\int_0^t \frac{\sin a\tau}{\tau}\,d\tau\right\}$.

(a) We know
$$\mathcal{L}\{t^n e^{-at}\} = \frac{n!}{(s+a)^{n+1}}.$$
It follows from (3.6.8) that
$$\mathcal{L}\left\{\int_0^t \tau^n e^{-a\tau}\,d\tau\right\} = \frac{n!}{s(s+a)^{n+1}}.$$

(b) Using (3.6.8) and Example 3.6.2(a), we obtain
$$\mathcal{L}\left\{\int_0^t \frac{\sin a\tau}{\tau}\,d\tau\right\} = \frac{1}{s}\tan^{-1}\left(\frac{a}{s}\right).$$

3.7 The Inverse Laplace Transform and Examples

It has already been demonstrated that the Laplace transform $\bar{f}(s)$ of a given function $f(t)$ can be calculated by direct integration. We now look at the inverse problem. Given a Laplace transform $\bar{f}(s)$ of an unknown function $f(t)$, how can we find $f(t)$? This is essentially concerned with the solution of the integral equation
$$\int_0^\infty e^{-st} f(t)\,dt = \bar{f}(s). \tag{3.7.1}$$
At this stage, is is rather difficult to handle the problem as it is. However, in simple cases, we can find the inverse transform from Table B-4 of Laplace transforms. For example

$$\mathcal{L}^{-1}\left\{\frac{1}{s}\right\} = 1, \quad \mathcal{L}^{-1}\left\{\frac{s}{s^2+a^2}\right\} = \cos at.$$

In general, the inverse Laplace transform can be determined by using four methods: (i) Partial Fraction Decomposition, (ii) the Convolution Theorem, (iii) Contour Integration of the Laplace Inversion Integral and (iv) Heaviside's Expansion Theorem.

(i) *Partial Fraction Decomposition Method*

If

$$\bar{f}(s) = \frac{\bar{p}(s)}{\bar{q}(s)}, \qquad (3.7.2)$$

where $\bar{p}(s)$ and $\bar{q}(s)$ are polynomials in s, and the degree of $\bar{p}(s)$ is less than that of $\bar{q}(s)$, the method of partial fractions may be used to express $\bar{f}(s)$ as the sum of terms which can be inverted by using a table of Laplace transforms. We illustrate the method by means of simple examples.

Example 3.7.1 To find

$$\mathcal{L}^{-1}\left\{\frac{1}{s(s-a)}\right\},$$

where a is a constant, we write

$$\mathcal{L}^{-1}\left\{\frac{1}{s(s-a)}\right\} = \mathcal{L}^{-1}\left[\frac{1}{a}\left\{\frac{1}{s-a} - \frac{1}{s}\right\}\right]$$

$$= \frac{1}{a}\left[\mathcal{L}^{-1}\left\{\frac{1}{s-a}\right\} - \mathcal{L}^{-1}\left\{\frac{1}{s}\right\}\right]$$

$$= \frac{1}{a}\left(e^{at} - 1\right).$$

Example 3.7.2 Show that

$$\mathcal{L}^{-1}\left\{\frac{1}{(s^2+a^2)(s^2+b^2)}\right\} = \frac{1}{b^2-a^2}\left(\frac{\sin at}{a} - \frac{\sin bt}{b}\right).$$

We write

$$\mathcal{L}^{-1}\left\{\frac{1}{(s^2+a^2)(s^2+b^2)}\right\} = \frac{1}{b^2-a^2}\left[\mathcal{L}^{-1}\left\{\frac{1}{s^2+a^2} - \frac{1}{s^2+b^2}\right\}\right]$$

$$= \frac{1}{(b^2-a^2)}\left(\frac{\sin at}{a} - \frac{\sin bt}{b}\right).$$

Example 3.7.3 Find

$$\mathcal{L}^{-1}\left\{\frac{s+7}{s^2+2s+5}\right\}.$$

We have

$$\mathcal{L}^{-1}\left\{\frac{s+7}{(s+1)^2+4}\right\} = \mathcal{L}^{-1}\left\{\frac{s+1+6}{(s+1)^2+2^2}\right\}$$

$$= \mathcal{L}^{-1}\left\{\frac{s+1}{(s+1)^2+2^2}\right\} + 3\mathcal{L}^{-1}\left\{\frac{2}{(s+1)^2+2^2}\right\}$$

$$= e^{-t}\cos 2t + 3e^{-t}\sin 2t.$$

Example 3.7.4 Evaluate the following inverse Laplace transform

$$\mathcal{L}^{-1}\left\{\frac{2s^2+5s+7}{(s-2)(s^2+4s+13)}\right\}.$$

We have

$$\mathcal{L}^{-1}\left\{\frac{2s^2+5s+7}{(s-2)(s^2+4s+13)}\right\} = \mathcal{L}^{-1}\left\{\frac{1}{s-2} + \frac{s+2}{(s+2)^2+3^2} + \frac{1}{(s+2)^2+3^2}\right\}$$

$$= \mathcal{L}^{-1}\left\{\frac{1}{s-2}\right\} + \mathcal{L}^{-1}\left\{\frac{s+2}{(s+2)^2+3^2}\right\} + \frac{1}{3}\mathcal{L}^{-1}\left\{\frac{3}{(s+2)^2+3^2}\right\}$$

$$= e^{2t} + e^{-2t}\cos 3t + \frac{1}{3}e^{-2t}\sin 3t.$$

(ii) Convolution Theorem

We shall apply the convolution theorem for calculation of inverse Laplace transforms.

Example 3.7.5 $\mathcal{L}^{-1}\left\{\dfrac{1}{s(s-a)}\right\} = 1 * e^{at} = \int_0^t e^{a\tau}d\tau = \dfrac{(e^{at}-1)}{a}.$

Example 3.7.6 $\mathcal{L}^{-1}\left\{\dfrac{1}{s^2(s^2+a^2)}\right\} = t * \dfrac{\sin at}{a}$

$$= \frac{1}{a}\int_0^t (t-\tau)\sin a\tau\, d\tau$$

$$= \frac{t}{a}\int_0^t \sin a\tau\, d\tau - \frac{1}{a}\int_0^t \tau\sin a\tau\, d\tau$$

$$= \frac{1}{a^2}\left(t - \frac{1}{a}\sin at\right).$$

Example 3.7.7 $\mathcal{L}^{-1}\left\{\dfrac{1}{(s^2+a^2)^2}\right\} = \dfrac{\sin at}{a} * \dfrac{\sin at}{a}$

Integral Transforms and Their Applications

$$= \frac{1}{a^2} \int_0^t \sin a\tau \, \sin a(t-\tau) d\tau$$

$$= \frac{1}{2a^3} (\sin at - at \cos at).$$

Example 3.7.8

$$\mathcal{L}^{-1}\left\{\frac{1}{\sqrt{s}(s-a)}\right\} = \frac{1}{\sqrt{\pi t}} * e^{at}, \quad (a > 0)$$

$$= \frac{1}{\sqrt{\pi}} \int_0^t \frac{1}{\sqrt{\tau}} e^{a(t-\tau)} d\tau$$

$$= \frac{2e^{at}}{\sqrt{\pi a}} \int_0^{\sqrt{at}} e^{-x^2} dx, \left(\text{putting } \sqrt{a\tau} = x\right)$$

$$= \frac{e^{at}}{\sqrt{a}} \operatorname{erf}\left(\sqrt{at}\right). \tag{3.7.3}$$

Example 3.7.9 Show that

$$\mathcal{L}^{-1}\left\{\frac{1}{s} e^{-a\sqrt{s}}\right\} = \operatorname{erfc}\left(\frac{a}{2\sqrt{t}}\right). \tag{3.7.4}$$

In view of Example 3.6.2(b), and the Convolution Theorem 3.5.1, we obtain

$$\mathcal{L}^{-1}\left\{\frac{1}{s} e^{-a\sqrt{s}}\right\} = 1 * \frac{a}{2} \frac{e^{-a^2/4t}}{\sqrt{\pi t^3}}$$

$$= \frac{a}{2\sqrt{\pi}} \int_0^t \frac{e^{-a^2/4\tau}}{\tau^{3/2}} d\tau,$$

which is, by putting $\frac{a}{2\sqrt{\tau}} = x$,

$$= \frac{2}{\sqrt{\pi}} \int_{\frac{a}{2\sqrt{t}}}^{\infty} e^{-x^2} dx = \operatorname{erfc}\left(\frac{a}{2\sqrt{t}}\right).$$

Example 3.7.10 Show that

$$\mathcal{L}^{-1}\left\{\frac{1}{\sqrt{s}+a}\right\} = \frac{1}{\sqrt{\pi t}} - a \exp(ta^2) \operatorname{erfc}(a\sqrt{t}). \tag{3.7.5}$$

We have

$$\mathcal{L}^{-1}\left\{\frac{1}{\sqrt{s}+a}\right\} = \mathcal{L}^{-1}\left\{\frac{1}{\sqrt{s}} - \frac{a}{\sqrt{s}(\sqrt{s}+a)}\right\}$$

$$= \mathcal{L}^{-1}\left\{\frac{1}{\sqrt{s}}\right\} - a\mathcal{L}^{-1}\left\{\frac{\sqrt{s}-a}{\sqrt{s}(s-a^2)}\right\}$$

$$= \mathcal{L}^{-1}\left\{\frac{1}{\sqrt{s}}\right\} - a\mathcal{L}^{-1}\left\{\frac{1}{s-a^2}\right\} + a^2\mathcal{L}^{-1}\left\{\frac{1}{\sqrt{s}(s-a^2)}\right\}$$

$$= \frac{1}{\sqrt{\pi t}} - a\exp(a^2 t) + a\exp(a^2 t)\, erf(a\sqrt{t}), \qquad \text{by (3.7.3)}$$

$$= \frac{1}{\sqrt{\pi t}} - a\exp(a^2 t)\, erfc(a\sqrt{t}).$$

Example 3.7.11 If $f(t) = \mathcal{L}^{-1}\{\bar{f}(s)\}$, then

$$\mathcal{L}^{-1}\left\{\frac{1}{s}\bar{f}(s)\right\} = \int_0^t f(x)\,dx. \qquad (3.7.6)$$

We have, by the Convolution Theorem with $g(t) = 1$ so that $\bar{g}(s) = \frac{1}{s}$,

$$\mathcal{L}^{-1}\left\{\frac{1}{s}\bar{f}(s)\right\} = \int_0^t f(t-\tau)\,d\tau,$$

which is, by putting $t - \tau = x$,

$$= \int_0^t f(x)\,dx.$$

(iii) *Contour Integration of the Laplace Inversion Integral*

In Section 3.2, the inverse Laplace transform is defined by the complex integral formula

$$\mathcal{L}^{-1}\{\bar{f}(s)\} = f(t) = \frac{1}{2\pi i}\int_{c-i\infty}^{c+i\infty} e^{st}\bar{f}(s)\,ds, \qquad (3.7.7)$$

where c is a suitable real constant and $\bar{f}(s)$ is an analytic function of the complex variable s in the right half-plane $\mathrm{Re}\, s > a$.

The details of evaluation of (3.7.7) depend on the nature of the singularities of $\bar{f}(s)$. Usually, $\bar{f}(s)$ is a single valued function with a finite or enumerably infinite number of polar singularities. Often it has branch points. The path of integration is the straight line L (see Figure 3.4(a)) in the complex s-plane with equation $s = c + iR$, $-\infty < R < \infty$, $\mathrm{Re}\, s = c$ being chosen so that all the singularities of the integrand of (3.7.7) lie to the left of the line L. This line is called the *Bromwich Contour*. In practice, the Bromwich Contour is closed by an arc of a circle of radius R as shown in Figure 3.4(a), and then the limit as

$R \to \infty$ is taken to expand the contour of integration to infinity so that all the singularities of $\bar{f}(s)$ lie inside the contour of integration.

When $\bar{f}(s)$ has a branch point at the origin, we draw the modified contour of integration by making a cut along the negative real axis and a small semicircle γ surrounding the origin as shown in Figure 3.4(b).

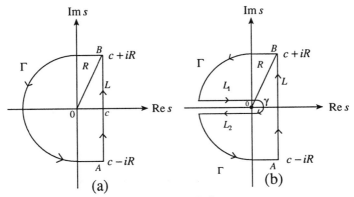

Figure 3.4
The Bromwich Contour and the contour of integration.

In either case, the Cauchy Residue Theorem is used to evaluate the integral

$$\int_L e^{st}\bar{f}(s)\,ds + \int_\Gamma e^{st}\bar{f}(s)\,ds = \int_C e^{st}\bar{f}(s)\,ds$$

$= 2\pi i \times$ [Sum of the residues of $e^{st}\bar{f}(s)$ at the poles inside C]. (3.7.8)

Letting $R \to \infty$, the integral over Γ tends to zero, and this is true in most problems of interest. Consequently, result (3.7.7) reduces to the form

$$\lim_{R\to\infty} \frac{1}{2\pi i} \int_{c-iR}^{c+iR} e^{st}\bar{f}(s)\,ds = \text{Sum of the residues of } e^{st}\bar{f}(s) \text{ at the poles of } \bar{f}(s).$$

(3.7.9)

We illustrate the above method of evaluation by simple examples.

Example 3.7.12 If $\bar{f}(s) = \dfrac{s}{s^2 + a^2}$, show that

$$f(t) = \frac{1}{2\pi i} \int_{c-i\infty}^{c+i\infty} e^{st}\bar{f}(s)\,ds = \cos at.$$

Clearly, the integrand has two simple poles at $s = \pm ia$ and the residues at these poles are

R_1 = Residue of $e^{st}\bar{f}(s)$ at $s = ia$

$$= \lim_{s \to ia}(s - ia)\frac{se^{st}}{(s^2 + a^2)} = \frac{1}{2}e^{iat}.$$

R_2 = Residue of $e^{st}\bar{f}(s)$ at $s = -ia$

$$= \lim_{s \to -ia}(s + ia)\frac{se^{st}}{(s^2 + a^2)} = \frac{1}{2}e^{-iat}.$$

Hence

$$f(t) = \frac{1}{2\pi i}\int_{c-i\infty}^{c+i\infty} e^{st}\bar{f}(s)\,ds = R_1 + R_2 = \frac{1}{2}(e^{iat} + e^{-iat}) = \cos at,$$

as obtained earlier.

If $\bar{g}(s) = e^{st}\bar{f}(s)$ has a pole of order n at $s = z$, then the residue R_1 of $\bar{g}(s)$ at this pole is given by the formula

$$R_1 = \lim_{s \to a}\frac{1}{(n-1)!}\frac{d^{n-1}}{ds^{n-1}}\left[(s-z)^n\bar{g}(s)\right]. \qquad (3.7.10)$$

This is obviously true for a simple pole $(n = 1)$ and for a double pole $(n = 2)$.

Example 3.7.13 Evaluate

$$\mathcal{L}^{-1}\left\{\frac{s}{(s^2 + a^2)^2}\right\}.$$

Clearly

$$\bar{g}(s) = e^{st}\bar{f}(s) = \frac{se^{st}}{(s^2 + a^2)^2}$$

has double poles at $s = \pm ia$. The residue formula (3.7.10) for double poles gives

$$R_1 = \lim_{s \to ia}\frac{d}{ds}\left[(s - ia)^2\frac{se^{st}}{(s^2 + a^2)^2}\right]$$

$$= \lim_{s \to ia}\frac{d}{ds}\left[\frac{se^{st}}{(s + ia)^2}\right] = \frac{te^{iat}}{2ia}.$$

Similarly, the residue at the double pole at $s = -ia$ is $(-te^{-iat})/2ia$.

Thus

$$f(t) = \text{Sum of the residues} = \frac{t}{2ia}(e^{iat} - e^{-iat}) = \frac{t}{2a}\sin at, \qquad (3.7.11)$$

as given in Table B-4 of Laplace transforms.

Example 3.7.14 Evaluate

$$\mathcal{L}^{-1}\left\{\frac{\cosh(\alpha x)}{s\cosh(\alpha \ell)}\right\}, \qquad \alpha = \sqrt{\frac{s}{a}}.$$

We have

$$f(t) = \frac{1}{2\pi i} \int_{c-i\infty}^{c+i\infty} e^{st} \frac{\cosh(\alpha x)}{\cosh(\alpha \ell)} \frac{ds}{s}.$$

Clearly, the integrand has simple poles at $s = 0$ and $s = s_n = -(2n+1)^2 \frac{a\pi^2}{4\ell^2}$, where $n = 0, 1, 2, \ldots$.

R_1 = Residue at the pole $s = 0$ is 1.
R_n = Residue at the pole $s = s_n$ is

$$\frac{\exp(-s_n t) \cosh\left\{i(2n+1)\frac{\pi x}{2\ell}\right\}}{\left[s \frac{d}{ds}\left\{\cosh \ell \sqrt{\frac{s}{a}}\right\}\right]_{s=s_n}}$$

$$= \frac{4(-1)^{n+1}}{(2n+1)\pi} \exp\left[-\left\{\frac{(2n+1)\pi}{2\ell}\right\}^2 at\right] \cosh\left\{(2n+1)\frac{\pi x}{2\ell}\right\}.$$

Thus
$f(t)$ = Sum of the residues at the poles

$$= 1 + \frac{4}{\pi} \sum_{n=0}^{\infty} \frac{(-1)^{n+1}}{(2n+1)} \exp\left[-(2n+1)^2 \frac{\pi^2 at}{4\ell^2}\right] \cosh\left\{(2n+1)\frac{\pi x}{2\ell}\right\}, \quad (3.7.12)$$

as given later by the Heaviside Expansion Theorem.

Example 3.7.15 Show that

$$f(t) = \mathcal{L}^{-1}\left\{\frac{e^{-a\sqrt{s}}}{s}\right\} = \frac{1}{2\pi i} \int_{c-i\infty}^{c+i\infty} \frac{1}{s} \exp(st - a\sqrt{s}) ds = \operatorname{erfc}\left(\frac{a}{2\sqrt{t}}\right). \quad (3.7.13)$$

The integrand has a branch point at $s = 0$. We use the contour of integration as shown in Figure 3.4(b) which excludes the branch point at $s = 0$. Thus the Cauchy Fundamental Theorem gives

$$\frac{1}{2\pi i}\left[\int_L + \int_\Gamma + \int_{L_1} + \int_{L_2} + \int_\gamma\right] \exp(st - a\sqrt{s}) \frac{ds}{s} = 0. \quad (3.7.14)$$

It is shown that the integral on Γ tends to zero as $R \to \infty$, and that on L gives the Bromwich integral. We now evaluate the remaining three integrals in (3.7.14). On L_1, we have $s = re^{i\pi} = -r$ and

$$\int_{L_1} \exp(st - a\sqrt{s}) \frac{ds}{s} = \int_{-\infty}^{0} \exp(st - a\sqrt{s}) \frac{ds}{s} = -\int_0^\infty \exp\{-(rt + ia\sqrt{r})\} \frac{dr}{r}.$$

On L_2, $s = re^{-i\pi} = -r$ and

$$\int_{L_2} \exp(st - a\sqrt{s}) \frac{ds}{s} = \int_0^{-\infty} \exp(st - a\sqrt{s}) \frac{ds}{s} = \int_0^\infty \exp\{-rt + ia\sqrt{r}\} \frac{dr}{r}.$$

Thus the integrals along L_1 and L_2 combined yield

$$-2i\int_0^\infty e^{-rt} \sin(a\sqrt{r}) \frac{dr}{r} = -4i \int_0^\infty e^{-x^2 t} \frac{\sin ax}{x} dx, \quad (r = x^2). \quad (3.7.15)$$

Integrating the following standard integral with respect to β

$$\int_0^\infty e^{-x^2\alpha^2} \cos(2\beta x) dx = \frac{\sqrt{\pi}}{2\alpha} \exp\left(-\frac{\beta^2}{\alpha^2}\right), \quad (3.7.16)$$

we obtain

$$\frac{1}{2}\int_0^\infty e^{-x^2\alpha^2} \frac{\sin 2\beta x}{x} dx = \frac{\sqrt{\pi}}{2\alpha} \int_0^\beta \exp\left(-\frac{\beta^2}{\alpha^2}\right) d\beta$$

$$= \frac{\sqrt{\pi}}{2} \int_0^{\beta/\alpha} e^{-u^2} du, \quad (\beta = \alpha u)$$

$$= \frac{\pi}{4} \text{erf}\left(\frac{\beta}{\alpha}\right). \quad (3.7.17)$$

In view of (3.7.17), result (3.7.15) becomes

$$-4i\int_0^\infty \exp(-tx^2) \frac{\sin ax}{x} dx = -2\pi i \text{ erf}\left(\frac{a}{2\sqrt{t}}\right). \quad (3.7.18)$$

Finally, on γ, we have $s = re^{i\theta}$, $ds = i re^{i\theta} d\theta$, and

$$\int_\gamma \left|\exp(st - a\sqrt{s})\right| \frac{ds}{s} = i\int_\pi^{-\pi} \exp\left(rt \cos\theta - a\sqrt{r} \cos\frac{\theta}{2}\right) d\theta = i\int_{-\pi}^\pi d\theta = 2\pi i, \quad (3.7.19)$$

in which the limit as $r \to 0$ is used and integration from π to $-\pi$ is interchanged to make γ in the counterclockwise direction.

Thus the final result follows from (3.7.14), (3.7.18), and (3.7.19) in the form

$$\mathcal{L}^{-1}\left\{\frac{e^{-a\sqrt{s}}}{s}\right\} = \frac{1}{2\pi i} \int_{c-i\infty}^{c+i\infty} \exp(st - a\sqrt{s}) \frac{ds}{s} = \left[1 - \text{erf}\left(\frac{a}{2\sqrt{t}}\right)\right] = \text{erfc}\left(\frac{a}{2\sqrt{t}}\right).$$

(iv) Heaviside's Expansion Theorem

Suppose $\bar{f}(s)$ is the Laplace transform of $f(t)$, which has a Maclaurin power series expansion in the form

$$f(t) = \sum_{r=0}^\infty a_r \frac{t^r}{r!}. \quad (3.7.20)$$

Taking the Laplace transform, it is possible to write formally

$$\bar{f}(s) = \sum_{r=0}^\infty \frac{a_r}{s^{r+1}}. \quad (3.7.21)$$

Conversely, we can derive (3.7.20) from a given expansion (3.7.21). This kind of expansion is useful for determining the behavior of the solution for small time. Further, it provides an alternating way to prove the Tauberian theorems.

Theorem 3.7.1 (*Heaviside's Expansion Theorem*). If $\bar{f}(s) = \dfrac{\bar{p}(s)}{\bar{q}(s)}$, where $\bar{p}(s)$ and $\bar{q}(s)$ are polynomials in s and the degree of $\bar{q}(s)$ is higher than that of $\bar{p}(s)$, then

$$\mathscr{L}^{-1}\left\{\frac{\bar{p}(s)}{\bar{q}(s)}\right\} = \sum_{k=1}^{n} \frac{\bar{p}(\alpha_k)}{\bar{q}'(\alpha_k)} \exp(t\alpha_k), \qquad (3.7.22)$$

where α_k are the distinct roots of the equation $\bar{q}(s) = 0$.

Proof. Without loss of generality, we can assume that the leading coefficient of $\bar{q}(s)$ is unity so that

$$\bar{q}(s) = (s - \alpha_1)(s - \alpha_2)\cdots(s - \alpha_k)\cdots(s - \alpha_n). \qquad (3.7.23)$$

Using the rules of partial fraction decomposition, we can write

$$\bar{f}(s) = \frac{\bar{p}(s)}{\bar{q}(s)} = \sum_{k=1}^{n} \frac{A_k}{(s - \alpha_k)}, \qquad (3.7.24)$$

where A_k are arbitrary constants to be determined. In view of (3.7.23), we find

$$\bar{p}(s) = \sum_{k=1}^{n} A_k (s - \alpha_1)(s - \alpha_2)\cdots(s - \alpha_{k-1})(s - \alpha_{k+1})\cdots(s - \alpha_n).$$

Substitution of $s = \alpha_k$ gives

$$\bar{p}(\alpha_k) = A_k (\alpha_k - \alpha_1)(\alpha_k - \alpha_2)\cdots(\alpha_k - \alpha_{k+1})\cdots(\alpha_k - \alpha_n), \qquad (3.7.25)$$

where $k = 1, 2, 3, \ldots, n$.

Differentiation of (3.7.23) yields

$$\bar{q}'(s) = \sum_{k=1}^{n}(s - \alpha_1)(s - \alpha_2)\cdots(s - \alpha_{k-1})(s - \alpha_{k+1})\cdots(s - \alpha_n),$$

whence it follows that

$$\bar{q}'(\alpha_k) = (\alpha_k - \alpha_1)(\alpha_k - \alpha_2)\cdots(\alpha_k - \alpha_{k-1})(\alpha_k - \alpha_{k+1})\cdots(\alpha_k - \alpha_n). \qquad (3.7.26)$$

From (3.7.25) and (3.7.26), we find

$$A_k = \frac{\bar{p}(\alpha_k)}{\bar{q}'(\alpha_k)},$$

and hence

$$\frac{\bar{p}(s)}{\bar{q}(s)} = \sum_{k=1}^{n} \frac{\bar{p}(\alpha_k)}{\bar{q}'(\alpha_k)} \frac{1}{(s - \alpha_k)}. \qquad (3.7.27)$$

Inversion gives immediately

$$\mathscr{L}^{-1}\left\{\frac{\bar{p}(s)}{\bar{q}(s)}\right\} = \sum_{k=1}^{n} \frac{\bar{p}(\alpha_k)}{\bar{q}'(\alpha_k)} \exp(t\alpha_k).$$

This proves the theorem. We give some examples of this theorem.

Example 3.7.16 We consider

$$\mathscr{L}^{-1}\left\{\frac{s}{s^2 - 3s + 2}\right\}.$$

Here $\bar{p}(s) = s$, and $\bar{q}(s) = s^2 - 3s + 2 = (s-1)(s-2)$. Hence

$$\mathcal{L}^{-1}\left\{\frac{s}{s^2 - 3s + 2}\right\} = \frac{\bar{p}(2)}{\bar{q}'(2)}e^{2t} + \frac{\bar{p}(1)}{\bar{q}'(1)}e^t = 2e^{2t} - e^t.$$

Example 3.7.17 Use Heaviside's power series expansion to evaluate

$$\mathcal{L}^{-1}\left\{\frac{1}{s}\frac{\sinh x\sqrt{s}}{\sinh \sqrt{s}}\right\}, \qquad 0 < x < 1, \qquad s > 0.$$

We have

$$\frac{1}{s}\frac{\sinh x\sqrt{s}}{\sinh \sqrt{s}} = \frac{1}{s}\frac{e^{x\sqrt{s}} - e^{-x\sqrt{s}}}{e^{\sqrt{s}} - e^{-\sqrt{s}}}$$

$$= \frac{1}{s}\frac{e^{-(1-x)\sqrt{s}} - e^{-(1+x)\sqrt{s}}}{1 - e^{-2\sqrt{s}}}$$

$$= \frac{1}{s}\left[e^{-(1-x)\sqrt{s}} - e^{-(1+x)\sqrt{s}}\right]\left(1 - e^{-2\sqrt{s}}\right)^{-1}$$

$$= \frac{1}{s}\left[e^{-(1-x)\sqrt{s}} - e^{-(1+x)\sqrt{s}}\right]\sum_{n=0}^{\infty}\exp(-2n\sqrt{s})$$

$$= \frac{1}{s}\sum_{n=0}^{\infty}\left[\exp\{-(1-x+2n)\sqrt{s}\} - \exp\{-(1+x+2n)\sqrt{s}\}\right].$$

Hence

$$\mathcal{L}^{-1}\left\{\frac{1}{s}\frac{\sinh x\sqrt{s}}{\sinh \sqrt{s}}\right\}$$

$$= \mathcal{L}^{-1}\left\{\frac{1}{s}\sum_{n=0}^{\infty}\left[\exp\{-(1-x+2n)\sqrt{s}\} - \exp\{-(1+x+2n)\sqrt{s}\}\right]\right\}$$

$$= \sum_{n=0}^{\infty}\left[\mathrm{erfc}\left(\frac{1-x+2n}{2\sqrt{t}}\right) - \mathrm{erfc}\left(\frac{1+x+2n}{2\sqrt{t}}\right)\right], \qquad \text{by (3.7.4)}.$$

Example 3.7.18 If $\alpha = \sqrt{\frac{s}{a}}$, show that

$$\mathcal{L}^{-1}\left[\frac{\cosh \alpha x}{s \cosh \alpha \ell}\right] = 1 - \frac{4}{\pi}\sum_{k=0}^{\infty}\frac{(-1)^k \cos\left\{\left(k + \frac{1}{2}\right)\frac{\pi x}{\ell}\right\}\exp\left[-(2k+1)^2\frac{a\pi^2 t}{4\ell^2}\right]}{(2k+1)}. \qquad (3.7.28)$$

Proof. In this case, we write

$$\mathcal{L}^{-1}\{\bar{f}(s)\} = \mathcal{L}^{-1}\left\{\frac{\bar{p}(s)}{\bar{q}(s)}\right\} = \mathcal{L}^{-1}\left\{\frac{\cosh \alpha x}{s \cosh \alpha \ell}\right\}.$$

Clearly, the zeros of $\bar{f}(s)$ are at $s = 0$ and at the roots of $\cosh \alpha \ell = 0$, that is, at $s = s_k = a\left(k + \frac{1}{2}\right)^2\left(\frac{\pi i}{\ell}\right)^2$, $k = 0, 1, 2, \ldots$. Thus

$$\alpha_k = \sqrt{\frac{s_k}{a}} = \left(k+\frac{1}{2}\right)\frac{\pi i}{\ell}, \quad k = 0,1,2,\ldots$$

Here $\bar{p}(s) = \cosh(\alpha x)$, $\bar{q}(s) = s\cosh(\alpha\ell)$. In order to apply the Heaviside Expansion Theorem, we need

$$\bar{q}'(s) = \frac{d}{ds}(s\cosh\alpha\ell) = \cosh(\alpha\ell) + \frac{1}{2}\alpha\ell\sinh(\alpha\ell).$$

For the zero $s = 0$, $\bar{q}'(0) = 1$, and for the zeros at $s = s_k$,

$$\bar{q}'(s_k) = \frac{1}{2}\left(k+\frac{1}{2}\right)\pi i \cdot \sinh\left[\left(k+\frac{1}{2}\right)\pi i\right]$$

$$= (2k+1)\frac{\pi i}{4} \cdot i\sin\left[\left(k+\frac{1}{2}\right)\pi\right]$$

$$= -(2k+1)\frac{\pi}{4}\cdot\cos k\pi = (-1)^{k+1}(2k+1)\frac{\pi}{4}.$$

Thus

$$\mathcal{L}^{-1}\left\{\frac{\cosh\alpha x}{s\cosh\alpha\ell}\right\} = 1 + \frac{4}{\pi}\sum_{k=0}^{\infty}\frac{(-1)^{k+1}}{(2k+1)}\cosh\left[(2k+1)\frac{\pi i x}{2\ell}\right]\exp(ts_k)$$

$$= 1 - \frac{4}{\pi}\sum_{k=0}^{\infty}\frac{(-1)^k}{(2k+1)}\cos\left[(2k+1)\frac{\pi x}{2\ell}\right]\exp\left[-\left(k+\frac{1}{2}\right)^2\frac{\pi^2 at}{\ell^2}\right].$$

3.8 Tauberian Theorems and Watson's Lemma

These theorems give the behavior of object functions in terms of the behavior of transform functions. Particularly, they determine the value of the object functions $f(t)$ for large and small values of time t. Tauberian theorems are extremely useful and have frequent applications.

Theorem 3.8.1 (*The Initial Value Theorem*). If $\mathcal{L}\{f(t)\} = \bar{f}(s)$ exists, then

$$\lim_{s\to\infty}\bar{f}(s) = 0. \tag{3.8.1}$$

In addition, if $f(t)$ and its derivatives exist as $t \to 0$, we obtain the *Initial Value Theorem*:

(i) $$\lim_{s\to\infty}\left[s\bar{f}(s)\right] = \lim_{t\to 0}f(t) = f(0), \tag{3.8.2}$$

(ii) $$\lim_{s\to\infty}\left[s^2\bar{f}(s) - sf(0)\right] = \lim_{t\to 0}f'(t) = f'(0), \tag{3.8.3}$$

and

(iii) $$\lim_{s\to\infty}\left[s^{n+1}\bar{f}(s) - s^n\bar{f}(s) - \cdots - sf^{(n-1)}(0)\right] = f^{(n)}(0). \tag{3.8.4}$$

Results (3.8.2)-(3.8.4), which are true under fairly general conditions, determine the initial values $f(0), f'(0), \ldots, f^{(n)}(0)$ of the function $f(t)$ and its derivatives from the Laplace transform $\bar{f}(s)$.

Proof. To prove (3.8.1), we use the fact that the Laplace integral (3.2.5) is uniformly convergent with respect to the parameter s. Hence it is permissible to take the limit $s \to \infty$ under the sign of integration so that

$$\lim_{s \to \infty} \bar{f}(s) = \int_0^\infty \left(\lim_{s \to \infty} e^{-st} \right) f(t) \, dt = 0.$$

Next, we use the same argument to obtain

$$\lim_{s \to \infty} \mathcal{L}\{f'(t)\} = \int_0^\infty \left(\lim_{s \to \infty} e^{-st} \right) f'(t) \, dt = 0.$$

Then it follows from result (3.4.10) that

$$\lim_{s \to \infty} \left[s \bar{f}(s) - f(0) \right] = 0,$$

and hence we obtain (3.8.2), that is,

$$\lim_{s \to \infty} \left[s \bar{f}(s) \right] = f(0) = \lim_{t \to 0} f(t).$$

A similar argument combined with Theorem 3.4.2 leads to (3.8.3) and (3.8.4).

Example 3.8.1 Verify the truth of Theorem 3.8.1 for $\bar{f}(s) = (n+1)! \, s^{-(n+1)}$ where n is a positive integer. Clearly, $f(t) = t^n$. Thus we have

$$\lim_{s \to \infty} \bar{f}(s) = \lim_{s \to \infty} \frac{(n+1)!}{s^{n+1}} = 0,$$

$$\lim_{s \to \infty} s \bar{f}(s) = 0 = f(0).$$

Example 3.8.2 Find $f(0)$ and $f'(0)$ when

(a) $\bar{f}(s) = \dfrac{1}{s(s^2 + a^2)}$, (b) $\bar{f}(s) = \dfrac{2s}{s^2 - 2s + 5}$.

(a) It follows from (3.8.2) and (3.8.3) that

$$f(0) = \lim_{s \to \infty} \left[s \bar{f}(s) \right] = \lim_{s \to \infty} \frac{1}{s^2 + a^2} = 0.$$

$$f'(0) = \lim_{s \to \infty} \left[s^2 \bar{f}(s) - s f(0) \right] = \lim_{s \to \infty} \frac{s}{s^2 + a^2} = 0.$$

(b) $$f(0) = \lim_{s \to \infty} \frac{2s^2}{s^2 + 2s + 5} = 2.$$

$$f'(0) = \lim_{s \to \infty} \left[s^2 \bar{f}(s) - s f(0) \right] = \lim_{s \to \infty} \left[\frac{2s^3}{s^2 + 2s + 5} - 2s \right] = -4.$$

Theorem 3.8.2 (The Final Value Theorem). If $\bar{f}(s) = \dfrac{\bar{p}(s)}{\bar{q}(s)}$, where $\bar{p}(s)$ and $\bar{q}(s)$ are polynomials in s, and the degree of $\bar{p}(s)$ is less than that of $\bar{q}(s)$, and if all roots of $\bar{q}(s) = 0$ have negative real parts with the possible exception of one root which may be at $s = 0$, then

(i) $$\lim_{s \to 0} \bar{f}(s) = \int_0^\infty f(t)\, dt, \qquad (3.8.5)$$

and

(ii) $$\lim_{s \to 0}\left[s\,\bar{f}(s)\right] = \lim_{t \to \infty} f(t), \qquad (3.8.6)$$

provided the limits exist.

Result (3.8.6) is true under more general conditions, and known as the *Final Value Theorem*. This theorem determines the final value of $f(t)$ at infinity from its Laplace transform at $s = 0$. However, if $\bar{f}(s)$ is more general than the rational function as stated above, a statement of a more general theorem is needed with appropriate conditions under which it is valid.

Proof. To prove (i), we use the same argument as employed in Theorem 3.8.1 and find

$$\lim_{s \to 0} \bar{f}(s) = \int_0^\infty \left(\lim_{s \to 0} \exp(-st)\right) f(t)\, dt = \int_0^\infty f(t)\, dt.$$

As before, we can use result (3.4.10) to obtain

$$\lim_{s \to 0} \mathscr{L}\{f'(t)\} = \lim_{s \to 0}\left[s\,\bar{f}(s) - f(0)\right] = \int_0^\infty \left(\lim_{s \to 0}\exp(-st)\right)f'(t)\, dt$$

$$= \int_0^\infty f'(t)\, dt = f(\infty) - f(0) = \lim_{t \to \infty}\left[f(t) - f(0)\right].$$

Thus it follows immediately that

$$\lim_{s \to 0}\left[s\,\bar{f}(s)\right] = \lim_{t \to \infty} f(t) = f(\infty).$$

Example 3.8.3 Find $f(\infty)$, if it exists, from the following functions:

(a) $\bar{f}(s) = \dfrac{1}{s(s^2 + 2s + 2)}$, (b) $\bar{f}(s) = \dfrac{1}{s - a}$,

(c) $\bar{f}(s) = \dfrac{s + a}{s^2 + b^2}$, $(b \ne 0)$, (d) $\bar{f}(s) = \dfrac{s}{s - 2}$.

(a) Clearly, $\bar{q}(s) = 0$ has roots at $s = 0$ and $s = -1 \pm i$, and the conditions of Theorem 3.8.2 are satisfied. Thus

$$\lim_{s \to 0}\left[s\,\bar{f}(s)\right] = \lim_{s \to 0} \frac{1}{s^2 + 2s + 2} = \frac{1}{2} = f(\infty).$$

(b) Here $\bar{q}(s) = 0$ has a real positive root at $s = a$ if $a > 0$, and a real negative root if $a < 0$. Thus when $a < 0$

$$\lim_{s \to 0}\left[s\,\bar{f}(s)\right] = \lim_{s \to 0} \frac{s}{s - a} = 0 = f(\infty).$$

If $a > 0$, the Final Value Theorem does not apply. In fact

$$f(t) = \mathcal{L}^{-1}\left\{\frac{1}{s-a}\right\} = e^{at} \to \infty \quad \text{as } t \to \infty.$$

(c) Here $\bar{q}(s) = 0$ has purely imaginary roots at $s = \pm ib$ which do not have negative real parts. The Final Value Theorem does not apply. In fact, $f(t) = \cos bt + \dfrac{a}{b}\sin bt$ and $\lim_{t \to \infty} f(t)$ does not exist. However, $f(t)$ is bounded and oscillatory for all $t > 0$.

(d) The Final Value Theorem does not apply as $\bar{q}(s) = 0$ has a positive root at $s = 2$.

Watson's Lemma. If (i) $f(t) = O(e^{at})$ as $t \to \infty$, that is, $|f(t)| \le K \exp(at)$ for $t > T$ where K and T are constants, and (ii) $f(t)$ has the expansion

$$f(t) = t^{\alpha}\left[\sum_{r=0}^{n} a_r t^r + R_{n+1}(t)\right] \quad \text{for } 0 < t < T \text{ and } \alpha > -1, \qquad (3.8.7)$$

where $|R_{n+1}(t)| < A t^{n+1}$ for $0 < t < T$ and A is a constant, then the Laplace transform $\bar{f}(s)$ has the *asymptotic expansion*

$$\bar{f}(s) \sim \sum_{r=0}^{n} a_r \frac{\Gamma(\alpha+r+1)}{s^{\alpha+r+1}} + O\left(\frac{1}{s^{\alpha+n+2}}\right) \quad \text{as } s \to \infty. \qquad (3.8.8)$$

Proof. We have, for $s > a$,

$$\bar{f}(s) = \int_0^T e^{-st} f(t)\, dt + \int_T^{\infty} e^{-st} f(t)\, dt$$

$$= \int_0^T e^{-st} t^{\alpha}\left[\sum_{r=0}^{n} a_r t^r\right] dt + \int_0^T e^{-st} t^{\alpha} R_{n+1}(t)\, dt + \int_T^{\infty} e^{-st} f(t)\, dt. \qquad (3.8.9)$$

The general term of the first integral in (3.8.9) can be written as

$$\int_0^T e^{-st} a_r\, t^{\alpha+r}\, dt = \int_0^{\infty} e^{-st} a_r\, t^{\alpha+r}\, dt - \int_T^{\infty} e^{-st} a_r\, t^{\alpha+r}\, dt$$

$$= a_r \frac{\Gamma(\alpha+r+1)}{s^{\alpha+r+1}} + O(e^{-Ts}). \qquad (3.8.10)$$

As $s \to \infty$, the second integral in (3.8.9) is less in magnitude than

$$A \int_0^T e^{-st} t^{\alpha+n+1}\, dt = O\left(\frac{1}{s^{\alpha+n+2}}\right), \qquad (3.8.11)$$

and the magnitude of the third integral in (3.8.9) is

$$\left|\int_T^{\infty} e^{-st} f(t)\, dt\right| \le K \int_T^{\infty} e^{-(s-a)t}\, dt = K \exp[-(s-a)T], \qquad (3.8.12)$$

which is exponentially small as $s \to \infty$.

Finally, combining (3.8.10), (3.8.11), and (3.8.12), we obtain

$$\bar{f}(s) \sim \sum_{r=0}^{n} a_r \frac{\Gamma(\alpha+r+1)}{s^{\alpha+r+1}} + O\left(\frac{1}{s^{\alpha+n+2}}\right) \quad \text{as } s \to \infty.$$

This completes the proof of Watson's lemma.

This lemma is one of the most widely used methods for finding asymptotic expansions. In order to further expand its applicability, this lemma has subsequently been generalized and its converse has also been proved. The reader is referred to Erdélyi (1956), Copson (1965), Wyman (1964), Watson (1981), Ursell (1990), and Wong (1989).

Example 3.8.4 Find the asymptotic expansion of the *parabolic cylinder function* $D_v(s)$, which is valid for $\text{Re}(v) < 0$, given by

$$D_v(s) = \frac{\exp\left(-\frac{s^2}{4}\right)}{\Gamma(-v)} \int_0^\infty \exp\left[-\left(st + \frac{t^2}{2}\right)\right] \frac{dt}{t^{v+1}}. \quad (3.8.13)$$

To find the asymptotic behavior of $D_v(s)$ as $s \to \infty$, we expand $\exp\left(-\frac{1}{2}t^2\right)$ as a power series in t in the form

$$\exp\left(-\frac{1}{2}t^2\right) = \sum_{n=0}^{\infty} (-1)^n \frac{t^{2n}}{2^n n!}. \quad (3.8.14)$$

According to Watson's lemma, as $s \to \infty$,

$$D_v(s) \sim \frac{\exp\left(-\frac{s^2}{4}\right)}{\Gamma(-v)} \sum_{n=0}^{\infty} \frac{(-1)^n}{2^n n!} \int_0^\infty t^{2n-v-1} e^{-st} dt$$

$$= \frac{\exp\left(-\frac{s^2}{4}\right)}{\Gamma(-v)} \sum_{n=0}^{\infty} \frac{(-1)^n}{2^n n!} \frac{\Gamma(2n-v)}{s^{2n-v}}. \quad (3.8.15)$$

This result is also valid for $\text{Re}(v) \geq 0$.

3.9 Laplace Transforms of Fractional Integrals and Fractional Derivatives

The *Riemann-Liouville fractional integral* is usually defined by

$$D^{-\alpha} f(t) = {}_0 D_t^{-\alpha} f(t) = \frac{1}{\Gamma(\alpha)} \int_0^t (t-x)^{\alpha-1} f(x) dx, \quad \text{Re } \alpha > 0. \quad (3.9.1)$$

Clearly, $D^{-\alpha}$ is a linear integral operator.

A simple change of variable $(t-x)^\alpha = u$ in (3.9.1) allows us to prove the following result

$$D[D^{-\alpha} f(t)] = D^{-\alpha}[D f(t)] + f(0) \frac{t^{\alpha-1}}{\Gamma(\alpha)}. \quad (3.9.2)$$

Clearly, the integral in (3.9.1) is a convolution, and hence the Laplace transform of (3.9.1) gives

$$\mathcal{L}\{D^{-\alpha}f(t)\} = \frac{1}{\Gamma(\alpha)}\mathcal{L}\{f(t)*g(t)\} = \mathcal{L}\{f(t)\}\mathcal{L}\{g(t)\}, \quad (3.9.3)$$

$$= s^{-\alpha}\bar{f}(s), \quad \alpha > 0. \quad (3.9.4)$$

where $g(t) = \dfrac{t^{\alpha-1}}{\Gamma(\alpha)}$ and $\bar{g}(s) = s^{-\alpha}$.

The result (3.9.4) is also valid for $\alpha = 0$, and

$$\lim_{\alpha \to 0} \mathcal{L}\left\{\frac{t^{\alpha-1}}{\Gamma(\alpha)}\right\} = \lim_{\alpha \to 0} s^{-\alpha} = 1. \quad (3.9.5)$$

Using (3.9.4), it can readily be verified that the fractional integral operator satisfies the laws of exponents

$$D^{-\alpha}\left[D^{-\beta}f(t)\right] = D^{-(\beta+\alpha)}f(t) = D^{-\beta}\left[D^{-\alpha}f(t)\right]. \quad (3.9.6)$$

Formula (3.9.4) can be used for evaluating the fractional integral of a given function using the inverse Laplace transform. The following examples illustrate this point.

$$\mathcal{L}\{D^{-\alpha}t^{\beta}\} = \frac{\Gamma(\beta+1)}{s^{\alpha+\beta+1}}, \quad \beta > -1. \quad (3.9.7)$$

Or equivalently,

$$D^{-\alpha}t^{\beta} = \mathcal{L}^{-1}\left\{\frac{\Gamma(\beta+1)}{s^{\alpha+\beta+1}}\right\} = \frac{\Gamma(\beta+1)}{\Gamma(\alpha+\beta+1)}t^{\alpha+\beta}. \quad (3.9.8)$$

In particular, if $\alpha = \dfrac{1}{2}$ and $\beta(=n)$ is an integer, then (3.9.8) gives

$$D^{-1/2}t^n = \frac{\Gamma(n+1)}{\Gamma\left(n+\dfrac{1}{2}+1\right)} \cdot t^{n+\frac{1}{2}}, \quad n > -1. \quad (3.9.9)$$

It also follows from (3.9.4) that

$$\mathcal{L}\{D^{-\alpha}e^{at}\} = \frac{1}{s^{\alpha}(s-a)}, \quad a > 0. \quad (3.9.10)$$

Or

$$D^{-\alpha}e^{at} = \mathcal{L}^{-1}\left\{\frac{1}{s^{\alpha}(s-a)}\right\} \quad (3.9.11)$$

$$= \mathcal{L}^{-1}\left\{\frac{1}{s^{\alpha+1}}\left(1+\frac{a}{s-a}\right)\right\}$$

$$= \frac{t^{\alpha}}{\Gamma(\alpha+1)} + aE(t,\alpha+1,a)$$

$$= E(t,\alpha,a), \quad (3.9.12)$$

where $E(t,\alpha,a)$ is defined by

$$E(t, \alpha, a) = \frac{1}{\Gamma(\alpha)} \int_0^t \xi^{\alpha-1} \exp\{a(t-\xi)\} d\xi. \tag{3.9.13}$$

In particular, if $\alpha = \frac{1}{2}$ then

$$D^{-1/2} e^{at} = \mathcal{L}^{-1}\left\{\frac{1}{\sqrt{s}(s-a)}\right\} = \frac{1}{\sqrt{\pi t}} * e^{at},$$

which is, by Example 3.7.8,

$$= \frac{e^{at}}{\sqrt{a}} erf(\sqrt{at}). \tag{3.9.14}$$

The following results follow readily from (3.9.4):

$$\mathcal{L}\{D^{-\alpha} \sin at\} = \frac{a}{s^\alpha (s^2 + a^2)}, \quad \alpha > 0. \tag{3.9.15}$$

$$\mathcal{L}\{D^{-\alpha} \cos at\} = \frac{s}{s^\alpha (s^2 + a^2)}, \quad \alpha > 0. \tag{3.9.16}$$

$$\mathcal{L}\{D^{-\alpha} e^{at} t^{\beta-1}\} = \frac{\Gamma(\beta)}{s^\alpha (s-a)^\beta}, \quad \alpha > 0, \; \beta > 0. \tag{3.9.17}$$

The inverse Lapalce transforms of these results combined with the Convolution Theorem lead to fractional integrals of functions involved.

One of the consequences of (3.9.1) is that

$$\lim_{\alpha \to 0} D^{-\alpha} f(t) = f(t). \tag{3.9.18}$$

This follows from the inverse Laplace transform of (3.9.4) combined with the limit as $\alpha \to 0$.

We now evaluate the Laplace transform of the fractional integral of the derivative and then the Laplace transform of the derivative of the integral. In view of (3.9.4), it follows that

$$\mathcal{L}\{D^{-\alpha}[Df(t)]\} = s^{-\alpha} \mathcal{L}\{Df(t)\}$$

$$= s^{-\alpha}[s \bar{f}(s) - f(0)], \quad \alpha > 0. \tag{3.9.19}$$

Although this result is proved for $\alpha > 0$, it is valid even if $\alpha = 0$.

On the other hand, the Laplace transform of (3.9.2) gives

$$\mathcal{L}\{D[D^{-\alpha} f(t)]\} = \mathcal{L}\{D^{-\alpha} D[f(t)]\} + f(0) \mathcal{L}\left\{\frac{t^{\alpha-1}}{\Gamma(\alpha)}\right\}$$

$$= s^{-\alpha}[s \bar{f}(s) - f(0)] + s^{-\alpha} f(0)$$

$$= s^{1-\alpha} \bar{f}(s), \quad \alpha \geq 0. \tag{3.9.20}$$

Obviously, if $\alpha = 0$, this result does not agree with that obtained from (3.9.19) as $\alpha \to 0$. This disagreement is due to the fact that "$\mathcal{L}$" and "lim" do not commute, as is seen from (3.9.5).

Another consequence of (3.9.1) is that the *fractional derivative* $D^\alpha f(t)$ can be defined as the solution $\phi(t)$ of the integral equation

$$D^{-\alpha}\phi(t) = f(t). \tag{3.9.21}$$

The Laplace transform of this result gives the solution for $\overline{\phi}(s)$ as

$$\overline{\phi}(s) = s^\alpha \overline{f}(s). \tag{3.9.22}$$

Inversion gives the fractional derivative of $f(t)$ as

$$\phi(t) = D^\alpha f(t) = \mathcal{L}^{-1}\{s^\alpha \overline{f}(s)\} \tag{3.9.23}$$

leading to the result

$$D^\alpha f(t) = \frac{1}{\Gamma(-\alpha)} \int_0^t (t-x)^{-\alpha-1} f(x) dx, \quad \alpha > 0. \tag{3.9.24}$$

This is the *Cauchy integral formula*, which is often used to define the fractional derivative. However, formula (3.9.23) can be used for finding the fractional derivatives. If $f(t) = t^\beta$, it is seen from (3.9.23) that

$$D^\alpha t^\beta = \mathcal{L}^{-1}\left\{\frac{\Gamma(\beta+1)}{s^{\beta-\alpha+1}}\right\} = \frac{\Gamma(\beta+1)}{\Gamma(\beta-\alpha+1)} t^{\beta-\alpha}. \tag{3.9.25}$$

In particular, if $\alpha = \frac{1}{2}$ and $\beta(=n)$ is an integer,

$$D^{1/2} t^n = \frac{\Gamma(n+1)}{\Gamma\left(n+\frac{1}{2}\right)} t^{n-\frac{1}{2}}, \quad n > -1. \tag{3.9.26}$$

$$D^{1/2} e^{at} = \mathcal{L}^{-1}\left\{\frac{\sqrt{s}}{s-a}\right\} = \mathcal{L}^{-1}\left\{\frac{1}{\sqrt{s}} + \frac{a}{\sqrt{s}(s-a)}\right\}$$

which is, by (3.2.21) and (3.7.3),

$$= \frac{1}{\sqrt{\pi t}} + \sqrt{a}\exp(at)\,\mathrm{erf}\left(\sqrt{at}\right). \tag{3.9.27}$$

In view of (3.9.23), it can easily be verified that the operator D^α is linear and satisfies the laws of exponents

$$D^\alpha\left[D^\beta f(t)\right] = D^{\alpha+\beta} f(t) = D^\beta\left[D^\alpha f(t)\right]. \tag{3.9.28}$$

Example 3.9.1 Show that

$$D^{-\alpha} J_0\left(a\sqrt{t}\right) = \left(\frac{2}{a}\right)^\alpha t^{\alpha/2} J_\alpha\left(a\sqrt{t}\right). \tag{3.9.29}$$

We apply the Laplace transform to the left hand side of (3.9.29) and use (3.9.22) to obtain

$$\mathcal{L}\left\{D^{-\alpha} J_0\left(a\sqrt{t}\right)\right\} = s^{-\alpha}\,\mathcal{L}\left\{J_0\left(a\sqrt{t}\right)\right\}$$

$$= s^{-(1+\alpha)} \exp\left(-\frac{a^2}{4s}\right).$$

The inverse Laplace transform gives

$$D^{-\alpha} J_0(a\sqrt{t}) = \mathcal{L}^{-1}\left\{s^{-(1+\alpha)} \exp\left(-\frac{a^2}{4s}\right)\right\}$$

$$= \left(\frac{2}{a}\right)^{\alpha} t^{\alpha/2} J_\alpha(a\sqrt{t}).$$

Example 3.9.2. Solve the Abel integral equation

$$g(t) = \int_0^t (t-x)^{-\alpha} f(x)\,dx, \quad 0 < \alpha < 1. \tag{3.9.30}$$

Clearly, it follows from (3.9.1) that

$$g(t) = \Gamma(1-\alpha) D^{\alpha-1} f(t).$$

Or

$$D^{1-\alpha} g(t) = \Gamma(1-\alpha) f(t).$$

Hence

$$f(t) = \frac{1}{\Gamma(1-\alpha)} D \cdot D^{-\alpha} g(t)$$

$$= \frac{1}{\Gamma(1-\alpha)} \cdot \frac{1}{\Gamma(\alpha)} \cdot D \int_0^t (t-x)^{\alpha-1} g(x)\,dx$$

$$= \frac{1}{\Gamma(\alpha)\Gamma(1-\alpha)} \cdot \frac{d}{dt} \int_0^t (t-x)^{\alpha-1} g(x)\,dx. \tag{3.9.31}$$

3.10 Exercises

1. Find the Laplace transforms of the following functions:
   (a) $2t + a\sin at$, (b) $(1-2t)\exp(-2t)$, (c) $t\cos at$,
   (d) $t^{3/2}$, (e) $H(t-3)\exp(t-3)$, (f) $H(t-a)\sinh(t-a)$,
   (g) $(t-3)^2 H(t-3)$, (h) $tH(t-a)$, (i) $(1+2at)t^{-\frac{1}{2}} \exp(at)$.

2. If n is a positive integer, show that $\mathcal{L}\{t^{-n}\}$ does not exist.

3. Use result (3.4.10) to find (a) $\mathcal{L}\{\cos at\}$ and (b) $\mathcal{L}\{\sin at\}$.

4. Use the Maclaurin series for $\sin at$ and $\cos at$ to find the Laplace transforms of these functions.

5. Show that $\mathcal{L}\left[\frac{1}{t}\{\exp(-at) - \exp(-bt)\}\right] = \log\left(\frac{s+b}{s+a}\right).$

6. Show that $\mathcal{L}\left\{\int_0^t \frac{f(u)}{u}\,du\right\} = \frac{1}{s}\int_s^\infty \bar{f}(x)\,dx.$

7. Obtain the inverse Laplace transforms of the following functions:

   (a) $\dfrac{s}{(s^2+a^2)(s^2+b^2)}$, (b) $\dfrac{1}{s^2(s^2+c^2)}$, (c) $\dfrac{1}{s^2}\exp(-as)$,

   (d) $\dfrac{1}{(s-1)^2(s-2)}$, (e) $\dfrac{1}{s^2+2s+5}$, (f) $\dfrac{1}{s^2(s+1)(s+2)}$.

8. Use the Convolution Theorem to find the inverse Laplace transforms of the following functions:

   (a) $\dfrac{s^2}{(s^2+a^2)^2}$, (b) $\dfrac{1}{s\sqrt{s+4}}$, (c) $\dfrac{\bar{f}(s)}{s}$,

   (d) $\dfrac{s}{(s^2+a^2)^2}$, (e) $\left(\dfrac{\omega}{s^2+\omega^2}\right)\bar{f}(s)$,

   (f) $\dfrac{s}{(s-a)(s^2+b^2)}$, (g) $\dfrac{1}{(s+1)^2}$, (h) $\dfrac{1}{s}\exp(-a\sqrt{s})$.

9. Show that

   (a) $\mathscr{L}\{\exp(-t^2)\} = \dfrac{\sqrt{\pi}}{2}\exp\left(\dfrac{s^2}{4}\right)\left(1-\operatorname{erf}\dfrac{s}{2}\right)$,

   (b) $\mathscr{L}^{-1}\left\{\dfrac{1}{\sqrt{s}-\sqrt{a}}\right\} = \sqrt{a}\,\exp(at) + \dfrac{1}{\sqrt{\pi t}} + \sqrt{a}\,\exp(at)\operatorname{erf}(\sqrt{at})$,

   (c) $\mathscr{L}^{-1}\left\{\dfrac{\sinh\left(\dfrac{sx}{a}\right)}{s^2\cosh\left(\dfrac{sb}{2a}\right)}\right\} = \dfrac{x}{a} + \sum_{n=0}^{\infty}(-1)^{n+1}\left(\dfrac{4b}{a\pi^2}\right)(2n+1)^{-2}$

   $\times\left[\sin\left\{(2n+1)\dfrac{\pi x}{b}\right\}\cos\left\{(2n+1)\dfrac{\pi at}{b}\right\}\right].$

10. Show that

    (a) $\mathscr{L}\left\{\dfrac{1}{t}(\sin at - at\cos at)\right\} = \tan^{-1}\left(\dfrac{a}{s}\right) - \dfrac{as}{s^2+a^2}$,

    (b) $\mathscr{L}\left\{\int_0^t \dfrac{1}{\tau}(\sin a\tau - a\tau\cos a\tau)d\tau\right\} = \dfrac{1}{s}\left[\tan^{-1}\left(\dfrac{a}{s}\right) - \dfrac{as}{s^2+a^2}\right].$

11. Using the Heaviside power series expansion, evaluate the inverse Laplace transforms of the following functions:

    (a) $\dfrac{1}{\sqrt{s^2+a^2}}$, (b) $\tan^{-1}\left(\dfrac{a}{s}\right)$, (c) $\sinh^{-1}\left(\dfrac{1}{s}\right)$,

    (d) $\dfrac{1}{s}\operatorname{cosech}(x\sqrt{s})$, (e) $\dfrac{1}{s}\exp\left(-\dfrac{1}{s}\right)$.

12. If $\mathscr{L}\{f(t)\} = \bar{f}(s)$, show that

(i) $\mathscr{L}^{-1}\left\{\dfrac{\bar{f}(s)}{s}\right\} = \displaystyle\int_0^t f(\tau)\,d\tau,$

(ii) $\mathscr{L}^{-1}\left\{\dfrac{\bar{f}(s)}{s^2}\right\} = \displaystyle\int_0^t\left[\int_0^{t_1} f(\tau)\,d\tau\right]dt_1 = \int_0^t (t-\tau)f(\tau)\,d\tau,$

(iii) $\mathscr{L}^{-1}\left\{\dfrac{\bar{f}(s)}{s^3}\right\} = \displaystyle\int_0^t\int_0^{t_1}\int_0^{t_2} f(\tau)\,d\tau\,dt_1\,dt_2 = \int_0^t \dfrac{1}{2}(t-\tau)^2 f(\tau)\,d\tau,$

and in general

(iv) $\mathscr{L}^{-1}\left\{\dfrac{\bar{f}(s)}{s^n}\right\} = \displaystyle\int_0^t\int_0^{t_1}\int_0^{t_2}\cdots\int_0^{t_{n-1}} f(\tau)\,d\tau\,dt_1\cdots dt_{n-1} = \int_0^t \dfrac{(t-\tau)^{n-1}}{(n-1)!} f(\tau)\,d\tau.$

13. The staircase function $f(t) = [t]$ represents the greatest integer less than or equal to t. Find its Laplace transform.

14. Use the convolution theorem to prove
$$\int_0^t J_0(\tau) J_0(t-\tau)\,d\tau = \sin t.$$

15. Show that

(a) $\mathscr{L}\{t\,H(t-a)\} = \left(\dfrac{1}{s^2} + \dfrac{a}{s}\right)\exp(-sa),$

(b) $\mathscr{L}\{t^n \exp(at)\} = n!(s-a)^{-(n+1)}.$

16. If $\mathscr{L}\{f(t)\} = \bar{f}(s)$ and $f(t)$ has a finite discontinuity at $t = a$, show that
$$\mathscr{L}\{f'(t)\} = s\bar{f}(s) - f(0) - \exp(-sa)[f]_a,$$
where $[f]_a = f(a+0) - f(a-0).$

17. If $f(t) = H\left(t - \dfrac{\pi}{2}\right)\sin t$, find its Laplace transform.

18. Establish the following results:

(a) $\mathscr{L}\{\sin^2 at\} = \dfrac{2a^2}{s(s^2+4a^2)},$

(b) $\mathscr{L}\{I_0(x)\} = \dfrac{1}{\sqrt{s^2+a^2}},$

(c) $\mathscr{L}\{|\sin at|\} = \dfrac{a}{s^2+a^2}\coth\left(\dfrac{\pi s}{2a}\right),\ s>0.$

(d) $\mathscr{L}\left\{\displaystyle\int_0^t \dfrac{\sin ax}{x}\,dx\right\} = \dfrac{1}{s}\tan^{-1}\left(\dfrac{a}{s}\right).$

(e) $\mathscr{L}\left\{\dfrac{d}{dt}(f*g)\right\} = g(0)\bar{f}(s) + \mathscr{L}\{f*g'\} = s\bar{f}(s)\,\bar{g}(s).$

19. Establish the following results:

    (a) $\mathscr{L}\{t^2 f''(t)\} = s^2 \dfrac{d^2}{ds^2}\bar{f}(s) + 4s\dfrac{d}{ds}\bar{f}(s) + 2\bar{f}(s),$

    (b) $\mathscr{L}\{t^m f^{(n)}(t)\} = (-1)^m \dfrac{d^m}{ds^m}\left[s^n \bar{f}(s) - s^{n-1} f(0) - s^{n-2} f'(0) - \cdots - f^{(n-1)}(0)\right].$

20. (a) Show that $f(t) = \sin\!\left(a\sqrt{t}\right)$ satisfies the differential equation
    $$4t\,f''(t) + 2f'(t) + a^2 f(t) = 0.$$
    Use this differential equation to show that

    (b) $\mathscr{L}\{\sin\sqrt{t}\} = \dfrac{1}{2}\Gamma\!\left(\dfrac{1}{2}\right) s^{-3/2} \exp\!\left(-\dfrac{1}{4s}\right),\quad s > 0,$

    (c) $\mathscr{L}\!\left\{\dfrac{\cos\sqrt{t}}{\sqrt{t}}\right\} = \Gamma\!\left(\dfrac{1}{2}\right)\dfrac{1}{\sqrt{s}}\exp\!\left(-\dfrac{1}{4s}\right),\quad s > 0.$

21. Establish the following results:

    (a) $\mathscr{L}\!\left\{\displaystyle\int_t^\infty \dfrac{f(x)}{x}\,dx\right\} = \dfrac{1}{s}\displaystyle\int_0^s \bar{f}(x)\,dx,$

    (b) $\mathscr{L}\!\left\{\displaystyle\int_0^\infty \dfrac{f(x)}{x}\,dx\right\} = \dfrac{1}{s}\displaystyle\int_0^\infty \bar{f}(x)\,dx.$

22. Use exercise 21(a) to find the Laplace transform of

    (a) the *cosine integral* defined by
    $$Ci(t) = \int_\infty^t \dfrac{\cos x}{x}\,dx,\quad t > 0,$$

    (b) the *exponential integral* defined by
    $$Ei(t) = \int_t^\infty \dfrac{e^{-x}}{x}\,dx,\quad t > 0.$$

23. Show that

    (a) $\mathscr{L}\{t e^{-bt} \cos at\} = \dfrac{(s+b)^2 - a^2}{\left[(s+b)^2 + a^2\right]^2},$

    (b) $\mathscr{L}\!\left\{\dfrac{\cos at - \cos bt}{t}\right\} = \dfrac{1}{2}\log\!\left(\dfrac{s^2 + a^2}{s^2 + b^2}\right),$

    (c) $\mathscr{L}\{L_n(t)\} = \dfrac{1}{s}\!\left(\dfrac{s-1}{s}\right)^n$, where $L_n(t)$ are the Laguerre polynomials.

24. If $\mathscr{L}\{f(t)\} = \bar{f}(s)$ and $\mathscr{L}\{g(x,t)\} = \bar{h}(s)\exp\{-x\bar{h}(s)\}$, prove that

(a) $\mathcal{L}\left\{\int_0^\infty g(x,t) f(x) dx\right\} = \bar{h}(s) \, \bar{f}\{\bar{h}(s)\}.$

(b) $\mathcal{L}\left\{\int_0^\infty J_0(2\sqrt{xt}) f(x) dx\right\} = \frac{1}{s} \bar{f}\left(\frac{1}{s}\right),$ when $g(x,t) = J_0(2\sqrt{xt}).$

25. Use Exercise 24(b) to show that

(a) $\int_0^\infty J_0(2\sqrt{xt}) \sin\left(\frac{x}{a}\right) dx = a \cos at,$ $(a \neq 0),$

(b) $\int_0^\infty J_0(2\sqrt{xt}) e^{-x} x^n \, dx = n! \, e^{-t} L_n(t).$

26. Find the Laplace transform of the *triangular wave* function defined over $(0, 2a)$ by

$$f(t) = \begin{cases} t, & 0 < t < a \\ 2a - t, & a < t < 2a \end{cases}.$$

27. Use the Initial Value Theorem to find $f(0)$, and $f'(0)$ from the following functions:

(a) $\bar{f}(s) = \dfrac{s}{s^2 - 5s + 12},$

(b) $\bar{f}(s) = \dfrac{1}{s(s^2 + a^2)},$

(c) $\bar{f}(s) = \dfrac{\exp(-sa)}{s^2 + 3s + 5},$ $a > 0,$

(d) $\bar{f}(s) = \dfrac{s^2 - 1}{(s^2 + 1)}.$

28. Use the Final Value Theorem to find $f(\infty)$, if it exists, from the following functions:

(a) $\bar{f}(s) = \dfrac{1}{s(s^2 + as + b)},$

(b) $\bar{f}(s) = \dfrac{s+2}{s^2 + 4},$

(c) $\bar{f}(s) = \dfrac{1}{1+as},$

(d) $\bar{f}(s) = \dfrac{3}{(s^2+4)^2}.$

29. If $\mathcal{L}\{f(t)\} = \bar{f}(s)$ and $\mathcal{L}\{g(t)\} = \bar{g}(s)$, establish Duhamel's integrals:

$$\mathcal{L}^{-1}\{s \bar{f}(s) \, \bar{g}(s)\} = \begin{cases} f(0)g(t) + \int_0^t f'(\tau) g(t - \tau) d\tau \\ g(0)f(t) + \int_0^t g'(\tau) f(t - \tau) d\tau \end{cases}.$$

30. Using Watson's lemma, find the asymptotic expansion of

(a) $\bar{f}(s) = \int_0^\infty (1 + t^2)^{-1} \exp(-st) dt,$ as $s \to \infty;$

(b) $K_0(s) = \int_1^\infty (t^2 - 1)^{-\frac{1}{2}} \exp(-st) dt$, as $s \to \infty$,

where $K_0(s)$ is the *modified Bessel* function.

31. Find the asymptotic expansion of $\bar{f}(s)$ as $s \to \infty$ when $f(t)$ is given by
   (a) $(1+t)^{-1}$,
   (b) $\sin 2\sqrt{t}$,
   (c) $\log(1+t)$,
   (d) $J_0(at)$.

Chapter 4

Applications Of Laplace Transforms

4.1 Introduction

Many problems of physical interest are described by ordinary or partial differential equations with appropriate initial or boundary conditions. These problems are usually formulated as *initial value problems, boundary value problems,* or *initial-boundary value problems* that seem to be mathematically more rigorous and physically realistic in applied and engineering sciences. The Laplace transform method is particularly useful for finding solutions of these problems. The method is very effective for the solution of the response of a linear system governed by an ordinary differential equation to the *initial data* and/or to an *external disturbance* (or *external input function*). More precisely, we seek the solution of a linear system for its state at subsequent time $t > 0$ due to the initial state at $t = 0$ and/or to the disturbance applied for $t > 0$.

This chapter deals with the solutions of ordinary and partial differential equations that arise in mathematical, physical, and engineering sciences. The applications of Laplace transforms to the solutions of certain integral equations and boundary value problems are also discussed in this chapter. It is shown by examples that the Laplace transform can also be used effectively for evaluating certain definite integrals. We also give a few examples of solutions of difference and differential equations using the Laplace transform technique. Some examples of applications of fractional derivatives and integrals are included. It is shown how the Laplace transform method is used in solving fractional differential equations with constant coefficients. The effective use of the joint Laplace and Fourier transform is illustrated by solving several initial-boundary value problems. Application of Laplace transforms to the problem of summation of infinite series in closed form is presented with examples. Finally, it is noted that the examples given in this chapter are only representative of a wide variety of problems which can be solved by the use of the Laplace transform method.

4.2 Solutions of Ordinary Differential Equations

As stated in the introduction of this chapter, the Laplace transform can be used as an effective tool for analyzing the basic characteristics of a linear system governed by the differential equation in response to initial data and/or to an external disturbance. The following examples illustrate the use of the Laplace transform in solving certain initial value problems described by ordinary differential equations.

Example 4.2.1 (*Initial Value Problem*). We consider the first-order ordinary differential equation

$$\frac{dx}{dt} + px = f(t), \qquad t > 0, \tag{4.2.1}$$

with the initial condition

$$x(t=0) = a, \tag{4.2.2}$$

where p and a are constants and $f(t)$ is an external input function so that its Laplace transform exists.

Application of the Laplace transform $\bar{x}(s)$ of the function $x(t)$ gives

$$s\bar{x}(s) - x(0) + p\bar{x}(s) = \bar{f}(s),$$

or

$$\bar{x}(s) = \frac{a}{s+p} + \frac{\bar{f}(s)}{s+p}. \tag{4.2.3}$$

The inverse Laplace transform together with the Convolution Theorem leads to the solution

$$x(t) = ae^{-pt} + \int_0^t f(t-\tau)e^{-p\tau}d\tau. \tag{4.2.4}$$

Thus the solution naturally splits into two terms—the first term corresponds to the response of the initial condition and the second term is entirely due to the external input function $f(t)$.

In particular, if $f(t) = q =$ constant, then the solution (4.2.4) becomes

$$x(t) = \frac{q}{p} + \left(a - \frac{q}{p}\right)e^{-pt}. \tag{4.2.5}$$

The first term of this solution is independent of time t and is usually called the *steady-state solution*. The second term depends on time t and is called the *transient solution*. In the limit as $t \to \infty$, the transient solution decays to zero if $p > 0$ and the steady-state solution is attained. On the other hand, when $p < 0$, the transient solution grows exponentially as $t \to \infty$, and the solution becomes unstable.

Equation (4.2.1) describes the law of natural growth or decay process with an external forcing function $f(t)$ according as $p > 0$ or < 0. In particular, if $f(t) = 0$ and $p > 0$, the resulting equation (4.2.1) occurs very frequently in chemical kinetics. Such an equation describes the rate of chemical reactions.

Example 4.2.2 (*Second Order Ordinary Differential Equation*). The second order linear ordinary differential equation has the general form

$$\frac{d^2x}{dt^2} + 2p\frac{dx}{dt} + qx = f(t), \qquad t > 0. \tag{4.2.6}$$

The initial conditions are

Integral Transforms and Their Applications

$$x(t) = a, \quad \frac{dx}{dt} = \dot{x}(t) = b \quad \text{at } t = 0, \tag{4.2.7ab}$$

where p, q, a and b are constants.

Application of the Laplace transform to this general initial value problem gives

$$s^2\bar{x}(s) - sx(0) - \dot{x}(0) + 2p\{s\bar{x}(s) - x(0)\} + q\bar{x}(s) = \bar{f}(s).$$

The use of (4.2.7ab) leads to the solution for $\bar{x}(s)$ as

$$\bar{x}(s) = \frac{(s+p)a + (b+pa) + \bar{f}(s)}{(s+p)^2 + n^2}, \quad n^2 = q - p^2. \tag{4.2.8}$$

The inverse transform gives the solution in three distinct forms depending on $q > = < p^2$, and they are

$$x(t) = ae^{-pt}\cos nt + \frac{1}{n}(b+pa)e^{-pt}\sin nt$$

$$+ \frac{1}{n}\int_0^t f(t-\tau)e^{-p\tau}\sin n\tau\, d\tau, \quad \text{when } n^2 = q - p^2 > 0, \tag{4.2.9}$$

$$x(t) = ae^{-pt} + (b+pa)te^{-pt} + \int_0^t f(t-\tau)\tau e^{-p\tau}d\tau, \text{ when } n^2 = q - p^2 = 0, \tag{4.2.10}$$

$$x(t) = ae^{-pt}\cosh mt + \frac{1}{m}(b+pa)e^{-pt}\sinh mt$$

$$+ \frac{1}{m}\int_0^t f(t-\tau)e^{-p\tau}\sinh m\tau\, d\tau, \quad \text{when } m^2 = p^2 - q > 0, \tag{4.2.11}$$

Example 4.2.3 (*Higher Order Ordinary Differential Equations*). We solve the linear equation of order n with constant coefficients as

$$f(D)\{x(t)\} \equiv D^n x + a_1 D^{n-1}x + a_2 D^{n-2}x + \cdots + a_n x = \phi(t), \quad t > 0, \tag{4.2.12}$$

with the initial conditions

$$x(t) = x_0, \quad Dx(t) = x_1, \quad D^2 x(t) = x_2, \ldots, D^{n-1}x(t) = x_{n-1}, \text{ at } t = 0, \tag{4.2.13}$$

where $D = \dfrac{d}{dt}$ is the differential operator and $x_0, x_1, \cdots, x_{n-1}$ are constants.

We take the Laplace transform of (4.2.12) to get

$$\left(s^n \bar{x} - s^{n-1}x_0 - s^{n-2}x_1 - \cdots - sx_{n-2} - x_{n-1}\right)$$

$$+ a_1\left(s^{n-1}\bar{x} - s^{n-2}x_0 - s^{n-3}x_1 - \cdots - x_{n-2}\right)$$

$$+ a_2\left(x^{n-2}\bar{x} - s^{n-3}x_0 - \cdots - x_{n-3}\right)$$

$$+ \cdots + a_{n-1}(s\bar{x} - x_0) + a_n \bar{x} = \bar{\phi}(s). \tag{4.2.14}$$

Or

$$(s^n + a_1 s^{n-1} + a_2 s^{n-2} + \cdots + a_n) \bar{x}(s)$$
$$= \bar{\phi}(s) + (s^{n-1} + a_1 s^{n-2} + \cdots + a_{n-1}) x_0$$
$$+ (s^{n-2} + a_1 s^{n-3} + \cdots + a_{n-2}) x_1 + \cdots + (s + a_1) x_{n-2} + x_{n-1}$$
$$= \bar{\phi}(s) + \bar{\psi}(s), \qquad (4.2.15)$$

where $\bar{\psi}(s)$ is made up of all terms on the right hand side of (4.2.15) except $\bar{\phi}(s)$, and is a polynomial in s of degree $(n-1)$.

Hence
$$\bar{f}(s) \bar{x}(s) = \bar{\phi}(s) + \bar{\psi}(s),$$
where
$$\bar{f}(s) = s^n + a_1 s^{n-1} + \cdots + a_n.$$

Or
$$\bar{x}(s) = \frac{\bar{\phi}(s) + \bar{\psi}(s)}{\bar{f}(s)}. \qquad (4.2.16)$$

Inversion yields
$$x(t) = \mathcal{L}^{-1}\left\{\frac{\bar{\phi}(s)}{\bar{f}(s)}\right\} + \mathcal{L}^{-1}\left\{\frac{\bar{\psi}(s)}{\bar{f}(s)}\right\}. \qquad (4.2.17)$$

The inverse operation on the right can be carried out by partial fraction decomposition, by the Heaviside Expansion Theorem, or by contour integration.

Example 4.2.4 (*Third Order Ordinary Differential Equations*). We solve
$$(D^3 + D^2 - 6D) x(t) = 0, \quad D \equiv \frac{d}{dt}, \quad t > 0, \qquad (4.2.18)$$

with the initial data
$$x(0) = 1, \ \dot{x}(0) = 0, \text{ and } \ddot{x}(0) = 5. \qquad (4.2.19)$$

The Laplace transform of equation (4.2.18) gives
$$[s^3 \bar{x} - s^2 x(0) - s \dot{x}(0) - \ddot{x}(0)] + [s^2 \bar{x} - s x(0) - \dot{x}(0)] - 6[s \bar{x} - x(0)] = 0.$$

In view of the initial conditions, we find
$$\bar{x}(s) = \frac{s^2 + s - 1}{s(s^2 + s - 6)} = \frac{s^2 + s - 1}{s(s+3)(s-2)}.$$

Or
$$\bar{x}(s) = \frac{1}{6} \cdot \frac{1}{s} + \frac{1}{3} \cdot \frac{1}{s+3} + \frac{1}{2} \cdot \frac{1}{s-2}.$$

Inverting gives the solution
$$x(t) = \frac{1}{6} + \frac{1}{3} e^{-3t} + \frac{1}{2} e^{2t}. \qquad (4.2.20)$$

Example 4.2.5 (*System of First Order Ordinary Differential Equations*). Consider the system

$$\frac{dx_1}{dt} = a_{11}x_1 + a_{12}x_2 + b_1(t) \left.\begin{matrix}\\\\\end{matrix}\right\}$$
$$\frac{dx_2}{dt} = a_{21}x_1 + a_{22}x_2 + b_2(t)$$

(4.2.21ab)

with the initial data

$$x_1(0) = x_{10} \text{ and } x_2(0) = x_{20};$$

(4.2.22ab)

where $a_{11}, a_{12}, a_{21}, a_{22}$ are constants.

Introducing the matrices

$$x \equiv \begin{pmatrix} x_1 \\ x_2 \end{pmatrix}, \quad \frac{dx}{dt} \equiv \begin{pmatrix} \frac{dx_1}{dt} \\ \frac{dx_2}{dt} \end{pmatrix}, \quad A \equiv \begin{pmatrix} a_{11} & a_{12} \\ a_{21} & a_{22} \end{pmatrix},$$

$$b(t) \equiv \begin{pmatrix} b_1(t) \\ b_2(t) \end{pmatrix} \text{ and } x_0 \equiv \begin{pmatrix} x_{10} \\ x_{20} \end{pmatrix},$$

we can write the above system in a matrix differential system as

$$\frac{dx}{dt} = Ax + b(t), \qquad x(0) = x_0.$$

(4.2.23ab)

We take the Laplace transform of the system with the initial conditions to get

$$(s - a_{11})\bar{x}_1 - a_{12}\bar{x}_2 = x_{10} + \bar{b}_1(s),$$
$$-a_{21}\bar{x}_1 + (s - a_{22})\bar{x}_2 = x_{20} + \bar{b}_2(s).$$

The solutions of this algebraic system are

$$\bar{x}_1(s) = \frac{\begin{vmatrix} x_{10} + \bar{b}_1(s) & -a_{12} \\ x_{20} + \bar{b}_2(s) & s - a_{22} \end{vmatrix}}{\begin{vmatrix} s - a_{11} & -a_{12} \\ -a_{21} & s - a_{22} \end{vmatrix}}, \quad \bar{x}_2(s) = \frac{\begin{vmatrix} s - a_{11} & x_{10} + \bar{b}_1(s) \\ -a_{21} & x_{20} + \bar{b}_2(s) \end{vmatrix}}{\begin{vmatrix} s - a_{11} & -a_{12} \\ -a_{21} & s - a_{22} \end{vmatrix}}.$$

(4.2.24a,b)

Expanding these determinants, results for $\bar{x}_1(s)$ and $\bar{x}_2(s)$ can readily be inverted, and the solutions for $x_1(t)$ and $x_2(t)$ can be found in closed forms.

Example 4.2.6 Solve the matrix differential system

$$\frac{dx}{dt} = Ax, \qquad x(0) = \begin{pmatrix} 0 \\ 1 \end{pmatrix},$$

(4.2.25)

where

$$x = \begin{pmatrix} x_1 \\ x_2 \end{pmatrix} \text{ and } A = \begin{pmatrix} 0 & 1 \\ -2 & 3 \end{pmatrix}.$$

This system is equivalent to

$$\frac{dx_1}{dt} - x_2 = 0,$$

$$\frac{dx_2}{dt} + 2x_1 - 3x_2 = 0,$$

with
$$x_1(0) = 0 \text{ and } x_2(0) = 1.$$
Taking the Laplace transform of the system with the initial data, we find
$$s\bar{x}_1 - \bar{x}_2 = 0,$$
$$2\bar{x}_1 + (s-3)\bar{x}_2 = 1.$$
This system has the solutions
$$\bar{x}_1(s) = \frac{1}{s^2 - 3s + 2} = \frac{1}{s-2} - \frac{1}{s-1},$$
$$\bar{x}_2(s) = \frac{s}{s^2 - 3s + 2} = \frac{2}{s-2} - \frac{1}{s-1}.$$
Inverting these results, we obtain
$$x_1(t) = e^{2t} - e^t, \quad x_2(t) = 2e^{2t} - e^t.$$
In matrix notation, the solution is
$$x(t) = \begin{pmatrix} e^{2t} - e^t \\ 2e^{2t} - e^t \end{pmatrix}. \tag{4.2.26}$$

Example 4.2.7 (*Second Order Differential System*). Solve the system
$$\left. \begin{array}{l} \dfrac{d^2 x_1}{dt^2} - 3x_1 - 4x_2 = 0 \\[6pt] \dfrac{d^2 x_2}{dt^2} + x_1 + x_2 = 0 \end{array} \right\} \quad t > 0, \tag{4.2.27}$$
with the initial conditions
$$x_1(t) = x_2(t) = 0; \quad \frac{dx_1}{dt} = 2 \text{ and } \frac{dx_2}{dt} = 0 \text{ at } t = 0. \tag{4.2.28}$$
The use of the Laplace transform to (4.2.27) with (4.2.28) gives
$$(s^2 - 3)\bar{x}_1 - 4\bar{x}_2 = 2$$
$$\bar{x}_1 + (s^2 + 1)\bar{x}_2 = 0.$$
Then
$$\bar{x}_1(s) = \frac{2(s^2+1)}{(s^2-1)^2} = \frac{(s+1)^2 + (s-1)^2}{(s^2-1)^2} = \frac{1}{(s-1)^2} + \frac{1}{(s+1)^2}.$$
Hence the inversion yields
$$x_1(t) = t(e^t + e^{-t}). \tag{4.2.29}$$
$$\bar{x}_2(s) = \frac{-2}{(s^2-1)^2} = \frac{1}{2}\left[\frac{1}{s-1} - \frac{1}{s+1} - \frac{1}{(s-1)^2} - \frac{1}{(s+1)^2}\right],$$
which can be readily inverted to find
$$x_2(t) = \frac{1}{2}(e^t - e^{-t} - te^t - te^{-t}). \tag{4.2.30}$$

Example 4.2.8 (The Harmonic Oscillator in a Non-Resisting Medium). The differential equation of the oscillator in the presence of an external driving force $F f(t)$ is

$$\frac{d^2 x}{dt^2} + \omega^2 x = F f(t), \qquad (4.2.31)$$

where ω is the frequency and F is a constant.

The initial conditions are

$$x(t) = a, \quad \dot{x}(t) = U \quad \text{at } t = 0, \qquad (4.2.32)$$

where a and U are constants.

Taking the Laplace transform of (4.2.31) with the initial conditions, we obtain

$$(s^2 + \omega^2)\bar{x}(s) = sa + U + F \bar{f}(s).$$

Or

$$\bar{x}(s) = \frac{as}{s^2 + \omega^2} + \frac{U}{s^2 + \omega^2} + \frac{F \bar{f}(s)}{s^2 + \omega^2}. \qquad (4.2.33)$$

Inversion together with the convolution theorem yields

$$x(t) = a \cos \omega t + \frac{U}{\omega} \sin \omega t + \frac{F}{\omega} \int_0^t f(t-\tau) \sin \omega \tau \, d\tau \qquad (4.2.34)$$

$$= A \cos(\omega t - \phi) + \frac{F}{\omega} \int_0^t f(t-\tau) \sin \omega \tau \, d\tau, \qquad (4.2.35)$$

where $A = \left(a^2 + \frac{U^2}{\omega^2}\right)^{1/2}$ and $\phi = \tan^{-1}\left(\frac{U}{\omega a}\right)$.

The solution (4.2.35) consists of two terms. The first term represents the response to the initial data, and it describes *free oscillations* with amplitude A, phase ϕ, and frequency ω, which is called the *natural frequency* of the oscillator. The second term arises in response to the external force, and hence it represents the *forced oscillations*. In order to investigate some interesting features of solution (4.2.35), we select the following cases of interest:

(i) Zero Forcing Function.

In this case, solution (4.2.35) reduces to

$$x(t) = A \cos(\omega t - \phi). \qquad (4.2.36)$$

This represents simple harmonic motion with amplitude A, frequency ω and phase ϕ. Evidently, the motion is oscillatory.

(ii) Steady Forcing Function, that is, $f(t) = 1$.

In this case, solution (4.2.35) becomes

$$x - \frac{F}{\omega^2} = A \cos(\omega t - \phi) - \frac{F}{\omega^2} \cos \omega t. \qquad (4.2.37)$$

In particular, when the particle is released from rest, $U = 0$, (4.2.37) takes the form

$$x - \frac{F}{\omega^2} = \left(a - \frac{F}{\omega^2}\right)\cos \omega t. \tag{4.2.38}$$

This corresponds to free oscillations with the natural frequency ω and displays a shift in the equilibrium position from the origin to the point $\frac{F}{\omega^2}$.

(iii) Periodic Forcing Function, that is, $f(t) = \cos \omega_0 t$.

The transform solution can readily be found from (4.2.33) in the form

$$\bar{x}(s) = \frac{as}{s^2 + \omega^2} + \frac{U}{s^2 + \omega^2} + \frac{Fs}{(s^2 + \omega_0^2)(s^2 + \omega^2)}$$

$$= \frac{as}{s^2 + \omega^2} + \frac{U}{s^2 + \omega^2} + \frac{Fs}{(\omega_0^2 - \omega^2)}\left(\frac{1}{s^2 + \omega^2} - \frac{1}{s^2 + \omega_0^2}\right). \tag{4.2.39}$$

Inversion yields the solution

$$x(t) = a \cos \omega t + \frac{U}{\omega}\sin \omega t + \frac{F}{(\omega_0^2 - \omega^2)}(\cos \omega t - \cos \omega_0 t) \tag{4.2.40}$$

$$= A \cos(\omega t - \phi) + \frac{F}{(\omega_0^2 - \omega^2)}\cos \omega_0 t, \tag{4.2.41}$$

where $A = \left\{\left(a + \frac{F}{\omega_0^2 - \omega^2}\right)^2 + \frac{U^2}{\omega^2}\right\}^{1/2}$ and $\tan \phi = \frac{U}{\omega} \div \left(a + \frac{F}{\omega_0^2 - \omega^2}\right)$.

It is noted that solution (4.2.41) consists of free oscillations of period $\frac{2\pi}{\omega}$ and forced oscillations of period $\frac{2\pi}{\omega_0}$, which is the same as that of the external periodic force. If $\omega_0 < \omega$, the phase of the forced oscillations is the same as that of the external periodic force. If $\omega_0 > \omega$, the forced term suffers from a phase change by an amount π. In other words, the forced motion is in phase or 180° out of phase with the external force according as $\omega >$ or $< \omega_0$.

When $\omega = \omega_0$, result (4.2.40) can be written as

$$x(t) = a \cos \omega t + \frac{U}{\omega}\sin \omega t + \frac{Ft}{(\omega_0 + \omega)}\left[\frac{\sin\left\{\frac{1}{2}(\omega - \omega_0)t\right\}\sin\left\{\frac{1}{2}(\omega + \omega_0)t\right\}}{\frac{1}{2}(\omega_0 - \omega)t}\right]$$

$$= a \cos \omega t + \frac{U}{\omega}\sin \omega t + \frac{Ft}{2\omega}\sin \omega t$$

$$= A \cos(\omega t - \phi) + \frac{Ft}{2\omega}\sin \omega t, \tag{4.2.42}$$

where

$$A^2 = \left(a^2 + \frac{U^2}{\omega^2}\right) \quad \text{and} \quad \tan\phi = \frac{U}{a\omega}.$$

This solution clearly shows that the amplitude of the forced motion increases with t. Thus, if the natural frequency is equal to the forcing frequency, the oscillations become unbounded, which is physically undesirable. This phenomenon is usually called *resonance*, and the corresponding frequency $\omega = \omega_0$ is referred to as the *resonant frequency* of the system. It may be emphasized that at the resonant frequency, the solution of the problem becomes mathematically invalid for large times, and hence it is physically unrealistic. In most dynamical systems, this kind of situation is resolved by including dissipating and/or nonlinear effects.

Example 4.2.9 (*Harmonic Oscillator in a Resisting Medium*). The differential equation of the oscillator in a resisting medium where the resistance is proportional to velocity is given by

$$\frac{d^2x}{dt^2} + 2k\frac{dx}{dt} + \omega^2 x = F f(t), \tag{4.2.43}$$

where $2k(>0)$ is a constant of proportionality and the right hand side represents the external driving force. The initial state of the system is

$$x(t) = a, \quad \frac{dx}{dt} = U \quad \text{at } t = 0. \tag{4.2.44}$$

In view of the initial conditions, the Laplace transform solution of equation (4.2.43) is obtained as

$$\bar{x}(s) = \frac{a(s+2k)+U+F\bar{f}(s)}{(s^2+2ks+\omega^2)}$$

$$= \frac{a(s+k)+(U+ak)+\bar{F}(s)}{(s+k)^2+n^2}, \tag{4.2.45}$$

where $n^2 = \omega^2 - k^2$.

Three possible cases deserve attention:

(i) $k < \omega$ (*small damping*).

In this case, $n^2 = \omega^2 - k^2 > 0$ and the inversion of (4.2.45) along with the Convolution Theorem yields

$$x(t) = a e^{-kt}\cos nt + \frac{(U+ak)}{n}e^{-kt}\sin nt + \frac{F}{n}\int_0^t f(t-\tau)e^{-k\tau}\sin n\tau \, d\tau. \tag{4.2.46}$$

This is the most general solution of the problem for an arbitrary form of the external driving force.

(ii) $k = \omega$ (*critical damping*) so that $n^2 = 0$.

The solution for this case can readily be obtained from (4.2.45) by inversion and has the form

$$x(t) = a e^{-kt} + (U+ak)t e^{-kt} + F\int_0^t f(t-\tau)\tau e^{-k\tau} d\tau. \tag{4.2.47}$$

(iii) $k > \omega$ (*large damping*).

Set $n^2 = -(k^2 - \omega^2) = -m^2$ where $m^2 = k^2 - \omega^2 > 0$.

The transformed solution (4.2.45) assumes the form

$$\bar{x}(s) = \frac{a(s+k)+(U+ak)+F\bar{f}(s)}{(s+k)^2 - m^2}. \tag{4.2.48}$$

After inversion, it turns out that

$$x(t) = ae^{-kt}\cosh mt + \left(\frac{U+ak}{m}\right)e^{-kt}\sinh mt$$

$$+ \frac{F}{m}\int_0^t f(t-\tau)e^{-k\tau}\sinh m\tau\, d\tau. \tag{4.2.49}$$

In order to examine the characteristic features of the problem, it is necessary to specify the nature and functional form of $f(t)$ involved in the external force term. Suppose the external driving force is zero. The solution can readily be written down in all three cases.

For $0 < k < \omega$, the solution is

$$x(t) = e^{-kt}\left(a\cos nt + \frac{U+ak}{n}\sin nt\right) = Ae^{-kt}\cos(nt - \phi), \tag{4.2.50}$$

where $A = \left\{a^2 + \frac{(U+ak)^2}{n^2}\right\}^{1/2}$ and $\phi = \tan^{-1}\left(\frac{U+ak}{an}\right)$.

Like the harmonic oscillator in a vacuum, the motion is oscillatory with the time-dependent amplitude Ae^{-kt} and the modified frequency

$$n = (\omega^2 - k^2)^{1/2} = \omega\left(1 - \frac{1}{2}\frac{k^2}{\omega^2} + \cdots\right), \qquad 0 < k < \omega.$$

This means that, when the resistance is small, the modified frequency (or the undamped natural frequency) is obviously smaller than the natural frequency, ω. Although the small resistance produces an insignificant effect on the frequency, the amplitude is radically modified. It should also be noted that the amplitude decays exponentially to zero as time $t \to \infty$. The phase of the motion is also changed by the small resistance. Thus the motion is called the *damped oscillatory motion*, and depicted by Figure 4.1.

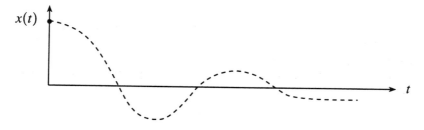

Figure 4.1 Damped oscillatory motion.

At the critical case, $\omega = k$, and hence $n = 0$. The solution can readily be found from (4.2.47) with $F = 0$, and has the form
$$x(t) = ae^{-kt} + (ak + U)te^{-kt}. \tag{4.2.51}$$
The motion ceases to be oscillatory and decays very rapidly as $t \to \infty$.

If damping is large with no external force, the solution (4.2.49) reduces to
$$x(t) = ae^{-kt} \cosh mt + \left(\frac{ak + U}{m}\right) e^{-kt} \sinh mt. \tag{4.2.52}$$

Using $\begin{matrix}\cosh\\ \sinh\end{matrix} mt = \frac{1}{2}(e^{mt} \pm e^{-mt})$, we can write down the solution as
$$x(t) = Ae^{-(k-m)t} + Be^{-(k+m)t}, \tag{4.2.53}$$
where $A = \frac{1}{2}\left(a + \frac{ak+U}{m}\right)$ and $B = \frac{1}{2}\left(a - \frac{ak+U}{m}\right)$.

The above solution suggests that the motion is no longer oscillatory and in fact, it decays very rapidly as $t \to \infty$.

Example 4.2.10 (*Harmonic Oscillator in a Resisting Medium with an External Periodic Force*). The motion is governed by the equation
$$\frac{d^2x}{dt^2} + 2k\frac{dx}{dt} + \omega^2 x = F\cos\omega_0 t, \qquad k > 0 \tag{4.2.54}$$
with the initial data
$$x(0) = a \quad \text{and} \quad \dot{x}(0) = U.$$
The transformed solution for the case of small damping $(k < \omega)$ is
$$\bar{x}(s) = \frac{a(s+k) + (U+ak)}{(s+k)^2 + n^2} + \frac{Fs}{\{(s+k)^2 + n^2\}(s^2 + \omega_0^2)}$$
$$= \frac{a(s+k) + (U+ak)}{(s+k)^2 + n^2} + F\left[\frac{As - B}{(s+k)^2 + n^2} - \frac{As - C}{s^2 + \omega_0^2}\right], \tag{4.2.55}$$
where
$$A = \frac{\omega_0^2 - \omega^2}{\left(\omega^2 - \omega_0^2\right)^2 + 4k^2\omega_0^2}, \qquad B = \frac{2k\omega^2}{\left(\omega^2 - \omega_0^2\right)^2 + 4k^2\omega_0^2},$$
and
$$C = \frac{2k\omega_0^2}{\left(\omega^2 - \omega_0^2\right)^2 + 4k^2\omega_0^2} \qquad \text{with } \omega^2 = n^2 + k^2.$$
The expression for $\bar{x}(s)$ can be inverted to obtain the solution
$$x(t) = (a + FA)e^{-kt}\cos nt + \frac{1}{n}(U + ak - FAk - FB)e^{-kt}\sin nt$$
$$- AF\cos\omega_0 t + \frac{CF}{\omega_0}\sin\omega_0 t. \tag{4.2.56}$$
It is convenient to write it in the form
$$x(t) = A_1 \cos(\omega_0 t - \phi_1) + A_2 e^{-kt}\cos(nt - \phi_2), \tag{4.2.57}$$

where

$$A_1^2 = F^2\left(A^2 + \frac{C^2}{\omega_0^2}\right) = \frac{F^2}{\left(\omega^2 - \omega_0^2\right)^2 + 4k^2\omega_0^2}, \quad (4.2.58)$$

$$\tan\phi_1 = -\frac{C}{A\omega_0} = \frac{2k\omega_0}{\omega^2 - \omega_0^2}, \quad (4.2.59)$$

$$A_2^2 = (a + FA)^2 + \frac{1}{n^2}(U + ak - kFA - FB)^2, \quad (4.2.60)$$

and

$$\tan\phi_2 = \frac{U + ak - kFA - FB}{n(a + FA)}. \quad (4.2.61)$$

This form of solution (4.2.57) lends itself to some interesting physical interpretations. First, the displacement field $x(t)$ essentially consists of the steady state and the transient terms, which are independently modified by the damping and driving forces involved in the equation of motion. In the limit as $t \to \infty$, the latter decays exponentially to zero. Consequently, the ultimate steady state is attained in the limit, and represented by the first term of (4.2.57). In fact, the steady-state solution is denoted by $x_{st}(t)$ and given by

$$x_{st}(t) = A_1 \cos(\omega_0 t - \phi_1), \quad (4.2.62)$$

where A_1 is the amplitude, ω_0 is the frequency, and ϕ_1 represents the phase lag given by

$$\phi_1 = \tan^{-1}\left\{\frac{2k\omega_0}{\left(\omega^2 - \omega_0^2\right)}\right\} \quad \text{when } \omega_0 < \omega,$$

$$= \pi - \tan^{-1}\left\{\frac{2k\omega_0}{\left(\omega_0^2 - \omega^2\right)}\right\} \quad \text{when } \omega_0 > \omega,$$

$$= \frac{\pi}{2} \quad \text{as } \omega_0 \to \omega.$$

It should be noted that the frequency of the steady-state solution is the same as that of the external driving force, but the amplitude and the phase are modified by the parameters ω, k and ω_0. It is of interest to examine the nature of the amplitude and the phase with respect to the forcing frequency ω_0. For a low frequency $(\omega_0 \to 0)$, $A_1 = \frac{F}{\omega^2}$ and $\phi_1 = 0$. As $\omega_0 \to \omega$, the amplitude of the motion is still bounded and equal to $\left(\frac{F}{2k\omega}\right)$ if $k \neq 0$. The displacement suffers from a phase lag of $\pi/2$. Further, we note that

$$\frac{dA_1}{d\omega_0} = \frac{2\omega_0 F\left(\omega^2 - \omega_0^2 - 2k^2\right)}{\left\{\left(\omega^2 - \omega_0^2\right)^2 + 4k^2\omega_0^2\right\}^{3/2}}. \quad (4.2.63)$$

It follows that A_1 has a minimum at $\omega_0 = 0$ with minimum value $\dfrac{F}{\omega^2}$, and a maximum at $\omega_0 = (\omega^2 - 2k^2)^{1/2}$ with maximum value $\dfrac{F}{2k(\omega^2 - 2k^2)^{1/2}}$ provided $2k^2 < \omega^2$. If $2k^2 > \omega^2$, A_1 has no maximum and gradually decreases. The non-dimensional amplitude $A^* = \left(\dfrac{2A_1\omega^2}{F}\right)$ is plotted against the non-dimensional frequency $\dfrac{\omega_0}{\omega}$ for a given value of $\dfrac{k}{\omega}(<1)$ in Figure 4.2.

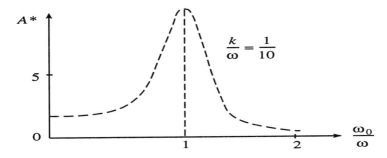

Figure 4.2 Amplitude versus frequency *with* damping.

In the absence of the damping term, the amplitude A_1 becomes
$$A_1 = \dfrac{F}{|\omega^2 - \omega_0^2|},$$
which is unbounded at $\omega_0 = \omega$ and shown in Figure 4.3.

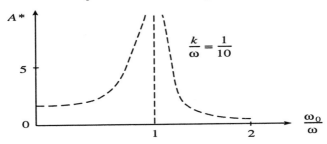

Figure 4.3 Amplitude versus frequency *without* damping.

This situation has already been encountered earlier, and the frequency $\omega_0 = \omega$ was defined as the *resonant frequency*. The difficulty for the resonant case has been resolved by the inclusion of small damping effect.

At the critical case $(k^2 = \omega^2)$, the solution is found from (4.2.55) by inversion and has the form
$$x(t) = A_1 \cos(\omega_0 t - \phi) + (a + FA)e^{-kt} + t(U + ak - FAk - FB)e^{-kt}. \qquad (4.2.64)$$
The transient term of this solution decays as $t \to \infty$ and the steady state is attained.

The solution for the case of high damping $(k^2 > \omega^2)$ is obtained from (4.2.55) as
$$x(t) = (a + FA)e^{-kt} \cosh mt + \frac{1}{m}(U + ak - FAk - FB)e^{-kt} \sinh mt$$
$$- AF \cos \omega_0 t + \frac{CF}{\omega_0} \sin \omega_0 t \qquad (4.2.65)$$
where $m^2 = -n^2 = k^2 - \omega^2 > 0$. This result is somewhat similar to that of (4.2.56) or (4.2.57) with the exception that the transient term decays very rapidly as $t \to \infty$. Like previous cases, the steady state is reached in the limit.

Example 4.2.11 Obtain the solution of the Bessel equation
$$t\frac{d^2 x}{dt^2} + \frac{dx}{dt} + a^2 t\, x(t) = 0, \qquad x(0) = 1. \qquad (4.2.66)$$
Application of the Laplace transform gives
$$\mathcal{L}\left\{t\frac{d^2 x}{dt^2}\right\} + \mathcal{L}\left\{\frac{dx}{dt}\right\} + a^2\, \mathcal{L}\{t\, x(t)\} = 0.$$
Or
$$-\frac{d}{ds}\left[\mathcal{L}\left\{\frac{d^2 x}{dt^2}\right\}\right] + s\bar{x}(s) - x(0) - a^2 \frac{d\bar{x}}{ds} = 0.$$
Or
$$-\frac{d}{ds}\left[s^2 \bar{x} - s x(0) - \dot{x}(0)\right] + s\bar{x}(s) - 1 - a^2 \frac{d\bar{x}}{ds} = 0.$$
Thus
$$(s^2 + a^2)\frac{d\bar{x}}{ds} + s\bar{x} = 0.$$
Or
$$\frac{d\bar{x}}{\bar{x}} = -\frac{s\, ds}{s^2 + a^2}.$$
Integration gives the solution for $\bar{x}(s)$
$$\bar{x}(s) = \frac{A}{\sqrt{s^2 + a^2}},$$
where A is an integrating constant. By the inverse transformation, we obtain the solution
$$x(t) = A\, J_0(at).$$

Example 4.2.12 Find the solution of the initial value problem
$$\frac{d^2x}{dt^2} + t\frac{dx}{dt} - 2x = 2, \quad x(0) = \dot{x}(0) = 0.$$
Taking the Laplace transform, we obtain
$$\mathcal{L}\left\{\frac{d^2x}{dt^2}\right\} + \mathcal{L}\left\{t\frac{dx}{dt}\right\} - 2\bar{x}(s) = \frac{2}{s}.$$
Or
$$s^2\bar{x} - \frac{d}{ds}\{s\bar{x}(s)\} - 2\bar{x} = \frac{2}{s}$$
$$\frac{d\bar{x}}{ds} + \left(\frac{3}{s} - s\right)\bar{x} = -\frac{2}{s^2}.$$
This is a first order linear equation which can be solved by the method of the integrating factor. The integrating factor is $s^3 \exp\left(-\frac{1}{2}s^2\right)$. Multiplying the equation by the integrating factor and integrating, it turns out that
$$\bar{x}(s) = \frac{2}{s^3} + \frac{A}{s^3}\exp\left(\frac{s^2}{2}\right),$$
where A is an integrating constant. As $\bar{x}(s) \to \infty$ as $s \to \infty$, we must have $A \equiv 0$. Thus $\bar{x}(s) = \frac{2}{s^3}$. Inverting, we get the solution
$$x(t) = t^2.$$

Example 4.2.13 (*Current and Charge in a Simple Electric Circuit*). The current in a circuit (see Figure 4.4) containing a inductance L, resistance R, and capacitance C with an applied voltage $E(t)$ is governed by the equation
$$L\frac{dI}{dt} + RI + \frac{1}{C}\int_0^t I\,dt = E(t), \quad (4.2.67)$$
where L, R, and C are constants and $I(t)$ is the current which is related to the accumulated charge Q on the condenser at time t by
$$Q(t) = \int_0^t I(t)\,dt \quad \text{so that} \quad \frac{dQ}{dt} = I(t). \quad (4.2.68)$$

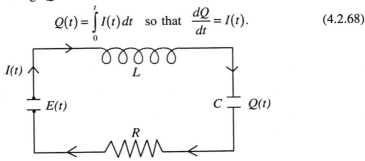

Figure 4.4 Simple electric circuit.

If the circuit is without a condenser $(C \to \infty)$, equation (4.2.67) reduces to

$$L\frac{dI}{dt} + RI = E(t), \quad t > 0. \tag{4.2.69}$$

This can easily be solved with the initial condition $I(t=0) = I_0$. However, we solve the system (4.2.67)-(4.2.68) with the initial data

$$I(t=0) = 0, \quad Q(t=0) = 0. \tag{4.2.70}$$

Then, in the limit $C \to \infty$, the solution of the system reduces to that of (4.2.69).

Application of the Laplace transform to (4.2.67) with (6.2.70) gives

$$\bar{I}(s) = \frac{1}{L} \frac{s\bar{E}(s)}{\left(s^2 + \frac{R}{L}s + \frac{1}{CL}\right)} = \frac{1}{L} \cdot \frac{(s+k-k)\bar{E}(s)}{(s+k)^2 + n^2}, \tag{4.2.71}$$

where $k = \dfrac{R}{2L}$, $\omega^2 = \dfrac{1}{LC}$ and $n^2 = \omega^2 - k^2$.

Inversion of (4.2.71) gives the current field for three cases:

$$I(t) = \frac{1}{L}\int_0^t E(t-\tau)\left[\cos n\tau - \frac{k}{n}\sin n\tau\right]e^{-k\tau}d\tau, \quad \text{if } \omega^2 > k^2, \tag{4.2.72}$$

$$= \frac{1}{L}\int_0^t E(t-\tau)(1-k\tau)e^{-k\tau}d\tau, \quad \text{if } \omega^2 = k^2, \tag{4.2.73}$$

$$= \frac{1}{L}\int_0^t E(t-\tau)\left[\cosh m\tau - \frac{k}{m}\sinh m\tau\right]e^{-k\tau}d\tau \quad \text{if } k^2 > \omega^2, \tag{4.2.74}$$

where $m^2 = -n^2$.

In particular, if $E(t) = \text{constant} = E_0$, then the solution can be obtained directly from (4.2.71) by inversion as

$$I(t) = \frac{E_0}{nL}\exp\left(-\frac{Rt}{2L}\right)\sin nt, \quad \text{if } n^2 = \frac{1}{CL} - \left(\frac{R}{2L}\right)^2 > 0, \tag{4.2.75}$$

$$= \frac{E_0}{L}t\exp\left(-\frac{Rt}{2L}\right), \quad \text{if } \left(\frac{R}{2L}\right)^2 = \frac{1}{CL}, \tag{4.2.76}$$

$$= \frac{E_0}{mL}\exp\left(-\frac{Rt}{2L}\right)\sinh mt, \quad \text{if } m^2 = \left(\frac{R}{2L}\right)^2 - \frac{1}{CL} > 0. \tag{4.2.77}$$

It may be observed that the solution for the case of low resistance $(R^2C < 4L)$, or small damping, describes a damped sinusoidal current with slowly decaying amplitude. In fact, the rate of damping is proportional to $\dfrac{R}{L}$, and when this quantity is large, the attenuation of the current is very rapid. The frequency of the oscillating current field is

$$n = \left(\frac{1}{CL} - \frac{R^2}{4L^2}\right)^{1/2},$$

which is called the *natural frequency* of the current field. If $\dfrac{R^2}{4L^2} \ll \dfrac{1}{CL}$, the frequency n is approximately equal to

$$n \sim \frac{1}{\sqrt{CL}}.$$

The case, $\dfrac{R^2}{4L^2} = \dfrac{1}{CL}$, corresponds to *critical damping*, and the solution for this case decays exponentially with time.

The last case, $R^2 C > 4L$, corresponds to high resistance or high damping. The current related to this case has the form

$$I(t) = \frac{E_0}{2mL}\left[e^{-\left(\frac{R}{2L}-m\right)t} - e^{-\left(\frac{R}{2L}+m\right)t} \right]. \tag{4.2.78}$$

It may be recognized that the solution is no longer oscillatory and decays exponentially to zero as $t \to \infty$. This is really expected in an electrical circuit with a very high resistance. If $C \to \infty$, the circuit is free from a condenser and $m \to \dfrac{R}{2L}$. Consequently, solution (4.2.77) reduces to

$$I(t) = \frac{E_0}{R}\left[1 - \exp\left(-\frac{Rt}{L}\right) \right]. \tag{4.2.79}$$

This is identical with the solution of equation (4.2.69).

We consider another special case where the alternating voltage is applied to the circuit so that

$$E(t) = E_0 \sin \omega_0 t. \tag{4.2.80}$$

The transformed solution for $\bar{I}(s)$ follows from (4.2.71) as

$$\bar{I}(s) = \frac{E_0 \omega_0}{L} \frac{s}{\{(s+k)^2 + n^2\}(s^2 + \omega_0^2)}. \tag{4.2.81}$$

Using the rules of partial fractions, it turns out that

$$\bar{I}(s) = \frac{E_0 \omega_0}{L}\left[\frac{As - B}{(s+k)^2 + n^2} - \frac{As - C}{s^2 + \omega_0^2} \right], \tag{4.2.82}$$

where $(A, B, C) \equiv \dfrac{\left(\omega_0^2 - \omega^2,\ 2k\omega^2,\ 2k\omega_0^2\right)}{\left(\omega^2 - \omega_0^2\right)^2 + 4k^2 \omega_0^2}$.

The inversion of (4.2.82) can be completed by Table B-4 of Laplace transforms, and the solution for $I(t)$ assumes three distinct forms according as $\omega^2 > = < k^2$.

The solution for the case of low resistance $(\omega^2 > k^2)$ is

$$I(t) = \frac{E_0 \omega_0}{L}\left[Ae^{-kt}\cos nt - \frac{1}{n}(Ak+B)e^{-kt}\sin nt \right.$$
$$\left. -A\cos\omega_0 t + \frac{C}{\omega_0}\sin\omega_0 t \right], \qquad (4.2.83)$$

which has the equivalent form
$$I(t) = A_1 \sin(\omega_0 t - \phi_1) + A_2 e^{-kt}\cos(nt - \phi_2), \qquad (4.2.84)$$

where
$$A_1^2 = \frac{E_0^2}{L^2}(A^2\omega_0^2 + C^2) = \frac{E_0^2\omega_0^2}{L^2\{(\omega^2 - \omega_0^2)^2 + 4k^2\omega_0^2\}}, \quad \tan\phi_1 = \left(\frac{A\omega_0}{C}\right), \quad (4.2.85)$$

$$A_2^2 = \frac{E_0^2\omega_0^2}{L^2}\left[A^2 + \frac{1}{n^2}(Ak+B)^2\right] \text{ and } \tan\phi_2 = -\frac{(Ak+B)}{An}. \qquad (4.2.86)$$

The current field consists of the steady-state and transient components. The latter decays exponentially in a time scale of the order $\frac{L}{R}$. Consequently, the steady current field is established in the electric circuit and describes the sinusoidal current with constant amplitude and phase lagging by an angle ϕ_1. The frequency of the steady oscillating current is the same as that of the applied voltage.

In the critical situation $(\omega^2 = k^2)$, the current field is derived from (4.2.82) by inversion and has the form
$$I(t) = A_1 \sin(\omega_0 t - \phi_1) + \frac{E_0\omega_0}{L}\left[Ae^{-kt} - (Ak+B)te^{-kt}\right]. \qquad (4.2.87)$$

This result suggests that the transient component of the current dies out exponentially in the limit as $t \to \infty$. Eventually, the steady oscillating current is set up in the circuit and described by the first term of (4.2.87). Finally, the solution related to the case of high resistance $(\omega^2 < k^2)$ can be found by direct inversion of (4.2.82) and is given by
$$I(t) = A_1 \sin(\omega_0 t - \phi_1) + \frac{E_0\omega_0}{L}\left[A\cosh mt\right.$$
$$\left. -\frac{1}{m}(Ak+B)\sinh mt\right]e^{-kt}. \qquad (4.2.88)$$

This solution is somewhat similar to (4.2.84) with the exception of the form of the transient term which, of course, decays very rapidly as $t \to \infty$. Consequently, the steady current field is established in the circuit and has the same value as in (4.2.84).

Finally, we close this example by suggesting a similarity between this electric circuit system and the mechanical system as described in Example 4.2.9. Differentiation of (4.2.67) with respect to t gives a second order equation for the current field as

$$L\frac{d^2I}{dt^2} + R\frac{dI}{dt} + \frac{I}{C} = \frac{dE}{dt}. \qquad (4.2.89)$$

Also, an equation for the charge field $Q(t)$ can be found from (4.2.67) and (4.2.68) as

$$L\frac{d^2Q}{dt^2} + R\frac{dQ}{dt} + \frac{Q}{C} = E(t). \qquad (4.2.90)$$

Writing $2k = \frac{R}{L}$ and $\omega^2 = \frac{1}{LC}$, the above equations can be put into the form

$$\left(\frac{d^2}{dt^2} + 2k\frac{d}{dt} + \omega^2\right)\binom{I}{Q} = \frac{1}{L}\binom{\frac{dE}{dt}}{E}. \qquad (4.2.91ab)$$

These equations are very similar to (4.2.43).

Example 4.2.14 (*Current and Charge in an Electrical Network*). An electrical network is a combination of several interrelated simple electric circuits. Consider a more general network consisting of two electric circuits coupled by the mutual inductance M with resistances R_1 and R_2, capacitances C_1 and C_2, self-inductances L_1 and L_2 as shown in Figure 4.5. A time-dependent voltage $E(t)$ is applied to the first circuit at time $t = 0$, when charges and currents are zero.

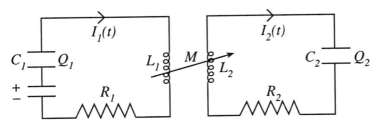

Figure 4.5 Two coupled electric circuits.

The charge and current fields in the network are governed by the system of ordinary differential equations

$$L_1\frac{dI_1}{dt} + R_1I_1 + M\frac{dI_2}{dt} + \frac{Q_1}{C_1} = E(t), \qquad t > 0 \qquad (4.2.92)$$

$$M\frac{dI_1}{dt} + L_2\frac{dI_2}{dt} + R_2I_2 + \frac{Q_2}{C_2} = 0, \qquad t > 0 \qquad (4.2.93)$$

with

$$\frac{dQ_1}{dt} = I_1 \text{ and } \frac{dQ_2}{dt} = I_2.$$

The initial conditions are
$$I_1 = 0, \; Q_1 = 0, \; I_2 = 0, \; Q_2 = 0 \qquad \text{at } t = 0. \qquad (4.2.94)$$

Eliminating the currents from (4.2.92) and (4.2.93), we obtain

$$\left(L_1 D^2 + R_1 D + \frac{1}{C_1}\right) Q_1 + M D^2 Q_2 = E(t), \qquad (4.2.95)$$

$$M D^2 Q_1 + \left(L_2 D^2 + R_2 D + \frac{1}{C_2}\right) Q_2 = 0, \qquad (4.2.96)$$

where $D \equiv \dfrac{d}{dt}$.

The Laplace transform can be used to solve this system for Q_1 and Q_2. Similarly, we can find solutions for the current fields I_1 and I_2 independently or from the charge fields. We leave it as an exercise for the reader.

In the absence of the external voltage $(E = 0)$ with $R_1 = R_2 = 0$, $L_1 = L_2 = L$ and $C_1 = C_2 = C$, addition and subtraction of (4.2.95) and (4.2.96) give

$$\ddot{Q}_+ + \alpha^2 Q_+ = 0, \quad \ddot{Q}_- + \beta^2 Q_- = 0, \qquad (4.2.97\text{ab})$$

where $Q_+ = Q_1 + Q_2$, $Q_- = Q_1 - Q_2$, $\alpha^2 = [C(L-M)]^{-1}$, and $\beta^2 = [C(L+M)]^{-1}$.

Clearly, the system executes uncoupled simple harmonic oscillations with frequencies α and β. Hence the normal modes can be generated in this freely oscillatory electrical system.

Finally, in the absence of capacitances $(C_1 \to \infty, C_2 \to \infty)$, the above network reduces to a simple one which consists of two electric circuits coupled by the mutual inductance M with inductances L_1 and L_2, and resistances R_1 and R_2. As shown in Figure 4.6, an external voltage is applied to the first circuit at time $t = 0$.

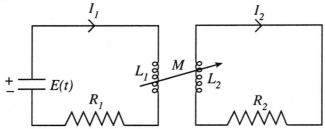

Figure 4.6 Two coupled electric circuits without capacitances.

The current fields in the network are governed by a pair of coupled ordinary differential equations

$$L_1 \frac{dI_1}{dt} + R_1 I_1 + M \frac{dI_2}{dt} = E(t), \qquad t > 0, \qquad (4.2.98)$$

$$M \frac{dI_1}{dt} + L_2 \frac{dI_2}{dt} + R_2 I_2 = 0, \qquad t > 0, \qquad (4.2.99)$$

where $I_1(t)$ and $I_2(t)$ are the currents in the first and the second circuits respectively. The initial conditions are

$$I_1(0) = I_2(0) = 0. \qquad (4.2.100)$$

We shall not pursue the problem further because the transform method of solution is a simple exercise.

Example 4.2.15 (*Linear Dynamical Systems and Signals*). In physical and engineering sciences, a large number of linear dynamical systems with a time dependent *input signal* $x(t)$ that generates an *output signal* $y(t)$ can be described by the ordinary differential equation with constant coefficients

$$\left(D^n + a_{n-1} D^{n-1} + \cdots + a_0\right) y(t) = \left(D^m + b_{m-1} D^{m-1} + \cdots + b_0\right) x(t), \qquad (4.2.101)$$

where $D \equiv \dfrac{d}{dt}$ is the differential operator, a_r and b_r are constants.

We apply the Laplace transform to find the output $y(t)$ so that (4.2.101) becomes

$$\overline{p}_n(s)\,\overline{y}(s) - \overline{R}_{n-1} = \overline{q}_m(s)\,\overline{x}(s) - \overline{S}_{m-1}, \qquad (4.2.102)$$

where

$$\overline{p}_n(s) = s^n + a_{n-1} s^{n-1} + \cdots + a_0, \quad \overline{q}_m(s) = s^m + a_{m-1} s^{m-1} + \cdots + b_0,$$

$$\overline{R}_{n-1} = \sum_{r=0}^{n-1} s^{n-r-1} y^{(r)}(0), \quad \overline{S}_{m-1} = \sum_{r=0}^{m-1} s^{m-r-1} x^{(r)}(0).$$

It is convenient to express (4.2.102) in the form

$$\overline{y}(s) = \overline{\phi}(s)\,\overline{x}(s) + \overline{\psi}(s), \qquad (4.2.103)$$

where

$$\overline{\phi}(s) = \frac{\overline{q}_m(s)}{\overline{p}_n(s)} \quad \text{and} \quad \overline{\psi}(s) = \frac{\left(\overline{R}_{n-1} - \overline{S}_{m-1}\right)}{\overline{p}_n(s)}, \qquad (4.2.104\text{a,b})$$

and $\overline{\phi}(s)$ is usually called the *transfer function*.

The inverse Laplace transform combined with the Convolution Theorem leads to the formal solution

$$y(t) = \int_0^t x(t-\tau)\,\phi(\tau)\,d\tau + \psi(t). \qquad (4.2.105)$$

When the initial data are zero, $\overline{\psi}(s) = 0$, the transfer function takes the simple form

$$\overline{\phi}(s) = \frac{\overline{y}(s)}{\overline{x}(s)}. \qquad (4.2.106)$$

If $x(t) = \delta(t)$ so that $\overline{x}(s) = 1$, then the output function is

$$y(t) = \int_0^t \delta(t-\tau)\,\phi(\tau)\,d\tau = \phi(t), \qquad (4.2.107)$$

and $\phi(t)$ is known as the *impulse response*.

Example 4.2.16 (*Delay Differential Equations*). In many problems, the derivatives of the unknown function $x(t)$ are related to its value at different time $t - \tau$. This leads us to consider differential equations of the form

$$\frac{dx}{dt} + a\,x(t-\tau) = f(t), \qquad (4.2.108)$$

where a is a constant and $f(t)$ is a given function. Equations of this type are called *delay differential equations*. In general, initial value problems for these equations involve the specification of $x(t)$ in the interval $t_0 - \tau \leq t < t_0$, and this information combined with the equation itself is sufficient to determine $x(t)$ for $t > t_0$.

We show how equation (4.2.108) can be solved by the Laplace transform when $t_0 = 0$ and $x(t) = x_0$ for $t \leq 0$. In view of the initial condition, we can write
$$x(t-\tau) = x(t-\tau) H(t-\tau)$$
so equation (4.2.108) is equivalent to
$$\frac{dx}{dt} + a\, x(t-\tau) H(t-\tau) = f(t). \tag{4.2.109}$$
Application of the Laplace transform to (4.2.109) gives
$$s\, \bar{x}(s) - x_0 + a \exp(-\tau s)\, \bar{x}(s) = \bar{f}(s).$$
Or
$$\bar{x}(s) = \frac{x_0 + \bar{f}(s)}{\{s + a \exp(-\tau s)\}} \tag{4.2.110}$$
$$= \frac{1}{s}\{x_0 + \bar{f}(s)\}\left[1 + \frac{a}{s}\exp(-\tau s)\right]^{-1}$$
$$= \frac{1}{s}\{x_0 + \bar{f}(s)\} \sum_{n=0}^{\infty}(-1)^n \left(\frac{a}{s}\right)^n \exp(-n\tau s). \tag{4.2.111}$$
The inverse Laplace transform gives the formal solution
$$x(t) = \mathscr{L}^{-1}\left[\frac{1}{s}\{x_0 + \bar{f}(s)\}\sum_{n=0}^{\infty}(-1)^n \left(\frac{a}{s}\right)^n \exp(-n\tau s)\right]. \tag{4.2.112}$$
In order to write an explicit solution, we choose $x_0 = 0$ and $f(t) = t$, and hence (4.2.112) becomes
$$x(t) = \mathscr{L}^{-1}\left[\frac{1}{s^3}\sum_{n=0}^{\infty}(-1)^n \left(\frac{a}{s}\right)^n \exp(-n\tau s)\right]$$
$$= \sum_{n=0}^{\infty}(-1)^n a^n \frac{(t-n\tau)^{n+2}}{(n+2)!} H(t-n\tau),\ t > 0. \tag{4.2.113}$$

Example 4.2.17 (*The Renewal Equation in Statistics*). The random function $X(t)$ of time t represents the number of times some event has occurred between time 0 and time t, and is usually referred to as a *counting process*. A random variable X_n that records the time it assumes for X to get the value n from the $n-1$ is referred to as an *inter-arrival time*. If the random variables $X_1, X_2, X_3, \ldots$ are independent and identically distributed, then the counting process $X(t)$ is called a *renewal process*. We represent their common probability distribution function by $F(t)$ and the density function by $f(t)$ so that $F'(t) = f(t)$. The *renewal function* is defined by the expected number of times the event being counted occurs by time t and is denoted by $r(t)$ so that

$$r(t) = E\{X(t)\} = \int_0^\infty E\{X(t)|X_1 = x\} f(x) dx, \qquad (4.2.114)$$

where $E\{X(t)|X_1 = x\}$ is the conditional expected value of $X(t)$ under the condition that $X_1 = x$ and has the value

$$E\{X(t)|X_1 = x\} = [1 + r(t-x)] H(t-x). \qquad (4.2.115)$$

Thus

$$r(t) = \int_0^t \{1 + r(t-x)\} f(x) dx.$$

Or

$$r(t) = F(t) + \int_0^t r(t-x) f(x) dx. \qquad (4.2.116)$$

This is called the *renewal equation* in mathematical statistics. We solve the equation by taking the Laplace transform with respect to t, and the Laplace transformed equation is

$$\bar{r}(s) = \bar{F}(s) + \bar{r}(s) \bar{f}(s),$$

Or

$$\bar{r}(s) = \frac{\bar{F}(s)}{1 - \bar{f}(s)}. \qquad (4.2.117)$$

The inverse Laplace transform gives the formal solution of the renewal function

$$r(t) = \mathscr{L}^{-1}\left\{\frac{\bar{F}(s)}{1 - \bar{f}(s)}\right\}. \qquad (4.2.118)$$

Example 4.2.18 *(Fractional Ordinary Differential Equations).* We first define a *fractional differential equation* with constant coefficients of order (n, q) as

$$\left[D^{n\alpha} + a_{n-1} D^{(n-1)\alpha} + \cdots + a_0 D^0\right] x(t) = 0, \quad t \geq 0, \qquad (4.2.119)$$

where $\alpha = \dfrac{1}{q}$. If $q = 1$, then $\alpha = 1$ and this equation is simply an ordinary differential equation of order n. Symbolically, we write (4.2.119) as

$$f(D^\alpha) x(t) = 0, \qquad (4.2.120)$$

where $f(D^\alpha)$ is a fractional differential operator.

We next use the Laplace transform method to solve a simple fractional differential equation with constant coefficients of order (2, 2) in the form

$$f\left(D^{\frac{1}{2}}\right)x(t) = \left(D^1 + a_1 D^{\frac{1}{2}} + a_0 D^0\right)x(t) = 0. \qquad (4.2.121)$$

Application of the Laplace transform to this equation gives

$$[s\,\bar{x}(s) - x(0)] + a_1\left[\sqrt{s}\,\bar{x}(s) - D^{-\frac{1}{2}}x(0)\right] + a_0\,\bar{x}(s) = 0.$$

Or

$$\bar{x}(s) = \frac{x(0) + a_1 D^{-\frac{1}{2}}x(0)}{\left(s + a_1\sqrt{s} + a_0\right)} = \frac{A}{f(\sqrt{s})}, \qquad (4.2.122)$$

where $f(x) = x^2 + a_1 x + a_0$ is an associated *indicial equation* and A is assumed to be a non-zero finite constant defined by

$$A = x(0) + a_1 D^{-\frac{1}{2}}\{x(0)\}. \qquad (4.2.123)$$

We next write the following partial fractions for the right hand side of (4.2.122) so that

$$\bar{x}(s) = \frac{A}{a-b}\left(\frac{1}{\sqrt{s}-a} - \frac{1}{\sqrt{s}-b}\right)$$

$$= \frac{A}{a-b}\left(\frac{\sqrt{s}}{s-a^2} + \frac{a}{s-a^2} - \frac{\sqrt{s}}{s-b^2} - \frac{b}{s-b^2}\right), \qquad (4.2.124)$$

where a and b are two distinct roots of $f(x) = 0$.

Using the inverse Laplace transform formula (3.9.11) with $\alpha = -\frac{1}{2}$ and $\alpha = 0$, we invert (4.2.124) to obtain the formal solution

$$x(t) = \frac{A}{a-b}\left[E\left(t, -\frac{1}{2}, a^2\right) + a E(t, 0, a^2) - b\left(t, -\frac{1}{2}, b^2\right) - E(t, 0, b^2)\right]. \qquad (4.2.125)$$

For equal roots $(a = b)$ of $f(x) = 0$, we find

$$\bar{x}(s) = \frac{A}{\left(\sqrt{s}-a\right)^2} = A\left[\frac{\sqrt{s}}{\left(s-a^2\right)} + \frac{a}{\left(s-a^2\right)}\right]^2. \qquad (4.2.126)$$

In view of the result

$$\mathcal{L}^{-1}\left\{\frac{1}{s^\alpha(s-a)^2}\right\} = t\,E(t, \alpha, a) - \alpha\,E(t, \alpha+1, a), \qquad (4.2.127)$$

the inverse Laplace transform of (4.2.126) gives the solution as

$$x(t) = A\left[\left(1 + 2a^2 t\right) E(t, 0, a^2) + a E\left(t, \frac{1}{2}, a^2\right) + 2at\,E\left(t, -\frac{1}{2}, a^2\right)\right]. \qquad (4.2.128)$$

This example is taken from a recent book by Miller and Ross (1993). The reader is referred to this book for more information on fractional differential equations.

4.3 Partial Differential Equations, Initial and Boundary Value Problems

The Laplace transform method is very useful in solving a variety of partial differential equations with assigned initial and boundary conditions. The following examples illustrate the use of the Laplace transform method.

Example 4.3.1 (*First-Order Initial-Boundary Value Problem*). Solve the equation
$$u_t + x u_x = x, \quad x > 0, \quad t > 0, \quad (4.3.1)$$
with the initial and boundary conditions
$$u(x,0) = 0 \quad \text{for } x > 0, \quad (4.3.2)$$
$$u(0,t) = 0 \quad \text{for } t > 0. \quad (4.3.3)$$
We apply the Laplace transform of $u(x,t)$ with respect to t to obtain
$$s\bar{u}(x,s) + x\frac{d\bar{u}}{dx} = \frac{x}{s}, \quad \bar{u}(0,s) = 0.$$
Using the integrating factor x^s, the solution of this transformed equation is
$$\bar{u}(x,s) = A x^{-s} + \frac{x}{s(s+1)},$$
where A is a constant of integration. Since $\bar{u}(0,s) = 0$, $A = 0$ for a bounded solution. Consequently,
$$\bar{u}(x,s) = \frac{x}{s(s+1)} = x\left(\frac{1}{s} - \frac{1}{s+1}\right).$$
The inverse Laplace transform gives the solution
$$u(x,t) = x(1 - e^{-t}). \quad (4.3.4)$$

Example 4.3.2 Find the solution of the equation
$$x u_t + u_x = x, \quad x > 0, \quad t > 0 \quad (4.3.5)$$
with the same initial and boundary conditions (4.3.2) and (4.3.3).

Application of the Laplace transform with respect to t to (4.3.5) with the initial condition gives
$$\frac{d\bar{u}}{dx} + xs\bar{u} = \frac{x}{s}.$$
Using the integrating factor $\exp\left(\frac{1}{2}x^2 s\right)$ gives the solution
$$\bar{u}(x,s) = \frac{1}{s^2} + A\exp\left(-\frac{1}{2}sx^2\right),$$
where A is an integrating constant. Since $\bar{u}(0,s) = 0$, $A = -\frac{1}{s^2}$ and hence the solution is
$$\bar{u}(x,s) = \frac{1}{s^2}\left[1 - \exp\left(-\frac{1}{2}x^2 s\right)\right]. \quad (4.3.6)$$

Finally, we obtain the solution by inversion
$$u(x,t) = t - \left(t - \frac{1}{2}x^2\right) H\left(t - \frac{x^2}{2}\right). \tag{4.3.7}$$

Or equivalently
$$u(x,t) = \begin{cases} t, & 2t < x^2 \\ \frac{1}{2}x^2, & 2t > x^2 \end{cases}. \tag{4.3.8}$$

Example 4.3.3 (*The Heat Conduction Equation in a Semi-Infinite Medium and Fractional Derivatives*). Solve the equation
$$u_t = \kappa u_{xx}, \qquad x > 0, \quad t > 0 \tag{4.3.9}$$
with the initial and boundary conditions
$$u(x,0) = 0, \qquad x > 0 \tag{4.3.10}$$
$$u(0,t) = f(t), \qquad t > 0 \tag{4.3.11}$$
$$u(x,t) \to 0 \qquad \text{as } x \to \infty, \quad t > 0. \tag{4.3.12}$$

Application of the Laplace transform with respect to t to (4.3.9) gives
$$\frac{d^2 \bar{u}}{dx^2} - \frac{\kappa}{s} \bar{u} = 0. \tag{4.3.13}$$

The general solution of this equation is
$$\bar{u}(x,s) = A \exp\left(-x\sqrt{\frac{s}{\kappa}}\right) + B \exp\left(x\sqrt{\frac{s}{\kappa}}\right), \tag{4.3.14}$$
where A and B are integrating constants. For bounded solutions, $B \equiv 0$, and using $\bar{u}(0,s) = \bar{f}(s)$, we obtain the solution
$$\bar{u}(x,s) = \bar{f}(s) \exp\left(-x\sqrt{\frac{s}{\kappa}}\right). \tag{4.3.15}$$

The inversion theorem gives the solution
$$u(x,t) = \frac{x}{2\sqrt{\pi\kappa}} \int_0^t f(t-\tau)\, \tau^{-3/2} \exp\left(-\frac{x^2}{4\kappa\tau}\right) d\tau, \tag{4.3.16}$$

which is, by putting $\lambda = \frac{x}{2\sqrt{\kappa\tau}}$ or $d\lambda = -\frac{x}{4\sqrt{\kappa}} \tau^{-3/2} d\tau$,
$$= \frac{2}{\sqrt{\pi}} \int_{\frac{x}{2\sqrt{\kappa t}}}^{\infty} f\left(t - \frac{x^2}{4\kappa\lambda^2}\right) e^{-\lambda^2} d\lambda. \tag{4.3.17}$$

This is the formal solution of the problem.

In particular, if $f(t) = T_0 = $ constant, solution (4.3.17) becomes
$$u(x,t) = \frac{2T_0}{\sqrt{\pi}} \int_{\frac{x}{2\sqrt{\kappa t}}}^{\infty} e^{-\lambda^2} d\lambda = T_0 \, \mathrm{erfc}\left(\frac{x}{2\sqrt{\kappa t}}\right). \tag{4.3.18}$$

Clearly, the temperature distribution tends asymptotically to the constant value T_0 as $t \to \infty$.

We consider another physical problem that is concerned with the determination of the temperature distribution in a semi-infinite solid when the rate of flow of heat is prescribed at the end $x = 0$. Thus the problem is to solve diffusion equation (4.3.9) subject to conditions (4.3.10) and (4.3.12)

$$-k\left(\frac{\partial u}{\partial x}\right) = g(t) \quad \text{at } x = 0, \quad t > 0, \tag{4.3.19}$$

where k is a constant that is called *thermal conductivity*.

Application of the Laplace transform gives the solution of the transformed problem

$$\bar{u}(x,s) = \frac{1}{k}\sqrt{\frac{\kappa}{s}}\ \bar{g}(s)\exp\left(-x\sqrt{\frac{s}{\kappa}}\right). \tag{4.3.20}$$

The inverse Laplace transform yields the solution

$$u(x,t) = \frac{1}{k}\sqrt{\frac{\kappa}{\pi}}\int_0^t g(t-\tau)\,\tau^{-\frac{1}{2}}\exp\left(-\frac{x^2}{4\kappa\tau}\right)d\tau, \tag{4.3.21}$$

which is, by the change of variable $\lambda = \dfrac{x}{2\sqrt{\kappa\tau}}$,

$$= \frac{x}{k\sqrt{\pi}}\int_{\frac{x}{2\sqrt{\kappa t}}}^{\infty} g\left(t - \frac{x^2}{4\kappa\lambda^2}\right)\lambda^{-2}e^{-\lambda^2}d\lambda. \tag{4.3.22}$$

In particular, if $g(t) = T_0 = $ constant, the solution becomes

$$u(x,t) = \frac{T_0 x}{k\sqrt{\pi}}\int_{\frac{x}{2\sqrt{\kappa t}}}^{\infty} \lambda^{-2}e^{-\lambda^2}d\lambda.$$

Integrating this result by parts gives

$$u(x,t) = \frac{T_0}{\kappa}\left[2\sqrt{\frac{\kappa t}{\pi}}\exp\left(-\frac{x^2}{4\kappa t}\right) - x\,\mathrm{erfc}\left(\frac{x}{2\sqrt{\kappa t}}\right)\right]. \tag{4.3.23}$$

Alternatively, the heat conduction problem (4.3.9)-(4.3.12) can be solved by using fractional derivatives (see Debnath, 1978). We recall (4.3.15) and rewrite it

$$\frac{\partial \bar{u}}{\partial x} = -\sqrt{\frac{s}{\kappa}}\,\bar{u}. \tag{4.3.24}$$

In view of (3.9.21), this can be expressed in terms of fractional derivative of order $\dfrac{1}{2}$ as

$$\frac{\partial u}{\partial x} = -\frac{1}{\sqrt{\kappa}}\,\mathscr{L}^{-1}\left\{\sqrt{s}\,\bar{u}(x,s)\right\} = -\frac{1}{\sqrt{\kappa}}\,_0D_t^{\frac{1}{2}} u(x,t). \tag{4.3.25}$$

Thus the heat flux is expressed in terms of the fractional derivative. In particular, when $u(0,t) = $ constant $= T_0$, then the heat flux at the surface is

$$-k\left(\frac{\partial u}{\partial x}\right)_{x=0} = \frac{k}{\sqrt{\kappa}} D_t^{\frac{1}{2}} T_0 = \frac{kT_0}{\sqrt{\pi\kappa\tau}}. \tag{4.3.26}$$

Example 4.3.4 (*Diffusion Equation in a Finite Medium*). Solve the diffusion equation

$$u_t = \kappa u_{xx}, \quad 0 < x < a, \quad t > 0, \tag{4.3.27}$$

with the initial and boundary conditions

$$u(x,0) = 0, \quad 0 < x < a, \tag{4.3.28}$$
$$u(0,t) = U, \quad t > 0, \tag{4.3.29}$$
$$u_x(a,t) = 0, \quad t > 0, \tag{4.3.30}$$

where U is a constant.

We introduce the Laplace transform of $u(x,t)$ with respect to t to obtain

$$\frac{d^2\bar{u}}{dx^2} - \frac{s}{\kappa}\bar{u} = 0, \quad 0 < x < a, \tag{4.3.31}$$

$$\bar{u}(0,s) = \frac{U}{s}, \quad \left(\frac{d\bar{u}}{dx}\right)_{x=a} = 0. \tag{4.3.32ab}$$

The general solution of (4.3.31) is

$$\bar{u}(x,s) = A \cosh\left(x\sqrt{\frac{s}{\kappa}}\right) + B \sinh\left(x\sqrt{\frac{s}{\kappa}}\right), \tag{4.3.33}$$

where A and B are constants of integration. Using (4.3.32ab), we obtain the values of A and B so that the solution (4.3.33) becomes

$$\bar{u}(x,s) = \frac{U}{s} \cdot \frac{\cosh\left[(a-x)\sqrt{\frac{s}{\kappa}}\right]}{\cosh\left(a\sqrt{\frac{s}{\kappa}}\right)}. \tag{4.3.34}$$

The inverse Laplace transform gives the solution

$$u(x,t) = U \mathcal{L}^{-1}\left\{\frac{\cosh(a-x)\sqrt{\frac{s}{\kappa}}}{s \cosh\sqrt{a\frac{s}{\kappa}}}\right\}. \tag{4.3.35}$$

The inversion can be carried out by the Cauchy Residue Theorem to obtain the solution

$$u(x,t) = U\left[1 + \frac{4}{\pi}\sum_{n=1}^{\infty}\frac{(-1)^n}{(2n-1)}\cos\left\{\frac{(2n-1)(a-x)\pi}{2a}\right\}\right.$$
$$\left. \times \exp\left\{-(2n-1)^2\left(\frac{\pi}{2a}\right)^2 \kappa t\right\}\right], \tag{4.3.36}$$

which is, by expanding the cosine term,

$$= U\left[1 - \frac{4}{\pi}\sum_{n=1}^{\infty}\frac{1}{(2n-1)}\sin\left\{\left(\frac{2n-1}{2a}\right)\pi x\right\}\right.$$
$$\left.\times \exp\left\{-(2n-1)^2\left(\frac{\pi}{2a}\right)^2 \kappa t\right\}\right]. \quad (4.3.37)$$

This result can be obtained by solving the problem by the method of separation of variables.

Example 4.3.5 (*Diffusion in a Finite Medium*). Solve the one-dimensional diffusion equation in a finite medium $0 < z < a$ where the concentration function $C(z,t)$ satisfies the equation
$$C_t = \kappa C_{zz}, \quad 0 < z < a, \quad t > 0, \quad (4.3.38)$$
and the initial and boundary data
$$C(z,0) = 0 \quad \text{for } 0 < z < a, \quad (4.3.39)$$
$$C(z,t) = C_0 \quad \text{for } z = a, \quad t > 0, \quad (4.3.40)$$
$$\frac{\partial C}{\partial z} = 0 \quad \text{for } z = 0, \quad t > 0, \quad (4.3.41)$$
where C_0 is a constant.

Application of the Laplace transform of $C(z,t)$ with respect to t gives
$$\frac{d^2\overline{C}}{dz^2} - \left(\frac{s}{\kappa}\right)\overline{C} = 0, \quad 0 < z < a,$$
$$\overline{C}(a,s) = \frac{C_0}{s}, \quad \left(\frac{d\overline{C}}{dz}\right)_{z=0} = 0.$$

The solution of this system is
$$\overline{C}(z,s) = \frac{C_0 \cosh\left(z\sqrt{\frac{s}{\kappa}}\right)}{s \cosh\left(a\sqrt{\frac{s}{\kappa}}\right)}, \quad (4.3.42)$$

which is, by writing $\alpha = \sqrt{\frac{s}{\kappa}}$,
$$= \frac{C_0}{s}\frac{\left(e^{\alpha z} + e^{-\alpha z}\right)}{\left(e^{\alpha a} + e^{-\alpha a}\right)}$$
$$= \frac{C_0}{s}\left[\exp\{-\alpha(a-z)\} + \exp\{-\alpha(a+z)\}\right]\sum_{n=0}^{\infty}(-1)^n \exp(-2n\alpha a)$$
$$= \frac{C_0}{s}\left\{\sum_{n=0}^{\infty}(-1)^n \exp\left[-\alpha\{(2n+1)a - z\}\right]\right.$$
$$\left. + \sum_{n=0}^{\infty}(-1)^n \exp\left[-\alpha\{(2n+1)a + z\}\right]\right\}. \quad (4.3.43)$$

Using the result (3.7.4), we obtain the final solution
$$C(z,t) = C_0 \left\{ \sum_{n=0}^{\infty} (-1)^n \left[erfc\left\{ \frac{(2n+1)a-z}{2\sqrt{\kappa t}} \right\} + erfc\left\{ \frac{(2n+1)a+z}{2\sqrt{\kappa t}} \right\} \right] \right\}. \quad (4.3.44)$$

This solution represents an infinite series of complementary error functions. The successive terms of this series are in fact the concentrations at depths $a-z$, $a+z$, $3a-z$, $3a+z$,... in the medium. The series converges rapidly for all except large values of $\left(\frac{\kappa t}{a^2} \right)$.

Example 4.3.6 (*The Wave Equation for the Transverse Vibration of a Semi-Infinite String*). Find the displacement of a semi-infinite string which is initially at rest in its equilibrium position. At time $t=0$, the end $x=0$ is constrained to move so that the displacement is $u(0,t) = Af(t)$ for $t \geq 0$. The problem is to solve the one-dimensional wave equation
$$u_{tt} = c^2 u_{xx}, \quad 0 \leq x < \infty, \quad t > 0, \quad (4.3.45)$$
with the boundary and initial conditions
$$u(x,t) = Af(t) \quad \text{at } x = 0, \ t \geq 0, \quad (4.3.46)$$
$$u(x,t) \to 0 \quad \text{as } x \to \infty, \ t \geq 0, \quad (4.3.47)$$
$$u(x,t) = 0 = \frac{\partial u}{\partial t} \quad \text{at } t = 0 \quad \text{for } 0 < x < \infty. \quad (4.3.48ab)$$

Application of the Laplace transform of $u(x,t)$ with respect to t gives
$$\frac{d^2 \bar{u}}{dx^2} - \frac{s^2}{c^2} \bar{u} = 0, \quad \text{for } 0 \leq x < \infty,$$
$$\bar{u}(x,s) = A\bar{f}(s) \quad \text{at } x = 0,$$
$$\bar{u}(x,s) \to 0 \quad \text{as } x \to \infty.$$

The solution of this differential system is
$$\bar{u}(x,s) = A\bar{f}(s) \exp\left(-\frac{xs}{c}\right). \quad (4.3.49)$$

Inversion gives the solution
$$u(x,t) = Af\left(t - \frac{x}{c}\right) H\left(t - \frac{x}{c}\right). \quad (4.3.50)$$

In other words, the solution is
$$u(x,t) = \begin{bmatrix} Af\left(t - \frac{x}{c}\right), & t > \frac{x}{c} \\ 0, & t < \frac{x}{c} \end{bmatrix}. \quad (4.3.51)$$

This solution represents a wave propagating at a velocity c with the characteristic $x = ct$.

Example 4.3.7 (*Potential and Current in an Electric Transmission Line*). We consider a transmission line which is a model of co-axial cable containing resistance R, inductance L, capacitance C, and leakage conductance G. The current $I(x,t)$ and potential $V(x,t)$ at a point x and time t in the line satisfy the coupled equations

$$L\frac{\partial I}{\partial t} + RI = -\frac{\partial V}{\partial x}, \qquad (4.3.52)$$

$$C\frac{\partial V}{\partial t} + GV = -\frac{\partial I}{\partial x}. \qquad (4.3.53)$$

If I or V is eliminated from these equations, both I and V satisfy the same equation in the form

$$\frac{1}{c^2}u_{tt} - u_{xx} + au_t + bu = 0 \qquad (4.3.54)$$

where $c^2 = (LC)^{-1}$, $a = LG + RC$, and $b = RG$. Equation (4.3.54) is called the *telegraph equation*.

For a lossless transmission line, $R = 0$ and $G = 0$, I or V satisfies the classical wave equation

$$u_{tt} = c^2 u_{xx}. \qquad (4.3.55)$$

The solution of this equation with the initial and boundary data is obtained from Example 4.3.6 using the boundary conditions in the potential $V(x,t)$:

(i) $\qquad V(x,t) = V_0 f(t) \qquad$ at $x = 0$, $t > 0$. $\qquad (4.3.56)$

This corresponds to a signal at the end $x = 0$ for $t > 0$, and $V(x,t) \to 0$ as $x \to \infty$ for $t > 0$.

A special case, $f(t) = H(t)$, is also of interest. The solution for this special case is given by

$$V(x,t) = V_0 f\left(t - \frac{x}{c}\right) H\left(t - \frac{x}{c}\right). \qquad (4.3.57)$$

This represents a wave propagating at a speed c with the characteristic $x = ct$.

Similarly, the solution associated with the boundary data

(ii) $\qquad V(x,t) = V_0 \cos \omega t \qquad$ at $x = 0$ for $t > 0 \qquad (4.3.58)$

$\qquad\qquad V(x,t) \to 0 \qquad$ as $x \to \infty$ for $t > 0 \qquad (4.3.59)$

can readily be obtained from Example 4.3.6.

For ideal submarine cable (or the *Kelvin ideal cable*), $L = 0$ and $G = 0$ equation (4.3.54) reduces to the classical diffusion equation

$$u_t = \kappa u_{xx}, \qquad (4.3.60)$$

where $\kappa = a^{-1} = (RC)^{-1}$.

The method of solution is similar to that discussed in Example 4.3.3. Using the boundary data (i), the solution for the potential $V(x,t)$ is given by

$$V(x,t) = V_0 \operatorname{erfc}\left(\frac{x}{2\sqrt{\kappa t}}\right). \qquad (4.3.61)$$

The current field is given by

$$I(x,t) = -\frac{1}{R}\left(\frac{\partial V}{\partial x}\right) = \frac{V_0}{R}(\pi \kappa t)^{-1/2} \exp\left(-\frac{x^2}{4\kappa t}\right). \qquad (4.3.62)$$

For very large x, the asymptotic representation of the complementary error function is

$$erfc(x) \sim \frac{1}{x\sqrt{\pi}} \exp(-x^2), \qquad x \to \infty. \qquad (4.3.63)$$

In view of this asymptotic representation, (4.3.61) becomes

$$V(x,t) \sim \frac{2V_0}{x} \left(\frac{\kappa t}{\pi}\right)^{1/2} \exp\left(-\frac{x^2}{4\kappa t}\right). \qquad (4.3.64)$$

For any $t > 0$, no matter how small, solution (4.3.61) reveals that $V(x,t) > 0$ for all $x > 0$, even though $V(x,t) \to 0$ as $x \to \infty$. Thus, the signal applied at $t = 0$ propagates with the infinite speed although its amplitude is very small for large x. Physically, the infinite speed is unrealistic and is essentially caused by the neglect of the first term in equation (4.3.54). In a real cable, the presence of some inductance would set a limit to the speed of propagation.

Instead of the Kelvin cable, a non-inductive leaky cable $(L = 0$ and $G \neq 0)$ is of interest. The equation for this case is obtained from (4.3.54) in the form

$$V_{xx} - a\, V_t - b\, V = 0, \qquad (4.3.65)$$

with zero initial conditions, and with the boundary data

$$V(0,t) = H(t) \text{ and } V(x,t) \to 0 \text{ as } x \to \infty. \qquad (4.3.66ab)$$

The Laplace transformed problem is

$$\frac{d^2 \overline{V}}{dx^2} = (sa+b)\overline{V}, \qquad (4.3.67)$$

$$\overline{V}(0,s) = \frac{1}{s}, \qquad \overline{V}(x,s) \to 0 \quad \text{as } x \to \infty. \qquad (4.3.68ab)$$

Thus the solution is given by

$$\overline{V}(x,s) = \frac{1}{s} \exp\left[-x(sa+b)^{1/2}\right]. \qquad (4.3.69)$$

With the aid of a standard table of the inverse Laplace transform, the solution is given by

$$V(x,t) = \frac{1}{2} e^{x\sqrt{b}} erfc\left\{\frac{x}{2}\sqrt{\frac{a}{t}} + \sqrt{\frac{bt}{a}}\right\} + \frac{1}{2} e^{-x\sqrt{b}} erfc\left\{\frac{x}{2}\sqrt{\frac{a}{t}} - \sqrt{\frac{bt}{a}}\right\}. \qquad (4.3.70)$$

When $G = 0$ $(b = 0)$, the solution becomes identical with (4.3.61).

For the Heaviside distortionless cable, $\frac{R}{L} = \frac{G}{C} = k = $ constant, the potential $V(x,t)$ and the current $I(x,t)$ satisfies the same equation

$$u_{tt} + 2k u_t + k^2 u = c^2 u_{xx}, \qquad 0 \leq x < \infty, \quad t > 0. \qquad (4.3.71)$$

We solve this equation with the initial data (4.3.48ab) and the boundary condition (4.3.56). Application of the Laplace transform with respect to t to (4.3.71) gives

$$\frac{d^2 \overline{V}}{dx^2} = \left(\frac{s+k}{c}\right)^2 \overline{V}. \qquad (4.3.72)$$

The solution for $\overline{V}(x,s)$ with the transformed boundary condition (4.3.56) is

$$\overline{V}(x,s) = V_0 \bar{f}(s)\exp\left[-\left(\frac{s+k}{c}\right)x\right]. \tag{4.3.73}$$

This can easily be inverted to obtain the final solution

$$V(x,t) = V_0 \exp\left(-\frac{kx}{c}\right) f\left(t-\frac{x}{c}\right) H\left(t-\frac{x}{c}\right). \tag{4.3.74}$$

This solution represents the signal that propagates with velocity $c = (LC)^{-1/2}$ with exponentially decaying amplitude, but with no distortion. Thus the signals can propagate along the Heaviside distortionless line over long distances if appropriate boosters are placed at regular intervals in order to increase the strength of the signal so as to counteract the effects of attenuation.

Example 4.3.8 Find the bounded solution of the axisymmetric heat conduction equation

$$u_t = \kappa\left(u_{rr} + \frac{1}{r} u_r\right), \quad 0 \leq r < a, \ t > 0, \tag{4.3.75}$$

with the initial and boundary data

$$u(r,0) = 0 \quad \text{for } 0 < r < a, \tag{4.3.76}$$

$$u(r,t) = f(t) \quad \text{at } r = a \text{ for } t > 0, \tag{4.3.77}$$

where κ and T_0 are constants.

Application of the Laplace transform to (4.3.75) gives

$$\frac{d^2\bar{u}}{dr^2} + \frac{1}{r}\frac{d\bar{u}}{dr} - \frac{s}{\kappa}\bar{u} = 0.$$

Or

$$r^2 \frac{d^2\bar{u}}{dr^2} + r\frac{d\bar{u}}{dr} - r^2\left(\frac{s}{\kappa}\right)\bar{u} = 0. \tag{4.3.78}$$

This is the standard Bessel equation with the solution

$$\bar{u}(r,s) = A I_0\left(r\sqrt{\frac{s}{\kappa}}\right) + B K_0\left(r\sqrt{\frac{s}{\kappa}}\right), \tag{4.3.79}$$

where A and B are constants of integration and $I_0(x)$ and $K_0(x)$ are the modified Bessel functions of zero order.

Since $K_0(\alpha r)$ is unbounded at $r = 0$, for the bounded solution $B \equiv 0$, and hence the solution is

$$\bar{u}(r,s) = A I_0(kr), \quad k = \sqrt{\frac{s}{\kappa}}.$$

In view of the transformed boundary condition $\bar{u}(a,s) = \bar{f}(s)$, we obtain the solution

$$\bar{u}(r,s) = \bar{f}(s)\frac{I_0(kr)}{I_0(ka)} = \bar{f}(s)\bar{g}(s), \tag{4.3.80}$$

where $\bar{g}(s) = \dfrac{I_0(kr)}{I_0(ka)}$.

By Convolution Theorem 3.5.1, the solution takes the form

$$u(r,t) = \int_0^t f(t-\tau)g(\tau)\,d\tau, \qquad (4.3.81)$$

where

$$g(t) = \dfrac{1}{2\pi i} \int_{c-i\infty}^{c+i\infty} e^{st} \dfrac{I_0(kr)}{I_0(ka)}\,ds. \qquad (4.3.82)$$

This complex integral can be evaluated by the theory of residues where the poles of the integrand are at the points $s = s_n = -\kappa\alpha_n^2$, $n = 1,2,3,\ldots$ and α_n are the roots of $J_0(a\alpha) = 0$. The residue at pole $s = s_n$ is

$$\left(\dfrac{2i\kappa\alpha_n}{a}\right)\dfrac{I_0(ir\alpha_n)}{I_0'(ia\alpha_n)}\exp(-\kappa t\alpha_n^2) = \left(\dfrac{2\kappa\alpha_n}{a}\right)\dfrac{J_0(r\alpha_n)}{J_1(a\alpha_n)}\exp(-\kappa t\alpha_n^2),$$

so that

$$g(t) = \left(\dfrac{2\kappa}{a}\right)\sum_{n=1}^{\infty} \dfrac{\alpha_n J_0(r\alpha_n)}{J_1(a\alpha_n)} \exp(-\kappa t\alpha_n^2).$$

Thus solution (4.3.81) becomes

$$u(r,t) = \left(\dfrac{2\kappa}{a}\right)\sum_{n=1}^{\infty} \dfrac{\alpha_n J_0(r\alpha_n)}{J_1(a\alpha_n)} \int_0^t f(t-\tau)\exp(-\kappa\tau\alpha_n^2)\,d\tau, \qquad (4.3.83)$$

where the summation is taken over the positive roots of $J_0(a\alpha) = 0$.

In particular, if $f(t) = T_0$, then the solution (4.3.83) reduces to

$$u(r,t) = \dfrac{2T_0}{a}\sum_{n=1}^{\infty} \dfrac{J_0(r\alpha_n)}{\alpha_n J_1(a\alpha_n)}\left(1 - e^{-\kappa t\alpha_n^2}\right)$$

$$= T_0\left[1 - \dfrac{2}{a}\sum_{n=1}^{\infty} \dfrac{J_0(r\alpha_n)}{\alpha_n J_1(a\alpha_n)} e^{-\kappa t\alpha_n^2}\right]. \qquad (4.3.84)$$

Example 4.3.9 (*Inhomogeneous Partial Differential Equation*). We solve the inhomogeneous problem

$$u_{xt} = -\omega \sin \omega t, \quad t > 0 \qquad (4.3.85)$$

$$u(x,0) = x, \quad u(0,t) = 0. \qquad (4.3.86\text{ab})$$

Application of the Laplace transform with respect to t gives

$$\dfrac{d\bar{u}}{dx} = \dfrac{s}{s^2 + \omega^2}$$

which has the general solution

$$\bar{u}(x,s) = \dfrac{sx}{s^2 + \omega^2} + A,$$

where A is a constant. Since $\bar{u}(0,s) = 0$, $A = 0$ and hence the solution is obtained by inversion as

$$u(x,t) = x \cos \omega t. \qquad (4.3.87)$$

Example 4.3.10 (*Inhomogeneous Wave Equation*). Find the solution of

$$\frac{1}{c^2} u_{tt} - u_{xx} = k \sin\left(\frac{\pi x}{a}\right), \quad 0 < x < a, \ t > 0, \qquad (4.3.88)$$

$$u(x,0) = 0 = u_t(x,0), \quad 0 < x < a, \qquad (4.3.89)$$

$$u(0,t) = 0 = u(a,t), \quad t > 0, \qquad (4.3.90)$$

where c, k, and a are constants.

Application of the Laplace transforms gives

$$\frac{d^2 \bar{u}}{dx^2} - \frac{s^2}{c^2} \bar{u} = -\frac{k}{s} \sin\left(\frac{\pi x}{a}\right), \qquad (4.3.91)$$

$$\bar{u}(0,s) = 0 = \bar{u}(a,s). \qquad (4.3.92)$$

The general solution of equation (4.3.91) is

$$\bar{u}(x,s) = A \exp\left(\frac{sx}{c}\right) + B \exp\left(-\frac{sx}{c}\right) + \frac{k \sin\left(\frac{\pi x}{a}\right)}{a^2 s \left(s^2 + \frac{\pi^2 c^2}{a^2}\right)}. \qquad (4.3.93)$$

In view of (4.3.92), $A = B = 0$, and hence the solution (4.3.93) becomes

$$\bar{u}(x,s) = \frac{k}{\pi^2 c^2} \sin\left(\frac{\pi x}{a}\right) \left[\frac{1}{s} - \frac{s}{s^2 + \frac{\pi^2 c^2}{a^2}}\right], \qquad (4.3.94)$$

which, by inversion, gives the solution,

$$u(x,t) = \frac{k}{(\pi c)^2} \left[1 - \cos\left(\frac{\pi c t}{a}\right)\right] \sin\left(\frac{\pi x}{a}\right). \qquad (4.3.95)$$

Example 4.3.11 (*The Stokes Problem and the Rayleigh Problem in Fluid Dynamics*). Solve the Stokes problem, which is concerned with the unsteady boundary layer flows induced in a semi-infinite viscous fluid bounded by an infinite horizontal disk at $z = 0$ due to non-torsional oscillations of the disk in its own plane with a given frequency ω.

We solve the boundary layer equation in fluid dynamics

$$u_t = v u_{zz}, \quad z > 0, \ t > 0, \qquad (4.3.96)$$

with the boundary and initial conditions

$$u(z,t) = U_0 e^{i\omega t} \quad \text{on } z = 0, \ t > 0, \qquad (4.3.97)$$

$$u(z,t) \to 0 \quad \text{as } z \to \infty, \ t > 0, \qquad (4.3.98)$$

$$u(z,0) = 0 \quad \text{at } t \leq 0 \text{ for all } z > 0, \qquad (4.3.99)$$

where $u(z,t)$ is the velocity of fluid of kinematic viscosity v and U_0 is a constant.

The Laplace transform solution of the problem with the transformed boundary conditions is

$$\bar{u}(z,s) = \frac{U_0}{(s-i\omega)} \exp\left(-z\sqrt{\frac{s}{v}}\right). \qquad (4.3.100)$$

Using a standard table of inverse Laplace transforms, we obtain the solution

$$u(z,t) = \frac{U_0}{2} e^{i\omega t}\left[\exp(-\lambda z) \, erfc\left(\zeta - \sqrt{i\omega t}\right) + \exp(\lambda z) \, erfc\left(\zeta + \sqrt{i\omega t}\right)\right], \quad (4.3.101)$$

where $\zeta = z/\left(2\sqrt{vt}\right)$ is called the *similarity variable* of the viscous boundary layer theory and $\lambda = (i\omega/v)^{\frac{1}{2}}$. The result (4.3.101) describes the unsteady boundary layer flow.

In view of the asymptotic formula for the complementary error function

$$erfc\left(\zeta \mp \sqrt{i\omega t}\right) \sim (2,0) \quad \text{as} \quad t \to \infty \qquad (4.3.102)$$

the above solution for $u(z,t)$ has the asymptotic representation

$$u(z,t) \sim U_0 \exp(i\omega t - \lambda z) = U_0 \exp\left[i\omega t - \left(\frac{\omega}{2v}\right)^{\frac{1}{2}}(1+i)z\right]. \qquad (4.3.103)$$

This is called the *Stokes steady-state solution*. This represents the propagation of shear waves which spread out from the oscillating disk with velocity $(\omega/k) = \sqrt{2v\omega}$ and exponentially decaying amplitude. The boundary layer associated with the solution has thickness of the order $(v/\omega)^{\frac{1}{2}}$ in which the shear oscillations imposed by the disk decay exponentially with distance z from the disk. This boundary layer is called the *Stokes layer*. In other words, the thickness of the Stokes layer is equal to the depth of penetration of vorticity which is essentially confined to the immediate vicinity of the disk for high frequency ω.

The Stokes problem with $\omega = 0$ becomes the *Rayleigh problem*. In other words, the motion is generated in the fluid from rest by moving the disk impulsively in its own plane with constant velocity U_0. In this case, the Laplace transformed solution is

$$\bar{u}(z,s) = \frac{U_0}{s} \exp\left(-z\sqrt{\frac{s}{v}}\right). \qquad (4.3.104)$$

Hence the inversion gives the Rayleigh solution

$$u(z,t) = U_0 \, erfc\left(\frac{z}{2\sqrt{vt}}\right). \qquad (4.3.105)$$

This describes the growth of a boundary layer adjacent to the disk. The associated boundary layer is called the *Rayleigh layer* of thickness of the order $\delta \sim \sqrt{vt}$ which grows with increasing time. The rate of growth is of the order $d\delta/dt \sim \sqrt{v/t}$, which diminishes with increasing time.

The vorticity of the unsteady flow is given by

$$\frac{\partial u}{\partial z} = \frac{U_0}{\sqrt{\pi v t}} \exp(-\zeta^2), \qquad (4.3.106)$$

which decays exponentially to zero as $z \gg \delta$.

Note that the vorticity is everywhere zero at $t = 0$. This implies that it is generated at the disk and diffuses outward within the Rayleigh layer. The total viscous diffusion time is $T_d \sim \delta^2/v$.

Another physical quantity related to the Stokes and Rayleigh problems is the *skin friction* on the disk defined by

$$\tau_0 = \mu \left(\frac{\partial u}{\partial z}\right)_{z=0}, \qquad (4.3.107)$$

where $\mu = v\rho$ is the dynamic viscosity and ρ is the density of the fluid. The skin friction can readily be calculated from the flow field given by (4.3.103) or (4.3.105).

4.4 Solutions of Integral Equations.

4.4.1 Definition. An equation in which the unknown function occurs under an integral is called an *integral equation*.

An equation of the form

$$f(t) = h(t) + \lambda \int_a^b k(t, \tau) f(\tau) d\tau, \qquad (4.4.1)$$

in which f is the unknown function, $h(t)$, $k(t, \tau)$; and the limits of integration a and b are known; and λ is a constant, is called the *linear integral equation* of *the second kind* or *the linear Volterra integral equation*. The function $k(t, \tau)$ is called the *kernel* of the equation. Such an equation is said to be *homogeneous* or *inhomogeneous* according as $h(t) = 0$ or $h(t) \neq 0$. If the kernel of the equation has the form $k(t, \tau) = g(t - \tau)$, the equation is referred to as the *convolution integral equation*.

In this section, we show how the Laplace transform method can be applied successfully to solve the convolution integral equations. This method is simple and straightforward, and can be illustrated by examples.

To solve the convolution equation of the form

$$f(t) = h(t) + \lambda \int_0^t g(t - \tau) f(\tau) d\tau, \qquad (4.4.2)$$

we take the Laplace transform of this equation to obtain

$$\bar{f}(s) = \bar{h}(s) + \lambda \mathscr{L}\left\{\int_0^t g(t - \tau) f(\tau) d\tau\right\},$$

which is, by the Convolution Theorem,

$$\bar{f}(s) = \bar{h}(s) + \lambda \bar{f}(s) \bar{g}(s).$$

Or

$$\bar{f}(s) = \frac{\bar{h}(s)}{1 - \lambda \bar{g}(s)}. \tag{4.4.3}$$

Inversion gives the formal solution

$$f(t) = \mathscr{L}^{-1}\left\{\frac{\bar{h}(s)}{1 - \lambda \bar{g}(s)}\right\}. \tag{4.4.4}$$

In many simple cases, the right hand side can be inverted by using partial fractions or the theory of residues. Hence the solution can readily be found.

Example 4.4.1 Solve the integral equation

$$f(t) = a + \lambda \int_0^t f(\tau) d\tau. \tag{4.4.5}$$

We take the Laplace transform of (4.4.5) to find

$$\bar{f}(s) = \frac{a}{s - \lambda},$$

whence, by inversion, it follows that

$$f(t) = a \exp(\lambda t). \tag{4.4.6}$$

Example 4.4.2 Solve the equation

$$f(t) = a \sin t + 2 \int_0^t f'(\tau) \sin(t - \tau) d\tau, \quad f(0) = 0. \tag{4.4.7}$$

Taking the Laplace transform, we obtain

$$\bar{f}(s) = \frac{a}{s^2 + 1} + 2 \mathscr{L}\{f'(t)\} \mathscr{L}\{\sin t\}$$

Or

$$\bar{f}(s) = \frac{a}{s^2 + 1} + 2 \frac{\{s\bar{f}(s) - f(0)\}}{s^2 + 1}.$$

Hence, by the initial condition,

$$\bar{f}(s) = \frac{a}{(s - 1)^2}.$$

Inversion yields the solution

$$f(t) = at \exp(t). \tag{4.4.8}$$

Example 4.4.3 Solve the *Abel integral equation*

$$g(t) = \int_0^t f'(t)(t - \tau)^{-\alpha} d\tau, \quad 0 < \alpha < 1. \tag{4.4.9}$$

Application of the Laplace transform gives

$$\bar{g}(s) = \mathscr{L}\{f'(t)\} \mathscr{L}\{t^{-\alpha}\}.$$

Or

$$\bar{f}(s) = \frac{f(0)}{s} + \frac{\bar{g}(s)}{\Gamma(1 - \alpha)} s^{-\alpha}.$$

Inverting, we find

$$f(t) = f(0) + \frac{1}{\Gamma(\alpha)\Gamma(1-\alpha)} \int_0^t g(\tau)(t-\tau)^{\alpha-1} d\tau. \qquad (4.4.10)$$

This is the required solution of Abel's equation.

Example 4.4.4 (*Abel's Problem of Tautochronous Motion*). The problem is to determine the form of a frictionless plane curve through the origin in a vertical plane along which a particle of mass m can fall in a time which does not depend on the initial position.

Suppose the particle is placed on a curve at the point (ξ, η), where η is measured positive upward. Let the particle be allowed to fall to the origin under the action of gravity. Suppose (x, y) is any position of the particle during its descent as shown in Figure 4.7. According to the principle of conservation of energy, the sum of the kinetic and potential energies is constant, that is,

$$\frac{1}{2}mv^2 + mgy = mg\eta = \text{constant}. \qquad (4.4.11)$$

This gives the velocity of the particle at any position (x, y)

$$v^2 = \left(\frac{ds}{dt}\right)^2 = 2g(\eta - y), \qquad (4.4.12)$$

where s is the length of the arc of the curve measured from the origin and t is the time.

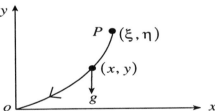

Figure 4.7 Abel's problem.

Integrating (4.4.12) from $y = \eta$ to $\eta = 0$ gives

$$\int_0^T dt = -\frac{1}{\sqrt{2g}} \int_\eta^0 \frac{ds}{\sqrt{\eta - y}} = \frac{1}{\sqrt{2g}} \int_0^\eta \frac{f'(y) dy}{\sqrt{\eta - y}},$$

where $s = f(y)$ represents the equation of the curve with $f(0) = 0$. Thus we obtain

$$\sqrt{2g}\, T = \int_0^\eta \frac{f'(y) dy}{\sqrt{\eta - y}}. \qquad (4.4.13)$$

This is the Abel integral equation (4.4.9) with $\alpha = \frac{1}{2}$ and $g(\eta) = T\sqrt{2g} = \text{constant}$. Thus the solution (4.4.10) becomes

$$f(y) = \frac{T\sqrt{2g}}{\pi} \int_0^\eta (\eta-y)^{\frac{1}{2}-1} dy,$$

which is, putting $\eta = y\sin^2\theta$ and $a = \frac{gT^2}{\pi^2}$,

$$f(y) = 2\sqrt{2ay}. \tag{4.4.14}$$

If ψ is the angle made by the tangent to the curve at a point (x,y), then $\frac{dy}{ds} = \sin\psi$ and $\frac{dx}{ds} = \cos\psi$ so that

$$\operatorname{cosec}\psi = \frac{ds}{dy} = f'(y) = \sqrt{\frac{2a}{y}}.$$

Or

$$s = f(y) = 2\sqrt{2ay} = 4a\sin\psi.$$

This is the equation of the curve and represents the cycloid with the vertex at the origin and the tangent at the vertex as the x-axis.

Alternatively, equation (4.4.13) can be expressed in terms of a fractional integral defined by (3.9.1) as

$$T\sqrt{2g} = \Gamma\left(\frac{1}{2}\right) D_\eta^{-\frac{1}{2}} f'(\eta). \tag{4.4.15}$$

We multiply this result by $D^{1/2}$ and use (3.9.26) to obtain

$$f'(\eta) = T\sqrt{\frac{2g}{\pi}} D_\eta^{1/2} \cdot 1 = T\sqrt{\frac{2g}{\pi}} \frac{1}{\sqrt{\pi\eta}} = \sqrt{\frac{2a}{\eta}}. \tag{4.4.16}$$

Or

$$f(\eta) = 2\sqrt{2a\eta}. \tag{4.4.17}$$

This is identical with (4.4.14) when $\eta = y$. However, solution (4.4.17) is obtained by using fractional derivatives. This is a simple application of fractional derivatives.

Example 4.4.5 (*Abel's Equation and Fractional Derivatives in a Problem of Fluid Flow*). We consider the flow of water along the x direction through a symmetric dam in a vertical yz plane with the y-axis along the face of the dam. The problem is to determine the form $y = f(z)$ of the opening of the dam so that the quantity of water per unit time is proportional to a given power of the depth of the stream. It follows from Bernoulli's equation of fluid mechanics that the fluid velocity v at a given height z above the base of the dam is given by

$$v^2 = 2g(h-z). \tag{4.4.18}$$

The volume flux dQ through an elementary cross section dA of the opening of the dam is $dQ = vdA = 2\sqrt{2g}\,(h-z)^{1/2} f(z)dz$ so that the total volume flux is

$$Q(h) = a\int_0^h (h-z)^{\frac{1}{2}} f(z)dz, \quad a = 2\sqrt{2g}. \tag{4.4.19}$$

This is the *Abel integral equation* and hence can be written as the fractional integral

$$Q(h) = a\Gamma\left(\frac{3}{2}\right)D_h^{-3/2}f(h). \qquad (4.4.20)$$

Multiplying this result by $D_h^{3/2}$ with a given $Q(h) = h^\beta$, we obtain, by (3.9.25),

$$f(h) = \frac{1}{a\Gamma\left(\frac{3}{2}\right)} D_h^{3/2} h^\beta = \frac{2a}{\sqrt{\pi}} \frac{\Gamma(\beta+1)}{\Gamma\left(\beta-\frac{1}{2}\right)} h^{\beta-3/2}, \qquad (4.4.21)$$

where $\beta > -1$. In particular, the shape $y = f(z)$ is either a parabola or a rectangle depending on whether, $\beta = \frac{7}{2}$ or $\frac{3}{2}$. There are also other shapes of the opening of the dam depending on the value of β.

4.5 Solutions of Boundary Value Problems

The Laplace transform technique is also very useful in finding solutions of certain simple boundary value problems which arise in many areas of applied mathematics and engineering sciences. We illustrate the method by solving boundary value problems in the theory of deflection of elastic beams.

A horizontal beam experiences a vertical deflection due to the combined effect of its own weight and the applied load on the beam. We consider a beam of length ℓ and its equilibrium position is taken along the horizontal x-axis.

Example 4.5.1 (*Deflection of Beams*). The differential equation for the vertical deflection $y(x)$ of a uniform beam under the action of a transverse load $W(x)$ per unit length at a distance x from the origin on the x-axis of the beam is

$$EI\frac{d^4y}{dx^4} = W(x), \qquad \text{for } 0 < x < \ell, \qquad (4.5.1)$$

where E is Young's modulus, I is the moment of inertia of the cross section about an axis normal to the plane of bending and EI is called the *flexural rigidity* of the beam.

Some physical quantities associated with the problem are $y'(x)$, $M(x) = EIy''(x)$ and $S(x) = M'(x) = EIy'''(x)$, which respectively represent the slope, bending moment, and shear at a point.

It is of interest to find the solution of (4.5.1) subject to a given loading function and simple boundary conditions involving the deflection, slope, bending moment and shear. We consider the following cases:

(i) Concentrated load on a clamped beam of length ℓ, that is,
$W(x) \equiv W\delta(x - a)$,
$y(0) = y'(0) = 0$ and $y(\ell) = y'(\ell) = 0$,
where W is a constant and $0 < a < \ell$.

(ii) Distributed load on a uniform beam of length ℓ clamped at $x = 0$ and unsupported at $x = \ell$, that is, $W(x) = W\,H(x - a)$, $y(0) = y'(0) = 0$, and $M(\ell) = S(\ell) = 0$.

(iii) A uniform semi-infinite beam freely hinged at $x = 0$ resting horizontally on an elastic foundation and carrying a load W per unit length.

In order to solve the problem, we use the Laplace transform $\bar{y}(s)$ of $y(x)$ defined by

$$\bar{y}(s) = \int_0^\infty e^{-sx} y(x)\, dx. \tag{4.5.2}$$

In view of this transformation, equation (4.5.1) becomes

$$EI\left[s^4\,\bar{y}(s) - s^3 y(0) - s^2 y'(0) - s y''(0) - y'''(0)\right] = \overline{W}(s). \tag{4.5.3}$$

The solution of the transformed deflection function $\bar{y}(s)$ for case (i) is

$$\bar{y}(s) = \frac{y''(0)}{s^3} + \frac{y'''(0)}{s^4} + \frac{W}{EI}\frac{e^{-as}}{s^4}. \tag{4.5.4}$$

Inversion gives

$$y(x) = y''(0)\frac{x^2}{2} + y'''(0)\frac{x^3}{6} + \frac{W}{6EI}(x - a)^3 H(x - a), \tag{4.5.5}$$

$$y'(x) = y''(0) x + \frac{1}{2} x^2 y'''(0) + \frac{W}{2EI}(x - a)^2 H(x - a). \tag{4.5.6}$$

The conditions $y(\ell) = y'(\ell) = 0$ require that

$$y''(0)\frac{\ell^2}{2} + y'''(0)\frac{\ell^3}{6} + \frac{W}{6EI}(\ell - a)^3 = 0,$$

$$y''(0)\ell + y'''(0)\frac{\ell^2}{2} + \frac{W}{2EI}(\ell - a)^2 = 0.$$

These algebraic equations determine the value of $y''(0)$ and $y'''(0)$. Solving these equations, it turns out that

$$y''(0) = \frac{Wa(\ell - a)^2}{EI\,\ell^2} \quad \text{and} \quad y'''(0) = -\frac{W(\ell - a)^2(\ell + 2a)}{EI\,\ell^3}.$$

Thus the final solution for case (i) is

$$y(x) = \frac{W}{2EI}\left[\frac{a(\ell - a)^2 x^2}{\ell^2} - \frac{(\ell - a)^2(\ell + 2a)x^3}{3\ell^3} + \frac{(x - a)^3 H(x - a)}{3}\right]. \tag{4.5.7}$$

It is now possible to calculate the bending moment and shear at any point of the beam, and, in particular, at the ends.

The solution for case (ii) follows directly from (4.5.3) in the form

$$\bar{y}(s) = \frac{y''(0)}{s^3} + \frac{y'''(0)}{s^4} + \frac{W}{EI}\frac{e^{-as}}{s^5}. \tag{4.5.8}$$

The inverse transformation yields

$$y(x) = \frac{1}{2} y''(0) x^2 + \frac{1}{6} y'''(0) x^3 + \frac{W}{24 EI}(x - a)^4 H(x - a), \tag{4.5.9}$$

where $y''(0)$ and $y'''(0)$ are to be determined from the remaining boundary conditions $M(\ell) = S(\ell) = 0$, that is, $y''(\ell) = y'''(\ell) = 0$.

From (4.5.9) with $y''(\ell) = y'''(\ell) = 0$, it follows that

$$y''(0) + y'''(0)\ell + \frac{W}{2EI}(\ell - a)^2 = 0$$

$$y'''(0) + \frac{W}{EI}(\ell - a) = 0$$

which give

$$y''(0) = \frac{W(\ell - a)(\ell + a)}{2EI} \quad \text{and} \quad y'''(0) = -\frac{W}{EI}(\ell - a).$$

Hence the solution for $y(x)$ for case (ii) is

$$y(x) = \frac{W}{2EI}\left[\frac{(\ell^2 - a^2)x^2}{2} - (\ell - a)\frac{x^3}{3} + \frac{W}{12}(x-a)^4 H(x-a)\right]. \qquad (4.5.10)$$

The shear, S, and the bending moment, M, at the origin, can readily be calculated from the solution.

The differential equation for case (iii) takes the form

$$EI\frac{d^4y}{dx^4} + ky = W, \qquad x > 0, \qquad (4.5.11)$$

where the second term on the left hand side represents the effect of elastic foundation and k is a positive constant.

Writing $\dfrac{k}{EI} = 4\omega^4$, equation (4.5.11) becomes

$$\left(\frac{d^4}{dx^4} + 4\omega^4\right)y(x) = \frac{W}{EI}, \qquad x > 0. \qquad (4.5.12)$$

This has to be solved subject to the boundary conditions

$$y(0) = y''(0) = 0, \qquad (4.5.13)$$
$$y(x) \text{ is finite as } x \to \infty. \qquad (4.5.14)$$

Using the Laplace transform with respect to x to (4.5.12), we obtain

$$(s^4 + 4\omega^4)\bar{y}(s) = \left(\frac{W}{EI}\right)\frac{1}{s} + sy''(0) + y'''(0). \qquad (4.5.15)$$

In view of the Tauberian Theorem 3.8.2 (ii), that is,

$$\lim_{s \to 0} s\bar{y}(s) = \lim_{x \to \infty} y(x),$$

it follows that $\bar{y}(s)$ must be of the form

$$\bar{y}(s) = \frac{W}{EI}\frac{1}{s(s^4 + 4\omega^4)}, \qquad (4.5.16)$$

which gives

$$\lim_{x \to \infty} y(x) = \frac{W}{k}. \qquad (4.5.17)$$

We now write (4.5.16) as

$$\bar{y}(s) = \frac{W}{EI} \frac{1}{4\omega^4} \left[\frac{1}{s} - \frac{s^3}{s^4 + 4\omega^4} \right]. \qquad (4.5.18)$$

Using the standard table of inverse Laplace transforms, we obtain

$$y(x) = \frac{W}{k} (1 - \cos \omega x \cosh \omega x)$$

$$= \frac{W}{k} \left[1 - \frac{1}{2} e^{-\omega x} \cos \omega x - \frac{1}{2} e^{\omega x} \cos \omega x \right]. \qquad (4.5.19)$$

In view of (4.5.17), the final solution is

$$y(x) = \frac{W}{k} \left(1 - \frac{1}{2} e^{-\omega x} \cos \omega x \right). \qquad (4.5.20)$$

4.6 Evaluation of Definite Integrals

The Laplace transform can be employed to evaluate easily certain definite integrals containing a parameter. Although the method of evaluation may not be very rigorous, it is quite simple and straightforward. The method is essentially based upon the permissibility of interchange of the order of integration, that is,

$$\mathcal{L} \int_a^b f(t,x) \, dx = \int_a^b \mathcal{L} f(t,x) \, dx, \qquad (4.6.1)$$

and may be well described by considering some important integrals.

Example 4.6.1 Evaluate the integral

$$f(t) = \int_0^\infty \frac{\sin tx}{x(1+x^2)} \, dx. \qquad (4.6.2)$$

We take the Laplace transform of (4.6.2) with respect to t and interchange the order of integration, which is permissible due to uniform convergence, to obtain

$$\bar{f}(s) = \int_0^\infty \frac{dx}{x(1+x^2)} \int_0^\infty e^{-st} \sin tx \, dt$$

$$= \int_0^\infty \frac{dx}{(1+x^2)(x^2+s^2)}$$

$$= \frac{1}{s^2-1} \int_0^\infty \left(\frac{1}{1+x^2} - \frac{1}{x^2+s^2} \right) dx$$

$$= \frac{1}{s^2-1} \left(1 - \frac{1}{s} \right) \frac{\pi}{2}$$

$$= \frac{\pi}{2} \frac{1}{s(s+1)} = \frac{\pi}{2} \left(\frac{1}{s} - \frac{1}{s+1} \right).$$

Inversion gives the value of the given integral

$$f(t) = \frac{\pi}{2}\left(1 - e^{-t}\right). \tag{4.6.3}$$

Example 4.6.2 Evaluate the integral

$$f(t) = \int_0^\infty \frac{\sin^2 tx}{x^2}\, dx. \tag{4.6.4}$$

A procedure similar to the above integral with $2\sin^2 tx = 1 - \cos(2tx)$ gives

$$\bar{f}(s) = \frac{1}{2}\int_0^\infty \frac{1}{x^2}\left(\frac{1}{s} - \frac{s}{4x^2 + s^2}\right) dx$$

$$= \frac{2}{s}\int_0^\infty \frac{dx}{4x^2 + s^2}$$

$$= \frac{1}{s}\int_0^\infty \frac{dy}{y^2 + s^2} = \frac{1}{s^2}\left[\tan^{-1}\frac{y}{s}\right]_0^\infty = \pm\frac{\pi}{2s^2}$$

according as $s > $ or < 0. The inverse transform yields

$$f(t) = \frac{\pi t}{2}\, \text{sgn}\, t. \tag{4.6.5}$$

Example 4.6.3 Show that

$$\int_0^\infty \frac{x\sin xt}{x^2 + a^2}\, dx = \frac{\pi}{2} e^{-at}, \quad (a, t > 0), \tag{4.6.6}$$

Suppose

$$f(t) = \int_0^\infty \frac{x\sin xt}{x^2 + a^2}\, dx.$$

Taking the Laplace transform with respect to t gives

$$\bar{f}(s) = \int_0^\infty \frac{x^2\, dx}{(x^2 + a^2)(x^2 + s^2)}$$

$$= \int_0^\infty \frac{dx}{x^2 + s^2} - \frac{a^2}{s^2 - a^2}\int_0^\infty \left(\frac{1}{x^2 + a^2} - \frac{1}{x^2 + s^2}\right) dx$$

$$= \frac{\pi}{2s}\left(1 - \frac{a}{s + a}\right) = \frac{\pi}{2}\frac{1}{(s + a)}.$$

Taking the inverse transform, we obtain

$$f(t) = \frac{\pi}{2} e^{-at}.$$

4.7 Solutions of Difference and Differential-Difference Equations

Like differential equations, the difference and differential-difference equations describe mechanical, electrical, and electronic systems of interest. These equations also arise frequently in problems of economics and business, and particularly in problems concerning interest, annuities, amortization, loans, and mortgages. Thus for the study of the above systems or problems, it is often necessary to solve difference or differential-difference equations with prescribed initial data. This section is essentially devoted to the solution of simple difference and differential-difference equations by the Laplace transform technique.

Suppose $\{u_r\}_{r=1}^{\infty}$ is a given sequence. We introduce the difference operators $\Delta, \Delta^2, \Delta^3, \ldots, \Delta^n$ defined by

$$\Delta u_r = u_{r+1} - u_r, \tag{4.7.1}$$

$$\Delta^2 u_r = \Delta(\Delta u_r) = \Delta(u_{r+1} - u_r) = u_{r+2} - 2u_{r+1} + u_r, \tag{4.7.2}$$

$$\Delta^3 u_r = \Delta^2(u_{r+1} - u_r) = u_{r+3} - 3u_{r+2} + 3u_{r+1} - u_r. \tag{4.7.3}$$

More generally

$$\Delta^n u_r = \Delta^{n-1}(u_{r+1} - u_r) = \sum_{k=0}^{n}(-1)^k \binom{n}{k} u_{r+n-k}. \tag{4.7.4}$$

These expressions are usually called the *first, second, third*, and *nth finite differences* respectively. Any equation expressing a relation between finite differences is called a *difference equation*. The highest order finite difference involved in the equation is referred to as its *order*. A difference equation containing the derivatives of the unknown function is called the *differential-difference equation*. Thus the differential-difference equation has two distinct orders—one is related to the highest order finite difference and the other is associated with the highest order derivatives. Equations

$$\Delta u_r - u_r = 0, \tag{4.7.5}$$

$$\Delta^2 u_r - 2\Delta u_r = 0, \tag{4.7.6}$$

are the examples of difference equations of the first and second order respectively. The most general linear nth order difference equation has the form

$$a_0 \Delta^n u_r + a_1 \Delta^{n-1} u_r + \cdots + a_{n-1} \Delta u_r + a_n u_r = f(n), \tag{4.7.7}$$

where $a_0, a_1, \ldots, a_n$ and $f(n)$ are either constants or functions of non-negative integer n. Like ordinary differential equations, (4.7.7) is called a *homogeneous* or *inhomogeneous* according as $f(n) = 0$ or $\neq 0$.

The following equations

$$u'(t) - u(t-1) = 0, \tag{4.7.8}$$

$$u'(t) - au(t-1) = f(t), \tag{4.7.9}$$

are the examples of the differential-difference equations, where $f(t)$ is a given function of t. The study of the above equation is facilitated by introducing the function
$$S_n(t) = H(t-n) - H(t-n-1), \qquad n \le t < n+1, \qquad (4.7.10)$$
where n is a non-negative integer and $H(t)$ is the Heaviside unit step function.

The Laplace transform of $S_n(t)$ is given by
$$\overline{S}_n(s) = \mathcal{L}\{S_n(t)\} = \int_0^\infty e^{-st}\{H(t-n) - H(t-n-1)\}\, dt$$
$$= \int_n^{n+1} e^{-st}\, dt = \frac{1}{s}(1-e^{-s})e^{-ns}$$
$$= \overline{S}_0(s)\exp(-ns), \qquad (4.7.11)$$
where $\overline{S}_0(s)$ is equal to $\frac{1}{s}(1-e^{-s})$.

We next define the function $u(t)$ by a series
$$u(t) = \sum_{n=0}^\infty u_n S_n(t), \qquad (4.7.12)$$
where $\{u_n\}_{n=0}^\infty$ is a given sequence. It follows that $u(t) = u_n$ in $n \le t < n+1$ and represents a staircase function. Further
$$u(t+1) = \sum_{n=0}^\infty u_n S_n(t+1) = \sum_{n=0}^\infty u_n [H(t+1-n) - H(t-n)]$$
$$= \sum_{n=1}^\infty u_n S_{n-1}(t) = \sum_{n=0}^\infty u_{n+1} S_n(t), \qquad (4.7.13)$$

Similarly,
$$u(t+2) = \sum_{n=0}^\infty u_{n+2} S_n(t), \qquad (4.7.14)$$

More generally,
$$u(t+k) = \sum_{n=0}^\infty u_{n+k} S_n(t). \qquad (4.7.15)$$

The Laplace transform of $u(t)$ is given by
$$\overline{u}(s) = \mathcal{L}\{u(t)\} = \int_0^\infty e^{-st} u(t)\, dt$$
$$= \sum_{n=0}^\infty u_n \int_0^\infty e^{-st} S_n(t)\, dt$$
$$= \frac{1}{s}(1-e^{-s}) \sum_{n=0}^\infty u_n \exp(-ns).$$

Thus
$$\bar{u}(s) = \frac{1}{s}(1-e^{-s})\zeta(s) = \bar{S}_0(s)\zeta(s), \qquad (4.7.16)$$
where $\zeta(s)$ represents the *Dirichlet function* defined by
$$\zeta(s) = \sum_{n=0}^{\infty} u_n \exp(-ns). \qquad (4.7.17)$$
We thus deduce
$$u(t) = \mathcal{L}^{-1}\{\bar{S}_0(s)\zeta(s)\}. \qquad (4.7.18)$$
In particular, if $u_n = a^n$ is a geometric sequence, then
$$\zeta(s) = \sum_{n=0}^{\infty}(ae^{-s})^n = \frac{1}{1-ae^{-s}} = \frac{e^s}{e^s - a}. \qquad (4.7.19)$$
Thus we obtain from (4.7.16) that
$$\mathcal{L}\{a^n\} = \bar{S}_0(s)\zeta(s) = \bar{S}_0(s)\frac{e^s}{e^s - a}, \qquad (4.7.20)$$
so that
$$\mathcal{L}^{-1}\left\{\bar{S}_0(s)\frac{e^s}{e^s - a}\right\} = a^n. \qquad (4.7.21)$$
From the identity,
$$\sum_{n=0}^{\infty}(n+1)(ae^{-s})^n = (1-ae^{-s})^{-2}, \qquad (4.7.22)$$
it further follows that
$$\mathcal{L}\{(n+1)a^n\} = \bar{S}_0(s)(1-ae^{-s})^{-2} = \frac{e^{2s}\bar{S}_0(s)}{(e^s - a)^2}. \qquad (4.7.23)$$
Thus
$$\mathcal{L}^{-1}\left\{\frac{e^{2s}\bar{S}_0(s)}{(e^s - a)^2}\right\} = (n+1)a^n. \qquad (4.7.24)$$
We deduce from (4.7.22) that
$$\sum_{n=0}^{\infty} na^n e^{-ns} = \frac{ae^s}{(1-ae^{-s})^2}. \qquad (4.7.25)$$
Hence
$$\mathcal{L}\{na^n\} = \bar{S}_0(s)\frac{ae^s}{(e^s - a)^2}. \qquad (4.7.26)$$
Therefore
$$\mathcal{L}^{-1}\left\{\frac{a\bar{S}_0(s)e^s}{(e^s - a)^2}\right\} = na^n. \qquad (4.7.27)$$

Integral Transforms and Their Applications

Theorem 4.7.1 If $\bar{u}(s) = \mathcal{L}\{u(t)\}$, then
$$\mathcal{L}\{u(t+1)\} = e^s\left[\bar{u}(s) - u_0\,\bar{S}_0(s)\right], \tag{4.7.28}$$
where $u_0 = u(0)$.

Proof. We have
$$\mathcal{L}\{u(t+1)\} = \int_0^\infty e^{-st}u(t+1)\,dt = e^s \int_1^\infty e^{-s\tau}u(\tau)\,d\tau$$

$$= e^s\left[\bar{u}(s) - \int_0^1 e^{-s\tau}u(\tau)\,d\tau\right]$$

$$= e^s\left[\bar{u}(s) - u(0)\int_0^1 e^{-s\tau}\,d\tau\right]$$

$$= e^s\left[\bar{u}(s) - u_0\,\bar{S}_0(s)\right].$$

This proves the theorem.

In view of this theorem, we derive
$$\mathcal{L}\{u(t+2)\} = e^s\left[\mathcal{L}\{u(t+1)\} - u(1)\,\bar{S}_0(s)\right]$$
$$= e^{2s}\left[\bar{u}(s) - u(0)\,\bar{S}_0(s)\right] - e^s u_1 \bar{S}_0(s)$$
$$= e^{2s}\left[\bar{u}(s) - \left(u_0 + u_1 e^{-s}\right)\bar{S}_0(s)\right], \tag{4.7.29}$$
where $u(1) = u_1$.

Similarly,
$$\mathcal{L}\{u(t+3)\} = e^{3s}\left[\bar{u}(s) - \left(u_0 + u_1 e^{-s} + u_2 e^{-2s}\right)\bar{S}_0(s)\right]. \tag{4.7.30}$$

More generally, if k is an integer,
$$\mathcal{L}\{u(t+k)\} = e^{ks}\left[\bar{u}(s) - \bar{S}_0(s)\sum_{r=0}^{k-1} u_r e^{-rs}\right]. \tag{4.7.31}$$

Example 4.7.1 Solve the difference equation
$$\Delta u_n - u_n = 0, \tag{4.7.32}$$
with the initial condition $u_0 = 1$.

We take the Laplace transform of the equation to obtain
$$\mathcal{L}\{u_{n+1}\} - 2\,\mathcal{L}\{u_n\} = 0,$$
which is, by (4.7.28),
$$e^s\left[\bar{u}(s) - u_0\,\bar{S}_0(s)\right] - 2\bar{u}(s) = 0.$$
Thus
$$\bar{u}(s) = \frac{e^s\,\bar{S}_0(s)}{e^s - 2}.$$
Inversion with (4.7.21) gives the solution
$$u_n = 2^n. \tag{4.7.33}$$

Example 4.7.2 Show that the solution of the difference equation
$$\Delta^2 u_n - 2\Delta u_n = 0 \tag{4.7.34}$$
is
$$u_n = A + B\, 3^n, \tag{4.7.35}$$
where $A = \frac{1}{2}(3u_0 - u_1)$ and $B = \frac{1}{2}(u_1 - u_0)$.

The given equation is
$$u_{n+2} - 4u_{n+1} + 3u_n = 0.$$
Taking the Laplace transform, we obtain
$$e^{2s}\left[\bar{u}(s) - (u_0 + u_1 e^{-s})\bar{S}_0(s)\right] - 4e^s\left[\bar{u}(s) - u_0\bar{S}_0(s)\right] + 3\bar{u}(s) = 0$$
or
$$(e^{2s} - 4e^s + 3)\bar{u}(s) = \left[u_0(e^{2s} - 4e^s) + u_1 e^s\right]\bar{S}_0(s).$$
Hence
$$\bar{u}(s) = \bar{S}_0(s)\left[\frac{u_0(e^{2s} - 4e^s) + u_1 e^s}{(e^s - 1)(e^s - 3)}\right]$$
$$= \bar{S}_0(s)\left[\frac{(3u_0 - u_1)e^s}{2(e^s - 1)} + \frac{(u_1 - u_0)e^s}{2(e^s - 3)}\right].$$
The inverse Laplace transform combined with (4.7.21) gives
$$u_n = A + B\, 3^n.$$

Example 4.7.3 Solve the difference equation
$$u_{n+2} - 2\lambda u_{n+1} + \lambda^2 u_n = 0, \tag{4.7.36}$$
with $u_0 = 0$ and $u_1 = 1$.

The Laplace transformed equation is
$$e^{2s}\left[\bar{u}(s) - e^{-s}\bar{S}_0(s)\right] - 2\lambda\, \bar{u}(s)e^s + \lambda^2\, \bar{u}(s) = 0$$
or
$$\bar{u}(s) = \frac{e^s\, \bar{S}_0(s)}{(e^s - \lambda)^2}.$$
The inverse transform gives the solution
$$u_n = \frac{1}{\lambda}n\lambda^n = n\lambda^{n-1}. \tag{4.7.37}$$

Example 4.7.4 Solve the differential-difference equation
$$u'(t) = u(t-1),\quad u(0) = 1. \tag{4.7.38}$$
Application of the Laplace transform gives
$$s\bar{u}(s) - u(0) = e^{-s}\left[\bar{u}(s) - u(0)\bar{S}_0(s)\right]$$
or
$$\bar{u}(s)(s - e^{-s}) = 1 + \frac{e^{-s}}{s}(e^{-s} - 1)$$

or

$$\bar{u}(s) = \left\{\frac{1}{s-e^{-s}} - \frac{e^{-s}}{s(s-e^{-s})}\right\} + \frac{e^{-2s}}{s(s-e^{-s})}$$

$$= \frac{1}{s} + \frac{e^{-2s}}{s^2}\left(1 - \frac{e^{-s}}{s}\right)^{-1}$$

$$= \frac{1}{s} + \frac{e^{-2s}}{s^2} + \frac{e^{-3s}}{s^3} + \frac{e^{-4s}}{s^4} + \cdots + \frac{e^{-ns}}{s^n} + \cdots.$$

In view of the result

$$\mathcal{L}^{-1}\left\{\frac{e^{-as}}{s^n}\right\} = \frac{(t-a)^{n-1}}{\Gamma(n)} H(t-a), \qquad (4.7.39)$$

we obtain the solution

$$u(t) = 1 + \frac{(t-2)}{1!} + \frac{(t-3)^2}{2!} + \cdots + \frac{(t-n)^{n-1}}{(n-1)!}, \quad t > n. \qquad (4.7.40)$$

Example 4.7.5 Solve the differential-difference equation

$$u'(t) - \alpha u(t-1) = \beta, \qquad u(0) = 0. \qquad (4.7.41)$$

Application of the Laplace transform yields

$$s\bar{u}(s) - u(0) - \alpha e^{-s}\left[\bar{u}(s) - u(0)\bar{S}_0(s)\right] = \frac{\beta}{s}.$$

Or

$$\bar{u}(s) = \frac{\beta}{s(s-\alpha e^{-s})} = \frac{\beta}{s^2}\left(1 - \frac{\alpha}{s}e^{-s}\right)^{-1}$$

$$= \beta\left[\frac{1}{s^2} + \frac{\alpha e^{-s}}{s^3} + \frac{\alpha^2 e^{-2s}}{s^4} + \cdots + \frac{\alpha^n e^{-ns}}{s^{n+2}} + \cdots\right].$$

Inverting with the help of (4.7.39), we obtain the solution

$$u(t) = \beta\left[t + \frac{\alpha(t-1)^2}{\Gamma(3)} + \alpha^2\frac{(t-2)^3}{\Gamma(4)} + \cdots + \frac{\alpha^n(t-n)^{n+1}}{\Gamma(n+2)}\right], \quad t > n. \qquad (4.7.42)$$

4.8 Applications of the Joint Laplace and Fourier Transform

Example 4.8.1 *(The Cauchy Problem for the Wave Equation).* Use the joint Fourier and Laplace transform method to solve the Cauchy problem for the wave equation as stated in Example 2.7.4.

We define the joint Fourier and Laplace transform of $u(x,t)$ by

$$\bar{U}(k,s) = \frac{1}{\sqrt{2\pi}}\int_{-\infty}^{\infty} e^{-ikx}dx \int_0^{\infty} e^{-st}u(x,t)dt. \qquad (4.8.1)$$

The transformed Cauchy problem has the solution in the form

$$\overline{U}(k,s) = \frac{sF(k)+G(k)}{(s^2+c^2k^2)}. \qquad (4.8.2)$$

The joint inverse transform gives the solution as

$$u(x,t) = \frac{1}{\sqrt{2\pi}} \int_{-\infty}^{\infty} e^{ikx} \mathcal{L}^{-1}\left[\frac{sF(k)+G(k)}{s^2+c^2k^2}\right]dk$$

$$= \frac{1}{\sqrt{2\pi}} \int_{-\infty}^{\infty} \left[F(k)\cos ckt + \frac{G(k)}{ck}\sin ckt\right]e^{ikx}dk$$

$$= \frac{1}{2\sqrt{2\pi}} \int_{-\infty}^{\infty} F(k)\left(e^{ickt}+e^{-ickt}\right)e^{ikx}dk + \frac{1}{2\sqrt{2\pi}} \int_{-\infty}^{\infty} \frac{G(k)}{ick}\left(e^{ickt}-e^{-ickt}\right)e^{ikx}dk$$

$$= \frac{1}{2}\left[f(x-ct)+f(x+ct)\right] + \frac{1}{\sqrt{2\pi}}\frac{1}{2c} \int_{-\infty}^{\infty} G(k)dk \int_{x-ct}^{x+ct} e^{ik\xi}d\xi$$

$$= \frac{1}{2}\left[f(x-ct)+f(x+ct)\right] + \frac{1}{2c} \int_{x-ct}^{x+ct} g(\xi)d\xi. \qquad (4.8.3)$$

This is identical with the d'Alembert solution (2.7.4).

Example 4.8.2 (*Dispersive Long Water Waves in a Rotating Ocean*). We use the joint Laplace and Fourier transform to solve the linearized horizontal equations of motion and the continuity equation in a rotating inviscid ocean. These equations in a rotating coordinate system (see Proudman, 1953) are given by

$$\frac{\partial \mathbf{u}}{\partial t} + f\hat{\mathbf{k}}\times\mathbf{u} = -\frac{1}{\rho}\nabla p + \frac{1}{\rho h}\boldsymbol{\tau}, \qquad (4.8.4)$$

$$\nabla \cdot \mathbf{u} = -\frac{1}{h}\frac{\partial \zeta}{\partial t}, \qquad (4.8.5)$$

where $\mathbf{u} = (u, v)$ is the horizontal velocity field, $\hat{\mathbf{k}}$ is the unit vector normal to the horizontal plane, $f = 2\Omega\sin\phi$ is the constant Coriolis parameter, ρ is the constant density of water, $\zeta(x,t)$ is the vertical free surface elevation, $\boldsymbol{\tau} = (\tau^x, \tau^y)$ represents the components of wind stress in the x and y directions, and the pressure is given by the hydrostatic equation

$$p = p_0 + g\rho(\zeta - z), \qquad (4.8.6)$$

where z is the depth of water below the mean free surface and g is the acceleration due to gravity.

Equations (4.8.4)-(4.8.5) combined with (4.8.6) reduce to the form

$$\frac{\partial u}{\partial t} - fv = -g\frac{\partial \zeta}{\partial x} + \frac{\tau^x}{\rho h}, \qquad (4.8.7)$$

$$\frac{\partial v}{\partial t} + fu = -g\frac{\partial \zeta}{\partial y} + \frac{\tau^y}{\rho h}, \tag{4.8.8}$$

$$\frac{\partial u}{\partial x} + \frac{\partial v}{\partial y} = -\frac{1}{h}\frac{\partial \zeta}{\partial t}. \tag{4.8.9}$$

It follows from (4.8.7)-(4.8.8) that

$$Du = -g\left(\frac{\partial^2}{\partial x \partial t} + f\frac{\partial}{\partial y}\right)\zeta + \frac{1}{\rho h}\left(\frac{\partial \tau^x}{\partial t} + f\tau^y\right), \tag{4.8.10}$$

$$Dv = -g\left(\frac{\partial^2}{\partial y \partial t} - f\frac{\partial}{\partial x}\right)\zeta + \frac{1}{\rho h}\left(\frac{\partial \tau^y}{\partial t} - f\tau^x\right), \tag{4.8.11}$$

where the differential operator D is

$$D \equiv \left(\frac{\partial^2}{\partial t^2} + f^2\right). \tag{4.8.12}$$

Elimination of u and v from (4.8.9) - (4.8.11) gives

$$\left(\nabla^2 - \frac{1}{c^2}D\right)\zeta_t = E(x,y,t), \tag{4.8.13}$$

where $c^2 = gh$ and ∇^2 is the horizontal Laplacian, and $E(x,y,t)$ is a known forcing function given by

$$E(x,y,t) = \frac{1}{\rho c^2}\left[\frac{\partial^2 \tau^x}{\partial x \partial t} - \frac{\partial^2 \tau^y}{\partial y \partial t} + f\left(\frac{\partial \tau^y}{\partial x} - \frac{\partial \tau^x}{\partial y}\right)\right]. \tag{4.8.14}$$

Further, we assume that the conditions are uniform in the y direction and the wind stress acts only in the x direction so that τ^x and E are given functions of x and t only. Consequently, equation (4.8.13) becomes

$$\left[\frac{\partial^2}{\partial x^2} - \frac{1}{c^2}\left(\frac{\partial^2}{\partial t^2} + f^2\right)\right]\zeta_t = \frac{1}{\rho c^2}\left(\frac{\partial^2 \tau^x}{\partial x \partial t}\right).$$

Integrating this equation with respect to t gives

$$\left[\frac{\partial^2}{\partial x^2} - \frac{1}{c^2}\left(\frac{\partial^2}{\partial t^2} + f^2\right)\right]\zeta = \frac{1}{\rho c^2}\left(\frac{\partial \tau^x}{\partial x}\right). \tag{4.8.15}$$

Similarly, the velocity $u(x,t)$ satisfies the equation

$$\left[\frac{\partial^2}{\partial x^2} - \frac{1}{c^2}\left(\frac{\partial^2}{\partial t^2} + f^2\right)\right]u = -\frac{1}{\rho hc^2}\left(\frac{\partial \tau^x}{\partial t}\right). \tag{4.8.16}$$

If the right hand side of equations (4.8.15) and (4.8.16) is zero, these equations are known as the *Klein-Gordon equations* which have received extensive attention in quantum mechanics and in applied mathematics.

Equation (4.8.16) is to be solved subject to the following boundary and initial conditions

$$|\zeta| \text{ is bounded as } |x| \to \infty, \tag{4.8.17}$$

$$\zeta(x,t) = 0 \text{ at } t = 0 \text{ for all real } x. \tag{4.8.18}$$

Before we solve the initial value problem, we seek a plane wave solution of the homogeneous equation (4.8.15) in the form
$$\zeta(x,t) = A \exp\{i(\omega t - kx)\}, \qquad (4.8.19)$$
where A is a constant amplitude, ω is the frequency and k is the wavenumber. Such a solution exists provided the dispersion relation
$$\omega^2 = c^2 k^2 + f^2, \qquad (4.8.20)$$
is satisfied. Thus the phase and the group velocities of waves are given by
$$C_p = \frac{\omega}{k} = \left(c^2 + \frac{f^2}{k^2}\right)^{\frac{1}{2}}, \quad C_g = \frac{\partial \omega}{\partial k} = \frac{c^2 k}{\left(c^2 k^2 + f^2\right)^{\frac{1}{2}}}. \qquad (4.8.21\text{ab})$$

Thus the waves are dispersive in a rotating ocean $(f \neq 0)$. However, in a non-rotating ocean $(f = 0)$ all waves would propagate with constant velocity c, and they are non-dispersive shallow water waves. Further, $C_p C_g = c^2$ whence it follows that the phase velocity has a minimum of c and the group velocity a maximum. The short waves will be observed first at a given point, even though they have the smallest phase velocity.

Application of the joint Laplace and Fourier transform to (4.8.15) together with (4.8.17)-(4.8.18) give the transformed solution
$$\bar{\tilde{\zeta}}(k,s) = -\frac{Ac^2}{(s^2 + a^2)} \tilde{f}(k,s), \qquad a^2 = \left(c^2 k^2 + f^2\right), \qquad (4.8.22)$$
where
$$f(x,t) = \frac{1}{\rho c^2}\left(\frac{\partial \tau^x}{\partial x}\right) H(t). \qquad (4.8.23)$$
The inverse transforms combined with the Convolution Theorem of the Laplace transform lead to the formal solution
$$\zeta(x,t) = -\frac{Ac}{\sqrt{2\pi}} \int_{-\infty}^{\infty} \left(k^2 + \frac{f^2}{c^2}\right)^{-\frac{1}{2}} e^{ikx} \, dk \int_0^t \tilde{f}(k, t-\tau) \sin a\tau \, d\tau. \qquad (4.8.24)$$
In general, this integral cannot be evaluated unless $f(x,t)$ is prescribed. Even if some particular form of f is given, an exact evaluation of (4.8.24) is almost a formidable task. Hence it is necessary to resort to asymptotic methods (see Debnath and Kulchar, 1972).

To investigate the solution, we choose a particular form of the wind stress distribution
$$\frac{\tau^x}{\rho c^2} = A e^{i\omega t} H(t) H(-x), \qquad (4.8.25)$$
where A is a constant and ω is the frequency of the applied disturbance. Thus
$$\frac{1}{\rho c^2}\left(\frac{\partial \tau^x}{\partial x}\right) = -A e^{i\omega t} H(t) \delta(-x). \qquad (4.8.26)$$
In this case, solution (4.8.24) reduces to the form

$$\zeta(x,t) = \frac{Ac}{\sqrt{2\pi}} \int_0^t e^{i\omega(t-\tau)} H(t-\tau) \mathscr{F}^{-1}\left[\frac{\sin a\tau}{\sqrt{k^2 + \frac{f^2}{c^2}}}\right] d\tau$$

$$= \frac{Ac}{2} \int_0^t e^{i\omega(t-\tau)} H(t-\tau) J_0\left\{\frac{f}{c}(c^2\tau^2 - x^2)^{\frac{1}{2}}\right\} H(c\tau - |x|) \, d\tau, \quad (4.8.27)$$

where $J_0(z)$ is the zero-order Bessel function of the first kind.

When $\omega \equiv 0$, this solution is identical with that of Crease (1956) who obtained the solution using the Green's function method. In this case, the solution becomes

$$\zeta = \frac{Ac}{2} \int_0^t H(t-\tau) J_0\left[f\left\{\tau^2 - \frac{x^2}{c^2}\right\}^{\frac{1}{2}}\right] H\left(\tau - \frac{|x|}{c}\right) d\tau. \quad (4.8.28)$$

In terms of non-dimensional parameters $f\tau = \alpha$, $ft = a$, and $\frac{fx}{c} = b$, solution (4.8.28) assumes the form

$$\left(\frac{2f}{Ac}\right)\zeta = \int_0^a H(a-\alpha) J_0\left[(\alpha^2 - b^2)^{\frac{1}{2}}\right] H(\alpha - |b|) \, d\alpha. \quad (4.8.29)$$

Or equivalently

$$\left(\frac{2f}{Ac}\right)\zeta = \int_{|b|}^d J_0\left[(\alpha^2 - b^2)^{\frac{1}{2}}\right] d\alpha, \quad (4.8.30)$$

where $d = \max(|b|, a)$. This is the basic solution of the problem.

In order to find the solution of (4.8.16), we first choose

$$\frac{1}{\rho c^2}\left(\frac{\partial \tau^x}{\partial t}\right) = A\delta(t)H(-x), \quad (4.8.31)$$

so that the joint Laplace and Fourier transform of this result is $A\mathscr{F}\{H(-x)\}$. Thus the transformed solution of (4.8.16) is

$$\bar{\mathbf{u}}(k,s) = \frac{Ac^2}{h} \mathscr{F}\{H(-x)\} \frac{1}{(s^2 + \omega^2)}, \quad \omega^2 = (ck)^2 + f^2. \quad (4.8.32)$$

The inverse transforms combined with the Convolution Theorem lead to the solution

$$u(x,t) = \frac{Ac}{2h} \int_{-\infty}^{\infty} H(-\xi) J_0\left[f\left\{t^2 - \left(\frac{x-\xi}{c}\right)^2\right\}^{\frac{1}{2}}\right] H\left(t - \frac{(x-\xi)}{c}\right) d\xi, \quad (4.8.33)$$

which is, by the change of variable $(x-\xi)f = c\alpha$, with $a = ft$ and $b = fx/c$,

$$= \frac{Ac^2}{2hf} \int_b^\infty J_0\left[(a^2-\alpha^2)^{\frac{1}{2}}\right] H(a-|\alpha|) d\alpha. \tag{4.8.34}$$

For the case $b > 0$, solution (4.8.34) becomes

$$u(x,t) = \frac{Ac^2}{2hf} H(a-b) \int_b^a J_0\left\{(a^2-\alpha^2)^{\frac{1}{2}}\right\} d\alpha. \tag{4.8.35}$$

When $b < 0$, the velocity field is

$$u(x,t) = \frac{Ac^2}{2hf}\left[\int_{-a}^a J_0\left\{(a^2-\alpha^2)^{\frac{1}{2}}\right\} d\alpha - H(a-|b|)\int_{-a}^b J_0\left\{(a^2-\alpha^2)^{\frac{1}{2}}\right\} d\alpha\right]$$

$$= \frac{gA}{2f}\left[2\sin a - H(a-|b|)\int_{|b|}^a J_0\left\{(a^2-\alpha^2)^{\frac{1}{2}}\right\} d\alpha\right], \tag{4.8.36}$$

which is, for $a < |b|$,

$$u(x,t) = \frac{gA}{2f}\sin a. \tag{4.8.37}$$

Finally, it can be shown that the velocity transverse to the direction of propagation is

$$v = -\frac{gA}{2f}\int_0^a d\beta \int_b^\infty J_0\left\{(\beta^2-\alpha^2)^{\frac{1}{2}}\right\} H(\beta-|\alpha|) d\alpha. \tag{4.8.38}$$

If $b > 0$, that is, x is outside the generating region, then

$$\left(\frac{2f}{gA}\right) v = -H(a-b)\int_b^a d\beta \int_b^\beta J_0\left\{(\beta^2-\alpha^2)^{\frac{1}{2}}\right\} d\alpha,$$

which becomes, after some simplification,

$$= -\left[(1-\cos a) - \int_0^b d\alpha \int_\alpha^a J_0\left\{(\beta^2-\alpha^2)^{\frac{1}{2}}\right\}\right] H(a-b). \tag{4.8.39}$$

For $b < 0$, it is necessary to consider two cases: (i) $a < |b|$ and (ii) $a > |b|$. In the former case, (4.8.38) takes the form

$$\left(\frac{2f}{gA}\right) v = -\int_0^a d\beta \int_{-\beta}^\beta J_0\left\{(\beta^2-\alpha^2)^{\frac{1}{2}}\right\} d\alpha = -2(1-\cos b). \tag{4.8.40}$$

In the latter case, the final form of the solution is

$$\left(\frac{2f}{gA}\right) v = -(1-\cos b) + \int_0^{|b|} d\alpha \int_\alpha^a J_0\left\{(\beta^2-\alpha^2)^{\frac{1}{2}}\right\} d\beta. \tag{4.8.41}$$

Finally, the steady-state solutions are obtained in the limit as $t \to \infty$ $(b \to \infty)$ as

$$\zeta = \frac{Ac}{2f}\exp(-|b|),$$

$$u = \frac{Ag}{2f}\sin ft,$$

$$v = \frac{Ag}{2f}\begin{bmatrix}\cos ft - \exp(-b), & b > 0 \\ \cos ft + \exp(-|b|) - 2, & b < 0\end{bmatrix}. \qquad (4.8.42)$$

Thus the steady-state solutions are attained in a rotating ocean. This shows a striking contrast with the corresponding solutions in the non-rotating ocean where an ever-increasing free surface elevation is found. The terms $\sin ft$ and $\cos ft$ involved in the steady-state velocity field represent inertial oscillations with frequency f.

Example 4.8.3 (*One-Dimensional Diffusion Equation on a Half Line*). Solve
$$u_t = \kappa u_{xx}, \quad 0 < x < \infty, \quad t > 0 \qquad (4.8.43)$$
with the boundary data
$$\left.\begin{array}{l}u(x,t) = f(t) \text{ for } x = 0 \\ u(x,t) \to 0 \quad \text{as } x \to \infty\end{array}\right\} \quad t > 0 \qquad (4.8.44ab)$$
and the initial condition
$$u(x,t) = 0 \quad \text{at } t = 0 \quad \text{for } 0 < x < \infty. \qquad (4.8.45)$$
We use the joint Fourier sine and Laplace transform defined by
$$\bar{U}_s(k,s) = \sqrt{\frac{2}{\pi}} \int_0^\infty e^{-st} dt \int_0^\infty u(x,t)\sin kx\, dx, \qquad (4.8.46)$$
so that the solution of the transformed problem is
$$\bar{U}_s(k,s) = \sqrt{\frac{2}{\pi}}(\kappa k)\frac{\bar{f}(s)}{(s + k^2\kappa)}. \qquad (4.8.47)$$

The inverse transform yields the solution
$$u(x,t) = \frac{2\kappa}{\pi}\int_0^\infty k\sin kx\, dk \int_0^t f(t-\tau)\exp(-\kappa \tau k^2)d\tau.$$

In particular, if $f(t) = T_0 = $ constant, then the solution becomes
$$u(x,t) = \frac{2T_0}{\pi}\int_0^\infty \frac{\sin kx}{k}\left(1 - e^{-\kappa k^2 t}\right)dk. \qquad (4.8.48)$$

Making use of the integral (2.10.11) gives the solution
$$u(x,t) = \frac{2T_0}{\pi}\left[\frac{\pi}{2} - \frac{\pi}{2}\,\text{erf}\left(\frac{x}{2\sqrt{\kappa t}}\right)\right]$$
$$= T_0\,\text{erfc}\left(\frac{x}{2\sqrt{\kappa t}}\right). \qquad (4.8.49)$$

This is identical with (2.10.12).

Example 4.8.4 (*The Bernoulli-Euler Equation on an Elastic Foundation*) Solve the equation

$$EI \frac{\partial^4 u}{\partial x^4} + \kappa u + m \frac{\partial^2 u}{\partial t^2} = W \,\delta(t)\delta(x), \quad -\infty < x < \infty, \quad t > 0, \quad (4.8.50)$$

with the initial data

$$u(x,0) = 0 \text{ and } u_t(x,0) = 0. \quad (4.8.51)$$

We use the joint Laplace and Fourier transform (4.8.1) to find the solution of the transformed problem in the form

$$\overline{U}(k,s) = \frac{W}{m\sqrt{2\pi}} \frac{1}{\left(s^2 + a^2 k^4 + \omega^2\right)}, \quad (4.8.52)$$

where

$$a^2 = \frac{EI}{m} \text{ and } \omega^2 = \frac{\kappa}{m}.$$

The inverse Laplace transform gives

$$U(k,t) = \frac{W}{m\sqrt{2\pi}} \left(\frac{\sin \alpha t}{\alpha}\right), \quad \alpha = \left(a^2 k^4 + \omega^2\right)^{\frac{1}{2}}. \quad (4.8.53\text{ab})$$

Then the inverse Fourier transform yields the formal solution

$$u(x,t) = \frac{W}{2\pi m} \int_{-\infty}^{\infty} e^{ikx} \frac{\sin \alpha t}{\alpha} \, dk. \quad (4.8.54)$$

Example 4.8.5 (*The Cauchy Poisson Wave Problem in Fluid Dynamics*) We consider the two-dimensional Cauchy Poisson problem for an inviscid liquid of infinite depth with a horizontal free surface. We assume that the liquid has constant density ρ and negligible surface tension. Waves are generated on the surface of water initially at rest for time $t < 0$ by the prescribed free surface displacement at $t = 0$.

In terms of the velocity potential $\phi(x,z,t)$ and the free surface elevation $\eta(x,t)$, the linearized surface wave motion in Cartesian coordinates (x,y,z) is governed by the following equation and free surface and boundary conditions:

$$\nabla^2 \phi = \phi_{xx} + \phi_{zz} = 0, \quad -\infty < z \leq 0, \quad -\infty < x < \infty, \quad t > 0, \quad (4.8.55)$$

$$\left.\begin{array}{r}\phi_z - \eta_t = 0 \\ \phi_t + g\eta = 0\end{array}\right\} \text{ on } z = 0, \ t > 0, \quad (4.8.56\text{ab})$$

$$\phi_z \to 0 \quad \text{as } z \to -\infty. \quad (4.8.57)$$

The initial conditions are

$$\phi(x,0,0) = 0 \text{ and } \eta(x,0) = \eta_0(x), \quad (4.8.58)$$

where $\eta_0(x)$ is a given function with compact support.

We introduce the Laplace transform with respect to t and the Fourier transform with respect to x defined by

$$\left[\tilde{\overline{\phi}}(k,z,s), \ \tilde{\overline{\eta}}(k,s)\right] = \frac{1}{\sqrt{2\pi}} \int_{-\infty}^{\infty} e^{-ikx} dx \int_0^{\infty} e^{-st} [\phi, \eta] dt. \quad (4.8.59)$$

The use of joint transform to the above system gives

$$\tilde{\bar{\phi}}_{zz} - k^2 \tilde{\bar{\phi}} = 0, \quad -\infty < z \leq 0, \tag{4.8.60}$$

$$\left. \begin{array}{l} \tilde{\bar{\phi}}_z = s\tilde{\bar{\eta}} - \tilde{\eta}_0(k) \\ s\tilde{\bar{\phi}} + g\tilde{\bar{\eta}} = 0 \end{array} \right\} \quad \text{on } z = 0, \tag{4.8.61ab}$$

$$\tilde{\bar{\phi}}_z \to 0 \quad \text{as } z \to -\infty. \tag{4.8.62}$$

The bounded solution of (4.8.60) is

$$\tilde{\bar{\phi}}(k,s) = \overline{A} \exp(|k|z) \tag{4.8.63}$$

where $\overline{A} = \overline{A}(s)$ is an arbitrary function of s, and $\tilde{\eta}_0(k) = \mathcal{F}\{\eta_0(x)\}$.

Substituting (4.8.63) into (4.8.61ab) and eliminating $\tilde{\bar{\eta}}$ from the resulting equations gives $\overline{A}$. Hence the solutions for $\tilde{\bar{\phi}}$ and $\tilde{\bar{\eta}}$ are

$$\left[\tilde{\bar{\phi}}, \tilde{\bar{\eta}} \right] = \left[-\frac{g\, \tilde{\eta}_0 \exp(|k|z)}{s^2 + \omega^2}, \frac{s\, \tilde{\eta}_0}{s^2 + \omega^2} \right], \tag{4.8.64ab}$$

where

$$\omega^2 = g|k|. \tag{4.8.65}$$

The inverse Laplace and Fourier transforms give the solutions

$$\phi(x,z,t) = -\frac{g}{\sqrt{2\pi}} \int_{-\infty}^{\infty} \frac{\sin \omega t}{\omega} \exp(ikx + |k|z) \tilde{\eta}_0(k) dk \tag{4.8.66}$$

$$\eta(x,t) = \frac{1}{\sqrt{2\pi}} \int_{-\infty}^{\infty} \tilde{\eta}_0(k) \cos \omega t \, e^{ikx} dk$$

$$= \frac{1}{\sqrt{2\pi}} \int_0^{\infty} \tilde{\eta}_0(k) \left[e^{i(kx-\omega t)} + e^{i(kx+\omega t)} \right] dk, \tag{4.8.67}$$

in which $\tilde{\eta}_0(-k) = \tilde{\eta}_0(k)$ is assumed.

Physically, the first and second integrals of (4.8.67) represent waves traveling in the positive and negative directions of x respectively with phase velocity $\dfrac{\omega}{k}$. These integrals describe superposition of all such waves over the wavenumber spectrum $0 < k < \infty$.

For the classical Cauchy Poisson wave problem, $\eta(x) = a\, \delta(x)$ where $\delta(x)$ is the Dirac delta function so that $\tilde{\eta}_0(k) = a/\sqrt{2\pi}$. Thus solution (4.8.67) becomes

$$\eta(x,t) = \frac{a}{2\pi} \int_0^{\infty} \left[e^{i(kx-\omega t)} + e^{i(kx+\omega t)} \right] dk. \tag{4.8.68}$$

The wave integrals (4.8.66) and (4.8.67) represent the exact solution for the velocity potential ϕ and the free surface elevation η for all x and $t > 0$. However, they do not lend any physical interpretations. In general, the exact evaluation of these integrals is almost a formidable task. So it is necessary to resort to asymptotic methods. It would be sufficient for the determination of the principal features of the wave motions to investigate (4.8.67) or (4.8.68) asymptotically for large time t and large distance x with (x/t) held fixed. The asymptotic solution for this kind of problem is available in many standard books (for example, see Debnath, 1994, p 85). We state the stationary phase approximation of a typical wave integral, for $t \to \infty$,

$$\eta(x,t) = \int_a^b f(k) \exp\left[it W(k)\right] dk \qquad (4.8.69)$$

$$\sim f(k_1) \left[\frac{2\pi}{t|W''(k_1)|}\right]^{\frac{1}{2}} \exp\left[i\left\{tW(k_1) + \frac{\pi}{4} sgn\ W''(k_1)\right\}\right], \qquad (4.8.70)$$

where $W(k) = \dfrac{kx}{t} - \omega(k)$, $x > 0$ and $k = k_1$ is a stationary point that satisfies the equation

$$W'(k_1) = \frac{x}{t} - \omega'(k_1) = 0, \quad a < k_1 < b. \qquad (4.8.71)$$

Application of (4.8.70) to (4.8.67) shows that only the first integral in (4.8.67) has a stationary point for $x > 0$. Hence the stationary phase approximation gives the asymptotic solution, as $t \to \infty$, $x > 0$,

$$\eta(x,t) \sim \left[\frac{1}{t|\omega''(k_1)|}\right]^{\frac{1}{2}} \tilde{\eta}_0(k_1) \exp\left[i\left\{(k_1 x - t\omega(k_1))\right\} + \frac{i\pi}{4} sgn\{-\omega''(k_1)\}\right], (4.8.72)$$

where $k_1 = \left(gt^2/4x^2\right)$ is the root of the equation $\omega'(k) = \dfrac{x}{t}$.

On the other hand, when $x < 0$, only the second integral of (4.8.67) has a stationary point $k_1 = \left(gt^2/4x^2\right)$, and hence the same result (4.8.70) can be used to obtain the asymptotic solution for $t \to \infty$ and $x < 0$ as

$$\eta(x,t) \sim \left[\frac{1}{t|\omega''(k_1)|}\right]^{\frac{1}{2}} \tilde{\eta}_0(k_1) \exp\left[i\{t\omega(k_1) - k_1|x|\} + \frac{i\pi}{4} sgn\ \omega''(k_1)\right]. \quad (4.8.73)$$

In particular, for the classical Cauchy Poisson solution (4.8.68), the asymptotic representation for $\eta(x,t)$ follows from (4.8.73) in the form

$$\eta(x,t) \sim \frac{at}{2\sqrt{2\pi}} \frac{\sqrt{g}}{x^{3/2}} \cos\left(\frac{gt^2}{4x}\right), \quad gt^2 >> 4x \qquad (4.8.74)$$

and a similar result for $x < 0$ and $t \to \infty$.

4.9 Summation of Infinite Series

With the aid of Laplace transforms, Wheelon (1954) first developed a direct method to the problem of summing infinite series in closed form. His method is essentially based on the operation that is contained in the summation of both sides of a Laplace transform with respect to the transform variable s which is treated as the dummy index of summation n. This is followed by an interchange of summation and integration that leads to the desired sum as the integral of a geometric or exponential series which can be summed in closed form. We next discuss this procedure in some detail.

If $\bar{f}(s) = \mathcal{L}\{f(x)\}$, then

$$\sum_{n=1}^{\infty} a_n \bar{f}(n) = \sum_{n=1}^{\infty} a_n \int_0^{\infty} f(x) e^{-nx} dx. \tag{4.9.1}$$

In many cases, it is possible to interchange the order of summation and integration so that (4.9.1) gives

$$\sum_{n=1}^{\infty} a_n \bar{f}(n) = \int_0^{\infty} f(t) b(t) dt, \tag{4.9.2}$$

where

$$b(t) = \sum_{n=1}^{\infty} a_n \exp(-nt). \tag{4.9.3}$$

We now assume $f(t) = \dfrac{1}{\Gamma(p)} t^{p-1} \exp(-xt)$ so that $\bar{f}(n) = (n+x)^{-p}$. Consequently, (4.9.2) becomes

$$\sum_{n=1}^{\infty} \frac{a_n}{(n+x)^p} = \frac{1}{\Gamma(p)} \int_0^{\infty} b(t) t^{p-1} \exp(-xt) dt. \tag{4.9.4}$$

This shows that a general series has been expressed in terms of an integral. We next illustrate the method by simple examples.

Example 4.9.1 Show that the sum of the series

$$\sum_{n=1}^{\infty} \frac{1}{n^2} = \frac{\pi^2}{6}. \tag{4.9.5}$$

Putting $x = 0$, $p = 2$, and $a_n = 1$ for all n, we find, from (4.9.3) and (4.9.4),

$$b(t) = \sum_{n=1}^{\infty} \exp(-nt) = \frac{1}{e^t - 1}, \tag{4.9.6}$$

and

$$\sum_{n=1}^{\infty} \frac{1}{n^2} = \int_0^{\infty} \frac{t\, dt}{e^t - 1} = \zeta(2) = \frac{\pi^2}{6}, \tag{4.9.7}$$

in which the following standard result is used.

$$\int_0^\infty \frac{t^{p-1}}{e^{at}-1}\,dt = \frac{\Gamma(p)}{a^p}\,\zeta(p), \qquad (4.9.8)$$

where $\zeta(p)$ is the *Riemann zeta function* defined below by (4.9.10).

Similarly, we can show

$$\sum_{n=1}^\infty \frac{1}{n^3} = \frac{1}{\Gamma(3)}\int_0^\infty \frac{t^2\,dt}{e^t-1} = \zeta(3). \qquad (4.9.9)$$

More generally, we obtain, from (4.9.8),

$$\sum_{n=1}^\infty \frac{1}{n^p} = \frac{1}{\Gamma(p)}\int_0^\infty \frac{t^{p-1}\,dt}{e^t-1} = \zeta(p). \qquad (4.9.10)$$

Example 4.9.2 Show that

$$\sum_{n=1}^\infty \frac{1}{n}\exp(-an) = -\log(1-e^{-a}). \qquad (4.9.11)$$

We put $x=0$, $p=1$, and $a_n = \exp(-an)$ so that

$$b(t) = \sum_{n=1}^\infty \exp[-n(t+a)] = \frac{1}{e^{a+t}-1}. \qquad (4.9.12)$$

Then result (4.9.4) gives

$$\sum_{n=1}^\infty \frac{1}{n}\exp(-an) = \int_0^\infty \frac{dt}{e^{a+t}-1}$$

which is, by letting $\exp(-t) = x$,

$$= \int_0^1 \frac{dx}{e^a - x} = -\log(1-e^{-a}).$$

Example 4.9.3 Show that

$$\sum_{n=1}^\infty \frac{1}{(n^2+x^2)} = \frac{1}{2x^2}(\pi x \coth \pi x - 1). \qquad (4.9.13)$$

We set

$$f(t) = \frac{1}{x}\sin xt, \quad \bar{f}(n) = \frac{1}{n^2+x^2}, \quad \text{and} \quad a_n = 1 \quad \text{for all } n.$$

Clearly

$$b(t) = \sum_{n=1}^\infty \exp(-nt) = \frac{1}{e^t-1}.$$

Thus

$$\sum_{n=1}^\infty \frac{1}{(n^2+x^2)} = \frac{1}{x}\int_0^\infty \frac{\sin xt}{e^t-1}\,dt = \frac{1}{2x^2}(\pi x \coth \pi x - 1).$$

4.10 Exercises

1. Using the Laplace transform, solve the following initial value problems

    (a) $\dfrac{dx}{dt} + ax = e^{-bt}$, $t > 0$, $a \neq b$ with $x(0) = 0$.

    (b) $\dfrac{dx}{dt} - x = t^2$, $t > 0$, $x(0) = 0$.

    (c) $\dfrac{dx}{dt} + 2x = \cos t$, $t > 0$, $x(0) = 1$.

    (d) $\dfrac{dx}{dt} - 2x = 4$, $t > 0$, $x(0) = 0$.

2. Solve the intitial value problem for the radioactive decay of an element
$$\dfrac{dx}{dt} = -kx, \ (k > 0), \ t > 0, \ x(0) = x_0.$$
Prove that the half-lifetime T of the element, which is defined as the time taken for half a given amount of the element to decay, is
$$T = \dfrac{1}{k} \log 2.$$

3. Find the solutions of the following systems of equations with the initial data:

    (a) $\dfrac{dx}{dt} = x - 2y$, $\dfrac{dy}{dt} = y - 2x$, $x(0) = 1$, $y(0) = 0$.

    (b) $\dfrac{dx_1}{dt} = x_1 + 2x_2 + t$, $\dfrac{dx_2}{dt} = x_2 + 2x_1 + t$; $x_1(0) = 2$, $x_2(0) = 4$.

    (c) $\dfrac{dx}{dt} = 6x - 7y + 4z$, $\dfrac{dy}{dt} = 3x - 4y + 2z$, $\dfrac{dz}{dt} = -5x + 5y - 3z$,
    with $x(0) = 5$, $y(0) = z(0) = 0$.

    (d) $\dfrac{dx}{dt} = 2x - 3y$, $\dfrac{dy}{dt} = y - 2x$; $x(0) = 2$, $y(0) = 1$.

4. Solve the matrix differential system
$$\dfrac{dx}{dt} = Ax \text{ with } x(0) = \begin{pmatrix} x_0 \\ 0 \end{pmatrix},$$
where $x = \begin{pmatrix} x_1 \\ x_2 \end{pmatrix}$ and $A = \begin{pmatrix} -3 & -2 \\ 3 & 2 \end{pmatrix}$.

5. Find the solution of the autonomous system described by
$$\dfrac{dx}{dt} = x, \ \dfrac{dy}{dt} = x + 2y \text{ with } x(0) = x_0, \ y(0) = y_0.$$

6. Solve the differential system

$$\left.\begin{array}{l}\dfrac{d^2x}{dt^2} - 2k\dfrac{dy}{dt} + lx = 0 \\ \dfrac{d^2y}{dt^2} + 2k\dfrac{dx}{dt} + ly = 0\end{array}\right\} \quad t > 0$$

with the initial conditions
$$x(0) = a, \quad \dot{x}(0) = 0; \quad y(0) = 0, \quad \dot{y}(0) = v$$
where $k, l, a,$ and v are constants.

7. The glucose concentration in the blood during continuous intravenous injection of glucose is $C(t)$, which is in excess of the initial value at the start of the infusion. The function $C(t)$ satisfies the initial value problem
$$\frac{dC}{dt} + kC = \frac{\alpha}{V}, \quad t > 0, \quad C(0) = 0,$$
where k is the constant velocity of elimination, α is the rate of infusion (in mg/min), and V is the volume in which glucose is distributed. Solve this problem.

8. The blood is pumped into the aorta by the contraction of the heart. The pressure $p(t)$ in the aorta satisfies the initial value problem
$$\frac{dp}{dt} + \frac{c}{k}p = cA \sin \omega t, \quad t > 0; \quad p(0) = p_0$$
where $c, k, A,$ and p_0 are constants. Solve this initial value problem.

9. The zero-order chemical reaction satisfies the initial value problem
$$\frac{dc}{dt} = -k_0, \quad t > 0, \quad \text{with } c = c_0 \text{ at } t = 0$$
where k_0 is a positive constant and $c(t)$ is the concentration of a reacting substance at time t. Show that
$$c(t) = c_0 - k_0 t.$$

10. Solve the equation governing the first order chemical reaction
$$\frac{dc}{dt} = -k_1 c \quad \text{with } c(t) = c_0 \text{ at } t = 0 \quad (k_1 > 0).$$

11. Obtain the solutions of the systems of differential equations governing the consecutive chemical reactions of the first order
$$\frac{dc_1}{dt} = -k_1 c_1, \quad \frac{dc_2}{dt} = k_1 c_1 - k_2 c_2, \quad \frac{dc_3}{dt} = k_2 c_2, \quad t > 0,$$
with the initial conditions
$$c_1(0) = c_1, \quad c_2(0) = c_3(0) = 0,$$
where $c_1(t)$ is the concentration of a substance A at time t, which breaks down to form a new substance A_2 with concentration $c_2(t)$, and $c_3(t)$ is the concentration of a new element originated from A_2.

12. Solve the following initial value problems
   (a) $\ddot{x} + \omega^2 x = \cos nt, \quad (\omega \neq n) \qquad x(0) = 1, \quad \dot{x}(0) = 0.$
   (b) $\ddot{x} + x = \sin 2t, \quad x(0) = \dot{x}(0) = 0.$

(c) $\ddot{x}+x=\sin 2t$, $x(0)=\dot{x}(0)=0$.

(d) $\dfrac{d^3x}{dt^3}+\dfrac{d^2x}{dt^2}=3e^{-4t}$, $x(0)=0$, $\dot{x}(0)=-1$, $\ddot{x}(0)=1$.

(e) $\dfrac{d^4x}{dt^4}=16x$, $x(t)=\ddot{x}(t)=0$, $\dot{x}(t)=\dddot{x}(t)=1$ at $t=0$.

(f) $(D^4+2D^3-D^2-2D+10)x(t)=0$, $t>0$,
$x(0)=-1$, $\dot{x}(0)=3$, $\ddot{x}(0)=-1$, $\dddot{x}(0)=4$.

13. Solve the following systems of equations:

(a) $\ddot{x}-2\dot{y}-x=0$, $\ddot{y}+2\dot{x}-y=0$,
$x(t)=y(t)=0$, $\dot{x}(t)=\dot{y}(t)=1$ at $t=0$.

(b) $\ddot{x}_1+3\dot{x}_1-2x_1+\dot{x}_2-3x_2=2e^{-t}$,
$2\dot{x}_1-x_1+\dot{x}_2-2x_2=0$,
with $x_1(0)=\dot{x}_1(0)=0$ and $x_2(0)=4$.

14. With the aid of the Laplace transform, investigate the motion of a particle governed by the equations of motion $\ddot{x}-\omega\dot{y}=0$, $\ddot{y}+\omega\dot{x}=\omega^2 a$ and the initial conditions $x(0)=y(0)=\dot{x}(0)=\dot{y}(0)=0$.

15. Show that the solution of the equation
$$\frac{d^2y}{dx^2}+(a+b)\frac{dy}{dx}+aby=e^{-\lambda x}, \quad x>0$$
with the initial data $y(x)=\dfrac{1}{\lambda^2}$ and $\dfrac{dy}{dx}=0$ at $x=0$ is
$$y(x)=\frac{1}{a^2(a-b)}\left(ab^{-bx}-be^{-ax}-xa^2e^{-ax}\right)+\frac{e^{-bx}-e^{-ax}}{(a-b)^2}.$$

16. The motion of an electron of charge $-e$ in a static electric field $\mathbf{E}=(E,0,0)$ and a static magnetic field $\mathbf{H}=(0,0,H)$ is governed by the vector equation
$$m\ddot{\mathbf{r}}=-e\mathbf{E}+\frac{e}{c}(\dot{\mathbf{r}}\times\mathbf{H}), \quad t>0,$$
with zero initial velocity and displacement $(\mathbf{r}=\dot{\mathbf{r}}=0$ at $t=0)$ where $\mathbf{r}=(x,y,z)$ and c is the velocity of light. Show that the displacement fields are
$$x(t)=\frac{eE}{m\omega^2}(\cos\omega t-1), \quad y(t)=\frac{eE}{m\omega^2}(\sin\omega t-\omega t), \quad z(t)=0,$$
where $\omega=\dfrac{eH}{mc}$. Hence calculate the velocity field.

17. An electron of mass m and charge $-e$ is acted on by a periodic electric field $E\sin\omega_0 t$ along the x-axis and a constant magnetic field H along the z-axis. Initially, the electron is emitted at the origin with zero velocity. With the same ω as given in exercise 16, show that

$$x(t) = \frac{eE}{m\omega(\omega^2 - \omega_0^2)}(\omega_0 \sin \omega t - \omega \sin \omega_0 t),$$

$$y(t) = \frac{eE}{m\omega(\omega^2 - \omega_0^2)\omega_0}\{(\omega^2 - \omega_0^2) + (\omega_0^2 \cos \omega t - \omega^2 \cos \omega_0 t)\}.$$

18. The stress-strain relation and equation of motion for a viscoelastic rod in the absence of external force are

$$\frac{\partial e}{\partial t} = \frac{1}{E}\frac{\partial \sigma}{\partial t} + \frac{\sigma}{\eta}, \quad \frac{\partial \sigma}{\partial x} = \rho \frac{\partial^2 u}{\partial t^2},$$

where e is the strain, η is the coefficient of viscosity, and the displacement $u(x,t)$ is related to the strain by $e = \frac{\partial u}{\partial x}$. Prove that the stress $\sigma(x,t)$ satisfies the equation

$$\frac{\partial^2 \sigma}{\partial x^2} - \frac{\rho}{\eta}\frac{\partial \sigma}{\partial t} = \frac{1}{c^2}\frac{\partial^2 \sigma}{\partial t^2}.$$

Show that the stress distribution in a semi-infinite viscoelastic rod subject to the boundary and initial conditions

$$\dot{u}(0,t) = UH(t), \quad \sigma(x,t) \to 0 \text{ as } x \to \infty,$$
$$\sigma(x,0) = 0, \quad \dot{u}(x,0) = 0, \quad \text{for } 0 < x < \infty,$$

is given by

$$\sigma(x,t) = -U\rho c \exp\left(-\frac{Et}{2\eta}\right) I_0\left[\frac{E}{2\eta}\left(t^2 - \frac{x^2}{c^2}\right)^{1/2}\right] H\left(t - \frac{x}{c}\right).$$

19. An elastic string is stretched between $x = 0$ and $x = \ell$ and is initially at rest in the equilibrium position. Find the Laplace transform solution for the displacement subject to the boundary conditions $y(0,t) = f(t)$ and $y(l,t) = 0$, $t > 0$

20. The end $x = 0$ of a semi-infinite submarine cable is maintained at a potential $V_0 H(t)$. If the cable has no initial current and potential, determine the potential $V(x,t)$ at a point x and at time t.

21. A semi-infinite lossless transmission line has no initial current or potential. A time-dependent electromagetic force, $V_0(t)H(t)$ is applied at the end $x = 0$. Find the potential $V(x,t)$. Hence determine the potential for cases (i) $V_0(t) = V_0 = $ constant, and (ii) $V_0(t) = V_0 \cos \omega t$.

22. Solve the Blasius problem of an unsteady boundary layer flow in a semi-infinite body of viscous fluid enclosed by an infinite horizontal disk at $z = 0$. The governing equation and the boundary and initial conditions are

$$\frac{\partial u}{\partial t} = v\frac{\partial^2 u}{\partial z^2}, \quad z > 0, \quad t > 0,$$
$$u(z,t) = Ut \quad \text{on } z = 0, \quad t > 0,$$
$$u(z,t) \to 0 \quad \text{as } z \to \infty, \quad t > 0,$$

$$u(z,t) = 0 \quad \text{at } t \leq 0, \; z > 0.$$

Explain the significance of the solution.

23. Obtain the solution of the Stokes-Ekman problem of an unsteady boundary layer flow in a semi-infinite body of viscous fluid bounded by an infinite horizontal disk at $z = 0$, when both the fluid and the disk rotate with a uniform angular velocity Ω about the z-axis. The governing boundary layer equation, the boundary and the initial conditions are

$$\frac{\partial q}{\partial t} + 2\Omega i q = v \frac{\partial^2 q}{\partial z^2}, \quad z > 0,$$

$$q(z,t) = ae^{i\omega t} + be^{-i\omega t} \quad \text{on } z = 0, \; t > 0,$$

$$q(z,t) \to 0 \quad \text{as } z \to \infty, \; t > 0,$$

$$q(z,t) = 0 \quad \text{at } t \leq 0 \text{ for all } z > 0,$$

where $q = u + iv$, ω is the frequency of oscillations of the disk and a, b are complex constants. Hence deduce the steady-state solution and determine the structure of the associated boundary layers.

24. Show that, when $\omega = 0$ in exercise 23, the steady flow field is given by

$$q(z,t) \sim (a+b)\exp\left\{\left(-\frac{2i\Omega}{v}\right)^{1/2} z\right\}.$$

Hence determine the thickness of the Ekman layer.

25. Solve the following integral equations:

(a) $f(t) = \sin 2t + \int_0^t f(t-\tau)\sin \tau \, d\tau.$

(b) $f(t) = \frac{t}{2}\sin t + \int_0^t f''(\tau)\sin(t-\tau)d\tau, \quad f(0) = f'(0) = 0.$

(c) $\int_0^t f(\tau) J_0[a(t-\tau)] \, d\tau = \sin at.$

(d) $f(t) = \sin t + \int_0^t f(\tau)\sin\{2(t-\tau)\}d\tau.$

(e) $f(t) = t^2 + \int_0^t f'(t-\tau)\exp(-a\tau)d\tau, \quad f(0) = 0.$

26. Prove that the solution of the integral equation

$$f(t) = \frac{2}{\sqrt{\pi}}\left[\sqrt{t} + a\int_0^t (t-\tau)^{1/2} f'(\tau)d\tau\right], \quad f(0) = 0$$

is

$$f(t) = \frac{e^{at}}{\sqrt{a}}\left[1 + \operatorname{erf}\sqrt{at}\right] - \frac{1}{\sqrt{a}}.$$

27. Solve the integro-differential equation

$$\frac{d^2x}{dt^2} = \exp(-2t) - \int_0^t \exp\{-2(t-\tau)\}\left(\frac{dx}{d\tau}\right) d\tau.$$

with $x(0) = 0$ and $\dot{x}(0) = 0$.

28. Using the Laplace transform, evaluate the following integrals:

(a) $\int_0^\infty \frac{\sin tx}{x(x^2 + a^2)} dx$, $(a, t > 0)$,

(b) $\int_0^\infty \frac{\sin tx}{x} dx$,

(c) $\int_{-\infty}^\infty \frac{\cos tx}{x^2 + a^2} dx$, $(a, t > 0)$,

(d) $\int_{-\infty}^\infty \frac{x \sin xt}{x^2 + a^2} dx$, $(a, t > 0)$,

(e) $\int_0^\infty \exp(-tx^2) dx$, $t > 0$.

29. Show that

(a) $\int_0^\infty e^{-ax}\left(\frac{\cos px - \cos qx}{x}\right) dx = \frac{1}{2}\log\left(\frac{a^2 + q^2}{a^2 + p^2}\right)$, $(a > 0)$.

(b) $\int_0^\infty e^{-ax}\left(\frac{\sin qx - \sin px}{x}\right) dx = \tan^{-1}\left(\frac{q}{a}\right) - \tan^{-1}\left(\frac{p}{a}\right)$, $a > 0$.

30. Establish the following results:

(a) $\int_{-\infty}^\infty \frac{\cos tx \, dx}{(x^2 + a^2)(x^2 + b^2)} = \frac{\pi}{a^2 - b^2}\left(\frac{e^{-bt}}{b} - \frac{e^{-at}}{a}\right)$, $a, b, t > 0$.

(b) $\int_0^\infty \frac{\sin(\pi tx)}{x(1+x^2)} dx = \frac{\pi}{2}(1 - \cos \pi t)$, $t > 0$.

(c) $\int_0^\infty \cos(tu^2) du = \int_0^\infty \sin(tu^2) du = \frac{1}{2}\left(\frac{\pi}{2t}\right)^{1/2}$, $t > 0$.

31. In Example 4.5.1(i), write down the solution when the point load is applied at the midpoint of the beam.

32. A uniform horizontal beam of length 2ℓ is clamped at the end $x = 0$ and freely supported at $x = 2\ell$. It carries a distributed load of constant value W in $\frac{\ell}{2} < x < \frac{3\ell}{2}$ and zero elsewhere. Obtain the deflection of the beam wh. h satisfies the boundary value problem

$$EI \frac{d^4y}{dx^4} = W\left[H\left(x - \frac{\ell}{2}\right) - H\left(x - \frac{3\ell}{2}\right)\right], \quad 0 < x < 2\ell,$$

$$y(0) = 0 = y'(0), \quad y''(2\ell) = 0 = y'''(2\ell).$$

33. Solve exercise 32 if the beam carries a constant distributed load W per unit length in $0 < x < \ell$ and zero in $\ell < x < 2\ell$. Find the bending moment and shear at $x = \frac{\ell}{2}$.

34. A horizontal cantilever beam of length 2ℓ is deflected under the combined effect of its own constant weight W and a point load of magnitude P located at the midpoint. Obtain the deflection of the beam which satisfies the boundary value problem

$$EI \frac{d^4y}{dx^4} = W[H(x) - H(x-2\ell)] + P\,\delta(x-\ell), \quad 0 < x < 2\ell,$$
$$y(0) = 0 = y'(0), \quad y''(2\ell) = 0 = y'''(2\ell).$$

Find the bending moment and shear at $x = \dfrac{\ell}{2}$.

35. Using the Laplace transform, solve the following difference equations:
    (a) $\Delta u_n - 2u_n = 0, \quad u_0 = 1$,
    (b) $\Delta^2 u_n - 2u_{n+1} + 3u_n = 0, \quad u_0 = 0 \text{ and } u_1 = 1$,
    (c) $u_{n+2} - 4u_{n+1} + 4u_n = 0, \quad u_0 = 1 \text{ and } u_1 = 4$,
    (d) $u_{n+2} - 5u_{n+1} + 6u_n = 0, \quad u_0 = 1 \text{ and } u_1 = 4$,
    (e) $\Delta^2 u_n + 3u_n = 0, \quad u_0 = 0, \quad u_1 = 1$,
    (f) $u_{n+2} - 4u_{n+1} + 3u_n = 0$,
    (g) $u_{n+2} - 9u_n = 0, \quad u_0 = 1 \text{ and } u_1 = 3$,
    (h) $\Delta u_n - (a-1)u_n = 0, \quad u_0 = \text{constant}$.

36. Show that the solution of the difference equation
$$u_{n+2} + 4u_{n+1} + u_n = 0, \quad \text{with } u_0 = 0 \text{ and } u_1 = 1,$$
is
$$u_n = \frac{1}{2\sqrt{3}}\left[\left(\sqrt{3}-2\right)^n + (-1)^{n+1}\left(2+\sqrt{3}\right)^n\right].$$

37. Show that the solution of the differential-difference equation
$$\dot{u}(t) - u(t-1) = 2, \quad u(0) = 0$$
is
$$u(t) = 2\left[t - \frac{(t-1)^2}{2!} + \frac{(t-2)^3}{3!} + \cdots + \frac{(t-n)^{n+1}}{(n+1)!}\right], \quad t > n.$$

38. Obtain the solution of the differential-difference equation
$$\dot{u} = u(t-1), \quad u(0) = 1, \quad 0 < t < \infty \text{ with } u(t) = 1 \text{ when } -1 \le t < 0.$$

39. Use the Laplace transform to solve the initial-boundary value problem
$$u_{tt} - u_{xx} = k^2 u_{xxtt}, \quad 0 < x < \infty, \quad t > 0,$$
$$u(x,0) = 0, \quad \left(\frac{\partial u}{\partial x}\right)_{t=0} = 0, \quad \text{for } x > 0,$$
$$u(x,t) \to 0 \quad \text{as } x \to \infty, \quad t > 0,$$
$$u(0,t) = 1 \quad \text{for } t > 0.$$

Hence show that
$$\left(\frac{\partial u}{\partial x}\right)_{x=0} = -\frac{1}{k} J_0\!\left(\frac{t}{k}\right).$$

40. Solve the telegraph equation
$$u_{tt} - c^2 u_{xx} + 2au_t = 0, \quad -\infty < x < \infty, \quad t > 0,$$
$$u(x,0) = 0, \quad u_t(x,0) = g(x).$$
41. Use the joint Laplace and Fourier transform to solve Example 2.7.3 in Chapter 2.
42. Use the Laplace transform to solve the initial-boundary value problem
$$u_t = c^2 u_{xx}, \quad 0 < x < a, \quad t > 0,$$
$$u(x,0) = x + \sin\left(\frac{3\pi x}{a}\right) \quad \text{for } 0 < x < a,$$
$$u(0,t) = 0 = u(a,t) \quad \text{for } t > 0.$$
43. Solve the diffusion equation
$$u_t = k u_{xx}, \quad -a < x < a, \quad t > 0,$$
$$u(x,0) = 1 \quad \text{for } -a < x < a,$$
$$u(-a,t) = 0 = u(a,t) \quad \text{for } t > 0.$$
44. Use the joint Laplace and Fourier transform to solve the initial value problem for water waves which satisfies (see Debnath, 1994, p. 92)
$$\nabla^2 \phi = \phi_{xx} + \phi_{zz} = 0, \quad -\infty < z < 0, \quad -\infty < x < \infty, \quad t > 0$$
$$\left.\begin{array}{l}\phi_z = \eta_t \\ \phi_t + g\eta = -\dfrac{P}{\rho} p(x) e^{i\omega t}\end{array}\right\} \quad \text{on } z = 0, \quad t > 0,$$
$$\phi(x,z,0) = 0 = \eta(x,0) \quad \text{for all } x \text{ and } z,$$
where P and ρ are constants.

45. Show that

(a) $\displaystyle\sum_{n=0}^{\infty} \frac{a_n}{\sqrt{n^2 + x^2}} = \int_0^{\infty} b(t) J_0(xt)\, dt$, where $b(t)$ is given by (4.9.3).

(b) $\displaystyle\sum_{n=0}^{\infty} \frac{1}{n^2 - a^2} = \frac{1}{2a^2}(1 - \pi a \cot \pi a)$.

46. Show that

(a) $\displaystyle\sum_{n=1}^{\infty} \frac{(-1)^n \cos nx}{(n^2 - a^2)} = \frac{1}{2a^2}\left[1 - \frac{\pi a \cos ax}{\sin a\pi}\right]$.

(b) $\displaystyle\sum_{n=1}^{\infty} \log\left(1 + \frac{a^2}{n^2}\right) = \log\left(\frac{\sinh \pi a}{\pi a}\right)$.

Chapter 5

Hankel Transforms

5.1 Introduction

The Hankel transform involving Bessel functions as the kernel arises naturally in the discussion of axisymmetric problems formulated in cylindrical polar coordinates. This chapter deals with the definition and basic operational properties of the Hankel transform. A large number of axisymmetric problems in cylindrical polar coordinates are solved with the aid of the Hankel transform. The use of the joint Laplace and Hankel transforms is illustrated by several examples of applications to partial differential equations.

5.2 The Hankel Transform and Examples

We introduce the definition of the Hankel transform from the two-dimensional Fourier transform and its inverse given by

$$\mathscr{F}\{f(x,y)\} = F(k,l) = \frac{1}{2\pi} \int_{-\infty}^{\infty} \int_{-\infty}^{\infty} \exp\{-i(\boldsymbol{\kappa} \cdot \mathbf{r})\} f(x,y)\, dx\, dy, \qquad (5.2.1)$$

$$\mathscr{F}^{-1}\{F(k,l)\} = f(x,y) = \frac{1}{2\pi} \int_{-\infty}^{\infty} \int_{-\infty}^{\infty} \exp\{i(\boldsymbol{\kappa} \cdot \mathbf{r})\} F(k,l)\, dk\, dl, \qquad (5.2.2)$$

where $\mathbf{r} = (x,y)$ and $\boldsymbol{\kappa} = (k,l)$. Introducing polar coordinates $(x,y) = r(\cos\theta, \sin\theta)$ and $(k,l) = \kappa(\cos\phi, \sin\phi)$, we find $\boldsymbol{\kappa} \cdot \mathbf{r} = \kappa r \cos(\theta - \phi)$ and then

$$F(\kappa,\phi) = \frac{1}{2\pi} \int_0^{\infty} r\, dr \int_0^{2\pi} \exp[-i\kappa r \cos(\theta - \phi)] f(r,\theta)\, d\theta. \qquad (5.2.3)$$

We next assume $f(r,\theta) = \exp(in\theta) f(r)$, which is not a very severe restriction, and make a change of variable $\theta - \phi = \alpha - \dfrac{\pi}{2}$ to reduce (5.2.3) to the form

$$F(\kappa,\phi) = \frac{1}{2\pi} \int_0^{\infty} r f(r)\, dr \int_{\phi_o}^{2\pi + \phi_o} \exp\left[in\left(\phi - \frac{\pi}{2}\right) + i(n\alpha - \kappa r \sin\alpha)\right] d\alpha, \qquad (5.2.4)$$

where $\phi_0 = \dfrac{\pi}{2} - \phi$.

Using the integral representation of the Bessel function of order n

193

$$J_n(\kappa r) = \frac{1}{2\pi} \int_{\phi_0}^{2\pi+\phi_0} \exp[i(n\alpha - \kappa r \sin\alpha)] d\alpha \qquad (5.2.5)$$

integral (5.2.4) becomes

$$F(\kappa,\phi) = \exp\left[in\left(\phi - \frac{\pi}{2}\right)\right] \int_0^\infty r J_n(\kappa r) f(r) dr \qquad (5.2.6)$$

$$= \exp\left[in\left(\phi - \frac{\pi}{2}\right)\right] \tilde{f}_n(\kappa), \qquad (5.2.7)$$

where $\tilde{f}_n(\kappa)$ is called the *Hankel transform* of $f(r)$ and is defined formally by

$$\mathcal{H}_n\{f(r)\} = \tilde{f}_n(\kappa) = \int_0^\infty r J_n(\kappa r) f(r) dr. \qquad (5.2.8)$$

Similarly, in terms of the polar variables with the assumption $f(x,y) = f(r,\theta) = e^{in\theta} f(r)$ with (5.2.7), the inverse Fourier transform (5.2.2) becomes

$$e^{in\theta} f(r) = \frac{1}{2\pi} \int_0^\infty \kappa \, d\kappa \int_0^{2\pi} \exp[i\kappa r \cos(\theta - \phi)] F(\kappa,\phi) d\phi$$

$$= \frac{1}{2\pi} \int_0^\infty \kappa \tilde{f}_n(\kappa) d\kappa \int_0^{2\pi} \exp\left[in\left(\phi - \frac{\pi}{2}\right) + i\kappa r \cos(\theta - \phi)\right] d\phi,$$

which is, by the change of variables $\theta - \phi = -\left(\alpha + \frac{\pi}{2}\right)$ and $\theta_0 = -\left(\theta + \frac{\pi}{2}\right)$,

$$= \frac{1}{2\pi} \int_0^\infty \kappa \, \tilde{f}_n(\kappa) d\kappa \int_{\theta_0}^{2\pi+\theta_0} \exp[in(\theta + \alpha) - i\kappa r \sin\alpha] d\alpha$$

$$= e^{in\theta} \int_0^\infty \kappa J_n(\kappa r) \tilde{f}_n(\kappa) d\kappa, \qquad \text{by (5.2.5).} \qquad (5.2.9)$$

Thus the *inverse Hankel transform* is defined by

$$\mathcal{H}_n^{-1}[\tilde{f}_n(\kappa)] = f(r) = \int_0^\infty \kappa J_n(\kappa r) \tilde{f}_n(\kappa) d\kappa. \qquad (5.2.10)$$

Instead of $\tilde{f}_n(\kappa)$, we often simply write $\tilde{f}(\kappa)$ for the Hankel transform specifying the order. Integrals (5.2.8) and (5.2.10) exist for certain large classes of functions, which usually occur in physical applications.

Alternatively, the famous Hankel integral formula (Watson, 1944, p 453)

$$f(r) = \int_0^\infty \kappa J_n(\kappa r) d\kappa \int_0^\infty p J_n(\kappa p) f(p) dp, \qquad (5.2.11)$$

can be used to define the Hankel transform (5.2.8) and its inverse (5.2.10).

In particular, the Hankel transforms of the zero order $(n = 0)$ and of order one $(n = 1)$ are often useful for the solution of problems involving Laplace's equation in an axisymmetric cylindrical geometry.

Example 5.2.1 Obtain the zero-order Hankel transforms of

(a) $r^{-1} \exp(-ar)$, (b) $\dfrac{\delta(r)}{r}$, (c) $H(a-r)$,

where $H(r)$ is the Heaviside unit step function.

(a) $\tilde{f}(\kappa) = \mathcal{H}_0 \left\{ \dfrac{1}{r} \exp(-ar) \right\} = \int_0^\infty \exp(-ar) J_0(\kappa r) dr = \dfrac{1}{\sqrt{\kappa^2 + a^2}}.$

(b) $\tilde{f}(\kappa) = \mathcal{H}_0 \left\{ \dfrac{\delta(r)}{r} \right\} = \int_0^\infty \delta(r) J_0(\kappa r) dr = 1.$

(c) $\tilde{f}(\kappa) = \mathcal{H}_0 \{ H(a-r) \} = \int_0^a r J_0(\kappa r) dr = \dfrac{1}{\kappa^2} \int_0^{a\kappa} p J_0(p) dp$

$= \dfrac{1}{\kappa^2} [p J_1(p)]_0^{a\kappa} = \dfrac{a}{\kappa} J_1(a\kappa).$

Example 5.2.2 Find the first order Hankel transforms of

(a) $f(r) = e^{-ar}$, (b) $f(r) = \dfrac{1}{r} e^{-ar}$.

(a) $\tilde{f}(\kappa) = \mathcal{H}_1 \{ e^{-ar} \} = \int_0^\infty r e^{-ar} J_1(\kappa r) dr = \dfrac{\kappa}{(a^2 + \kappa^2)}.$

(b) $\tilde{f}(\kappa) = \mathcal{H}_1 \left\{ \dfrac{e^{-ar}}{r} \right\} = \int_0^\infty e^{-ar} J_1(\kappa r) dr = \dfrac{1}{\kappa} \left[1 - a(\kappa^2 + a^2)^{-\tfrac{1}{2}} \right].$

Example 5.2.3 Find the nth order Hankel transforms of

(a) $f(r) = r^n H(a-r)$, (b) $f(r) = r^n \exp(-ar^2)$.

(a) $\tilde{f}(\kappa) = \mathcal{H}_n [r^n H(a-r)] = \int_0^a r^{n+1} J_n(\kappa r) dr = \dfrac{a^{n+1}}{\kappa} J_{n+1}(a\kappa).$

(b) $\tilde{f}(\kappa) = \mathcal{H}_n [r^n \exp(-ar^2)]$

$= \int_0^\infty r^{n+1} J_n(\kappa r) \exp(-ar^2) dr = \dfrac{\kappa^n}{(2a)^{n+1}} \exp\left(-\dfrac{\kappa^2}{4a} \right).$

5.3 Operational Properties of the Hankel Transform

Theorem 5.3.1 (Scaling). If $\mathcal{H}_n \{ f(r) \} = \tilde{f}_n(\kappa)$, then

$$\mathcal{H}_n\{f(ar)\} = \frac{1}{a^2} \tilde{f}_n\left(\frac{\kappa}{a}\right), \qquad a > 0. \tag{5.3.1}$$

Proof. We have, by definition,

$$\mathcal{H}_n\{f(ar)\} = \int_0^\infty r J_n(\kappa r) f(ar) dr$$

$$= \frac{1}{a^2} \int_0^\infty s J_n\left(\frac{\kappa}{a} s\right) f(s) ds = \frac{1}{a^2} \tilde{f}_n\left(\frac{\kappa}{a}\right).$$

Theorem 5.3.2 (*Parseval's Relation*). If $\tilde{f}(\kappa) = \mathcal{H}_n\{f(r)\}$ and $\tilde{g}(\kappa) = \mathcal{H}_n\{g(r)\}$, then

$$\int_0^\infty r f(r) g(r) dr = \int_0^\infty \kappa \tilde{f}(\kappa) \tilde{g}(\kappa) d\kappa. \tag{5.3.2}$$

Proof. We proceed formally to obtain

$$\int_0^\infty \kappa \tilde{f}(\kappa) \tilde{g}(\kappa) d\kappa = \int_0^\infty \kappa \tilde{f}(\kappa) d\kappa \int_0^\infty r J_n(\kappa r) g(r) dr$$

which is, interchanging the order of integration,

$$= \int_0^\infty r g(r) dr \int_0^\infty \kappa J_n(\kappa r) \tilde{f}(\kappa) d\kappa$$

$$= \int_0^\infty r g(r) f(r) dr.$$

Theorem 5.3.3 (*Hankel Transforms of Derivatives*). If $\tilde{f}_n(k) = \mathcal{H}_n\{f(r)\}$, then

$$\mathcal{H}_n\{f'(r)\} = \frac{\kappa}{2n}\left[(n-1)\tilde{f}_{n+1}(\kappa) - (n+1)\tilde{f}_{n-1}(\kappa)\right], \qquad n \geq 1, \tag{5.3.3}$$

$$\mathcal{H}_1\{f'(r)\} = -\kappa \tilde{f}_0(\kappa), \tag{5.3.4}$$

provided $[r f(r)]$ vanishes as $r \to 0$ and $r \to \infty$.

Proof. We have, by definition,

$$\mathcal{H}_n\{f'(r)\} = \int_0^\infty r J_n(\kappa r) f'(r) dr$$

which is, by integrating by parts,

$$= [r f(r) J_n(\kappa r)]_0^\infty - \int_0^\infty f(r) \frac{d}{dr}[r J_n(\kappa r)] dr. \tag{5.3.5}$$

We now use the properties of the Bessel function

$$\frac{d}{dr}[r J_n(\kappa r)] = J_n(\kappa r) + r\kappa J_n'(\kappa r) = J_n(\kappa r) + r\kappa J_{n-1}(\kappa r) - n J_n(\kappa r)$$

$$= (1-n) J_n(\kappa r) + r\kappa J_{n-1}(\kappa r). \tag{5.3.6}$$

Integral Transforms and Their Applications

In view of the given condition, the first term of (5.3.5) vanishes as $r \to 0$ and $r \to \infty$, and the derivative within the integral in (5.3.5) can be replaced by (5.3.6) so that (5.3.5) becomes

$$\mathcal{H}_n\{f'(r)\} = (n-1)\int_0^\infty f(r) J_n(\kappa r)\, dr - \kappa\, \tilde{f}_{n-1}(\kappa). \tag{5.3.7}$$

We next use the standard recurrence relation for the Bessel function

$$J_n(\kappa r) = \frac{\kappa r}{2n}[J_{n-1}(\kappa r) + J_{n+1}(\kappa r)]. \tag{5.3.8}$$

Thus, (5.3.7) can be rewritten as

$$\mathcal{H}_n[f'(r)] = -\kappa \tilde{f}_{n-1}(\kappa) + \kappa\left(\frac{n-1}{2n}\right)\left[\int_0^\infty r f(r)\{J_{n-1}(\kappa r) + J_{n+1}(\kappa r)\} dr\right]$$

$$= -\kappa \tilde{f}_{n-1}(\kappa) + \kappa\left(\frac{n-1}{2n}\right)\left[\tilde{f}_{n-1}(\kappa) + \tilde{f}_{n+1}(\kappa)\right]$$

$$= \left(\frac{\kappa}{2n}\right)\left[(n-1)\tilde{f}_{n+1}(\kappa) - (n+1)\tilde{f}_{n-1}(\kappa)\right].$$

In particular, when $n=1$, (5.3.4) follows immediately.

Similarly, repeated applications of (5.3.3) lead to the following result

$$\mathcal{H}_n\{f''(r)\} = \frac{\kappa}{2n}\left[(n-1)\mathcal{H}_{n+1}\{f'(r)\} - (n+1)\mathcal{H}_{n-1}\{f'(r)\}\right]$$

$$= \frac{\kappa^2}{4}\left[\left(\frac{n+1}{n-1}\right)\tilde{f}_{n-2}(\kappa) - 2\left(\frac{n^2-3}{n^2-1}\right)\tilde{f}_n(\kappa) + \left(\frac{n-1}{n+1}\right)\tilde{f}_{n+2}(\kappa)\right]. \tag{5.3.9}$$

Theorem 5.3.4 If $\mathcal{H}_n\{f(r)\} = \tilde{f}_n(\kappa)$, then

$$\mathcal{H}_n\left\{\left(\nabla^2 - \frac{n^2}{r^2}\right) f(r)\right\} = \mathcal{H}_n\left\{\frac{1}{r}\frac{d}{dr}\left(r\frac{df}{dr}\right) - \frac{n^2}{r^2} f(r)\right\} = -\kappa^2\, \tilde{f}_n(\kappa), \tag{5.3.10}$$

provided both $r f'(r)$ and $r f(r)$ vanish as $r \to 0$ and $r \to \infty$.

Proof. We have, by definition (5.2.8),

$$\mathcal{H}_n\left\{\frac{1}{r}\frac{d}{dr}\left(r\frac{df}{dr}\right) - \frac{n^2}{r^2} f(r)\right\} = \int_0^\infty J_n(\kappa r)\left[\frac{d}{dr}\left(r\frac{df}{dr}\right)\right] dr - \int_0^\infty \frac{n^2}{r^2}[r J_n(\kappa r)] f(r)\, dr$$

which is, invoking integration by parts,

$$= \left[\left(r\frac{df}{dr}\right) J_n(\kappa r)\right]_0^\infty - \kappa\int_0^\infty r \frac{df}{dr} J_n'(\kappa r)\, dr - \int_0^\infty \frac{n^2}{r^2}[r J_n(\kappa r)] f(r)\, dr,$$

which is, by replacing the first term with zero because of the given assumption, and by invoking integration by parts again,

$$= -[\kappa r\, f(r) J_n'(\kappa r)]_0^\infty + \int_0^\infty \frac{d}{dr}[\kappa r\, J_n'(\kappa r)] f(r)\, dr - \int_0^\infty \frac{n^2}{r^2}[r J_n(\kappa r)] f(r)\, dr.$$

We use the given assumptions and Bessel's differential equation,

$$\frac{d}{dr}[\kappa r J_n'(\kappa r)] + r\left(\kappa^2 - \frac{n^2}{r^2}\right)J_n(\kappa r) = 0, \tag{5.3.11}$$

to obtain

$$\mathcal{H}_n\left\{\left(\nabla^2 - \frac{n^2}{r^2}\right)f(r)\right\} = -\int_0^\infty \left(\kappa^2 - \frac{n^2}{r^2}\right)r f(r) J_n(\kappa r)\, dr - \int_0^\infty \frac{n^2}{r^2}[r f(r)] J_n(\kappa r)\, dr$$

$$= -\kappa^2 \int_0^\infty r J_n(\kappa r) f(r)\, dr = -\kappa^2\, \mathcal{H}_n[f(r)] = -\kappa^2\, \tilde{f}_n(\kappa).$$

This proves the theorem.

In particular, when $n = 0$ and $n = 1$, we obtain

$$\mathcal{H}_0\left\{\frac{1}{r}\frac{d}{dr}\left(r\frac{df}{dr}\right)\right\} = -\kappa^2 \tilde{f}_0(\kappa), \tag{5.3.12}$$

$$\mathcal{H}_1\left\{\frac{1}{r}\frac{d}{dr}\left(r\frac{df}{dr}\right) - \frac{1}{r^2}f(r)\right\} = -\kappa^2 \tilde{f}_1(\kappa). \tag{5.3.13}$$

Results (5.3.10), (5.3.12), and (5.3.13) are widely used for finding solutions of partial differential equations in axisymmetric cylindrical configurations. We illustrate this point by considering several examples of applications.

5.4 Applications of Hankel Transforms to Partial Differential Equations

The Hankel transforms are extremely useful in solving a variety of partial differential equations in cylindrical polar coordinates. The following examples illustrate applications of the Hankel transforms. The examples given here are only representative of a whole variety of physical problems that can be solved in a similar way.

Example 5.4.1 (*Free Vibration of a Large Circular Membrane*). Obtain the solution of the free vibration of a large circular elastic membrane governed by the initial value problem

$$c^2\left(\frac{\partial^2 u}{\partial r^2} + \frac{1}{r}\frac{\partial u}{\partial r}\right) = \frac{\partial^2 u}{\partial t^2}, \quad 0 < r < \infty, \quad t > 0, \tag{5.4.1}$$

$$u(r,0) = f(r), \quad u_t(r,0) = g(r), \quad \text{for } 0 \le r < \infty, \tag{5.4.2ab}$$

where $c^2 = (T/\rho)$ = constant, T is the tension in the membrane, and ρ is the surface density of the membrane.

Application of the zero-order Hankel transform with respect to r

$$\tilde{u}(\kappa,t) = \int_0^\infty r J_0(\kappa r) u(r,t)\, dr, \tag{5.4.3}$$

to (5.4.1)-(5.4.2ab) gives

$$\frac{d^2\tilde{u}}{dt^2} + c^2\kappa^2 \tilde{u} = 0, \tag{5.4.4}$$

$$\tilde{u}(\kappa,0) = \tilde{f}(\kappa), \quad \tilde{u}_t(\kappa,0) = \tilde{g}(\kappa). \tag{5.4.5ab}$$

The general solution of this transformed system is

$$\tilde{u}(\kappa,t) = \tilde{f}(\kappa) \cos(c\kappa t) + \tilde{g}(\kappa) \sin(c\kappa t). \tag{5.4.6}$$

The inverse Hankel transform leads to the solution

$$u(r,t) = \int_0^\infty \kappa \, \tilde{f}(\kappa) \cos(c\kappa t) J_0(\kappa r) d\kappa$$

$$+ \int_0^\infty \kappa \, \tilde{g}(\kappa) \sin(c\kappa t) J_0(\kappa r) d\kappa. \tag{5.4.7}$$

In particular, we consider

$$u(r,0) = f(r) = Aa(r^2 + a^2)^{-\frac{1}{2}}, \quad u_t(r,0) = g(r) = 0, \tag{5.4.8ab}$$

so that $\tilde{g}(\kappa) \equiv 0$ and

$$\tilde{f}(\kappa) = Aa \int_0^\infty r(a^2 + r^2)^{-\frac{1}{2}} J_0(\kappa r) dr = \frac{Aa}{\kappa} e^{-a\kappa}, \text{ by Example 5.2.1(a)}.$$

Thus the formal solution (5.4.7) becomes

$$u(r,t) = Aa \int_0^\infty e^{-a\kappa} J_0(\kappa r) \cos(c\kappa t) d\kappa = Aa \, \text{Re} \int_0^\infty \exp[-\kappa(a + ict)] J_0(\kappa r) d\kappa$$

$$= Aa \, \text{Re} \left\{ r^2 + (a + ict)^2 \right\}^{-\frac{1}{2}}, \quad \text{by Example 5.2.1(a)}. \tag{5.4.9}$$

Example 5.4.2 (*Steady Temperature Distribution in a Semi-Infinite Solid with a Steady Heat Source*). Find the solution of the Laplace equation for the steady temperature distribution $u(r,z)$ with a steady and symmetric heat source $Q_o \, q(r)$:

$$u_{rr} + \frac{1}{r} u_r + u_{zz} = -Q_o q(r), \quad 0 < r < \infty, \quad 0 < z < \infty, \tag{5.4.10}$$

$$u(r,0) = 0, \quad 0 < r < \infty, \tag{5.4.11}$$

where Q_o is a constant. This boundary condition represents zero temperature at the boundary $z = 0$.

Application of the zero-order Hankel transform to (5.4.10) and (5.4.11) gives

$$\frac{d^2\tilde{u}}{dz^2} - \kappa^2 \tilde{u} = -Q_o \tilde{q}(\kappa), \quad \tilde{u}(\kappa,0) = 0.$$

The bounded general solution of this system is

$$\tilde{u}(\kappa,z) = A \, \exp(-\kappa z) + \frac{Q_o}{\kappa^2} \tilde{q}(\kappa),$$

where A is a constant to be determined from the transformed boundary condition. In this case

$$A = -\frac{Q_o}{\kappa^2}\tilde{q}(\kappa).$$

Thus the formal solution is

$$\tilde{u}(\kappa,z) = \frac{Q_o\,\tilde{q}(\kappa)}{\kappa^2}\left(1-e^{-\kappa z}\right). \tag{5.4.12}$$

The inverse Hankel transform yields the exact integral solution

$$u(r,z) = Q_o\int_0^\infty \frac{\tilde{q}(\kappa)}{\kappa}\left(1-e^{-\kappa z}\right)J_0(\kappa r)\,d\kappa. \tag{5.4.13}$$

Example 5.4.3 (*Axisymmetric Diffusion Equation*). Find the solution of the axisymmetric diffusion equation

$$u_t = \kappa\left(u_{rr} + \frac{1}{r}u_r\right), \qquad 0<r<\infty,\ t>0, \tag{5.4.14}$$

where $\kappa(>0)$ is a diffusivity constant and

$$u(r,0) = f(r), \qquad \text{for } 0<r<\infty. \tag{5.4.15}$$

We apply the zero-order Hankel transform defined by (5.4.3) to obtain

$$\frac{d\tilde{u}}{dt} + k^2\kappa\tilde{u} = 0, \qquad \tilde{u}(k,0) = \tilde{f}(k),$$

where k is the Hankel transform variable. The solution of this transformed system is

$$\tilde{u}(k,t) = \tilde{f}(k)\exp(-\kappa k^2 t). \tag{5.4.16}$$

Application of the inverse Hankel transform gives

$$u(r,t) = \int_0^\infty k\,\tilde{f}(k)\,J_0(kr)\,e^{-\kappa k^2 t}\,dk = \int_0^\infty k\left[\int_0^\infty l\,J_0(kl)f(l)\,dl\right]e^{-\kappa k^2 t}J_0(kr)\,dk$$

which is, interchanging the order of integration,

$$= \int_0^\infty l\,f(l)\,dl\int_0^\infty k\,J_0(kl)\,J_0(kr)\exp(-\kappa k^2 t)\,dk. \tag{5.4.17}$$

Using a standard table of integrals involving Bessel functions, we state

$$\int_0^\infty k\,J_0(kl)\,J_0(kr)\exp(-k^2\kappa t)\,dk = \frac{1}{2\kappa t}\exp\left[-\frac{(r^2+l^2)}{4\kappa t}\right]I_0\left(\frac{rl}{2\kappa t}\right), \tag{5.4.18}$$

where $I_0(x)$ is the modified Bessel function and $I_0(0)=1$. In particular, when $l=0$, $J_0(0)=1$ and integral (5.4.18) becomes

$$\int_0^\infty k\,J_0(kr)\exp(-k^2\kappa t)\,dk = \frac{1}{2\kappa t}\exp\left(-\frac{r^2}{4\kappa t}\right). \tag{5.4.19}$$

We next use (5.4.18) to rewrite (5.4.17) as

$$u(r,t) = \frac{1}{2\kappa t}\int_0^\infty l\,f(l)\,I_0\left(\frac{rl}{2\kappa t}\right)\exp\left[-\frac{(r^2+l^2)}{4\kappa t}\right]dl. \tag{5.4.20}$$

We now assume $f(r)$ to represent a heat source concentrated in a circle of radius a and allow $a \to 0$ so that the heat source is concentrated at $r=0$ and

$$\lim_{a \to 0} 2\pi \int_0^a r f(r) \, dr = 1.$$

Or equivalently

$$f(r) = \frac{1}{2\pi} \frac{\delta(r)}{r},$$

where $\delta(r)$ is the Dirac delta function.

Thus the final solution due to the concentrated heat source at $r=0$ is

$$u(r,t) = \frac{1}{4\pi\kappa t} \int_0^\infty \delta(l) I_0\left(\frac{rl}{2\kappa t}\right) \exp\left[-\frac{r^2+l^2}{4\kappa t}\right] dl$$

$$= \frac{1}{4\pi\kappa t} \exp\left(-\frac{r^2}{4\kappa t}\right). \tag{5.4.21}$$

Example 5.4.4 (*Axisymmetric Acoustic Radiation Problem*). Obtain the solution of the wave equation

$$c^2\left(u_{rr} + \frac{1}{r} u_r + u_{zz}\right) = u_{tt}, \quad 0 < r < \infty, \quad z > 0, \quad t > 0, \tag{5.4.22}$$

$$u_z = F(r,t) \quad \text{on } z = 0, \tag{5.4.23}$$

where $F(r,t)$ is a given function and c is a constant. We also assume that the solution is bounded and behaves as outgoing spherical waves.

We seek a steady-state solution for the acoustic radiation potential $u = e^{i\omega t}\phi(r,z)$ with $F(r,t) = e^{i\omega t} f(r)$, so that ϕ satisfies the Helmholtz equation

$$\phi_{rr} + \frac{1}{r}\phi_r + \phi_{zz} + \frac{\omega^2}{c^2}\phi = 0, \quad 0 < r < \infty, \quad z > 0, \tag{5.4.24}$$

with the boundary condition

$$\phi_z = f(r) \quad \text{on } z = 0, \tag{5.4.25}$$

where $f(r)$ is a given function of r.

Application of the Hankel transform $\mathcal{H}_0\{\phi(r,z)\} = \tilde{\phi}(k,z)$ to (5.4.24)-(5.4.25) gives

$$\tilde{\phi}_{zz} = \kappa^2 \tilde{\phi}, \quad z > 0,$$

$$\tilde{\phi}_z = \tilde{f}(k), \quad \text{on } z = 0,$$

where

$$\kappa = \left(k^2 - \frac{\omega^2}{c^2}\right)^{\frac{1}{2}}.$$

The solution of this differential system is

$$\tilde{\phi}(k,z) = -\frac{1}{\kappa} \tilde{f}(k) \exp(-\kappa z), \tag{5.4.26}$$

where κ is real and positive for $k > \omega/c$, and purely imaginary for $k < \omega/c$.
The inverse Hankel transform yields the formal solution

$$\phi(r,z) = -\int_0^\infty \frac{k}{\kappa} \tilde{f}(k) J_0(kr) \exp(-\kappa z) dk. \qquad (5.4.27)$$

Since the exact evaluation of this integral is difficult for an arbitrary $\tilde{f}(k)$, we choose a simple form of $f(r)$ as

$$f(r) = A H(a-r), \qquad (5.4.28)$$

where A is a constant, and hence $\tilde{f}(k) = \dfrac{Aa}{k} J_1(ak)$.

Thus the solution (5.4.27) takes the form

$$\phi(r,z) = -Aa \int_0^\infty \frac{1}{\kappa} J_1(ak) J_0(kr) \exp(-\kappa z) dk. \qquad (5.4.29)$$

For an asymptotic evaluation of this integral, it is convenient to express (5.4.29) in terms of R which is the distance from the z-axis so that $R^2 = r^2 + z^2$ and $z = R \cos\theta$. Using the asymptotic result for the Bessel function

$$J_0(kr) \sim \left(\frac{2}{\pi kr}\right)^{\frac{1}{2}} \cos\left(kr - \frac{\pi}{4}\right) \quad \text{as } r \to \infty, \qquad (5.4.30)$$

where $r = R\sin\theta$. Consequently, (5.4.29) combined with $u = \exp(i\omega t)\phi$ becomes

$$u \sim -\frac{Aa\sqrt{2} e^{i\omega t}}{\sqrt{\pi R \sin\theta}} \int_0^\infty \frac{1}{\kappa\sqrt{k}} J_1(ak) \cos\left(kR\sin\theta - \frac{\pi}{4}\right) \exp(-\kappa z) dk.$$

This integral can be evaluated asymptotically for $R \to \infty$ using the stationary phase approximation formula to obtain the final result

$$u \sim -\frac{Aac}{\omega R \sin\theta} J_1(ak_1) \exp\left[i\left(\omega t - \frac{\omega R}{c}\right)\right], \qquad (5.4.31)$$

where $k_1 = \omega/(c\sin\theta)$ is the stationary point. Physically, this solution represents outgoing spherical waves with constant velocity c and decaying amplitude as $R \to \infty$.

Example 5.4.5 (*Axisymmetric Biharmonic Equation*). We solve the axisymmetric boundary value problem

$$\nabla^4 u(r,z) = 0, \quad 0 \le r < \infty, \quad z > 0 \qquad (5.4.32)$$

with the boundary data

$$u(r,0) = f(r), \quad 0 \le r < \infty, \qquad (5.4.33)$$

$$\frac{\partial u}{\partial z} = 0 \quad \text{on } z = 0, \quad 0 \le r < \infty, \qquad (5.4.34)$$

$$u(r,z) \to 0 \quad \text{as } r \to \infty, \qquad (5.4.35)$$

where the axisymmetric biharmonic operator is

$$\nabla^4 = \nabla^2(\nabla^2) = \left(\frac{\partial^2}{\partial r^2} + \frac{1}{r}\frac{\partial}{\partial r} + \frac{\partial^2}{\partial z^2}\right)\left(\frac{\partial^2}{\partial r^2} + \frac{1}{r}\frac{\partial}{\partial r} + \frac{\partial^2}{\partial z^2}\right). \qquad (5.4.36)$$

The use of the Hankel transform $\mathcal{H}_0\{u(r,z)\} = \tilde{u}(k,z)$ to this problem gives

$$\left(\frac{d^2}{dz^2} - k^2\right)\tilde{u}(k,z) = 0, \quad z > 0, \tag{5.4.37}$$

$$\tilde{u}(k,0) = \tilde{f}(k), \quad \frac{d\tilde{u}}{dz} = 0 \quad \text{on } z = 0. \tag{5.4.38}$$

The bounded solution of (5.4.37) is
$$\tilde{u}(k,z) = (A + zB)\exp(-kz), \tag{5.4.39}$$
where A and B are integrating constants to be determined by (5.4.38) as $A = \tilde{f}(k)$ and $B = k\tilde{f}(k)$. Thus solution (5.4.39) becomes

$$\tilde{u}(k,z) = (1 + kz)\tilde{f}(k)\exp(-kz). \tag{5.4.40}$$

The inverse Hankel transform gives the formal solution

$$u(r,z) = \int_0^\infty (1 + kz)\,\tilde{f}(k)\,J_0(kr)\exp(-kz)\,dk. \tag{5.4.41}$$

Example 5.4.6 (*The Axisymmetric Cauchy-Poisson Water Wave Problem*). We consider the initial value problem for an inviscid water of finite depth h with a free horizontal surface at $z = 0$, and the z-axis positive upward. We assume that the liquid has constant density ρ with no surface tension. The surface waves are generated in water which is initially at rest for $t < 0$ by the prescribed free surface elevation. In cylindrical polar coordinates (r, θ, z), the axisymmetric water wave equations for the velocity potential $\phi(r, z, t)$ and the free surface elevation $\eta(r, t)$ are

$$\nabla^2 \phi = \phi_{rr} + \frac{1}{r}\phi_r + \phi_{zz} = 0, \quad 0 \le r < \infty, \quad -h \le z \le 0, \quad t > 0, \tag{5.4.42}$$

$$\left.\begin{array}{l}\phi_z - \eta_t = 0\\ \phi_t + g\eta = 0\end{array}\right\} \quad \text{on } z = 0, \quad t > 0, \tag{5.4.43ab}$$

$$\phi_z = 0 \quad \text{on } z = -h, \quad t > 0. \tag{5.4.44}$$

The initial conditions are
$$\phi(r,0,0) = 0 \quad \text{and} \quad \eta(r,0) = \eta_0(r), \quad \text{for } 0 \le r < \infty, \tag{5.4.45}$$
where g is the constant gravitational acceleration and $\eta_0(r)$ is the given free surface elevation.

We apply the joint Laplace and the zero-order Hankel transform defined by

$$\bar{\tilde{\phi}}(k,z,s) = \int_0^\infty e^{-st}dt\int_0^\infty r J_0(kr)\phi(r,z,t)\,dr, \tag{5.4.46}$$

to (5.4.42)-(5.4.44) so that these equations reduce to

$$\left(\frac{d^2}{dz^2} - k^2\right)\bar{\tilde{\phi}} = 0,$$

$$\left.\begin{array}{c}\dfrac{d\tilde{\bar{\phi}}}{dz} - s\tilde{\bar{\eta}} = -\tilde{\eta}_0(k) \\ s\tilde{\bar{\phi}} + g\tilde{\bar{\eta}} = 0 \end{array}\right\} \text{ on } z = 0,$$

$$\tilde{\bar{\phi}}_z = 0 \qquad \text{ on } z = -h,$$

where $\tilde{\eta}_0(k)$ is the Hankel transform of $\eta_0(r)$.

The solutions of this system are

$$\tilde{\bar{\phi}}(k,z,s) = -\frac{g\,\tilde{\eta}_0(k)}{(s^2 + \omega^2)}\,\frac{\cosh k(z+h)}{\cosh kh}, \qquad (5.4.47)$$

$$\tilde{\bar{\eta}}(k,s) = \frac{s\,\tilde{\eta}_0(k)}{(s^2 + \omega^2)}, \qquad (5.4.48)$$

where

$$\omega^2 = gk\,\tanh(kh), \qquad (5.4.49)$$

is the famous *dispersion relation* between frequency ω and wavenumber k for water waves in a liquid of depth h. Physically, this dispersion relation describes the interaction between the inertial and gravitational forces.

Application of the inverse transforms gives the integral solutions

$$\phi(r,z,t) = -g\int_0^\infty k\,J_0(kr)\,\tilde{\eta}_0(k)\left(\frac{\sin \omega t}{\omega}\right)\frac{\cosh k(z+h)}{\cosh kh}\,dk, \qquad (5.4.50)$$

$$\eta(r,t) = \int_0^\infty k\,J_0(kr)\,\tilde{\eta}_0(k)\cos \omega t\,dk. \qquad (5.4.51)$$

These wave integrals represent exact solutions for ϕ and η at any r and t, but the physical features of the wave motions cannot be described by them. In general, the exact evaluation of the integrals is almost a formidable task. In order to resolve this difficulty, it is necessary and useful to resort to asymptotic methods. It will be sufficient for the determination of the basic features of the wave motions to evaluate (5.4.50) or (5.4.51) asymptotically for a large time and distance with (r/t) held fixed. We now replace $J_0(kr)$ by its asymptotic formula (5.4.30) for $kr \to \infty$, so that (5.4.51) gives

$$\eta(r,t) \sim \left(\frac{2}{\pi r}\right)^{\frac{1}{2}} \int_0^\infty \sqrt{k}\,\tilde{\eta}_0(k)\,\cos\left(kr - \frac{\pi}{4}\right)\cos \omega t\,dk$$

$$= (2\pi r)^{-\frac{1}{2}}\,\text{Re}\int_0^\infty \sqrt{k}\,\tilde{\eta}_0(k)\exp\left[i\left(\omega t - kr + \frac{\pi}{4}\right)\right]dk. \qquad (5.4.52)$$

Application of the stationary phase method to (5.4.52) yields the solution

$$\eta(r,t) \sim \left[\frac{k_1}{rt|\omega''(k_1)|}\right]^{\frac{1}{2}} \tilde{\eta}_0(k_1)\,\cos\left[t\omega(k_1) - k_1 r\right], \qquad (5.4.53)$$

where the stationary point $k_1 = (gt^2/4r^2)$ is the root of the equation
$$\omega'(k) = \frac{r}{t}. \tag{5.4.54}$$
For sufficiently deep water, $kh \to \infty$, the dispersion relation becomes
$$\omega^2 = gk. \tag{5.4.55}$$
The solution of the axisymmetric Cauchy-Poisson problem is based on a prescribed initial displacement of unit volume that is concentrated at the origin, which means that $\eta_0(r) = (a/2\pi r)\delta(r)$ so that $\tilde{\eta}_0(k) = \frac{a}{2\pi}$. Thus, the asymptotic solution is obtained from (5.4.53) in the form
$$\eta(r,t) \sim \frac{agt^2}{4\pi\sqrt{2}\, r^3} \cos\left(\frac{gt^2}{4r}\right), \quad gt^2 \gg 4r. \tag{5.4.56}$$

It is noted that solution (5.4.53) is no longer valid when $\omega''(k_1) = 0$. This case can be handled by a modification of the asymptotic evaluation (see Debnath, 1994, p 91).

A wide variety of other physical problems solved by the Hankel transform, and/or by the joint Hankel and Laplace transform are given in books by Sneddon (1951, 1972) and by Debnath (1994), and in research papers by Debnath (1969, 1983, 1989), Mohanti (1979), and Debnath and Rollins (1992) listed in the Bibliography.

5.5 Exercises

1. Show that

   (a) $\mathcal{H}_0\{(a^2 - r^2)H(a - r)\} = \frac{4a}{\kappa^3} J_1(\kappa a) - \frac{2a^2}{\kappa^2} J_0(a\kappa),$

   (b) $\mathcal{H}_n\{r^n e^{-ar}\} = \frac{a}{\sqrt{\pi}} \cdot 2^{n+1} \Gamma\left(n + \frac{3}{2}\right) \kappa^n (a^2 + \kappa^2)^{-\left(n+\frac{3}{2}\right)},$

   (c) $\mathcal{H}_n\left\{\frac{2n}{r} f(r)\right\} = k\,\mathcal{H}_{n-1}\{f(r)\} + k\,\mathcal{H}_{n+1}\{f(r)\}.$

2. Show that the solution of the boundary value problem
   $$u_{rr} + \frac{1}{r} u_r + u_{zz} = 0, \quad 0 < r < \infty, \quad 0 < z < \infty,$$
   $$u(r,z) = \frac{1}{\sqrt{a^2 + r^2}} \quad \text{on } z = 0, \quad 0 < r < \infty,$$
   is
   $$u(r,z) = \int_0^\infty e^{-\kappa(z+a)} J_0(\kappa z)\, d\kappa = \frac{1}{\sqrt{(z+a)^2 + r^2}}.$$

3. (a) The axisymmetric initial value problem is governed by
$$u_t = \kappa\left(u_{rr} + \frac{1}{r}u_r\right) + \delta(t)f(r), \quad 0 < r < \infty, \quad t > 0,$$
$$u(r,0) = 0 \quad \text{for } 0 < r < \infty.$$
Show that the formal solution of this problem is
$$u(r,t) = \int_0^\infty k J_0(kr) \tilde{f}(k) \exp(-k^2 \kappa t) dk.$$

(b) For the special case when $f(r) = \dfrac{Q}{\pi a^2} H(a-r)$, show that the solution is
$$u(r,t) = \frac{Q}{\pi a^2} \int_0^\infty J_0(kr) J_1(ak) \exp(-k^2 \kappa t) dk.$$

4. If $f(r) = A(a^2 + r^2)^{-\frac{1}{2}}$ where A is a constant, show that the solution of the biharmonic equation described in Example 5.4.5 is
$$u(r,z) = A \frac{\{r^2 + (z+a)(2z+a)\}}{\left[r^2 + (z+a)^2\right]^{3/2}}.$$

5. Show that the solution of the boundary value problem
$$u_{rr} + \frac{1}{r}u_r + u_{zz} = 0, \quad 0 \le r < \infty, \quad z > 0,$$
$$u(r,0) = u_0 \quad \text{for } 0 \le r \le a, \quad u_0 \text{ is a constant,}$$
$$u(r,z) \to 0 \quad \text{as } z \to \infty,$$
is
$$u(r,z) = a u_0 \int_0^\infty J_1(ak) J_0(kr) \exp(-kz) dk.$$
Find the solution of the problem when u_0 is replaced by an arbitrary function $f(r)$, and a by infinity.

6. Solve the axisymmetric biharmonic equation for the free vibration of an elastic disk
$$b^2\left(\frac{\partial^2}{\partial r^2} + \frac{1}{r}\frac{\partial}{\partial r}\right)^2 u + u_{tt} = 0, \quad 0 < r < \infty, \quad t > 0,$$
$$u(r,0) = f(r), \quad u_t(r,0) = 0 \quad \text{for } 0 < r < \infty,$$
where $b^2 = \dfrac{D}{2\sigma h}$ is the ratio of the flexural rigidity of the disk and its mass $2h\sigma$ per unit area.

7. Show that the zero-order Hankel transform solution of the axisymmetric Laplace equation
$$u_{rr} + \frac{1}{r}u_r + u_{zz} = 0, \quad 0 < r < \infty, \quad -\infty < z < \infty,$$
with the boundary data
$$\lim_{r \to 0}\left(r^2 u\right) = 0, \quad \lim_{r \to 0}\left(2\pi r\right)u_r = -f(z), \quad -\infty < z < \infty,$$
is
$$\tilde{u}(k,z) = \frac{1}{4\pi k}\int_{-\infty}^{\infty} \exp\{-k|z-\zeta|\}f(\zeta)\,d\zeta.$$
Hence show that
$$u(r,z) = \frac{1}{4\pi}\int_{-\infty}^{\infty}\left\{r^2 + (z-\zeta)^2\right\}^{-\frac{1}{2}} f(\zeta)\,d\zeta.$$

8. Solve the non-homogeneous diffusion problem
$$u_t = \kappa\left(u_{rr} + \frac{1}{r}u_r\right) + Q(r,t), \quad 0 < r < \infty, \quad t > 0,$$
$$u(r,0) = f(r) \quad \text{for } 0 < r < \infty,$$
where κ is a constant.

9. Solve the problem of the electrified unit disk in the xy plane with center at the origin. The electric potential $u(r,z)$ is axisymmetric and satisfies the boundary value problem
$$u_{rr} + \frac{1}{r}u_r + u_{zz} = 0, \quad 0 < r < \infty, \quad 0 < z < \infty,$$
$$u(r,0) = u_0, \quad 0 \le r < a,$$
$$\frac{\partial u}{\partial z} = 0, \quad \text{on } z = 0 \quad \text{for } a < r < \infty,$$
$$u(r,z) \to 0 \quad \text{as } z \to \infty \quad \text{for all } r,$$
where u_0 is constant. Show that the solution is
$$u(r,z) = \frac{2au_0}{\pi}\int_0^\infty J_0(kr)\frac{\sin ak}{k}e^{-kz}\,dk.$$

10. Solve the axisymmetric surface wave problem in deep water due to an oscillatory surface pressure. The governing equations are
$$\nabla^2 \phi = \phi_{rr} + \frac{1}{r}\phi_r + \phi_{zz} = 0, \quad 0 \le r < \infty, \quad -\infty < z \le 0,$$
$$\left.\begin{array}{l}\phi_t + g\eta = -\dfrac{P}{\rho}p(r)\exp(i\omega t) \\ \phi_z - \eta_t = 0\end{array}\right\} \quad \text{on } z = 0, \quad t > 0,$$
$$\phi(r,z,0) = 0 = \eta(r,0), \quad \text{for } 0 \le r < \infty, \quad -\infty < z \le 0.$$

11. Solve the Neumann problem for the Laplace equation
$$u_{rr} + \frac{1}{r}u_r + u_{zz} = 0, \quad 0 < r < \infty, \quad 0 < z < \infty$$
$$u_z(r,0) = -\frac{1}{\pi a^2} H(a-r), \quad 0 < r < \infty$$
$$u(r,z) \to 0 \quad \text{as } z \to \infty \quad \text{for } 0 < r < \infty.$$
Show that
$$\lim_{a \to 0} u(r,z) = \frac{1}{2\pi}(r^2 + z^2)^{-\frac{1}{2}}.$$

12. Solve the Cauchy problem for the wave equation in a dissipating medium
$$u_{tt} + 2\kappa u_t = c^2\left(u_{rr} + \frac{1}{r}u_r\right), \quad 0 < r < \infty, \quad t > 0,$$
$$u(r,0) = f(r), \quad u_t(r,0) = g(r) \text{ for } 0 < r < \infty,$$
where κ is a constant.

13. Use the joint Laplace and Hankel transform to solve the initial-boundary value problem
$$c^2\left(u_{rr} + \frac{1}{r}u_r + u_{zz}\right) = u_{tt}, \quad 0 < r < \infty, \quad 0 < z < \infty, \quad t > 0,$$
$$u_z(r,0,t) = H(a-r)H(t), \quad 0 < r < \infty, \quad t > 0,$$
$$u(r,z,t) \to 0 \quad \text{as } r \to \infty \text{ and } u(r,z,t) \to 0 \quad \text{as } z \to \infty,$$
$$u(r,z,0) = 0 = u_t(r,z,0),$$
and show that
$$u_t(r,z,t) = -ac\, H\!\left(t - \frac{z}{c}\right) \int_0^\infty J_1(ak)\, J_0\!\left\{ck\sqrt{t^2 - \frac{z^2}{c^2}}\right\} J_0(kr)\, dk.$$

14. Find the steady temperature $u(r,z)$ in a beam $0 \le r < \infty$, $0 \le z \le a$ when the face $z = 0$ is kept at temperature $u(r,0) = 0$, and the face $z = a$ is insulated except that heat is supplied through a circular hole such that
$$u_z(r,a) = H(b-r).$$
The temperature $u(r,z)$ satisfies the axisymmetric equation
$$u_{rr} + \frac{1}{r}u_r + u_{zz} = 0, \quad 0 \le r < \infty, \quad 0 \le z \le a.$$

15. Find the integral solution of the initial-boundary value problem
$$u_{rr} + \frac{1}{r}u_r + u_{zz} = u_t, \quad 0 \le r < \infty, \quad 0 \le z < \infty, \quad t > 0,$$
$$u(r,z,0) = 0 \quad \text{for all } r \text{ and } z,$$
$$\left(\frac{\partial u}{\partial r}\right)_{r=0} = 0, \quad \text{for } 0 \le z < \infty, \quad t > 0,$$

$$\left(\frac{\partial u}{\partial z}\right)_{z=0} = -\frac{H(a-r)}{\sqrt{a^2-r^2}}, \quad \text{for } 0<r<\infty, \ 0<t<\infty,$$

$u(r,z,t) \to 0$ as $r \to \infty$ or $z \to \infty$.

16. Heat is supplied at a constant rate Q per unit area per unit time over a circular area of radius a in the plane $z=0$ to an infinite solid of thermal conductivity K, the rest of the plane is kept at zero temperature. Solve for the steady temperature field $u(r,z)$ that satisfies the Laplace equation

$$u_{rr} + \frac{1}{r}u_r + u_{zz} = 0, \quad 0<r<\infty, \ -\infty<z<\infty,$$

with the boundary conditions

$$u \to 0 \text{ as } r \to \infty, \quad u \to 0 \text{ as } |z| \to \infty,$$

$$-K u_z = \frac{1}{2}Q H(a-r) \text{ when } z=0.$$

17. The velocity potential $\phi(r,z)$ for the flow of an inviscid fluid through a circular aperture of unit radius in a plane rigid screen satisfies the Laplace equation

$$\phi_{rr} + \frac{1}{r}\phi_r + \phi_{zz} = 0, \ 0<r<\infty$$

with the boundary conditions

$$\left.\begin{array}{ll} \phi = 1 & \text{for } 0<r<1 \\ \phi_z = 0 & \text{for } r>1 \end{array}\right\} \text{ on } z=0.$$

Obtain the solution of this boundary value problem.

18. Solve the Cauchy-Poisson wave problem (Debnath, 1989) for a viscous liquid of finite or infinite depth governed by the equations, free surface, boundary, and initial conditions

$$\phi_{rr} + \frac{1}{r}\phi_r + \phi_{zz} = 0,$$

$$\psi_t = v\left(\psi_{rr} + \frac{1}{r}\psi_r - \frac{1}{r^2}\psi + \psi_{zz}\right),$$

where $\phi(r,z,t)$ and $\psi(r,z,t)$ represents the potential and stream functions respectively, $0 \le r < \infty$, $-h \le z \le 0$ (or $-\infty < z \le 0$) and $t>0$.

The free surface conditions are

$$\left.\begin{array}{l} \eta_t - w = 0 \\ \mu(u_z + w_r) = 0 \\ \phi_t + g\eta + 2v\, w_z = 0 \end{array}\right\} \text{on } z=0, \ t>0$$

where $\eta = \eta(r,t)$ is the free surface elevation, $u = \phi_r + \psi_z$ and $w = \phi_z - \frac{\psi}{r} - \psi_r$ are the radial and vertical velocity components of liquid particles, $\mu = \rho v$ is the dynamic viscosity, ρ is the density, and v is the kinematic viscosity of the liquid.

The boundary conditions at the rigid bottom are
$$\left. \begin{array}{l} u = \phi_r + \psi_z = 0 \\ w = \phi_z - \dfrac{1}{r}(r\psi)_r = 0 \end{array} \right\} \quad \text{on } z = -h.$$

The initial conditions are
$$\eta = a\frac{\delta(r)}{r}, \quad \phi = \psi = 0 \quad \text{at } t = 0,$$
where a is a constant and $\delta(r)$ is the Dirac delta function.

If the liquid is of infinite depth, the bottom boundary conditions are
$$(\phi, \psi) \to (0,0) \quad \text{as } z \to -\infty.$$

19. Use the joint Hankel and Laplace transform method to solve the initial-boundary value problem
$$u_{rr} + \frac{1}{r}u_r - u_{tt} - 2\varepsilon\, u_t = a\,\frac{\delta(r)}{r}\,\delta(t), \quad 0 < r < \infty, \quad t > 0,$$
$$u(r,t) \to 0 \quad \text{as } r \to \infty,$$
$$u(0,t) \text{ is finite for } t > 0,$$
$$u(r,0) = 0 = u_t(r,0) \quad \text{for } 0 < r < \infty.$$

20. Surface waves are generated in an inviscid liquid of infinite depth due to an explosion (Sen, 1963) above it which generates the pressure field $p(r,t)$. The velocity potential $\phi(r,z,t)$ satisfies the Laplace equation
$$u_{rr} + \frac{1}{r}u_r + u_{zz} = 0, \quad 0 < r < \infty, \quad t > 0,$$
and the free surface condition
$$u_{tt} + g\, u_z = \frac{1}{\rho}\left(\frac{\partial p}{\partial t}\right)\left[H(r) - H\{r, r_0(t)\}\right] \quad \text{on } z = 0,$$
where ρ is the constant density of the liquid, $r_0(t)$ is the extent of the blast, and the liquid is initially at rest.

Solve this problem.

Chapter 6
Mellin Transforms

6.1 Introduction

This chapter is concerned with the theory and applications of the Mellin transform. We derive the Mellin transform and its inverse from the complex Fourier transform. This is followed by several examples and the basic operational properties of Mellin transforms. We discuss several applications of Mellin transforms to boundary value problems and to summation of infinite series. The Weyl transform and the Weyl fractional derivatives with examples are also included.

Historically, Riemann (1876) first recognized the Mellin transform in his famous memoir on prime numbers. Its explicit formulation was given by Cahen (1894). Almost simultaneously, Mellin (1896, 1902) gave an elaborate discussion of the Mellin transform and its inversion formula.

6.2 Definition of the Mellin Transform and Examples

We derive the Mellin transform and its inverse from the complex Fourier transform and its inverse, which are defined respectively by

$$\mathscr{F}\{g(\xi)\} = G(k) = \frac{1}{\sqrt{2\pi}} \int_{-\infty}^{\infty} e^{-ik\xi} g(\xi) d\xi, \qquad (6.2.1)$$

$$\mathscr{F}^{-1}\{G(k)\} = g(\xi) = \frac{1}{\sqrt{2\pi}} \int_{-\infty}^{\infty} e^{ik\xi} G(k) dk. \qquad (6.2.2)$$

Making the changes of variables $\exp(\xi) = x$ and $ik = c - p$, where c is a constant, in results (6.2.1) and (6.2.2) we obtain

$$G(ip - ic) = \frac{1}{\sqrt{2\pi}} \int_{0}^{\infty} x^{p-c-1} g(\log x) dx, \qquad (6.2.3)$$

$$g(\log x) = \frac{1}{\sqrt{2\pi}} \int_{c-i\infty}^{c+i\infty} x^{c-p} G(ip - ic) dp. \qquad (6.2.4)$$

We now write $\frac{1}{\sqrt{2\pi}} x^{-c} g(\log x) \equiv f(x)$ and $G(ip - ic) \equiv \tilde{f}(p)$ to define the *Mellin transform* of $f(x)$ and its *inverse* as

$$\mathcal{M}\{f(x)\} = \tilde{f}(p) = \int_0^\infty x^{p-1} f(x) dx, \qquad (6.2.5)$$

$$\mathcal{M}^{-1}\{\tilde{f}(p)\} = f(x) = \frac{1}{2\pi i} \int_{c-i\infty}^{c+i\infty} x^{-p} \tilde{f}(p) dp, \qquad (6.2.6)$$

where $f(x)$ is a real valued function defined on $(0, \infty)$ and the Mellin transform variable p is a complex number. Sometimes, the Mellin transform of $f(x)$ is denoted explicitly by $\tilde{f}(p) = \mathcal{M}[f(x), p]$. Obviously, $\mathcal{M}$ and $\mathcal{M}^{-1}$ are linear integral operators.

Example 6.2.1 (a) If $f(x) = e^{-nx}$ where $n > 0$, then

$$\mathcal{M}\{e^{-nx}\} = \tilde{f}(p) = \int_0^\infty x^{p-1} e^{-nx} dx,$$

which is, by putting $nx = t$,

$$= \frac{1}{n^p} \int_0^\infty t^{p-1} e^{-t} dt = \frac{\Gamma(p)}{n^p}. \qquad (6.2.7)$$

(b) If $f(x) = \frac{1}{1+x}$, then

$$\mathcal{M}\left\{\frac{1}{1+x}\right\} = \tilde{f}(p) = \int_0^\infty x^{p-1} \cdot \frac{dx}{1+x},$$

which is, by substituting $x = \frac{t}{1-t}$ or $t = \frac{x}{1+x}$,

$$= \int_0^1 t^{p-1} (1-t)^{(1-p)-1} dt = B(p, 1-p) = \Gamma(p)\Gamma(1-p),$$

which is, by a well-known result for the gamma function,

$$= \pi \operatorname{cosec}(\pi p). \qquad (6.2.8)$$

(c) If $f(x) = (e^x - 1)^{-1}$, then

$$\mathcal{M}\left\{\frac{1}{e^x - 1}\right\} = \tilde{f}(p) = \int_0^\infty x^{p-1} \frac{1}{e^x - 1} dx,$$

which is, by using $\sum_{n=0}^\infty e^{-nx} = \frac{1}{1 - e^{-x}}$ and hence $\sum_{n=1}^\infty e^{-nx} = \frac{1}{e^x - 1}$,

$$= \sum_{n=1}^\infty \int_0^\infty x^{p-1} e^{-nx} dx = \sum_{n=1}^\infty \frac{\Gamma(p)}{n^p} = \Gamma(p)\zeta(p), \qquad (6.2.9)$$

where $\zeta(p) = \sum_{n=1}^\infty \frac{1}{n^p}$, (Re $p > 1$) is the famous *Riemann zeta function*.

(d) If $f(x) = \frac{2}{e^{2x} - 1}$, then

$$\mathcal{M}\left\{\frac{2}{e^{2x}-1}\right\} = \tilde{f}(p) = 2\int_0^\infty x^{p-1}\frac{dx}{e^{2x}-1} = 2\sum_{n=1}^\infty \int_0^\infty x^{p-1} e^{-2nx} dx$$

$$= 2\sum_{n=1}^\infty \frac{\Gamma(p)}{(2n)^p} = 2^{1-p}\,\Gamma(p)\sum_{n=1}^\infty \frac{1}{n^p} = 2^{1-p}\,\Gamma(p)\zeta(p). \quad (6.2.10)$$

(e) If $f(x) = \dfrac{1}{e^x+1}$, then

$$\mathcal{M}\left\{\frac{1}{e^x+1}\right\} = \left(1 - 2^{1-p}\right)\Gamma(p)\,\zeta(p). \quad (6.2.11)$$

This follows from the result

$$\left[\frac{1}{e^x-1} - \frac{1}{e^x+1}\right] = \frac{2}{e^{2x}-1}$$

combined with (6.2.9) and (6.2.10).

(f) If $f(x) = \dfrac{1}{(1+x)^n}$, then

$$\mathcal{M}\left\{\frac{1}{(1+x)^n}\right\} = \int_0^\infty x^{p-1}(1+x)^{-n}\,dx,$$

which is, by putting $x = \dfrac{t}{1-t}$ or $t = \dfrac{x}{1+x}$,

$$= \int_0^1 t^{p-1}(1-t)^{n-p-1}\,dt$$

$$= B(p, n-p) = \frac{\Gamma(p)\Gamma(n-p)}{\Gamma(n)}, \quad (6.2.12)$$

where $B(p,q)$ is the standard Beta function.
Hence

$$\mathcal{M}^{-1}\{\Gamma(p)\Gamma(n-p)\} = \frac{\Gamma(n)}{(1+x)^n}.$$

(g) Find the Mellin transform of $\cos kx$ and $\sin kx$.
It follows from Example 6.2.1(a) that

$$\mathcal{M}\left[e^{-ikx}\right] = \frac{\Gamma(p)}{(ik)^p} = \frac{\Gamma(p)}{k^p}\left(\cos\frac{p\pi}{2} - i\sin\frac{p\pi}{2}\right).$$

Separating real and imaginary parts, we find

$$\mathcal{M}[\cos kx] = k^{-p}\,\Gamma(p)\cos\left(\frac{\pi p}{2}\right), \quad (6.2.13)$$

$$\mathcal{M}[\sin kx] = k^{-p}\,\Gamma(p)\sin\left(\frac{\pi p}{2}\right). \quad (6.2.14)$$

These results can be used to calculate the Fourier cosine and Fourier sine transforms of x^{p-1}. Result (6.2.13) can be written as

$$\int_0^\infty x^{p-1} \cos kx \, dx = \frac{\Gamma(p)}{k^p} \cos\left(\frac{\pi p}{2}\right).$$

Or equivalently

$$\mathscr{F}_c\left\{\sqrt{\frac{\pi}{2}} \, x^{p-1}\right\} = \frac{\Gamma(p)}{k^p} \cos\left(\frac{\pi p}{2}\right).$$

Or

$$\mathscr{F}_c\{x^{p-1}\} = \sqrt{\frac{2}{\pi}} \, \frac{\Gamma(p)}{k^p} \cos\left(\frac{\pi p}{2}\right). \tag{6.2.15}$$

Similarly

$$\mathscr{F}_s\{x^{p-1}\} = \sqrt{\frac{2}{\pi}} \, \frac{\Gamma(p)}{k^p} \sin\left(\frac{\pi p}{2}\right). \tag{6.2.16}$$

6.3 Basic Operational Properties

If $\mathcal{M}\{f(x)\} = \tilde{f}(p)$, then the following operational properties hold:

(a) (*Scaling Property*).

$$\mathcal{M}\{f(ax)\} = a^{-p} \tilde{f}(p), \ a > 0. \tag{6.3.1}$$

Proof. By definition, we have,

$$\mathcal{M}\{f(ax)\} = \int_0^\infty x^{p-1} f(ax) \, dx,$$

which is, by substituting $ax = t$,

$$= \frac{1}{a^p} \int_0^\infty t^{p-1} f(t) \, dt = \frac{\tilde{f}(p)}{a^p}.$$

(b) (*Shifting Property*).

$$\mathcal{M}[x^a f(x)] = \tilde{f}(p + a). \tag{6.3.2}$$

Its proof follows from the definition.

(c)
$$\mathcal{M}\{f(x^a)\} = \frac{1}{a} \tilde{f}\left(\frac{p}{a}\right), \tag{6.3.3}$$

$$\mathcal{M}\left\{\frac{1}{x} f\left(\frac{1}{x}\right)\right\} = \tilde{f}(1 - p), \tag{6.3.4}$$

$$\mathcal{M}\{\log x \, f(x)\} = \frac{d}{dp} \tilde{f}(p). \tag{6.3.5}$$

The proofs of (6.3.3) and (6.3.4) are easy and hence left to the reader.
Result (6.3.5) can easily be proved by using the result

$$\frac{d}{dp} x^{p-1} = (\log x) x^{p-1}. \tag{6.3.6}$$

(d) (*Mellin Transforms of Derivatives*).
$$\mathcal{M}[f'(x)] = -(p-1)\tilde{f}(p-1), \tag{6.3.7}$$
provided $[x^{p-1}f(x)]$ vanishes at $x \to 0$ and as $x \to \infty$.
$$\mathcal{M}[f''(x)] = (p-1)(p-2)\tilde{f}(p-2). \tag{6.3.8}$$
More generally,
$$\mathcal{M}[f^{(n)}(x)] = (-1)^n \frac{\Gamma(p)}{\Gamma(p-n)}\tilde{f}(p-n)$$
$$= (-1)^n \frac{\Gamma(p)}{\Gamma(p-n)}\mathcal{M}[f(x), p-n], \tag{6.3.9}$$
provided $x^{p-r-1}f^{(r)}(x) = 0$ as $x \to 0$ for $r = 0, 1, 2, \ldots, (n-1)$.

Proof. We have, by definition,
$$\mathcal{M}[f'(x)] = \int_0^\infty x^{p-1} f'(x)\, dx,$$
which is, by integrating by parts,
$$= [x^{p-1} f(x)]_0^\infty - (p-1)\int_0^\infty x^{p-2} f(x)\, dx$$
$$= -(p-1)\tilde{f}(p-1).$$
The proofs of (6.3.8) and (6.3.9) are similar and left to the reader.

(e) If $\mathcal{M}\{f(x)\} = \tilde{f}(p)$, then
$$\mathcal{M}\{x f'(x)\} = -p\tilde{f}(p), \tag{6.3.10}$$
provided $x^p f(x)$ vanishes at $x = 0$ and as $x \to \infty$.
$$\mathcal{M}\{x^2 f''(x)\} = (-1)^2 p(p+1)\tilde{f}(p). \tag{6.3.11}$$
More generally
$$\mathcal{M}\{x^n f^{(n)}(x)\} = (-1)^n \frac{\Gamma(p+n)}{\Gamma(p)}\tilde{f}(p). \tag{6.3.12}$$

Proof. We have, by definition,
$$\mathcal{M}\{x f'(x)\} = \int_0^\infty x^p f'(x)\, dx,$$
which is, by integrating by parts,
$$= [x^p f(x)]_0^\infty - p\int_0^\infty x^{p-1} f(x)\, dx$$
$$= -p\tilde{f}(p).$$
Similar arguments can be used to prove results (6.3.11) and (6.3.12).

(f) *(Mellin Transforms of Differential Operators)*.
If $\mathcal{M}\{f(x)\} = \tilde{f}(p)$, then

$$\mathcal{M}\left[\left(x\frac{d}{dx}\right)^2 f(x)\right] = \mathcal{M}\left[x^2 f''(x) + x f'(x)\right] = (-1)^2 p^2 \tilde{f}(p), \quad (6.3.13)$$

and more generally,

$$\mathcal{M}\left[\left(x\frac{d}{dx}\right)^n f(x)\right] = (-1)^n p^n \tilde{f}(p). \quad (6.3.14)$$

Proof. We have, by definition,

$$\mathcal{M}\left[\left(x\frac{d}{dx}\right)^2 f(x)\right] = \mathcal{M}\left[x^2 f''(x) + x f'(x)\right]$$

$$= \mathcal{M}\left[x^2 f''(x)\right] + \mathcal{M}\left[x f'(x)\right]$$

$$= -p\tilde{f}(p) + p(p+1)\tilde{f}(p) \quad \text{by (6.3.10) and (6.3.11)}$$

$$= (-1)^2 p^2 \tilde{f}(p).$$

Similar arguments can be used to prove the general result (6.3.14).

(g) *(Mellin Transforms of Integrals)*.

$$\mathcal{M}\left\{\int_0^x f(t)\,dt\right\} = -\frac{1}{p}\tilde{f}(p+1). \quad (6.3.15)$$

In general,

$$\mathcal{M}\{I_n f(x)\} = \mathcal{M}\left\{\int_0^x I_{n-1} f(t)\,dt\right\} = (-1)^n \frac{\Gamma(p)}{\Gamma(p+n)} \tilde{f}(p+n), \quad (6.3.16)$$

where $I_n f(x)$ is the nth repeated integral of $f(x)$ defined by

$$I_n f(x) = \int_0^x I_{n-1} f(t)\,dt. \quad (6.3.17)$$

Proof. We write

$$F(x) = \int_0^x f(t)\,dt$$

so that $F'(x) = f(x)$ with $F(0) = 0$. Application of (6.3.7) with $F(x)$ as defined gives

$$\mathcal{M}\{f(x) = F'(x), p\} = -(p-1)\mathcal{M}\left\{\int_0^x f(t)\,dt, p-1\right\},$$

which is, replacing p by $p+1$,

$$\mathcal{M}\left\{\int_0^x f(t)\,dt, p\right\} = -\frac{1}{p}\mathcal{M}\{f(x), p+1\} = -\frac{1}{p}\tilde{f}(p+1).$$

An argument similar to this can be used to prove (6.3.16).

(h) (*Convolution Type Theorems*).
If $\mathcal{M}\{f(x)\} = \tilde{f}(p)$ and $\mathcal{M}\{g(x)\} = \tilde{g}(p)$, then

$$\mathcal{M}[f(x)*g(x)] = \mathcal{M}\left[\int_0^\infty f(\xi)g\left(\frac{x}{\xi}\right)\frac{d\xi}{\xi}\right] = \tilde{f}(p)\tilde{g}(p), \quad (6.3.18)$$

$$\mathcal{M}[f(x) \circ g(x)] = \mathcal{M}\left[\int_0^\infty f(x\xi)g(\xi)d\xi\right] = \tilde{f}(p)\tilde{g}(1-p). \quad (6.3.19)$$

Proof. We have, by definition,

$$\mathcal{M}[f(x)*g(x)] = \mathcal{M}\left[\int_0^\infty f(\xi)g\left(\frac{x}{\xi}\right)\frac{d\xi}{\xi}\right]$$

$$= \int_0^\infty x^{p-1}\,dx \int_0^\infty f(\xi)g\left(\frac{x}{\xi}\right)\frac{d\xi}{\xi}$$

$$= \int_0^\infty f(\xi)\frac{d\xi}{\xi} \int_0^\infty x^{p-1} g\left(\frac{x}{\xi}\right)dx,$$

which, is by letting, $\dfrac{x}{\xi} = \eta$

$$= \int_0^\infty f(\xi)\frac{d\xi}{\xi} \int_0^\infty (\xi\eta)^{p-1} g(\eta)\,\xi\,d\eta$$

$$= \int_0^\infty \xi^{p-1} f(\xi)d\xi \int_0^\infty \eta^{p-1} g(\eta)d\eta$$

$$= \tilde{f}(p)\tilde{g}(p).$$

Similarly, we have

$$\mathcal{M}[f(x)\circ g(x)] = \mathcal{M}\left[\int_0^\infty f(x\xi)g(\xi)d\xi\right]$$

$$= \int_0^\infty x^{p-1}\,dx \int_0^\infty f(x\xi)g(\xi)d\xi,$$

which is, by substituting $x\xi = \eta$,

$$= \int_0^\infty g(\xi)d\xi \int_0^\infty \eta^{p-1}\,\xi^{1-p} f(\eta)\,\frac{d\eta}{\xi}$$

$$= \int_0^\infty \xi^{1-p-1} g(\xi)d\xi \int_0^\infty \eta^{p-1} f(\eta)d\eta$$

$$= \tilde{g}(1-p)\tilde{f}(p).$$

Note that, in this case, the operation $\circ$ is not commutative.

Clearly,
$$\mathcal{M}^{-1}\{\tilde{f}(1-p)\tilde{g}(p)\} = \int_0^\infty g(st)f(t)\,dt.$$

Putting $g(t) = e^{-t}$ and $\tilde{g}(p) = \Gamma(p)$, we obtain the Laplace transform of $f(t)$

$$\mathcal{M}^{-1}\{\tilde{f}(1-p)\Gamma(p)\} = \int_0^\infty e^{-st} f(t)\,dt = \mathcal{L}\{f(t)\} = \bar{f}(s). \quad (6.3.20)$$

(i) (*Parseval's Type Property*).

If $\mathcal{M}\{f(x)\} = \tilde{f}(p)$ and $\mathcal{M}\{g(x)\} = \tilde{g}(p)$, then

$$\mathcal{M}[f(x)g(x)] = \frac{1}{2\pi i} \int_{c-i\infty}^{c+i\infty} \tilde{f}(s)\tilde{g}(p-s)\,ds. \quad (6.3.21)$$

Or equivalently

$$\int_0^\infty x^{p-1} f(x)g(x)\,dx = \frac{1}{2\pi i} \int_{c-i\infty}^{c+i\infty} \tilde{f}(s)\tilde{g}(p-s)\,ds. \quad (6.3.22)$$

In particular, when $p = 1$, we obtain the *Parseval formula* for the Mellin transform,

$$\int_0^\infty f(x)g(x)\,dx = \frac{1}{2\pi i} \int_{c-i\infty}^{c+i\infty} \tilde{f}(s)\tilde{g}(1-s)\,ds. \quad (6.3.23)$$

Proof. By definition, we have

$$\mathcal{M}[f(x)g(x)] = \int_0^\infty x^{p-1} f(x)g(x)\,dx$$

$$= \frac{1}{2\pi i} \int_0^\infty x^{p-1} g(x)\,dx \int_{c-i\infty}^{c+i\infty} x^{-s} \tilde{f}(s)\,ds$$

$$= \frac{1}{2\pi i} \int_{c-i\infty}^{c+i\infty} \tilde{f}(s)\,ds \int_0^\infty x^{p-s-1} g(x)\,dx$$

$$= \frac{1}{2\pi i} \int_{c-i\infty}^{c+i\infty} \tilde{f}(s)\tilde{g}(p-s)\,ds.$$

When $p = 1$, the above result becomes (6.3.23).

6.4 Applications of Mellin Transforms

Example 6.4.1 Obtain the solution of the boundary value problem
$$x^2 u_{xx} + x u_x + u_{yy} = 0, \quad 0 \le x < \infty, \quad 0 < y < 1 \quad (6.4.1)$$

$$u(x,0) = 0, \quad u(x,1) = \begin{cases} A, & 0 \le x \le 1 \\ 0, & x > 1 \end{cases}, \quad (6.4.2)$$

where A is a constant.

We apply the Mellin transform of $u(x,y)$ with respect to x defined by
$$\tilde{u}(p,y) = \int_0^\infty x^{p-1} u(x,y) dx$$
to reduce the given system into the form
$$\tilde{u}_{yy} + p^2 \tilde{u} = 0, \quad 0 < y < 1$$
$$\tilde{u}(p,0) = 0, \quad \tilde{u}(p,1) = A \int_0^1 x^{p-1} dx = \frac{A}{p}.$$
The solution of the transformed problem is
$$\tilde{u}(p,y) = \frac{A}{p} \frac{\sin py}{\sin p}, \quad 0 < \operatorname{Re} p < 1.$$
The inverse Mellin transform gives
$$u(x,y) = \frac{A}{2\pi i} \int_{c-i\infty}^{c+i\infty} \frac{x^{-p}}{p} \frac{\sin py}{\sin p} dp. \tag{6.4.3}$$
where $\tilde{u}(p,y)$ is analytic in the vertical strip $0 < \operatorname{Re}(p) < \pi$ and hence $0 < c < \pi$. The integrand of (6.4.3) has simple poles at $p = n\pi$, $n = 1, 2, 3, \ldots$ which lie inside a semi-circular contour in the right half plane. Evaluating (6.4.3) by the theory of residues gives the solution for $x > 1$ as
$$u(x,y) = \frac{A}{\pi} \sum_{n=1}^\infty \frac{1}{n} (-1)^n x^{-n\pi} \sin n\pi y. \tag{6.4.4}$$

Example 6.4.2 (*Potential in an Infinite Wedge*). Find the potential $\phi(r,\theta)$ that satisfies the Laplace equation
$$r^2 \phi_{rr} + r\phi_r + \phi_{\theta\theta} = 0 \tag{6.4.5}$$
in an infinite wedge $0 < r < \infty$, $-\alpha < \theta < \alpha$ with the boundary conditions
$$\phi(r,\alpha) = f(r), \quad \phi(r,-\alpha) = g(r), \quad 0 \le r < \infty, \tag{6.4.6ab}$$
$$\phi(r,\theta) \to 0 \text{ as } r \to \infty \text{ for all } \theta \text{ in } -\alpha < \theta < \alpha. \tag{6.4.7}$$

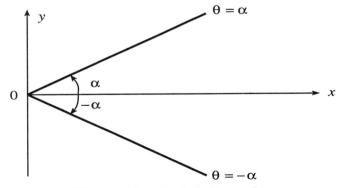

Figure 6.1 An infinite wedge.

We apply the Mellin transform of the potential $\phi(r,\theta)$ defined by
$$\mathcal{M}[\phi(r,\theta)] = \tilde{\phi}(p,\theta) = \int_0^\infty r^{p-1}\phi(r,\theta)\,dr$$
to the differential system (6.4.5)-(6.4.7) to obtain
$$\frac{d^2\tilde{\phi}}{d\theta^2} + p^2\tilde{\phi} = 0, \tag{6.4.8}$$
$$\tilde{\phi}(p,\alpha) = \tilde{f}(p), \quad \tilde{\phi}(p,-\alpha) = \tilde{g}(p). \tag{6.4.9ab}$$
The general solution of the transformed equation is
$$\tilde{\phi}(p,\theta) = A\cos p\theta + B\sin p\theta, \tag{6.4.10}$$
where A and B are functions of p and α. The boundary conditions (6.4.9ab) determine A and B, which satisfy
$$A\cos p\alpha + B\sin p\alpha = \tilde{f}(p),$$
$$A\cos p\alpha - B\sin p\alpha = \tilde{g}(p).$$
These give
$$A = \frac{\tilde{f}(p) + \tilde{g}(p)}{2\cos p\alpha}, \quad B = \frac{\tilde{f}(p) - \tilde{g}(p)}{2\sin p\alpha}.$$
Thus (6.4.10) becomes
$$\tilde{\phi}(p,\theta) = \tilde{f}(p)\cdot\frac{\sin p(\alpha+\theta)}{\sin(2p\alpha)} + \tilde{g}(p)\frac{\sin p(\alpha-\theta)}{\sin(2p\alpha)}$$
$$= \tilde{f}(p)\tilde{h}(p,\alpha+\theta) + \tilde{g}(p)\tilde{h}(p,\alpha-\theta), \tag{6.4.11}$$
where
$$\tilde{h}(p,\theta) = \frac{\sin p\theta}{\sin(2p\alpha)}.$$
Or equivalently
$$h(r,\theta) = \mathcal{M}^{-1}\left\{\frac{\sin p\theta}{\sin 2p\alpha}\right\} = \left(\frac{1}{2\alpha}\right)\frac{r^n \sin n\theta}{(1+2r^n \cos n\theta + r^{2n})}, \tag{6.4.12}$$
where
$$n = \frac{\pi}{2\alpha}\left(2\alpha = \frac{\pi}{n}\right).$$
Application of the inverse Mellin transform to (6.4.11) gives
$$\phi(r,\theta) = \mathcal{M}^{-1}\{\tilde{f}(p)\tilde{h}(p,\alpha+\theta)\} + \mathcal{M}^{-1}\{\tilde{g}(p)\tilde{h}(p,\alpha-\theta)\},$$
which is, by the convolution property (6.3.18),

$$\phi(r,\theta) = \frac{r^n \cos n\theta}{2\alpha} \left[\int_0^\infty \frac{\xi^{n-1} f(\xi) d\xi}{\xi^{2n} - 2(r\xi)^n \sin n\theta + r^{2n}} \right.$$

$$\left. + \int_0^\infty \frac{\xi^{n-1} g(\xi) d\xi}{\xi^{2n} + 2(r\xi)^n \sin n\theta + r^{2n}} \right], \quad |\alpha| < \frac{\pi}{2n}. \quad (6.4.13)$$

This is the formal solution of the problem.

In particular, when $f(r) = g(r)$, solution (6.4.11) becomes

$$\tilde{\phi}(p,\theta) = \tilde{f}(p) \frac{\cos p\theta}{\cos p\alpha} = \tilde{f}(p)\tilde{h}(p,\theta), \quad (6.4.14)$$

where

$$\tilde{h}(p,\theta) = \frac{\cos p\theta}{\cos p\alpha} = \mathcal{M}\{h(r,\theta)\}.$$

Application of the inverse Mellin transform to (6.4.14) combined with the convolution property (6.3.18) yields the solution

$$\phi(r,\theta) = \int_0^\infty f(\xi) \, h\!\left(\frac{r}{\xi},\theta\right) \frac{d\xi}{\xi}, \quad (6.4.15)$$

where

$$h(r,\theta) = \mathcal{M}^{-1}\left\{\frac{\cos p\theta}{\cos p\alpha}\right\} = \left(\frac{r^n}{\alpha}\right) \frac{(1+r^{2n})\cos(n\theta)}{(1+2r^{2n}\cos 2n\theta + r^{2n})}, \quad (6.4.16)$$

and $n = \dfrac{\pi}{2\alpha}$.

Some applications of the Mellin transform to boundary value problems are given by Sneddon (1951) and Tranter (1966).

Example 6.4.3 Solve the integral equation

$$\int_0^\infty f(\xi) k(x\xi) d\xi = f(x), \quad x > 0. \quad (6.4.17)$$

Application of the Mellin transform with respect to x to equation (6.4.17) combined with (6.3.19) gives

$$\tilde{f}(1-p)\tilde{k}(p) = \tilde{g}(p),$$

which gives, replacing p by $1-p$,

$$\tilde{f}(p) = \tilde{g}(1-p)\tilde{h}(p),$$

where

$$\tilde{h}(p) = \frac{1}{\tilde{k}(1-p)}.$$

The inverse Mellin transform combined with (6.3.19) leads to the solution

$$f(x) = \mathcal{M}^{-1}\{\tilde{g}(1-p)\tilde{h}(p)\} = \int_0^\infty g(\xi) h(x\xi) d\xi, \quad (6.4.18)$$

provided $h(x) = \mathcal{M}^{-1}\{\tilde{h}(p)\}$ exists. Thus the problem is formally solved.

If, in particular, $\tilde{h}(p) = \tilde{k}(p)$, then the solution of (6.4.18) becomes

$$f(x) = \int_0^\infty g(\xi) k(x\xi) d\xi, \qquad (6.4.19)$$

provided $\tilde{k}(p)\tilde{k}(1-p) = 1$.

Example 6.4.4 Solve the integral equation

$$\int_0^\infty f(\xi) g\left(\frac{x}{\xi}\right) \frac{d\xi}{\xi} = h(x), \qquad (6.4.20)$$

where $f(x)$ is unknown and $g(x)$ and $h(x)$ are given functions.

Applications of the Mellin transform with respect to x gives

$$\tilde{f}(p) = \tilde{h}(p)\tilde{k}(p), \qquad \tilde{k}(p) = \frac{1}{\tilde{g}(p)}.$$

Inversion, by the convolution property (6.3.18), gives the solution

$$f(x) = \mathcal{M}^{-1}\{\tilde{h}(p)\tilde{k}(p)\} = \int_0^\infty h(\xi) k\left(\frac{x}{\xi}\right) \frac{d\xi}{\xi}. \qquad (6.4.21)$$

6.5 Mellin Transforms of the Weyl Fractional Integral and the Weyl Fractional Derivative

Definition 6.5.1 The Mellin transform of the *Weyl fractional integral* of $f(x)$ defined by

$$W^{-\alpha}[f(x)] = \frac{1}{\Gamma(\alpha)} \int_x^\infty (t-x)^{\alpha-1} f(t) dt, \qquad 0 < \operatorname{Re} \alpha < 1, \, x > 0. \quad (6.5.1)$$

Often $_xW_\infty^{-\alpha}$ is used instead of $W^{-\alpha}$ to indicate the limits of integration. Result (6.5.1) can be interpreted as the *Weyl transform* of $f(t)$, defined by

$$W^{-\alpha}[f(t)] = F(x, \alpha) = \frac{1}{\Gamma(\alpha)} \int_x^\infty (t-x)^{\alpha-1} f(t) dt. \qquad (6.5.2)$$

We first give some simple examples of the Weyl transform.

If $f(t) = \exp(-at)$, $\operatorname{Re} a > 0$, then the Weyl transform of $f(t)$ is given by

$$W^{-\alpha}[\exp(-at)] = \frac{1}{\Gamma(\alpha)} \int_x^\infty (t-x)^{\alpha-1} \exp(-at) dt$$

which is, by the change of variable $t - x = y$,

$$= \frac{e^{-ax}}{\Gamma(\alpha)} \int_0^\infty y^{\alpha-1} \exp(-ay) dy$$

which is, by letting $ay = t$,

$$W^{-\alpha}[f(t)] = \frac{e^{-ax}}{a^\alpha} \frac{1}{\Gamma(\alpha)} \int_0^\infty t^{\alpha-1} e^{-t} \, dt = \frac{e^{-ax}}{a^\alpha}. \tag{6.5.3}$$

Similarly, it can be shown that

$$W^{-\alpha}[t^{-\mu}] = \frac{\Gamma(\mu - \alpha)}{\Gamma(\mu)} x^{\alpha - \mu}, \qquad 0 < \operatorname{Re} \alpha < \operatorname{Re} \mu. \tag{6.5.4}$$

Making reference to Grashteyn and Ryzhik (1980, p 424), we obtain

$$W^{-\alpha}[\sin at] = a^{-\alpha} \sin\left(ax + \frac{\pi\alpha}{2}\right), \tag{6.5.5}$$

$$W^{-\alpha}[\cos at] = a^{-\alpha} \cos\left(ax + \frac{\pi\alpha}{2}\right). \tag{6.5.6}$$

where $0 < \operatorname{Re} \alpha < 1$ and $a > 0$.

It can be shown that, for any two positive numbers α and β, the Weyl fractional integral satisfies the laws of exponents

$$W^{-\alpha}\left[W^{-\beta} f(x)\right] = W^{-(\beta + \alpha)}[f(x)] = W^{-\beta}\left[W^{-\alpha} f(x)\right]. \tag{6.5.7}$$

Invoking a change of variable $t - x = y$ in (6.5.1), we obtain

$$W^{-\alpha}[f(x)] = \frac{1}{\Gamma(\alpha)} \int_0^\infty y^{\alpha-1} f(x + y) \, dy. \tag{6.5.8}$$

We next differentiate (6.5.8) to obtain, $D = \dfrac{d}{dx}$,

$$D\left[W^{-\alpha} f(x)\right] = \frac{1}{\Gamma(\alpha)} \int_0^\infty t^{\alpha-1} \frac{\partial}{\partial x} f(x + t) \, dt$$

$$= \frac{1}{\Gamma(\alpha)} \int_0^\infty t^{\alpha-1} D f(x + t) \, dt$$

$$= W^{-\alpha}[Df(x)]. \tag{6.5.9}$$

A similar argument leads to a more general result

$$D^n\left[W^{-\alpha} f(x)\right] = W^{-\alpha}[D^n f(x)], \tag{6.5.10}$$

where n is a positive integer.

Or symbolically

$$D^n W^{-\alpha} = W^{-\alpha} D^n. \tag{6.5.11}$$

We now calculate the Mellin transform of the Weyl fractional integral by putting $h(t) = t^\alpha f(t)$ and $g\left(\dfrac{x}{t}\right) = \dfrac{1}{\Gamma(\alpha)}\left(1 - \dfrac{x}{t}\right)^{\alpha-1} H\left(1 - \dfrac{x}{t}\right)$ where $H\left(1 - \dfrac{x}{t}\right)$ is the Heaviside unit step function so that (6.5.1) becomes

$$F(x, \alpha) = \int_0^\infty h(t) \, g\left(\frac{x}{t}\right) \frac{dt}{t}, \tag{6.5.12}$$

which is, by the convolution property (6.3.18),

where
$$\tilde{F}(p, \alpha) = \tilde{h}(p)\, \tilde{g}(p),$$

$$\tilde{h}(p) = \mathcal{M}\{x^\alpha f(x)\} = \tilde{f}(p+\alpha),$$

and
$$\tilde{g}(p) = \mathcal{M}\left\{\frac{1}{\Gamma(\alpha)}(1-x)^{\alpha-1} H(1-x)\right\}$$
$$= \frac{1}{\Gamma(\alpha)} \int_0^1 x^{p-1}(1-x)^{\alpha-1}\, dx$$
$$= \frac{B(p, \alpha)}{\Gamma(\alpha)} = \frac{\Gamma(p)}{\Gamma(p+\alpha)}.$$

Consequently
$$\tilde{F}(p, \alpha) = \mathcal{M}\left[W^{-\alpha} f(x), p\right] = \frac{\Gamma(p)}{\Gamma(p+\alpha)} \tilde{f}(p+\alpha). \qquad (6.5.13)$$

It is important to note that this result is an obvious extension of result 7(b) in Exercise 6.6.

Definition 6.5.2 If β is a positive number and n is the smallest integer greater than β such that $n - \beta = \alpha > 0$, the *Weyl fractional derivative* of a function $f(x)$ is defined by
$$W^\beta[f(x)] = E^n\, W^{-(n-\beta)}[f(x)]$$
$$= \frac{(-1)^n}{\Gamma(n-\beta)} \frac{d^n}{dx^n} \int_x^\infty (t-x)^{n-\beta-1} f(t)\, dt, \qquad (6.5.14)$$

where $E = -D$.

Or symbolically
$$W^\beta = E^n\, W^{-\alpha} = E^n\, W^{-(n-\beta)}. \qquad (6.5.15)$$

It can be shown that, for any β,
$$W^{-\beta} W^\beta = I = W^\beta W^{-\beta}. \qquad (6.5.16)$$

And, for any β and γ, the Weyl fractional derivative satisfies the laws of exponents
$$W^\beta\left[W^\gamma f(x)\right] = W^{\beta+\gamma}[f(x)] = W^\gamma\left[W^\beta f(x)\right]. \qquad (6.5.17)$$

We now calculate the Weyl fractional derivative of some elementary functions.

If $f(x) = \exp(-ax), a > 0$, then the definition (6.5.14) gives
$$W^\beta e^{-ax} = E^n\left[W^{-(n-\beta)} e^{-ax}\right], \qquad (6.5.18)$$

Writing $n - \beta = \alpha > 0$ and using (6.5.3) yields
$$W^\beta e^{-ax} = E^n\left[W^{-\alpha} e^{-ax}\right] = E^n\left[a^{-\alpha} e^{-ax}\right]$$
$$= a^{-\alpha}\left(a^n e^{-ax}\right) = a^\beta e^{-ax}. \qquad (6.5.19)$$

Replacing β by $-\alpha$ in (6.5.19) leads to result (6.5.3) as expected.
Similarly, we obtain

$$W^\beta x^{-\mu} = \frac{\Gamma(\beta+\mu)}{\Gamma(\mu)} x^{-(\beta+\mu)}. \tag{6.5.20}$$

Finally, it is easy to see that

$$W^\beta(\cos ax) = E\left[W^{-(1-\beta)}\cos ax\right],$$

which is, by (6.5.6),

$$= a^\beta \cos\left(ax - \frac{1}{2}\pi\beta\right). \tag{6.5.21}$$

Similarly,

$$W^\beta(\sin ax) = a^\beta \sin\left(ax - \frac{1}{2}\pi\beta\right), \tag{6.5.22}$$

provided α and β lie between 0 and 1.

If β is replaced by $-\alpha$, results (6.5.20)-(6.5.22) reduce to (6.5.4)-(6.5.6) respectively.

Finally, we calculate the Mellin transform of the Weyl fractional derivative with the help of (6.3.9) and find

$$\mathcal{M}\left[W^\beta f(x)\right] = \mathcal{M}\left[E^n W^{-(n-\beta)} f(x)\right] = (-1)^n \mathcal{M}\left[D^n W^{-(n-\beta)} f(x)\right]$$

$$= \frac{\Gamma(p)}{\Gamma(p-n)} \mathcal{M}\left[W^{-(n-\beta)} f(x), p-n\right]$$

which is, by result (6.5.13),

$$= \frac{\Gamma(p)}{\Gamma(p-n)} \cdot \frac{\Gamma(p-n)}{\Gamma(p-\beta)} \tilde{f}(p-\beta)$$

$$= \frac{\Gamma(p)}{\Gamma(p-\beta)} M[f(x), p-\beta].$$

$$= \frac{\Gamma(p)}{\Gamma(p-\beta)} \tilde{f}(p-\beta). \tag{6.5.23}$$

Example 6.5.1 (*The Fourier Transform of the Weyl Fractional Integral*).

$$\mathcal{F}\{W^{-\alpha} f(x)\} = \exp\left(-\frac{\pi i \alpha}{2}\right) k^{-\alpha} \mathcal{F}\{f(x)\}. \tag{6.5.24}$$

We have

$$\mathcal{F}\{W^{-\alpha} f(x)\} = \frac{1}{\sqrt{2\pi}} \frac{1}{\Gamma(\alpha)} \int_{-\infty}^{\infty} e^{-ikx} dx \int_{x}^{\infty} (t-x)^{\alpha-1} f(t) dt$$

$$= \frac{1}{\sqrt{2\pi}} \int_{-\infty}^{\infty} f(t) dt \cdot \frac{1}{\Gamma(\alpha)} \int_{-\infty}^{t} \exp(-ikx)(t-x)^{\alpha-1} dx.$$

Thus

$$\mathcal{F}\{W^{-\alpha}f(x)\} = \frac{1}{\sqrt{2\pi}} \int_{-\infty}^{\infty} e^{-ikt}f(t)dt \cdot \frac{1}{\Gamma(\alpha)} \int_{0}^{\infty} e^{ik\tau}\tau^{\alpha-1}d\tau, \quad (t-x = \tau)$$

$$= \mathcal{F}\{f(x)\} \frac{1}{\Gamma(\alpha)} \mathcal{M}\{e^{ik\tau}\}$$

$$= \exp\left(-\frac{\pi i \alpha}{2}\right) k^{-\alpha} \mathcal{F}\{f(x)\}.$$

In the limit as $\alpha \to 0$

$$\lim_{\alpha \to 0} \mathcal{F}\{W^{-\alpha}f(x)\} = \mathcal{F}\{f(x)\}.$$

This implies that

$$W^{0}\{f(x)\} = f(x).$$

We conclude this section by proving a general property of the Riemann-Liouville fractional integral operator $D^{-\alpha}$ and the Weyl fractional integral operator $W^{-\alpha}$. It follows from the definition (3.9.1) that $D^{-\alpha}f(t)$ can be expressed as the convolution

$$D^{-\alpha}f(x) = g_{\alpha}(t) * f(t), \qquad (6.5.25)$$

where

$$g_{\alpha}(t) = \frac{t^{\alpha-1}}{\Gamma(\alpha)}, \quad t > 0.$$

Similarly, $W^{-\alpha}f(x)$ can also be written in terms of the convolution

$$W^{-\alpha}f(x) = g_{-\alpha}(-x) * f(x). \qquad (6.5.26)$$

Then, under suitable conditions,

$$\mathcal{M}[D^{-\alpha}f(x)] = \frac{\Gamma(1-\alpha-p)}{\Gamma(1-p)} \tilde{f}(p+\alpha), \qquad (6.5.27)$$

$$\mathcal{M}[W^{-\alpha}f(x)] = \frac{\Gamma(p)}{\Gamma(\alpha+p)} \tilde{f}(p+\alpha). \qquad (6.5.28)$$

Finally, a formal computation gives

$$\int_{0}^{\infty} \{D^{-\alpha}f(x)\}g(x)dx = \frac{1}{\Gamma(\alpha)} \int_{0}^{\infty} g(x)dx \int_{0}^{x} (x-t)^{\alpha-1}f(t)dt$$

$$= \int_{0}^{\infty} f(t)dt \cdot \frac{1}{\Gamma(\alpha)} \int_{t}^{\infty} (x-t)^{\alpha-1}g(x)dx$$

$$= \int_{0}^{\infty} f(t)[W^{-\alpha}g(t)]dt,$$

which is, using the inner product notation,

$$(D^{-\alpha}f, g) = (f, W^{-\alpha}g). \qquad (6.5.29)$$

This shows that $D^{-\alpha}$ and $W^{-\alpha}$ behave like adjoint operators. Obviously, this result can be used to define fractional integrals of distributions. This result is taken from Debnath and Grum (1988).

6.6 Application of Mellin Transforms to Summation of Series

In this section we discuss a method of summation of series which is particularly associated with the work of Macfarlane (1949).

Theorem 6.6.1 If $\mathcal{M}\{f(x)\} = \tilde{f}(p)$, then

$$\sum_{n=0}^{\infty} f(n+a) = \frac{1}{2\pi i} \int_{c-i\infty}^{c+i\infty} \tilde{f}(p)\zeta(p,a)\,dp, \qquad (6.6.1)$$

where $\xi(p,a)$ is the *Hurwitz zeta function* defined by

$$\xi(p,a) = \sum_{n=0}^{\infty} \frac{1}{(n+a)^p}, \qquad 0 \le a \le 1,\ \operatorname{Re}(p) > 1. \qquad (6.6.2)$$

Proof. It follows from the inverse Mellin transform that

$$f(n+a) = \frac{1}{2\pi i} \int_{c-i\infty}^{c+i\infty} \tilde{f}(p)(n+a)^{-p}\,dp. \qquad (6.6.3)$$

Summing this over all n gives

$$\sum_{n=0}^{\infty} f(n+a) = \frac{1}{2\pi i} \int_{c-i\infty}^{c+i\infty} \tilde{f}(p)\zeta(p,a)\,dp.$$

This completes the proof.

Similarly, the scaling property (6.3.1) gives

$$f(nx) = \mathcal{M}^{-1}\{n^{-p}\tilde{f}(p)\} = \frac{1}{2\pi i} \int_{c-i\infty}^{c+i\infty} x^{-p} n^{-p} \tilde{f}(p)\,dp.$$

Thus

$$\sum_{n=1}^{\infty} f(nx) = \frac{1}{2\pi i} \int_{c-i\infty}^{c+i\infty} x^{-p}\tilde{f}(p)\zeta(p)\,dp = \mathcal{M}^{-1}\{\tilde{f}(p)\zeta(p)\}, \qquad (6.6.4)$$

where $\zeta(p) = \sum_{n=1}^{\infty} n^{-p}$ is the *Riemann zeta function*.

When $x = 1$, (6.6.4) reduces to

$$\sum_{n=1}^{\infty} f(n) = \frac{1}{2\pi i} \int_{c-i\infty}^{c+i\infty} \tilde{f}(p)\zeta(p)\,dp. \qquad (6.6.5)$$

This can be obtained from (6.6.1) when $a = 0$.

Example 6.6.1 Show that

$$\sum_{n=1}^{\infty} (-1)^{n-1} n^{-p} = (1 - 2^{1-p})\zeta(p). \qquad (6.6.6)$$

Using Example 6.2.1(a), we can write the left hand side of (6.6.6) multiplied by t^n as

$$\sum_{n=1}^{\infty}(-1)^{n-1}n^{-p}t^n = \sum_{n=1}^{\infty}(-1)^{n-1}t^n \cdot \frac{1}{\Gamma(p)}\int_0^{\infty} x^{p-1}e^{-nx}dx$$

$$= \frac{1}{\Gamma(p)}\int_0^{\infty} x^{p-1}dx \sum_{n=1}^{\infty}(-1)^{n-1}t^n e^{-nx}$$

$$= \frac{1}{\Gamma(p)}\int_0^{\infty} x^{p-1} \cdot \frac{te^{-x}}{1+te^{-x}}dx$$

$$= \frac{1}{\Gamma(p)}\int_0^{\infty} x^{p-1} \cdot \frac{t}{e^x+t}dx.$$

In the limit as $t \to 1$, the above result gives

$$\sum_{n=1}^{\infty}(-1)^{n-1}n^{-p} = \frac{1}{\Gamma(p)}\int_0^{\infty} x^{p-1}\frac{1}{e^x+1}dx$$

$$= \frac{1}{\Gamma(p)}\mathcal{M}\left\{\frac{1}{e^x+1}\right\} = \left(1-2^{1-p}\right)\zeta(p),$$

in which result (6.2.11) is used.

Example 6.6.2 Show that

$$\sum_{n=1}^{\infty}\frac{\sin an}{n} = \frac{1}{2}(\pi-a), \qquad 0 < a < 2\pi. \tag{6.6.7}$$

Proof. The Mellin transform of $f(x) = \left(\dfrac{\sin ax}{x}\right)$ gives

$$\mathcal{M}\left[\frac{\sin ax}{x}\right] = \int_0^{\infty} x^{p-2}\sin ax\, dx$$

$$= \mathcal{F}_s\left\{\sqrt{\frac{\pi}{2}}\, x^{p-2}\right\}$$

$$= -\frac{\Gamma(p-1)}{a^{p-1}}\cos\left(\frac{\pi p}{2}\right).$$

Substituting this result into (6.6.5) gives

$$\sum_{n=1}^{\infty}\frac{\sin an}{n} = -\frac{1}{2\pi i}\int_{c-i\infty}^{c+i\infty}\frac{\Gamma(p-1)}{a^{p-1}}\zeta(p)\cos\left(\frac{\pi p}{2}\right)dp. \tag{6.6.8}$$

We next use the well-known functional equation for the zeta function

$$(2\pi)^p\zeta(1-p) = 2\Gamma(p)\zeta(p)\cos\left(\frac{\pi p}{2}\right) \tag{6.6.9}$$

in the integrand of (6.6.8) to obtain

$$\sum_{n=1}^{\infty} \frac{\sin an}{n} = -\frac{a}{2} \cdot \frac{1}{2\pi i} \int_{c-i\infty}^{c+i\infty} \left(\frac{2\pi}{a}\right)^p \frac{\zeta(1-p)}{p-1} \, dp.$$

The integral has two simple poles at $p = 0$ and $p = 1$ with residues 1 and $-\pi/a$ respectively, and the complex integral is evaluated by calculating the residues at these poles. Thus the sum of the series is

$$\sum_{n=1}^{\infty} \frac{\sin an}{n} = \frac{1}{2}(\pi - a).$$

6.7 Generalized Mellin Transforms

In order to extend the applicability of the classical Mellin transform, Naylor (1963) generalized the method of Mellin integral transform. This generalized Mellin transform is useful for finding solutions of boundary value problems in regions bounded by the natural coordinate surfaces of a spherical or cylindrical coordinate system. They can be used to solve boundary value problems in finite region or in infinite regions bounded internally.

The *generalized Mellin transform* of a function $f(r)$ defined in $a < r < \infty$ is introduced by the integral

$$\mathcal{M}_-\{f(r)\} = F_-(p) = \int_a^\infty \left(r^{p-1} - \frac{a^{2p}}{r^{p+1}}\right) f(r) \, dr. \quad (6.7.1)$$

The inverse transform is given by

$$\mathcal{M}_-^{-1}\{F_-(p)\} = f(r) = \frac{1}{2\pi i} \int_L r^{-p} F(p) \, dp, \quad r > a, \quad (6.7.2)$$

where L is the line Re $p = c$, and $F(p)$ is analytic in the strip $|\text{Re}(p)| < \gamma$ with $c < \gamma$.

By integrating by parts, we can show that

$$\mathcal{M}_-\left[r^2 \frac{\partial^2 f}{\partial r^2} + r \frac{\partial f}{\partial r}\right] = p^2 \, F_-(p) + 2p \, a^p \, f(a), \quad (6.7.3)$$

provided $f(r)$ is appropriately behaved at infinity. More precisely

$$\lim_{r \to \infty} \left[\left(r^p - a^{2p} \, r^{-p}\right) r f_r - p\left(r^p + a^{2p} \, r^{-p}\right) f\right] = 0. \quad (6.7.4)$$

Obviously, this generalized transform seems to be very useful for finding the solution of boundary value problems in which $f(r)$ is prescribed on the internal boundary at $r = a$.

On the other hand, if the derivative of $f(r)$ is prescribed at $r = a$, it is convenient to define the associated integral transform by

$$\mathcal{M}_+[f(r)] = F_+(p) = \int_a^\infty \left(r^{p-1} + \frac{a^{2p}}{r^{p+1}}\right) f(r) \, dr, \quad |\text{Re}(p)| < r, \quad (6.7.5)$$

and its inverse given by
$$\mathcal{M}_+^{-1}[f(p)] = f(r) = \frac{1}{2\pi i} \int_L r^{-p} F_+(p)\, dp, \quad r > a. \tag{6.7.6}$$
In this case, we can show by integration by parts that
$$\mathcal{M}_+\left[r^2 \frac{\partial^2 f}{\partial r^2} + r \frac{\partial f}{\partial r}\right] = p^2 F_+(p) - 2a^{p+1} f'(a), \tag{6.7.7}$$
where $f'(r)$ exists at $r = a$.

Theorem 6.7.1 (*Convolution*). If $\mathcal{M}_+\{f(r)\} = F_+(p)$, and $\mathcal{M}_+\{g(r)\} = G_+(p)$, then
$$\mathcal{M}_+\{f(r)\, g(r)\} = \frac{1}{2\pi i} \int_L F_+(\xi)\, G_+(p-\xi)\, d\xi. \tag{6.7.8}$$

Or equivalently
$$f(r) g(r) = \mathcal{M}_+^{-1}\left[\frac{1}{2\pi i} \int_L F_+(\xi)\, G_+(p-\xi)\, d\xi\right]. \tag{6.7.9}$$

Proof. We assume that $F_+(p)$ and $G_+(p)$ are analytic in some strip $|\operatorname{Re}(p)| < \gamma$. Then
$$\mathcal{M}_+\{f(r)\, g(r)\} = \int_a^\infty \left(r^{p-1} + \frac{a^{2p}}{r^{p+1}}\right) f(r) g(r)\, dr$$
$$= \int_a^\infty r^{p-1} f(r) g(r)\, dr + \int_a^\infty \frac{a^{2p}}{r^{p+1}} f(r) g(r)\, dr. \tag{6.7.10}$$
$$= \frac{1}{2\pi i} \int_L F_+(\xi)\, d\xi \int_a^\infty r^{p-\xi-1} g(r)\, dr$$
$$+ \frac{1}{2\pi i} \int_a^\infty \frac{a^{2p}}{r^{p+1}} g(r)\, dr \int_L r^{-\xi} F_+(\xi)\, d\xi. \tag{6.7.11}$$

Replacing ξ by $-\xi$ in the first integral term and using $F_+(\xi) = a^{2\xi} F_+(-\xi)$, which follows from the definition (6.7.5), we obtain
$$\int_L r^{-\xi} F_+(\xi)\, d\xi = \int_L r^\xi a^{-2\xi} F_+(\xi)\, d\xi. \tag{6.7.12}$$

The path of integration L, $\operatorname{Re}(\xi) = c$, becomes $\operatorname{Re}(\xi) = -c$, but these paths can be reconciled if $F(\xi)$ tends to zero for large $\operatorname{Im}(\xi)$.

In view of (6.7.11), we have rewritten
$$\int_a^\infty \frac{a^{2p}}{r^{p+1}} f(r) g(r)\, dr = \frac{1}{2\pi i} \int_L F_+(\xi)\, d\xi \int_a^\infty \frac{a^{2p-2\xi}}{r^{p-\xi+1}} g(r)\, dr. \tag{6.7.13}$$

This result is used to rewrite (6.7.10) as

$$\mathcal{M}_+\{f(r)g(r)\} = \int_a^\infty \left(r^{p-1} + \frac{a^{2p}}{r^{p+1}}\right) f(r)g(r)\,dr$$

$$= \int_a^\infty r^{p-1} f(r)g(r)\,dr + \int_a^\infty \frac{a^{2p}}{r^{p+1}} f(r)g(r)\,dr$$

$$= \frac{1}{2\pi i} \int_L F_+(\xi)\,d\xi \int_a^\infty r^{p-\xi-1} g(r)\,dr$$

$$\quad + \frac{1}{2\pi i} \int_L F_+(\xi)\,d\xi \int_a^\infty \frac{a^{2p-2\xi}}{r^{p-\xi+1}} g(r)\,dr$$

$$= \frac{1}{2\pi i} \int_L F_+(\xi)\, G_+(p-\xi)\,d\xi.$$

This completes the proof.

If the range of integration is finite, then we define the *generalized finite Mellin transform* by

$$\mathcal{M}_-^a\{f(r)\} = F_-^a(p) = \int_0^a \left(r^{p-1} - \frac{a^{2p}}{r^{p+1}}\right) f(r)\,dr, \qquad (6.7.14)$$

where Re $p < \gamma$.

The corresponding inverse transform is given by

$$f(r) = -\frac{1}{2\pi i} \int_L \left(\frac{r}{a^2}\right)^p F_-^a(p)\,dp, \quad 0 < r < a,$$

which is, by replacing p by $-p$ and using $F_-^a(-p) = -a^{-2p} F_-^a(p)$,

$$= \frac{1}{2\pi i} \int_L r^{-p} F_-^a(p)\,dp, \quad 0 < r < a, \qquad (6.7.15)$$

where the path L is Re $p = -c$ with $|c| < \gamma$.

It is easy to verify the result

$$\mathcal{M}_-^a\{r^2 f_{rr} + r f_r\} = \int_0^a \left(r^{p-1} - \frac{a^{2p}}{r^{p+1}}\right)\{r^2 f_{rr} + r f_r\}\,dr$$

$$= p^2 F_-^a(p) - 2p\, a^p\, f(a). \qquad (6.7.16)$$

This is a useful result for applications.

Similarly, we define the finite transform-pair by

$$\mathcal{M}_+^a\{f(r)\} = F_+^a(p) = \int_0^a \left(r^{p-1} + \frac{a^{2p}}{r^{p+1}}\right) f(r)\,dr, \qquad (6.7.17)$$

$$f(r) = \left(\mathcal{M}_+^a\right)^{-1}\left[F_+^a(p)\right] = \frac{1}{2\pi i} \int_L r^{-p} F_+^a(p)\,dp, \qquad (6.7.18)$$

where $|\text{Re } p| < \gamma$.

For this finite transform, we can also prove

$$\mathcal{M}_+^a\left[r^2 f_{rr} + r f_r\right] = \int_0^a \left(r^{p-1} + \frac{a^{2p}}{r^{p+1}}\right)\left(r^2 f_{rr} + r f_r\right) dr$$

$$= p^2 \, F_+^a(p) + 2a^{p-1} f'(a). \qquad (6.7.19)$$

This result seems to be useful for applications. The reader is referred to Naylor (1963) for applications of the above results to boundary value problems.

6.8 Exercises

1. Find the Mellin transform of each of the following functions:
    (a) $f(x) = H(a-x), \quad a > 0,$
    (b) $f(x) = x^m e^{-nx}, \quad m, n > 0,$
    (c) $f(x) = \dfrac{1}{1+x^2},$
    (d) $f(x) = J_0^2(x).$

2. Derive the Mellin transform-pairs from the bilateral Laplace transform and its inverse given by

$$\bar{g}(p) = \int_{-\infty}^{\infty} e^{-pt} g(t)\, dt, \qquad g(t) = \frac{1}{2\pi i}\int_{c-i\infty}^{c+i\infty} e^{pt}\, \bar{g}(p)\, dp.$$

3. Show that

$$\mathcal{M}\left[\frac{1}{e^x + e^{-x}}\right] = \Gamma(p) L(p),$$

where $L(p) = \dfrac{1}{1^p} - \dfrac{1}{3^p} + \dfrac{1}{5^p} - \cdots$ is the *Dirichlet L-function*.

4. Show that

$$\mathcal{M}\left\{\frac{1}{(1+ax)^n}\right\} = \frac{\Gamma(p)\Gamma(n-p)}{a^p\, \Gamma(n)}.$$

5. Show that

$$\mathcal{M}\{x^{-n} J_n(ax)\} = \frac{1}{2}\left(\frac{a}{2}\right)^{n-p} \frac{\Gamma\!\left(\dfrac{p}{2}\right)}{\Gamma\!\left(n - \dfrac{p}{2} + 1\right)}, \quad a > 0,\ n > -\frac{1}{2}.$$

6. Show that
    (a) $\mathcal{M}^{-1}\left[\cos\left(\dfrac{\pi p}{2}\right) \Gamma(p) \tilde{f}(1-p)\right] = \mathcal{F}_c\left\{\sqrt{\dfrac{\pi}{2}}\, f(x)\right\},$
    (b) $\mathcal{M}^{-1}\left[\sin\left(\dfrac{\pi p}{2}\right) \Gamma(p) \tilde{f}(1-p)\right] = \mathcal{F}_s\left\{\sqrt{\dfrac{\pi}{2}}\, f(x)\right\}.$

7. If $I_n^\infty f(x)$ denotes the nth repeated integral of $f(x)$ defined by

$$I_n^\infty f(x) = \int_x^\infty I_{n-1}^\infty f(t)\, dt,$$

show that

(a) $\mathcal{M}\left[\int_x^\infty f(t)\,dt, p\right] = \frac{1}{p}\tilde{f}(p+1)$,

(b) $\mathcal{M}[I_n^\infty f(x)] = \dfrac{\Gamma(p)}{\Gamma(p+n)}\tilde{f}(p+n)$.

8. Show that the integral equation

$$f(x) = h(x) + \int_0^\infty g(x\xi)\,f(\xi)\,d\xi$$

has the formal solution

$$f(x) = \frac{1}{2\pi i}\int_{c-i\infty}^{c+i\infty}\left[\frac{\tilde{h}(p)+\tilde{g}(p)\tilde{h}(1-p)}{1-\tilde{g}(p)\tilde{g}(1-p)}\right]x^{-p}\,dp.$$

9. Find the solution of the Laplace integral equation

$$\int_0^\infty e^{-x\xi}f(\xi)\,d\xi = \frac{1}{(1+x)^n}.$$

10. Show that the integral equation

$$f(x) = h(x) + \int_0^\infty f(\xi)\,g\!\left(\frac{x}{\xi}\right)\frac{d\xi}{\xi}$$

has the formal solution

$$f(x) = \frac{1}{2\pi i}\int_{c-i\infty}^{c+i\infty}\frac{x^{-p}\,\tilde{h}(p)}{1-\tilde{g}(p)}\,dp.$$

11. Show that the solution of the integral equation

$$f(x) = e^{-ax} + \int_0^\infty \exp\!\left(-\frac{x}{\xi}\right) f(\xi)\,\frac{d\xi}{\xi}$$

is

$$f(x) = \frac{1}{2\pi i}\int_{c-i\infty}^{c+i\infty}(ax)^{-p}\left\{\frac{\Gamma(p)}{1-\Gamma(p)}\right\}dp.$$

12. Assuming

$$\mathcal{M}[f(re^{i\theta})] = \int_0^\infty r^{p-1}f(re^{i\theta})\,dr$$

and putting $re^{i\theta} = \xi$, show that

(a) $\mathcal{M}[f(re^{i\theta})] = \exp(-ip\theta)\,\tilde{f}(p)$.

Hence deduce

(b) $\mathcal{M}^{-1}\{\tilde{f}(p)\cos p\theta\} = \mathrm{Re}[f(re^{i\theta})]$,

(c) $\mathcal{M}^{-1}\{\tilde{f}(p)\sin p\theta\} = -\mathrm{Im}[f(re^{i\theta})]$.

13. (a) If $\mathcal{M}[\exp(-r)] = \Gamma(p)$, show that
$$\mathcal{M}[\exp(-re^{i\theta})] = \Gamma(p)e^{-ip\theta},$$

(b) If $\mathcal{M}[\log(1+r)] = \dfrac{\pi}{p \sin \pi p}$, then show that
$$\mathcal{M}[\mathrm{Re}\,\log(1+re^{i\theta})] = \dfrac{\pi \cos p\theta}{p \sin \pi p}.$$

14. Use $\mathcal{M}^{-1}\left\{\dfrac{\pi}{\sin p\pi}\right\} = \dfrac{1}{1+x} = f(x)$, and Exercises 12(a) and 12(b) respectively to show that

(a) $\mathcal{M}^{-1}\left\{\dfrac{\pi \cos p\theta}{\sin p\pi}\right\} = \dfrac{1+r\cos\theta}{1+2r\cos\theta+r^2},$

(b) $\mathcal{M}^{-1}\left\{\dfrac{\pi \sin p\theta}{\sin p\pi}\right\} = \dfrac{r\sin\theta}{1+2r\cos\theta+r^2}.$

15. Find the inverse Mellin transforms of

(a) $\Gamma(p) \cos p\theta$, where $-\dfrac{\pi}{2} < \theta < \dfrac{\pi}{2}.$

(b) $\Gamma(p) \sin p\theta$.

16. Obtain the solution of Example 6.4.2 for the boundary data
$$\phi(r,\alpha) = \phi(r,-\alpha) = H(a-r).$$

17. Show that
$$\sum_{n=1}^{\infty} \dfrac{\cos kn}{n^2} = \dfrac{k^2}{4} - \dfrac{\pi k}{2} + \dfrac{\pi^2}{6}.$$

Hence deduce
$$\sum_{n=1}^{\infty} \dfrac{1}{n^2} = \dfrac{\pi^2}{6}.$$

18. If $f(x) = \sum_{n=1}^{\infty} a_n e^{-nx}$, show that
$$\mathcal{M}\{f(x)\} = \tilde{f}(p) = \Gamma(p)g(p),$$
where $g(p) = \sum_{n=1}^{\infty} a_n n^{-p}$ is the Dirichlet series. If $a_n = 1$ for all n, derive
$$\tilde{f}(p) = \Gamma(p)\zeta(p).$$
Show that
$$\mathcal{M}\left\{\dfrac{\exp(-ax)}{1-e^{-x}}\right\} = \Gamma(p)\zeta(p,a).$$

19. Show that

    (a) $\sum_{n=1}^{\infty} \frac{(-1)^{n-1}}{n^p} = (1-2^{1-p})\zeta(p).$

    Hence deduce

    (b) $\sum_{n=1}^{\infty} \frac{(-1)^{n-1}}{n^2} = \frac{\pi^2}{12},$ (c) $\sum_{n=1}^{\infty} \frac{(-1)^{n-1}}{n^4} = \left(\frac{7}{8}\right)\frac{\pi^4}{90}.$

20. Find the sum of the following series

    (a) $\sum_{n=1}^{\infty} \frac{(-1)^{n-1}}{n^2} \cos kn,$ (b) $\sum_{n=1}^{\infty} \frac{(-1)^{n-1}}{n} \sin kn.$

21. Show that the solution of the boundary value problem
    $$r^2 \phi_{rr} + r\phi_r + \phi_{\theta\theta} = 0, \quad 0 < r < \infty, \; 0 < \theta < \pi$$
    $$\phi(r,0) = \phi(r,\pi) = f(r),$$

    is

    $$\phi(r,\theta) = \frac{1}{2\pi i} \int_{c-i\infty}^{c+i\infty} r^{-p} \frac{\tilde{f}(p) \cos\left\{p\left(\theta - \frac{\pi}{2}\right)\right\}}{\cos\left(\frac{\pi p}{2}\right)} dp.$$

22. Evaluate

    $$\sum_{n=1}^{\infty} \frac{\cos an}{n^3} = \frac{1}{12}(a^3 - 3\pi a^2 + 2\pi^2 a).$$

23. Prove the following results:

    (a) $\mathcal{M}\left[\int_0^{\infty} \xi^n f(x\xi) g(\xi) d\xi\right] = \tilde{f}(p) \tilde{g}(1+n-p),$

    (b) $\mathcal{M}\left[\int_0^{\infty} \xi^n f\left(\frac{x}{\xi}\right) g(\xi) d\xi\right] = \tilde{f}(p) \tilde{g}(p+n+1).$

24. Show that

    (a) $W^{-\alpha}[e^{-x}] = e^{-x}, \quad \alpha > 0,$

    (b) $W^{\frac{1}{2}}\left[\frac{1}{\sqrt{x}} \exp(-\sqrt{x})\right] = \frac{K_1(\sqrt{x})}{\sqrt{\pi x}}, \quad x > 0,$

    where $K_1(x)$ is the modified Bessel function of the second kind and order one.

25. (a) Show that the integral (Wong, 1989, pp 186-187)

    $$I(x) = \int_0^{\pi/2} J_v^2(x \cos \theta) \, d\theta, \quad v > -\frac{1}{2},$$

    can be written as a Mellin convolution

$$I(x) = \int_0^\infty f(x\xi) g(\xi) d\xi,$$

where

$$f(\xi) = J_\nu^2(\xi) \text{ and}$$

$$g(\xi) = \begin{cases} (1-\xi^2)^{-\frac{1}{2}}, & 0 < \xi < 1 \\ 0, & \xi \geq 1 \end{cases}.$$

(b) Prove that the integration contour in the Parseval identity

$$I(x) = \frac{1}{2\pi i} \int_{c-i\infty}^{c+i\infty} x^{-p} \tilde{f}(p) \tilde{g}(1-p) dp, \quad -2\nu < c < 1,$$

cannot be shifted to the right beyond the vertical line Re $p = 2$.

26. If $f(x) = \int_0^\infty \exp(-x^2 t^2) \cdot \frac{\sin t}{t^2} J_1(t) dt$, show that

$$\mathcal{M}\{f(x)\} = \frac{\Gamma\left(p+\frac{3}{2}\right) \Gamma\left(\frac{1-p}{2}\right)}{p \, \Gamma(p+3)}.$$

Chapter 7

Hilbert and Stieltjes Transforms

7.1 Introduction

Both Hilbert and Stieltjes transforms arise in many problems in applied mathematics, mathematical physics, and engineering sciences. The former one plays an important role in fluid mechanics, signal processing, and electronics, while the latter arises in the moment problem. This chapter deals with definitions of Hilbert and Stieltjes transforms with examples. This is followed by a discussion of basic operational properties of these transforms. Finally, examples of applications of Hilbert and Stieltjes transforms to physical problems are discussed.

7.2 Definition of the Hilbert Transform and Examples

If $f(t)$ is defined on the real line $-\infty < t < \infty$, its *Hilbert transform*, denoted by $\hat{f}_H(x)$, is defined by

$$H\{f(t)\} = \hat{f}_H(x) = \frac{1}{\pi} \oint_{-\infty}^{\infty} \frac{f(t)}{t-x} \, dt, \qquad (7.2.1)$$

where x is real and the integral is treated as a Cauchy principal value, that is,

$$\oint_{-\infty}^{\infty} \frac{f(t)\,dt}{t-x} = \lim_{\varepsilon \to 0} \left[\int_{-\infty}^{x-\varepsilon} + \int_{x+\varepsilon}^{\infty} \right] \frac{f(t)\,dt}{t-x}. \qquad (7.2.2)$$

To derive the *inverse Hilbert transform*, we rewrite (7.2.1) as

$$\frac{1}{\sqrt{2\pi}} \int_{-\infty}^{\infty} f(t) g(x-t) \, dt = \hat{f}_H(x), \qquad (7.2.3)$$

where $g(x) = \sqrt{\frac{2}{\pi}} \left(-\frac{1}{x}\right)$. Application of the Fourier transform with respect to x gives

$$F(k) = \frac{\hat{F}_H(k)}{G(k)}, \quad G(k) = i \, \text{sgn} \, k. \qquad (7.2.4)$$

Taking the inverse Fourier transform, we obtain the solution for $f(x)$ as

$$f(x) = -\frac{1}{\sqrt{2\pi}} \int_{-\infty}^{\infty} (i \text{ sgn } k) \hat{F}_H(k) \exp(ikx) dk$$

which is, by the Convolution Theorem 2.4.5,

$$= \frac{1}{\pi} \oint_{-\infty}^{\infty} \frac{\hat{f}_H(\xi)}{x - \xi} d\xi = -H\{\hat{f}_H(\xi)\}. \tag{7.2.5}$$

Obviously, $-H[H\{f(t)\}] = f(x)$ and hence $H^{-1} = -H$. Thus the *inverse Hilbert transform* is given by

$$f(t) = H^{-1}\{\hat{f}_H(x)\} = -H\{\hat{f}_H(x)\} = -\frac{1}{\pi} \oint_{-\infty}^{\infty} \frac{\hat{f}_H(x) dx}{x - t}. \tag{7.2.6}$$

Example 7.2.1 Find the Hilbert transform of a rectangular pulse given by

$$f(t) = \begin{cases} 1, & \text{for } |t| < a \\ 0, & \text{for } |t| > a \end{cases}. \tag{7.2.7}$$

We have, by definition,

$$\hat{f}_H(x) = \frac{1}{\pi} \int_{-a}^{a} \frac{dt}{t - x}.$$

If $|t| < a$, the integrand has a singularity at $t = x$, and hence

$$\hat{f}_H(x) = \frac{1}{\pi} \lim_{\varepsilon \to 0} \left[\int_{-a}^{x-\varepsilon} \frac{dt}{t - x} + \int_{x+\varepsilon}^{a} \frac{dt}{t - x} \right]$$

$$= \frac{1}{\pi} \lim_{\varepsilon \to 0} \left\{ [\log|t - x|]_{-a}^{x-\varepsilon} + [\log|t - x|]_{x+\varepsilon}^{a} \right\}$$

$$= \frac{1}{\pi} \lim_{\varepsilon \to 0} \{\log|\varepsilon| - \log|a + x| + \log|a - x| - \log|\varepsilon|\}$$

$$= \frac{1}{\pi} \log\left|\frac{a - x}{a + x}\right| \quad \text{for } |x| < a.$$

On the other hand, if $|t| > a$, the integrand has no singularity in $-a < t < a$, and hence

$$\hat{f}_H(x) = \frac{1}{\pi} \int_{-a}^{a} \frac{dt}{t - x} = \frac{1}{\pi} [\log|t - x|]_{-a}^{a} = \frac{1}{\pi} \log\left|\frac{a - x}{a + x}\right| \quad \text{for } |x| > a.$$

Finally, we obtain the Hilbert transform of $f(t)$ defined by (7.2.7) as

$$\hat{f}_H(x) = \frac{1}{\pi} \log\left|\frac{a - x}{a + x}\right|. \tag{7.2.8}$$

Example 7.2.2 Find the Hilbert transform of

$$f(t) = \frac{t}{(t^2 + a^2)}, \quad a > 0. \tag{7.2.9}$$

We have, by definition,

$$\hat{f}_H(x) = \frac{1}{\pi} \oint_{-\infty}^{\infty} \frac{t\,dt}{(t^2+a^2)(t-x)}$$

$$= \frac{1}{\pi(a^2+x^2)} \oint_{-\infty}^{\infty} \left[\frac{a^2}{t^2+a^2} + \frac{x}{t-x} - \frac{xt}{t^2+a^2}\right] dt$$

$$= \frac{1}{\pi}\frac{1}{(a^2+x^2)} \left[a^2 \int_{-\infty}^{\infty} \frac{dt}{(t^2+a^2)} + x \oint_{-\infty}^{\infty} \frac{dt}{(t-x)} - x \int_{-\infty}^{\infty} \frac{t\,dt}{(t^2+a^2)}\right].$$

The second and third integrals as the Cauchy principal value vanish and hence only the first integral makes a non-zero contribution. Thus, we obtain

$$\hat{f}_H(x) = \frac{1}{\pi}\frac{1}{(a^2+x^2)} \cdot (a\pi) = \frac{a}{(a^2+x^2)}. \qquad (7.2.10)$$

Example 7.2.3 Find the Hilbert transform of
(a) $f(t) = \cos \omega t$ and (b) $f(t) = \sin \omega t$.

It follows from the definition of the Hilbert transform that

$$\hat{f}_H(x) = \frac{1}{\pi} \int_{-\infty}^{\infty} \frac{\cos \omega t}{(t-x)} dt$$

$$= \frac{1}{\pi} \int_{-\infty}^{\infty} \frac{\cos\{\omega(t-x) + \omega x\} dt}{(t-x)}$$

$$= \frac{1}{\pi} \int_{-\infty}^{\infty} (t-x)^{-1}\left[\cos \omega(t-x)\cos \omega x - \sin \omega(t-x)\sin \omega x\right] dt$$

$$= \frac{\cos \omega x}{\pi} \int_{-\infty}^{\infty} \frac{\cos \omega(t-x)}{t-x} dt - \frac{\sin \omega x}{\pi} \int_{-\infty}^{\infty} \frac{\sin \omega(t-x)\,dt}{t-x}$$

which is, in terms of the new variable $T = t - x$,

$$= \frac{\cos \omega x}{\pi} \int_{-\infty}^{\infty} \frac{\cos \omega T}{T} dT - \frac{\sin \omega x}{\pi} \int_{-\infty}^{\infty} \frac{\sin \omega T}{T} dT. \qquad (7.2.11)$$

Obviously, the first integral vanishes because its integrand is an odd function of T. On the other hand, the second integral makes a non-zero contribution so that (7.2.11) gives

$$H\{\cos \omega t\} = \hat{f}_H(x) = -\frac{\sin \omega x}{\pi} \cdot \pi = -\sin \omega x. \qquad (7.2.12)$$

Similarly, it can be shown that

$$H\{\sin \omega t\} = \cos \omega x. \qquad (7.2.13)$$

7.3 Basic Properties of Hilbert Transforms

Theorem 7.3.1 If $H\{f(t)\} = \hat{f}_H(x)$, then the following properties hold:

(a) $\quad H\{f(t+a)\} = \hat{f}_H(x+a),$ (7.3.1)

(b) $\quad H\{f(at)\} = \hat{f}_H(ax), \quad a > 0,$ (7.3.2)

(c) $\quad H\{f(-at)\} = -\hat{f}_H(-ax),$ (7.3.3)

(d) $\quad H\{f'(t)\} = \dfrac{d}{dx} \hat{f}_H(x),$ (7.3.4)

(e) $\quad H\{t\, f(t)\} = x\, \hat{f}_H(x) + \dfrac{1}{\pi} \int_{-\infty}^{\infty} f(t)\, dt,$ (7.3.5)

(f) $\quad \mathcal{F}[H\{f(t)\}] = (-i\, \text{sgn}\, k)\, \mathcal{F}\{f(x)\},$ (7.3.6)

(g) $\quad \|H\{f(t)\}\| = \|f(t)\|,$ (7.3.7)

where $\|\cdot\|$ denotes the norm in $L^2(R)$, that is (see Debnath and Mikusinski, 1990, p 197),

$$\|f\| = \left[\int_{-\infty}^{\infty} |f(x)|^2 \, dx\right]^{\frac{1}{2}}.$$ (7.3.8)

Proof. (a) We have, by definition,

$$H\{f(t+a)\} = \frac{1}{\pi} \oint_{-\infty}^{\infty} \frac{f(t+a)\, dt}{t-x} \quad (t+a = u)$$

$$= \frac{1}{\pi} \oint_{-\infty}^{\infty} \frac{f(u)\, du}{u-(x+a)} = \hat{f}_H(x+a).$$

(b) $\quad H\{f(at)\} = \dfrac{1}{\pi} \oint_{-\infty}^{\infty} \dfrac{f(at)\, dt}{t-x} \quad (at = u,\ a > 0)$

$$= \frac{1}{\pi} \oint_{-\infty}^{\infty} \frac{f(u)\, du}{u - ax} = \hat{f}_H(ax).$$

Similarly, result (c) can be proved.

(d) $\quad H\{f'(t)\} = \dfrac{1}{\pi} \oint_{-\infty}^{\infty} \dfrac{f'(t)\, dt}{t-x}$

which is, by integrating by parts,

$$= \frac{1}{\pi} \left[\frac{f(t)}{t-x}\right]_{-\infty}^{\infty} + \frac{1}{\pi} \oint_{-\infty}^{\infty} \frac{f(t)\, dt}{(t-x)^2}$$

$$= \frac{d}{dx} \hat{f}_H(x).$$

Proofs of (e)-(g) are similar and hence can be left to the reader.

Theorem 7.3.2 If $f(t)$ is an even function of t, then

$$\hat{f}_H(x) = \frac{x}{\pi} \int_{-\infty}^{\infty} \frac{f(t) - f(x)}{(t^2 - x^2)}\, dt.$$ (7.3.9)

Proof. As the Cauchy principal value, we have

$$\oint_{-\infty}^{\infty} \frac{dt}{t-x} = 0.$$

Consequently,

$$\hat{f}_H(x) = \frac{1}{\pi} \oint_{-\infty}^{\infty} \frac{f(t) - f(x)}{t-x} dt$$

$$= \frac{1}{\pi} \oint_{-\infty}^{\infty} \frac{(t+x)\{f(t) - f(x)\}}{(t^2 - x^2)} dt$$

$$= \frac{1}{\pi} \oint_{-\infty}^{\infty} \frac{t\{f(t) - f(x)\}}{(t^2 - x^2)} dx + \frac{x}{\pi} \oint_{-\infty}^{\infty} \frac{\{f(t) - f(x)\} dt}{(t^2 - x^2)}. \quad (7.3.10)$$

Since $f(t)$ is an even function, the integrand of the first integral of (7.3.10) is an odd function; hence the first integral vanishes, and (7.3.10) gives (7.3.9).

We close this section by adding a comment. A more rigorous mathematical treatment of classical Hilbert transforms can be found in a treatise by Titchmarsh (1959). Further results and references of related work on Hilbert transforms and their applications are given by Kober (1943a,b), Gakhov (1966), Newcomb (1962), and Muskhelishvili (1953).

Several authors including Okikiolu (1965) and Kober (1967) introduced the *modified Hilbert transform* of a function $f(t)$, which is defined by

$$H_\alpha[f(t)] = \hat{f}_{H\alpha}(x) = \frac{\operatorname{cosec}\left(\dfrac{\pi\alpha}{2}\right)}{2\Gamma(\alpha)} \oint_{-\infty}^{\infty} (t-x)^{\alpha-1} f(t) dt, \quad (7.3.11)$$

where x is real and $0 < \alpha < 1$, and the integral is treated as the Cauchy principal value. Obviously, $H_\alpha[f(t)]$ is closely related to the Weyl fractional integral $W^{-\alpha}$ defined by (6.4.22) so that

$$2 \sin\left(\frac{\pi\alpha}{2}\right) H_\alpha[f(t)] = W^{-\alpha}[f(t), x] - W^{-\alpha}[f(-t), -x]. \quad (7.3.12)$$

Several properties of $H_\alpha[f(t)]$ and $W^{-\alpha}[f(t)]$ are investigated by Kober (1967). He also proved the following results, which is stated below without proof.

Theorem 7.3.3 (*Parseval's Relation*). If $H_\alpha[f(t)] = \hat{f}_{H\alpha}(x)$, then

$$\int_{-\infty}^{\infty} H_\alpha[f(t), x] h(x) dx = -\int_{-\infty}^{\infty} H_\alpha[h(t), x] f(x) dx. \quad (7.3.13)$$

7.4 Hilbert Transforms in the Complex Plane

In communication and coherence problems in electrical engineering (see, for example, Tuttle, 1958), the Hilbert transform in the complex plane plays an important role. In order to define such a transform, we first consider the function $f_0(z)$ of a complex variable $z = x + iy$ given by

$$f_0(z) = \frac{1}{\pi} \oint_{-\infty}^{\infty} \frac{f(t)\,dt}{t-z}, \qquad y > 0. \tag{7.4.1}$$

Application of the Fourier transform defined by (2.4.28) and its inverse by (2.4.27) to (7.2.1) and (7.4.1) gives

$$\hat{F}_H(\omega) = i\,\text{sgn}(\omega)\,F(\omega), \tag{7.4.2}$$

$$F_0(\omega) = 2i\,\exp(-\omega y)\,H(\omega)\,F(\omega). \tag{7.4.3}$$

In view of (7.4.2), (7.4.3) can be written as

$$F_0(\omega) = 2\,\exp(-\omega y)\,H(\omega)\,\hat{F}_H(\omega). \tag{7.4.4}$$

Taking the inverse Fourier transform, we obtain

$$f_0(z) = \frac{i}{\pi} \oint_{-\infty}^{\infty} \frac{\hat{f}_H(t)}{t-z}\,dt.$$

Or

$$\oint_{-\infty}^{\infty} \frac{f(t)\,dt}{t-z} = i \oint_{-\infty}^{\infty} \frac{\hat{f}_H(t)\,dt}{t-z}, \qquad y > 0. \tag{7.4.5}$$

Since

$$\lim_{y \to 0+} \frac{1}{t-z} = \frac{1}{t-x} + \pi i\,\delta(t-x),$$

we have, from (7.4.5),

$$f_0(z) = \lim_{y \to 0+} \frac{1}{\pi} \oint_{-\infty}^{\infty} \frac{f(t)\,dt}{t-z} = \frac{1}{\pi} \oint_{-\infty}^{\infty} \frac{f(t)\,dt}{t-x} + i\,f(x) = \hat{f}_H(x) + i\,f(x).$$

This gives a relation between $f_0(z)$ and Hilbert transforms.

We now define a *complex analytic signal* $f_c(x)$ from a real signal $f(x)$ by

$$f_c(x) = \frac{1}{\pi} \int_{-\infty}^{\infty} F(\omega)\,H(\omega)\,\exp(i\omega x)\,d\omega. \tag{7.4.6}$$

Since

$$F(\omega)\,H(\omega) = \frac{1}{2}\left[F(\omega) + \text{sgn}(\omega)\,F(\omega)\right] = \frac{1}{2}\left[F(\omega) - i\hat{F}_H(\omega)\right],$$

$$f_c(x) = \frac{1}{2\pi} \int_{-\infty}^{\infty} \left[F(\omega) - i\hat{F}_H(\omega)\right] \exp(i\omega x)\,d\omega = f(x) - i\hat{f}_H(x). \tag{7.4.7}$$

Since $f(x)$ is real, $\text{Re}\{f_c(x)\} = f(x)$ and

$$\operatorname{Im}\{\hat{f}_c(x)\} = -\hat{f}_H(x) = \frac{1}{\pi} \int_{-\infty}^{\infty} \frac{f(t)\, dt}{x-t}.$$

Thus it follows from the inverse Hilbert transform that

$$f(t) = -\frac{1}{\pi} \oint_{-\infty}^{\infty} \frac{\operatorname{Im}\{\hat{f}_c(x)\}\, dx}{x-t} = \frac{1}{\pi} \oint_{-\infty}^{\infty} \frac{\hat{f}_H(x)\, dx}{x-t}. \qquad (7.4.8)$$

7.5 Applications of Hilbert Transforms

Example 7.5.1 (*Boundary Value Problems*). Solve the Laplace equation
$$u_{xx} + u_{yy} = 0, \quad -\infty < x < \infty, \quad y > 0, \qquad (7.5.1)$$
with the boundary conditions
$$u_x(x,y) = f(x) \quad \text{on } y=0, \text{ for } -\infty < x < \infty, \qquad (7.5.2)$$
$$u(x,y) \to 0 \quad \text{as } r = (x^2+y^2)^{\frac{1}{2}} \to \infty. \qquad (7.5.3)$$

Application of the Fourier transform defined by (2.3.1) with respect to x gives the solution for $U(k,y)$ as

$$U(k,y) = \frac{F(k)}{ik} \exp(-|k|y) = F(k)\, G(k), \qquad (7.5.4)$$

where $G(k) = (ik)^{-1} \exp(-|k|y)$ so that $g(x) = \sqrt{\frac{2}{\pi}}\, \tan^{-1}\left(\frac{x}{y}\right)$.

Using the Convolution Theorem 2.4.5 gives the formal solution

$$u(x,y) = \frac{1}{\sqrt{2\pi}} \int_{-\infty}^{\infty} f(t)\, g(x-t)\, dt$$

$$= \frac{1}{\pi} \int_{-\infty}^{\infty} f(t)\, \tan^{-1}\left(\frac{x-t}{y}\right) dt. \qquad (7.5.5)$$

Obviously, it follows from (7.5.5) that

$$u_y(x,0) = \frac{1}{\pi} \int_{-\infty}^{\infty} \frac{f(t)\, dt}{t-x} = H\{f(t)\}. \qquad (7.5.6)$$

Thus the Hilbert transform of the tangential derivative $u_x(x,0) = f(x)$ is the normal derivative $u_y(x,0)$ on the boundary at $y=0$.

Example 7.5.2 (*Nonlinear Internal Waves*). We consider a linear homogeneous partial differential equation with constant coefficients in the form

$$P\left(\frac{\partial}{\partial t}, \frac{\partial}{\partial x}, \frac{\partial}{\partial y}, \frac{\partial}{\partial z}\right) u(\mathbf{x},t) = 0, \qquad (7.5.7)$$

where P is a polynomial in partial derivatives, and $\mathbf{x} = (x,y,z)$ and time $t > 0$.

We seek a three-dimensional plane wave solution of (7.2.7) in the form

$$u(\mathbf{x},t) = a\, \exp\left[i(\mathbf{\kappa}\cdot\mathbf{x} - \omega t)\right], \qquad (7.5.8)$$

where a is the amplitude, $\kappa = (k, \ell, m)$ is the wavenumber vector, and ω is the frequency. If this solution (7.5.8) is substituted into (7.5.7), partial derivatives $\frac{\partial}{\partial t}, \frac{\partial}{\partial x}, \frac{\partial}{\partial y}$, and $\frac{\partial}{\partial z}$ will be replaced by $-i\omega$, ik, $i\ell$, and im respectively. Hence the solution exists provided the algebraic equation

$$P(-i\omega, ik, i\ell, im) = 0 \qquad (7.5.9)$$

is satisfied. This relation is universally known as the *dispersion relation*. Physically, this gives the frequency ω in terms of wavenumbers k, ℓ, and m. Further, the above analysis shows that there is a direct correspondence between the governing equation (7.5.7) and the dispersion relation (7.5.9) given by

$$\frac{\partial}{\partial t} \leftrightarrow -i\omega, \quad \left(\frac{\partial}{\partial x}, \frac{\partial}{\partial y}, \frac{\partial}{\partial z}\right) \leftrightarrow (ik, i\ell, im). \qquad (7.5.10)$$

Clearly, the dispersion relation can be derived from the governing equation and vice versa by using (7.5.10).

In many physical problems, the dispersion relation can be written explicitly in terms of the wavenumbers as

$$\omega = W(k, \ell, m). \qquad (7.5.11)$$

The phase and the group velocities of the waves are defined by

$$C_p(\kappa) = \frac{\omega}{\kappa} \hat{\kappa}, \quad C_g(\kappa) = \nabla_\kappa \omega, \qquad (7.5.12ab)$$

where $\hat{\kappa}$ is the unit vector in the direction of the wave vector κ. In the one-dimensional problem, (7.5.11)-(7.5.12ab) reduce to

$$\omega = W(k), \quad C_p = \frac{\omega}{k}, \quad C_g = \frac{d\omega}{dk}. \qquad (7.5.13abc)$$

Thus the one-dimensional waves given by (7.5.8) are called *dispersive* if the group velocity $C_g = \omega'(k)$ is not constant, that is, $\omega''(k) \neq 0$. Physically, as time progresses, the different waves disperse in the medium with the result that a single hump breaks into a series of wavetrains.

We consider a simple model of internal solitary waves in an inviscid, stably stratified two-fluid system between rigid horizontal planes at $z = h_1$ and $z = h_2$. The upper fluid of depth h_1 and density ρ_1 lies over the heavier lower fluid of depth h_2 and density $\rho_2(>\rho_1)$. Both fluids are subjected to a vertical gravitational force g, and the effects of surface tension are neglected. With $z = \eta(x, t)$ as the internal wave displacement field, the linear dispersion relation for the two-fluid system is

$$\omega^2 = \frac{gk(\rho_2 - \rho_1)}{(\rho_1 \coth kh_1 + \rho_2 \coth kh_2)}, \qquad (7.5.14)$$

where $\omega(k)$ and k are frequency and wavenumber for a small amplitude sinusoidal disturbance at the interface of the two fluids. Several important limiting cases of (7.5.14) are of interest.

Case (i): *Deep-Water Theory* (Benjamin 1967; Ono 1975).

Integral Transforms and Their Applications 245

In this case, the depth of the lower fluid is assumed to be infinite $(h_2 \to \infty)$, and waves are long compared with the depth h_1 of the upper fluid. This leads to the double limit in the form
$$\lim_{k \to 0} \lim_{h_2 \to \infty} \omega^2 = c_0^2 k^2 - 2\alpha c_0 k^3 (sgn\, k + \cdots), \qquad (7.5.15)$$
where $k \to 0$ is used with fixed h_1, and the limit $h_2 \to \infty$ is taken with k and h_1 fixed, and
$$c_0^2 = \left(\frac{\rho_2 - \rho_1}{\rho_1}\right) gh_1 \text{ and } \alpha = \left(\frac{\rho_2}{\rho_1}\right)\left(\frac{h_1 c_0}{2}\right). \qquad (7.5.16ab)$$

We consider internal waves propagating only in one direction and retain the first dispersive term so that the associated dispersion relation becomes
$$\omega = c_0 k - \alpha k |k|. \qquad (7.5.17)$$

This enables us to define the appropriate space and time scales associated with this limiting case as
$$\xi = \beta(x - c_0 t), \qquad \tau = \beta^2 t, \qquad (7.5.18ab)$$
where $\beta(\ll 1)$ is the longwave parameter defined as the ratio of the wave-guide scale to the wavelength.

The linear evolution equation associated with (7.5.17) is
$$\eta_t + c_o \eta_x + \alpha\, H\{\eta_{xx}\} = \beta^2 \left[\eta_\tau + \alpha H\{\eta_{\xi\xi}\}\right] = 0, \qquad (7.5.19)$$
where $H\{\eta(x',t)\}$ is the Hilbert transform of $\eta(x',t)$ defined by
$$H\{\eta(x',t)\} = \frac{1}{\pi} \oint_{-\infty}^{\infty} \frac{\eta(x',t)\, dx'}{(x' - x)}. \qquad (7.5.20)$$

Equation (7.5.19) is often called the *linear Benjamin-Ono equation*. Benjamin (1967) and Ono (1975) investigated nonlinear internal wave motion and discovered the following nonlinear equation
$$(\eta_t + c_o \eta_x) + c_1 \eta \eta_x + \alpha H\{\eta_{xx}\} = 0, \qquad (7.5.21)$$
where c_1 and α are constants which are the characteristics of specific flows. This equation is usually known as the *Benjamin-Ono equation*.

The solitary wave solution of (7.5.21) has the form (Benjamin, 1967)
$$\eta(x - ct) = \frac{a\lambda^2}{(x - ct)^2 + \lambda^2}, \qquad (7.5.22)$$
where $c = c_0 + \frac{1}{2} a c_1$ and $a\lambda = -\frac{4\alpha}{c_1}$.

It is noted here that the Benjamin-Ono equation is one of the model nonlinear evolution equations and it arises in a large variety of physical wave systems.

Case (ii): *Shallow-Water Theory* (Benjamin, 1966).

In this case, long wave $(k \to 0)$ disturbances with the length scale $h = (h_1 + h_2)$ fixed lead to the result
$$\lim_{k \to 0} (\omega^2) = c_0^2 k^2 - 2c_0 \gamma k^4, \qquad (7.5.23)$$

where
$$c_0^2 = \frac{g(\rho_2 - \rho_1) h_1 h_2}{(\rho_1 h_2 + \rho_2 h_1)} \quad \text{and} \quad \gamma = c_0 h_1 h_2 \left(\frac{\rho_1 h_1 + \rho_2 h_2}{\rho_1 h_2 + \rho_2 h_1}\right). \quad (7.5.24\text{ab})$$

If we retain only the first dispersive term in (7.5.23) and assume that the wave propagates only to the right, it turns out that
$$\omega(k) = c_0 k - \gamma k^3. \quad (7.5.25)$$

The evolution equation associated with (7.6.25) is the well-known linear KdV equation
$$\eta_t + c_0 \eta_x + \gamma \eta_{xxx} = 0. \quad (7.5.26)$$

In terms of a slow time scale τ and a slow spatial modulation ξ in a coordinate system moving at the linear wave velocity defined by $\xi = \beta(x - c_0 t)$ and $\tau = \beta^3 t$, the above equation (7.5.26) reduces to the linear KdV equation
$$\eta_\tau + \gamma \eta_{\xi\xi\xi} = 0. \quad (7.5.27)$$

The standard nonlinear KdV equation is given by
$$\eta_t + c_0 \eta_x + \alpha \eta \eta_x + \gamma \eta_{xxx} = 0. \quad (7.5.28)$$

It is well known that this equation admits the *soliton* solution in the form
$$\eta(x - ct) = a \operatorname{sech}^2 \left(\frac{x - ct}{\lambda}\right), \quad (7.5.29)$$

where $c = c_0 + \dfrac{a\alpha}{3}$ and $a\lambda^2 = \dfrac{12\gamma}{\alpha}$.

A similar argument can be employed to determine the integrodifferential nonlinear evolution equation associated with an arbitrary dispersion relation $\omega(k) = k\, c(k)$ in the form
$$\frac{\partial \eta}{\partial t} + c_1 \eta \eta_x + \int_{-\infty}^{\infty} K(x - \zeta) \left(\frac{\partial \eta}{\partial \zeta}\right) d\zeta = 0, \quad (7.5.30)$$

where the kernel $K(x)$ is a given function. The linearized version of (7.5.30) admits the wavelike solution
$$\eta(x, t) = A \exp[i(kx - \omega t)] \quad (7.5.31)$$

provided the following dispersion relation holds,
$$(-i\omega) \exp(ikx) + i \int_{-\infty}^{\infty} K(x - \zeta)\, k \exp(ik\zeta)\, d\zeta = 0.$$

Substituting $x - \zeta = \xi$, this can be rewritten in the form
$$\omega = k \int_{-\infty}^{\infty} K(\xi) \exp(-ik\xi)\, d\xi = k\, c(k), \quad (7.5.32)$$

where $c(k)$ is the Fourier transform of the given kernel $K(x)$ so that $K(x) = \mathscr{F}^{-1}\{c(k)\}$. This means that any phase velocity $c(k) = \mathscr{F}\{K(x)\}$ can be obtained by choosing the kernel $K(x)$.

In particular, if
$$K(x) = c_0\,\delta(x) + \gamma\,\delta''(x), \quad c(k) = c_0 + \gamma\,k^2 \tag{7.5.33}$$
equation (7.5.30) reduces to the linear KdV equation
$$\eta_t + c_0\,\eta_x + \gamma\,\eta_{xxx} = 0. \tag{7.5.34}$$
Combining the general dispersion of the integral form with typical nonlinearity, we obtain
$$\eta_t + c_0\,\eta_x + \alpha\,\eta\eta_x + \int_{-\infty}^{\infty} K(x-\zeta)\left(\frac{\partial \eta}{\partial \zeta}\right) d\zeta = 0. \tag{7.5.35}$$

Using (7.5.33) in (7.5.35), we can derive the KdV equation (7.5.28). On the other hand, if $c(k) = c_0(1 - \alpha|k|)$, we can deduce the Benjamin-Ono equation (7.5.21) from (7.5.30).

Case (iii): *Finite-Depth Theory* (Kubota et al, 1978).

In this case, $h_2 \gg h_1$, that is, $(h_1/h_2) = O(\beta)$, but $kh_1 = O(\beta)$ and $kh_2 = O(1)$. The dispersion relation appropriate for this case is
$$\omega = c_0\,k - \frac{1}{2}\left(\frac{\rho_2}{\rho_1}\right) c_0\,h_1\,k^2 \coth(k\,h_2), \tag{7.5.36}$$
where
$$c_0^2 = \left(\frac{\rho_2 - \rho_1}{\rho_1}\right) g\,h_1. \tag{7.5.37}$$

We can use (7.5.18ab) for the appropriate space and time scales to investigate this case. Thus the finite-depth evolution equation can be derived from (7.5.36) and has the form (see Kubota et al, 1978)
$$\eta_t + c_0\,\eta_x + c_1\,\eta\eta_x + c_2\,\frac{\partial^2}{\partial x^2}\left[\int_{-\infty}^{\infty} \eta(x',t)\left[\coth\frac{\pi(x-x')}{2h}\right.\right.$$
$$\left.\left. - \operatorname{sgn}\left(\frac{x-x'}{h}\right)\right]\right] dx'. \tag{7.5.38}$$

The solitary wave solution of this equation has been obtained by Joseph and Adams (1981).

It should be noted here that the finite-depth equation reduces to the Benjamin-Ono equation and the KdV equation in the deep- and shallow-water limits, respectively.

Finally, all of the above theories can be formulated in the framework of a generalized evolution equation usually known as the *Whitham equation* (Whitham, 1967) in the form
$$\frac{\partial \eta}{\partial t} + c_1\,\eta\eta_x + \frac{\partial}{\partial x}\left[\int_{-\infty}^{\infty} \eta(x',t)\,dx'\,\frac{1}{2\pi}\int_{-\infty}^{\infty} \exp\{ik(x-x')\}\,c(k)\,dk\right] = 0. \tag{7.5.39}$$

Subsequently, Maslowe and Redekopp (1980) have generalized the theory of long nonlinear waves in a stratified shear flows. They have obtained the governing nonlinear evolution equation, which involves the Hilbert transform.

In their analysis, the evolution equation contains a damping term describing energy loss by radiation which can be used to determine the persistence of solitary waves or nonlinear wave packets in physically realistic situations.

Finally, the *finite Hilbert transform* is defined by Tricomi (1951) as

$$H\{f(t), a, b\} = \hat{f}_H(x, a, b) = \frac{1}{\pi} \int_a^b \frac{f(t)}{t - x} dt. \qquad (7.5.40)$$

Such transforms arise naturally in aerodynamics. Tricomi (1951) studied the finite Hankel transform and its applications to airfoil theory. Subsequently, considerable attention has been given to the methods of solution of the singular integral equation for the unknown function $f(t)$ and known $\hat{f}_H(x)$ as

$$\frac{1}{\pi} \int_{-1}^1 \frac{f(t)}{t - x} dt = \hat{f}_H(x), \quad -1 < x < 1, \qquad (7.5.41)$$

where $f(x)$ and $\hat{f}_H(x)$ satisfy the Hölder conditions on $(-1,1)$. This equation arises in boundary value problems in elasticity and in other areas. Several authors including Muskhelishvili (1963), Gakhov (1966), Peters (1972), Chakraborty (1980, 1988), Chakraborty and Williams (1980), Williams (1978), Comninou (1977), Gautesen and Dunders (1987ab), and Pennline (1976) have studied the methods of the solution of (7.5.41) and its various generalizations. The readers are referred to these papers.

7.6 Asymptotic Expansions of One-Sided Hilbert Transforms

A two-sided Hilbert transform can be written as the sum of two one-sided transforms

$$\oint_{-\infty}^{\infty} \frac{f(t)}{t - x} dt = \oint_0^{\infty} \frac{f(t)}{t - x} dt - \int_0^{\infty} \frac{f(-t)}{t + x} dt, \qquad (7.6.1)$$

when $x > 0$ (with a similar expression for $x < 0$) where the second integral is actually a Stieltjes transform of $[-f(-t)]$ which has been defined by (7.7.4) in Section 7.7.

We examine the *one-sided Hilbert transform*, which is defined by

$$H^+\{f(t)\} = \hat{f}_H^+(x) = \int_0^{\infty} \frac{f(t)}{t - x} dt. \qquad (7.6.2)$$

The Mellin transform of $H^+\{f(t)\}$ is

$$\mathcal{M}\left[H^+\{f(t)\}\right] = \int_0^{\infty} x^{p-1} \left[\oint_0^{\infty} \frac{f(t)}{t - x} dt\right] dx$$

$$= \int_0^{\infty} f(t) \left[\oint_0^{\infty} \frac{x^{p-1}}{t - x} dx\right] dt$$

$$\mathcal{M}\{H^+\{f(t)\}\} = \pi \cot(\pi p) \int_0^\infty t^{p-1} f(t)\, dt$$
$$= \pi \cot(\pi p)\, \mathcal{M}\{f(t)\}.$$

Taking the inverse Mellin transform, we obtain
$$H^+\{f(t)\} = \hat{f}_H^+(x) = \frac{1}{2\pi i} \int_{c-i\infty}^{c+i\infty} x^{-p} \pi \cot(\pi p)\, \tilde{f}(p)\, dp. \tag{7.6.3}$$

Example 7.6.1 (*Asymptotic Expansion of One-Sided Hilbert Transforms*).

$$\oint_0^\infty \frac{\cos \omega t}{t-x}\, dt \sim -\pi \sin \omega x - \sum_{n=0}^\infty \frac{n!}{(\omega x)^{n+1}} \cos\left\{(n+1)\frac{\pi}{2}\right\}, \quad \text{as } x \to \infty, \tag{7.6.4}$$

$$\oint_0^\infty \frac{\sin \omega t}{t-x}\, dt \sim \pi \cos \omega x - \sum_{n=0}^\infty \frac{n!}{(\omega x)^{n+1}} \sin\left\{(n+1)\frac{\pi}{2}\right\}, \quad \text{as } x \to \infty. \tag{7.6.5}$$

We have
$$\oint_0^\infty \frac{\exp(i\omega t)}{t-x}\, dt = \pi i \exp(i\omega x) + \int_0^\infty \frac{\exp(i\omega t)}{t-x}\, dt,$$

where in the integral on the right the contour of integration passes above the pole $t = x$. The contour can be deformed into the positive imaginary axis on which $t = iu$ with $u > 0$. Thus
$$\int_0^\infty \frac{\exp(i\omega t)}{t-x}\, dt = -i \int_0^\infty \frac{\exp(-\omega u)}{x - iu}\, du$$

$$\sim -\sum_{n=0}^\infty \frac{i^{n+1}}{x^{n+1}} \int_0^\infty u^n \exp(-\omega u)\, du, \qquad \text{by Watson's lemma,}$$

$$= -\sum_{n=0}^\infty \frac{n!}{(\omega x)^{n+1}} \exp\left\{i(n+1)\frac{\pi}{2}\right\}. \tag{7.6.6}$$

Separating the real and imaginary parts, we obtain the desired results. These results are due to Ursell (1983).

Theorem 7.6.1 (Ursell, 1983). If $f(t)$ is analytic for real t, $0 \le t < \infty$, and if it has the asymptotic expansion in the form

$$f(t) \sim \sum_{n=1}^\infty \frac{a_n}{t^n} + \cos \omega t \sum_{n=1}^\infty \frac{A_n}{t^n} + \sin \omega t \sum_{n=1}^\infty \frac{B_n}{t^n} \qquad \text{as } t \to \infty, \tag{7.6.7}$$

where the coefficients a_n, A_n and B_n are known and $\omega > 0$, then the one-sided Hilbert transform $\hat{f}^+(x)$ has the following asymptotic expansion

$$\hat{f}_H^+(x) = H^+\{f(t)\} = \oint_0^\infty \frac{f(t)}{t-x}\, dt \sim \sum_1^\infty \frac{c_n}{x^n} - \log x \sum_1^\infty \frac{a_n}{x^n}$$

$$+ \left(\sum_1^\infty \frac{A_n}{x^n}\right) \oint_0^\infty \frac{\cos \omega t}{t-x}\, dt + \left(\sum_1^\infty \frac{B_n}{x^n}\right) \oint_0^\infty \frac{\sin \omega t}{t-x}\, dt \quad \text{as } x \to \infty, \tag{7.6.8}$$

where c_n is given by

$$c_n = d_n - \sum_{r=1}^{n-1} \frac{\Gamma(n-r)}{\omega^{n-r}} \left[A_r \cos\left\{\frac{\pi}{2}(n-r)\right\} + B_r \sin\left\{\frac{\pi}{2}(n-r)\right\} \right] \quad (7.6.9)$$

and

$$d_n = \lim_{p \to n} \left[\mathcal{M}\{f(t), p\} + \frac{a_n}{p-n} \right]. \quad (7.6.10)$$

Note that when $a_n = 0$, (7.6.10) becomes

$$d_n = \mathcal{M}\{f(t), n\} = \int_0^\infty t^{n-1} f(t) \, dt. \quad (7.6.11)$$

Substituting (7.6.9) in (7.6.8) and using (7.6.4) and (7.6.5), we obtain the following theorem:

Theorem 7.6.2 (Ursell, 1983). Under the same conditions of Theorem 7.6.1, the one-sided Hilbert transform $\hat{f}^+(x)$ has the asymptotic expansion

$$\hat{f}_H^+(x) = \int_0^\infty \frac{f(t)}{t-x} \, dt \sim -\sum_1^\infty \frac{d_n}{x^n} - \log x \sum_1^\infty \frac{a_n}{x^n} - (\pi \sin \omega x) \sum_1^\infty \frac{A_n}{x^n}$$

$$+ (\pi \cos \omega x) \sum_1^\infty \frac{B_n}{x^n} \qquad \text{as } x \to \infty, \quad (7.6.12)$$

where d_n is given by (7.6.10).

The reader is referred to Ursell (1983) for a detailed discussion of proof of Theorems 7.6.1 and 7.6.2.

7.7 Definition of the Stieltjes Transform and Examples

We use the Laplace transform of $\bar{f}(s) = \mathcal{L}\{f(t)\}$ with respect to s to define the Stieltjes transform of $f(t)$. Clearly

$$\mathcal{L}\{\bar{f}(s)\} = \tilde{f}(z) = \int_0^\infty e^{-sz} \bar{f}(s) \, ds$$

$$= \int_0^\infty e^{-sz} ds \int_0^\infty e^{-st} f(t) \, dt. \quad (7.7.1)$$

Interchanging the order of integration and evaluating the inner integral, we obtain

$$\tilde{f}(z) = \int_0^\infty \frac{f(t)}{t+z} \, dt. \quad (7.7.2)$$

The *Stieltjes transform* of a locally integrable function $f(t)$ on $0 \le t < \infty$ is denoted by $\tilde{f}(z)$ and defined by

Integral Transforms and Their Applications

$$\mathcal{S}\{f(t)\} = \tilde{f}(z) = \int_0^\infty \frac{f(t)}{t+z}\, dt, \tag{7.7.3}$$

where z is a complex variable in the cut plane $|\arg z| < \pi$.
If $z = x$ is real and positive, then

$$\mathcal{S}\{f(t)\} = \tilde{f}(x) = \int_0^\infty \frac{f(t)}{t+x}\, dt. \tag{7.7.4}$$

Differentiating (7.7.4) with respect to x, we obtain

$$\frac{d^n}{dx^n}\tilde{f}(x) = (-1)^n\, n! \int_0^\infty \frac{f(t)}{(t+x)^{n+1}}\, dt, \tag{7.7.5}$$

where $n = 1, 2, 3, \ldots$.

We now state the *inversion theorem* for the Stieltjes transform without proof.

Theorem 7.7.1 If $f(t)$ is absolutely integrable in $0 \le t \le T$ for every positive T and is such that the integral (7.7.4) converges for $x > 0$, then $\tilde{f}(z)$ exists for complex $z\ (z \ne 0)$ not lying on the negative real axis and

$$\lim_{\varepsilon \to 0+} \frac{1}{2\pi i}\left[\tilde{f}(-x-i\varepsilon) - \tilde{f}(-x+i\varepsilon)\right] = \frac{1}{2}\left[f(x+0) + f(x-0)\right] \tag{7.7.6}$$

for any positive x at which $f(x+0)$ and $f(x-0)$ exist.

For a rigorous proof of this theorem the reader is referred to Widder (1941, pp 340-341).

Example 7.7.1 Find the Stieltjes transform of each of the following functions:

$$\text{(a) } f(t) = (t+a)^{-1}, \quad \text{(b) } f(t) = t^{\alpha-1}.$$

(a) We have, by definition,

$$\tilde{f}(z) = \int_0^\infty \frac{dt}{(t+a)(t+z)} = \frac{1}{(a-z)}\left[\int_0^\infty \left(\frac{1}{t+z} - \frac{1}{t+a}\right) dt\right]$$

$$= \frac{1}{(a-z)} \log\left|\frac{a}{z}\right|. \tag{7.7.7}$$

(b) $$\tilde{f}(z) = \mathcal{S}\{t^{\alpha-1}\} = \int_0^\infty \frac{t^{\alpha-1}}{t+z}\, dt = z^{-1} \int_0^\infty \left(1 + \frac{t}{z}\right)^{-1} t^{\alpha-1}\, dt$$

which is, by the change of variable $\dfrac{t}{z} = x$,

$$= z^{\alpha-1} \int_0^\infty \frac{x^{\alpha-1}\, dx}{1+x} = z^{\alpha-1}\, \mathcal{M}\left\{\frac{1}{1+x}\right\}$$

which is, by Example 6.2.1(b),

$$= z^{\alpha-1}\, \pi\, \text{cosec}(\pi\alpha). \tag{7.7.8}$$

Example 7.7.2 Show that the Stieltjes transform of $J_\nu^2(t)$

$$\tilde{f}(x) = \mathscr{S}\{J_\nu^2(t)\} = \int_0^\infty \frac{J_\nu^2(t)\,dt}{t+x} \tag{7.7.9}$$

satisfies the *Parseval relation*

$$\tilde{f}(x) = \frac{1}{2\pi i}\int_{c-i\infty}^{c+i\infty} x^{-p}\,\tilde{f}(p)\,\tilde{g}(1-p)\,dp. \tag{7.7.10}$$

Proof. We write $t = xu$ so that (7.7.9) becomes

$$\tilde{f}(x) = \int_0^\infty f(xu)\,g(u)\,du, \tag{7.7.11}$$

where $f(u) = J_\nu^2(u)$ and $g(u) = (1+u)^{-1}$.

Taking the Mellin transform of (7.7.11) with respect to x, we obtain

$$\mathscr{M}\{\tilde{f}(x), p\} = \tilde{f}(p)\,\tilde{g}(1-p)$$

where, from Oberhettinger (1974, p 98),

$$\tilde{f}(p) = \frac{2^{p-1}\,\Gamma\!\left(\nu+\dfrac{p}{2}\right)\pi\,\mathrm{cosec}\,(\pi p)}{\left\{\Gamma\!\left(1-\dfrac{p}{2}\right)\right\}^2 \Gamma\!\left(1+\nu-\dfrac{p}{2}\right)\Gamma(p)},$$

$$\tilde{g}(1-p) = \pi\,\mathrm{cosec}\,(\pi p).$$

Thus the inverse Mellin transform gives the desired result.

Example 7.7.3 Show that

$$\mathscr{S}\{\sin(k\sqrt{t})\} = \pi\,\exp(-k\sqrt{z}), \qquad k > 0. \tag{7.7.12}$$

We have, by definition,

$$\mathscr{S}\{\sin(k\sqrt{t})\} = \int_0^\infty \frac{\sin(k\sqrt{t})}{t+z}\,dt$$

which is, by putting $\sqrt{t} = u$,

$$= 2\int_0^\infty \frac{u\,\sin ku}{(u^2+z)}\,du$$

$$= \pi\,\exp(-k\sqrt{z}) \qquad \text{by (2.8.6)}.$$

7.8 Basic Operational Properties of Stieltjes Transforms

The following results hold for the Stieltjes transform:

(a) $$\mathscr{S}\{f(t+a)\} = \tilde{f}(z-a) \tag{7.8.1}$$

(b) $$\mathscr{S}\{f(at)\} = \tilde{f}(az), \qquad a > 0 \tag{7.8.2}$$

(c) $$\mathscr{S}\{t\,f(t)\} = -z\,\tilde{f}(z) + \int_0^\infty f(t)\,dt \tag{7.8.3}$$

provided the integral on the right hand side exists.

(d) $$\mathscr{S}\left\{\frac{f(t)}{t+a}\right\} = \frac{1}{a-z}\left[\tilde{f}(z) - \tilde{f}(a)\right] \tag{7.8.4}$$

(e) $$\mathscr{S}\left\{\frac{1}{t}f\left(\frac{a}{t}\right)\right\} = \frac{1}{z}\tilde{f}\left(\frac{a}{z}\right), \qquad a > 0. \tag{7.8.5}$$

Proof. (a) We have, by definition,
$$\mathscr{S}\{f(t+a)\} = \int_0^\infty \frac{f(t+a)}{t+z}\,dt$$
which is, by the change of variable $t + a = \tau$,
$$\mathscr{S}\{f(t+a)\} = \int_0^\infty \frac{f(\tau)}{\tau + (z-a)}\,d\tau = \tilde{f}(z-a).$$

(b) We have, by definition,
$$\mathscr{S}\{f(at)\} = \int_0^\infty \frac{f(at)}{t+z}\,dt$$
which is, by writing $at = \tau$,
$$= \int_0^\infty \frac{f(\tau)}{\tau + az}\,d\tau = \tilde{f}(az).$$

(c) We have from the definition
$$\mathscr{S}\{t\,f(t)\} = \int_0^\infty \frac{t\,f(t)}{t+z}\,dt = \int_0^\infty \frac{(t+z-z)\,f(t)}{t+z}\,dt$$
$$= \int_0^\infty f(t)\,dt - z \int_0^\infty \frac{f(t)}{t+z}\,dt.$$

This gives the desired result.

(d) We have, by definition,
$$\mathscr{S}\left\{\frac{f(t)}{t+a}\right\} = \int_0^\infty \frac{f(t)}{(t+a)(t+z)}\,dt$$
$$= \frac{1}{a-z}\left[\int_0^\infty \left\{\frac{1}{t+z} - \frac{1}{t+a}\right\}f(t)\,dt\right]$$
$$= \frac{1}{a-z}\left[\tilde{f}(z) - \tilde{f}(a)\right].$$

(e) We have, by definition,

$$\mathscr{S}\left\{\frac{1}{t} f\left(\frac{a}{t}\right)\right\} = \int_0^\infty \frac{1}{t(t+z)} f\left(\frac{a}{t}\right) dt$$

which is, by substituting $\frac{a}{t} = \tau$,

$$= \frac{1}{z} \int_0^\infty \frac{f(\tau)}{\left(\tau + \frac{a}{z}\right)} d\tau = \frac{1}{z} \tilde{f}\left(\frac{a}{z}\right).$$

Theorem 7.8.1 (*Stieltjes Transforms of Derivatives*). If $\mathscr{S}\{f(t)\} = \tilde{f}(z)$, then

$$\mathscr{S}\{f'(t)\} = -\frac{1}{z} f(0) - \frac{d}{dz} \tilde{f}(z), \qquad (7.8.6)$$

$$\mathscr{S}\{f''(t)\} = -\left[\frac{1}{z} f'(0) + \frac{1}{z^2} f(0)\right] - \frac{d^2}{dz^2} \tilde{f}(z). \qquad (7.8.7)$$

More generally,

$$\mathscr{S}\{f^{(n)}(t)\} = -\left[\frac{1}{z} f^{(n-1)}(0) + \frac{1}{z^2} f^{(n-2)}(0) + \cdots + \frac{1}{z^n} f(0)\right] - \frac{d^n}{dz^n} \tilde{f}(z). \qquad (7.8.8)$$

Proof. Using the definition and integrating by parts, we obtain

$$\mathscr{S}\{f'(t)\} = \int_0^\infty \frac{f'(t)}{t+z} dt$$

$$= \left[\frac{f(t)}{t+z}\right]_0^\infty + \int_0^\infty \frac{f(t) \, dt}{(t+z)^2} = -\frac{1}{z} f(0) - \frac{d}{dz} \tilde{f}(z).$$

This proves result (7.8.6).

Similarly, other results can readily be proved.

7.9 Inversion Theorems for Stieltjes Transforms

We first introduce the following differential operator that can be used to establish inversion theorems for the Stieltjes transform.

A differential operator is defined for any real positive number t by the following equations:

$$L_{k,t}[f(x)] = (-1)^{k-1} c_k \, t^{k-1} \, D_t^{(2k-1)} [t^k \, f(t)], \qquad (7.9.1)$$

$$L_{0,t}[f(x)] = f(t), \qquad (7.9.2)$$

$$L_{1,t}[f(x)] = D_t [t \, f(t)], \qquad (7.9.3)$$

where $k = 2, 3, \ldots,$ $c_k = [k! \, (k-2)!]^{-1}$, $D_t \equiv \dfrac{d}{dt}$, and $f(x)$ has derivatives of all orders.

We state a basic theorem due to Widder (1941) without proof.

Theorem 7.9.1 If $\mathscr{S}\{f(t)\} = \tilde{f}(x)$ exists and is defined by

$$\tilde{f}(x) = \int_0^\infty \frac{f(t)}{t+x} \, dt \tag{7.9.4}$$

then, for all positive t,

(i) $\quad L_{k,t}\left[\tilde{f}(x)\right] = (2k-1)! \, c_k \, t^{k-1} \int_0^\infty \frac{u^k \, f(u)}{(t+u)^{2k}} \, du,\tag{7.9.5}$

(ii) $\quad \lim_{k\to\infty} L_{k,t}\left[\tilde{f}(x)\right] = f(t).\tag{7.9.6}$

Obviously,

$$\frac{t^k}{t+u} = \frac{t^k - (-u)^k}{t+u} + \frac{(-u)^k}{t+u} = t^{k-1} - u\, t^{k-2} + \cdots \pm u^{k-1} + \frac{(-u)^k}{t+u}.$$

In view of this result, we can find

$$L_{k,t}\left[\tilde{f}(x)\right] = (-1)^{k-1} \, c_k \, t^{k-1} \, D_t^{(2k-1)}\left[t^k \, \tilde{f}(t)\right]$$

$$= c_k \, t^{k-1} (2k-1)! \int_0^\infty \frac{u^k}{(t+u)^{2k}} \, f(u) \, du.$$

Theorem 7.9.2 If $\tilde{f}(x)$ is the Laplace transform of $\tilde{f}(s) = \mathscr{L}\{f(t)\}$ so that

$$\tilde{f}(x) = \mathscr{L}\{\tilde{f}(s)\} = \int_0^\infty \frac{f(t)}{t+x} \, dt \quad \text{for } s > 0,$$

then

(i) $\quad L_{k,x}\left[\tilde{f}(x)\right] = (-1)^k \, c_k \, x^{k-1} \int_0^\infty e^{-xt} \, t^{2k-1} \, f^{(k)}(t) \, dt,\tag{7.9.7}$

(ii) $\quad \lim_{k\to\infty} L_{k,x}\left[\tilde{f}(x)\right] = f(x), \quad \text{for all positive } x.\tag{7.9.8}$

Proof. We have, by definition of the operator defined by (7.9.1),

$$L_{k,x}\left[\tilde{f}(x)\right] = (-1)^{k-1} \, x^{k-1} \, c_k \, D_x^{(2k-1)}\left[x^k \, \tilde{f}(x)\right].\tag{7.9.9}$$

We use the result (see Widder, 1941, p 350)

$$x^{k-1} \, D_x^{(2k-1)}\left[x^k \, f(x)\right] = D_x^k\left[x^{2k-1} \, f^{(k-1)}(x)\right],\tag{7.9.10}$$

where $f(x)$ is any function which has derivatives of all orders. This can easily be verified by computing both sides of (7.9.10). Each of both sides is equal to

$$\sum_{n=0}^k \frac{(2k-1)!}{(2k-n-1)!} \binom{k}{n} x^{2k-n-1} f^{(2k-n-1)}(x).$$

In view of (7.9.10), result (7.9.9) becomes

$$L_{k,x}\left[\tilde{f}(x)\right] = (-1)^{k-1} \, c_k \, D_x^k\left[x^{2k-1} \, \tilde{f}^{(k-1)}(x)\right].\tag{7.9.11}$$

We next show that

$$(-1)^{k-1} \tilde{f}^{(k-1)}(x) = (-1)^{k-1} \frac{d^{k-1}}{dx^{k-1}} \int_0^\infty e^{-xt} f(t)\, dt$$

$$= (-1)^{2(k-1)} \int_0^\infty e^{-xt} t^{k-1} f(t)\, dt$$

$$= \frac{1}{x} \int_0^\infty e^{-u} \left(\frac{u}{x}\right)^{k-1} f\left(\frac{u}{x}\right) du. \qquad (7.9.12)$$

Using (7.9.12) in (7.9.11), we obtain

$$L_{k,x}\left[\tilde{f}(x)\right] = c_k \int_0^\infty e^{-u} u^{k-1}\, D_x^k \left\{ x^{k-1} f\left(\frac{u}{x}\right) \right\} du,$$

which is, due to Lemma 25 (Widder 1941, p 385),

$$= c_k\, x^{-(k+1)} \int_0^\infty e^{-u} u^{k-1} (-u)^k f^{(k)}\left(\frac{u}{x}\right) du. \qquad (7.9.13)$$

We again set $u = xt$ in (7.9.13) to obtain the desired result

$$L_{k,x}\left[\tilde{f}(x)\right] = c_k (-1)^k x^{k-1} \int_0^\infty e^{-xt} t^{2k-1} f^{(k)}(t)\, dt.$$

We next take the limit as $k \to \infty$ and use Widder's result (7.9.6) to derive (7.9.8). Thus the proof is complete.

It is important to note that result (7.9.7) depends on the values of all derivatives of $f(x)$ in the domain $(0, \infty)$. This seems to be a very severe restriction on the formula (7.9.7). This restriction can be eliminated by applying the operator $L_{k,x}$ to the Laplace integral directly. Then we prove the following theorem:

Theorem 7.9.3 (*Widder, 1941*). Under the same conditions of Theorem 7.9.2, the following results hold

(i)
$$L_{k,x}\left[\tilde{f}(x)\right] = \int_0^\infty e^{-xs} P_{2k-1}(xs)\, \tilde{f}(s)\, ds, \qquad (7.9.14)$$

(ii)
$$\lim_{k \to \infty} L_{k,x}\left[\tilde{f}(x)\right] = f(x). \qquad (7.9.15)$$

where

$$P_{2k-1}(t) = (-1)^{k-1} c_k (2k-1)! \sum_{n=0}^{k} \binom{k}{n} \frac{(-t)^{2k-n-1}}{(2k-n-1)!}. \qquad (7.9.16)$$

Proof. We apply the operator $L_{k,x}$ to the Laplace integral directly to obtain

$$L_{k,x}\left[\tilde{f}(x)\right] = L_{k,x}\left[\mathscr{L}\{\tilde{f}(s)\}\right]$$

$$= L_{k,x}\left[\int_0^\infty e^{-sx} \tilde{f}(s)\, ds\right]$$

$$L_{k,x}\left[\tilde{f}(x)\right] = \int_0^\infty L_{k,x}\left[e^{-sx}\right] \tilde{f}(s)\, ds$$

which is, after direct computation of $L_{k,x}[\exp(-sx)]$,

$$= \int_0^\infty e^{-xs} P_{2k-1}(xs)\, \tilde{f}(s)\, ds.$$

Taking the limit as $k \to \infty$ and using result (7.9.8), we obtain

$$\lim_{k \to \infty} L_{k,x}\left[\tilde{f}(x)\right] = \lim_{k \to \infty} \int_0^\infty e^{-xs} P_{2k-1}(xs)\, \tilde{f}(s)\, ds = f(x) \quad \text{for all } x > 0.$$

The significance of this result lies in the fact that the integral representation for $f(x)$ depends *only* on the values of $\tilde{f}(s)$ in $(0, \infty)$ and *not* on any of its derivatives.

7.10 Applications of Stieltjes Transforms

Example 7.9.1 (*Moment Problem*). If $f(t)$ has an exponential rate of decay as $t \to \infty$, then all of the *moments* exist and are given by

$$m_r = \int_0^\infty t^r f(t)\, dt, \quad r = 0, 1, 2, \ldots. \tag{7.10.1}$$

Then it can easily be shown from (7.7.4) that

$$\tilde{f}(x) = \sum_{r=0}^{n-1} (-1)^r\, m_r\, x^{-(r+1)} + \varepsilon_n(x), \tag{7.10.2}$$

where

$$|\varepsilon_n(x)| \leq x^{-(n+1)} \sup_{0 < t < \infty} \left|\int_0^t \tau^n f(\tau)\, d\tau\right|. \tag{7.10.3}$$

The Stieltjes transform is found to arise in the problems of moments for the semi-infinite interval. The reader is referred to Tamarkin and Shohat (1943).

Example 7.10.2 (*Solution of Integral Equations*). Find the solution of the integral equation

$$\lambda \int_0^\infty \frac{f(t)}{t+x}\, dt = f(x), \tag{7.10.4}$$

where λ is a real parameter.

Case (i): Suppose $\lambda \neq \dfrac{1}{\pi}$.

In this case, we show that the solution of (7.10.4) is

$$f(t) = A\, t^{-\alpha} + B\, t^{\alpha-1}, \tag{7.10.5}$$

where A and B are arbitrary constants and α is a root of the equation $\sin \alpha\pi = \lambda\pi$ between zero and unity if $\lambda < \dfrac{1}{\pi}$, and with real part $\dfrac{1}{2}$ if $\lambda > \dfrac{1}{\pi}$.

If $0 < \operatorname{Re} \alpha < 1$, then

$$\lambda \int_0^\infty \frac{t^{-\alpha}}{t+x} \, dt = \lambda \int_0^\infty \frac{t^{p-1}}{t+x} \, dt, \qquad p = 1 - \alpha$$

$$= \lambda \pi \, x^{p-1} \operatorname{cosec}(\pi p)$$

$$= x^{-\alpha} \left(\frac{\lambda \pi}{\sin \pi \alpha} \right) = x^{-\alpha}, \qquad (7.10.6)$$

so that $x^{-\alpha}$ is a solution of (7.10.4). Obviously, equation (7.10.6) holds if α is replaced by $1 - \alpha$, and hence $t^{\alpha-1}$ is also a solution. Thus (7.10.5) is a solution of equation (7.10.4).

Case (ii): $\lambda = \dfrac{1}{\pi}$.

In this case, we show that $\dfrac{1}{\sqrt{t}}$, and $\dfrac{1}{\sqrt{t}} \log t$ are solutions of (7.10.4).

$$f(x) = \frac{1}{\pi} \int_0^\infty \frac{dt}{\sqrt{t}\,(t+x)} = \frac{1}{\pi} \int_0^\infty \frac{1}{(t+x)} t^{\frac{1}{2}-1} \, dt$$

$$= \frac{1}{\pi} x^{\frac{1}{2}-1} \pi \operatorname{cosec}\left(\frac{\pi}{2}\right) \qquad \text{by Example 7.7.1(b)}$$

$$= \frac{1}{\sqrt{x}}.$$

Thus $\dfrac{1}{\sqrt{t}}$ is a solution of (7.10.4).

To show that $f(t) = \dfrac{1}{\sqrt{t}} \log t$ is a solution, we write

$$f(x) = \frac{1}{\pi} \int_0^\infty \frac{\log t}{\sqrt{t}\,(t+x)} \, dt \qquad (\log t = u)$$

$$= \frac{1}{\pi} \int_{-\infty}^\infty \frac{u}{(x+e^u)} \exp\left(\frac{u}{2}\right) du.$$

Replacing x by e^x and multiplying both sides by $\exp\left(\dfrac{x}{2}\right)$, we find

$$\exp\left(\frac{x}{2}\right) f(e^x) = \frac{1}{\pi} \int_{-\infty}^\infty \left(\frac{u}{e^x + e^u} \right) \exp\left(\frac{x+u}{2}\right) du$$

$$= \frac{1}{2\pi} \int_{-\infty}^\infty u \operatorname{sech}\left(\frac{x-u}{2}\right) du \qquad (x - u = t)$$

$$\exp\left(\frac{x}{2}\right) f(e^x) = \frac{1}{2\pi} \int_{-\infty}^\infty (x-t) \operatorname{sech}\left(\frac{t}{2}\right) dt = \frac{x}{2\pi} \int_{-\infty}^\infty \operatorname{sech}\left(\frac{t}{2}\right) dt = x.$$

Or
$$f(e^x) = x \exp\left(-\frac{x}{2}\right).$$

Thus
$$f(t) = \frac{1}{\sqrt{t}} \log t$$

is a solution, and hence $\frac{1}{\sqrt{t}}(A + B \log t)$ is also a solution of (7.10.4).

7.11 The Generalized Stieltjes Transform

The *generalized Stieltjes transform* of a function $f(t)$ is defined by

$$\mathcal{S}_g\{f(t)\} = \tilde{f}(z, p) = \int_0^\infty \frac{f(t)}{(t+z)^p} dt, \qquad (7.11.1)$$

provided the integral exists and $|\arg z| < \pi$.

Example 7.11.1 If Re $a > 0$, find the generalized Stieltjes transform of
  (a) t^{a-1},   (b) $\exp(-at)$.

(a) We have, by definition,

$$\mathcal{S}_g\{t^{a-1}\} = \int_0^\infty \frac{t^{a-1}}{(t+z)^p} dt$$

$$= z^{-p} \int_0^\infty t^{a-1} \left(1 + \frac{t}{z}\right)^{-p} dt, \qquad (t = zu),$$

$$= z^{a-p} \int_0^\infty u^{a-1}(1+u)^{-p} du. \qquad (7.11.2)$$

Substituting $x = \frac{u}{1+u}$ or $u = \frac{x}{1-x}$ into integral (7.11.2), we obtain

$$\mathcal{S}_g\{t^{a-1}\} = z^{a-p} \int_0^1 x^{a-1}(1-x)^{p-a-1} dx$$

$$= z^{a-p} B(a, p-a) = \frac{\Gamma(a)\,\Gamma(p-a)}{\Gamma(p)} z^{a-p}. \qquad (7.11.3)$$

(b) We have, by definition,

$$\mathcal{S}_g\{\exp(-at)\} = \int_0^\infty \frac{\exp(-at)}{(t+z)^p} dt$$

which is, by putting $t + z = u$,

$$\mathcal{S}_g\{\exp(-at)\} = \exp(-az) \int_0^\infty e^{-au} u^{-p} du.$$

Substituting $au = x$ into this integral, we obtain

$$\mathcal{S}_g\{\exp(-at)\} = a^{\rho-1}\exp(az)\int_0^\infty e^{-t}x^{1-\rho-1}dx$$

$$= a^{\rho-1}\exp(az)\Gamma(1-\rho). \qquad (7.11.4)$$

The reader is referred to Erdélyi *et al.* (1954, pp 234-235) where there is a table for generalized Stieltjes transforms.

7.12 Basic Properties of the Generalized Stieltjes Transform

The generalized Stieltjes transform satisfies the following properties:

(a) $\qquad \mathcal{S}_g\{f(at)\} = a^{\rho-1}\tilde{f}(az), \quad a>0,$ (7.12.1)

(b) $\qquad \mathcal{S}_g\{t\,f(t)\} = \tilde{f}(z,\rho-1) - z\,\tilde{f}(z,\rho),$ (7.12.2)

(c) $\qquad \mathcal{S}_g\{f'(t)\} = \rho\,\tilde{f}(z,\rho+1) - z^{-\rho}f(0),$ (7.12.3)

provided $f(t) \to 0$ as $t \to \infty$.

(d) $\qquad \mathcal{S}_g\left\{\int_0^t f(x)dx\right\} = (\rho-1)^{-1}\tilde{f}(z,\rho-1), \quad \text{Re } \rho > 1.$ (7.12.4)

Proof. (a) We have, by definition,

$$\mathcal{S}_g\{f(at)\} = \int_0^\infty \frac{f(at)\,dt}{(t+z)^\rho}$$

which is, by putting $at = x$,

$$= a^{\rho-1}\int_0^\infty \frac{f(x)\,dx}{(x+az)^\rho} = a^{\rho-1}\tilde{f}(az).$$

(b) It follows from the definition that

$$\mathcal{S}_g\{t\,f(t)\} = \int_0^\infty \frac{t\,f(t)}{(t+z)^\rho}\,dt = \int_0^\infty \frac{(t+z-z)\,f(t)}{(t+z)^\rho}\,dt$$

$$= \int_0^\infty \frac{f(t)}{(t+z)^{\rho-1}}\,dt - z\int_0^\infty \frac{f(t)\,dt}{(t+z)^\rho}$$

$$= \tilde{f}(z,\rho-1) - z\,\tilde{f}(z,\rho).$$

(c) We have, by definition,

$$\mathcal{S}_g\{f'(t)\} = \int_0^\infty \frac{f'(t)}{(t+z)^\rho}\,dt$$

which is, by integrating by parts,

Integral Transforms and Their Applications 261

$$\mathcal{S}_g\{f'(t)\} = \left[\frac{f(t)}{(t+z)^\rho}\right]_0^\infty + \rho \int_0^\infty \frac{f(t)}{(t+z)^{\rho+1}}\, dt$$

$$= \rho\, \tilde{f}(z, \rho+1) - z^{-\rho}\, f(0).$$

(d) We write
$$g(t) = \int_0^t f(x)\, dx$$

so that $g'(t) = f(t)$ and $g(0) = 0$.

Thus
$$\mathcal{S}_g\{f(t), \rho\} = \mathcal{S}_g\{g'(t), \rho\}$$
which is, by (7.12.3),
$$= \rho\, \tilde{g}(z, \rho+1) - z^{-\rho}\, g(0)$$
$$= \rho\, \mathcal{S}_g\{g(t), \rho+1\}.$$

Replacing ρ by $\rho-1$, we obtain (7.12.4).

7.13 Exercises

1. Find the Hilbert transform of each of the following functions:

   (a) $f(t) = \dfrac{1}{(a^2+t^2)}$, Re $a > 0$,

   (b) $f(t) = \dfrac{t^\alpha}{(t+a)}$, $|\text{Re}\,\alpha| < 1$,

   (c) $f(t) = \exp(-at)$,

   (d) $f(t) = t^{-\alpha} \exp(-at)$, Re $a > 0$, Re $\alpha < 1$,

   (e) $f(t) = \dfrac{1}{t}\sin(a\sqrt{t})$, $a > 0$,

   (f) $f(t) = \dfrac{\sin t}{t}$.

2. Show that

   (a) $\mathcal{S}\left\{\dfrac{1}{\sqrt{t}}\cos(a\sqrt{t})\right\} = \dfrac{\pi}{\sqrt{z}}\exp(-a\sqrt{z})$, $z > 0$.

   (b) $\mathcal{S}\{\sin(a\sqrt{t})\, J_0(b\sqrt{t})\} = \pi\,\exp(-a\sqrt{z})\, I_0(b\sqrt{z})$, $0 < b < a$.

3. If $\tilde{f}(z) = \mathcal{S}\{f(t)\}$ and $f(t) = \mathcal{S}\{g(u)\}$, then show that
$$\tilde{f}(z) = \int_0^\infty K(z,u)\, g(u)\, du,$$

where $K(z,u) = (z-u)^{-1} \log\left(\dfrac{z}{u}\right)$.

4. Show that

   (a) $\mathscr{S}_g\left\{t^{\rho-2} f\left(\dfrac{a}{t}\right)\right\} = a^{\rho-1} z^{-\rho} \tilde{f}\left(\dfrac{a}{z}\right)$, $a > 0$.

   (b) $\mathscr{S}_g\left\{\dfrac{1}{\Gamma(\alpha)} \displaystyle\int_0^t f(x)(t-x)^{\alpha-1} dx\right\} = \dfrac{\Gamma(\rho-\alpha)}{\Gamma(\rho)} \tilde{f}(z, \rho-\alpha)$,

   where $0 < \mathrm{Re}\,\alpha < \mathrm{Re}\,\rho$.

5. Show that the dispersion relation associated with the linearized Benjamin-Ono equation
   $$u_t + H\{u_{xx}\} = 0$$
   is
   $$\omega = -k|k|.$$

6. Find the Stieltjes transforms of each of the following functions:

   (a) $f(t) = \dfrac{t^{\alpha-1}}{t+a}$,

   (b) $f(t) = \dfrac{1}{t^2 + a^2}$,

   (c) $f(t) = \dfrac{t}{t^2 + a^2}$.

7. Show that

   (a) $\mathscr{S}\left[f(te^{i\pi}) - f(te^{-i\pi})\right] = 2\pi i\, \tilde{f}(z)$,

   (b) $\mathscr{S}\left[f(\sqrt{t})\right] = \tilde{f}(i\sqrt{z}) + \tilde{f}(-i\sqrt{z})$.

8. Suppose $f(t)$ is a locally integrable function on $(0, \infty)$ and has the asymptotic representation (see Wong, 1989)
   $$f(t) \sim \sum_{r=0}^{\infty} a_r t^{\alpha_r} \quad \text{as } t \to 0+$$
   where $\mathrm{Re}\,\alpha_r \uparrow +\infty$ as $r \to \infty$, $\mathrm{Re}\,\alpha_0 > -1$, and $f(t) = O(t^{-a})$, $a > 1$.
   Show that the generalized Stieltjes transform
   $$\tilde{f}(x) = \int_0^\infty \dfrac{f(t)\,dt}{(t+x)^\rho}, \quad \rho > 0$$
   has the asymptotic representation, as $x \to 0+$,
   $$\tilde{f}(x) \sim \sum_{r=0}^\infty a_r \dfrac{\Gamma(1+\alpha_r)\,\Gamma(\rho-1-\alpha_r)}{\Gamma(\rho)} x^{1+\alpha_r-\rho}$$
   $$+ \sum_{r=0}^\infty (-1)^r \dfrac{\Gamma(r+\rho)\,\mathcal{M}[f; 1-\rho-r]}{r!\,\Gamma(\rho)} x^r$$

provided $1+\alpha_r \neq p+n$ for all non-negative integers r and n.

9. Show that the one-sided Hilbert transform (involved in recent work on water waves by Hulme (1981))

$$\hat{f}_H(x) = \int_0^\infty \frac{J_0^2(t)\, dt}{t-x}$$

satisfies the Parseval relation

$$\hat{f}_H(x) = \frac{1}{2\pi i} \int_{c-i\infty}^{c+i\infty} x^{-p}\, \mathcal{M}\left[J_0^2(x); p\right] \pi \cot \pi p \, dp.$$

10. Prove the following asymptotic expansions (Ursell, 1983):

$$\oint_0^\infty \frac{J_0^2(t)}{t-x}\, dt \sim -\frac{1}{\pi x}(\log x + \gamma + 3\log 2) + \frac{1}{x}\cos 2x + \frac{1}{4x^2}\sin 2x$$

$$+ \frac{1}{8\pi x^3}\left(\log x + \gamma + 3\log 2 - \frac{5}{2}\right) - \frac{5}{32x^3}\cos 2x, \quad \text{as } x \to \infty,$$

and

$$\oint_0^\infty \frac{J_0^2(t)}{t-x}\, dt \sim -\frac{\pi}{2} J_0(x) Y_0(x)$$

$$-\sqrt{\pi}\sum_{r=0}^\infty \cos(\pi r) \frac{\Gamma(r+1) x^{2r+1}}{\left\{\Gamma\left(r+\frac{3}{2}\right)\right\}^3}, \quad \text{as } x \to 0.$$

11. If $\lambda = \frac{1}{\sqrt{\pi}}$, show that $f(t) = \frac{A}{\sqrt{t}}$, where A is a constant, is the only solution of the integral equation

$$f(s) = \lambda \int_0^\infty e^{-st} f(t)\, dt.$$

12. If $\lambda = -\frac{1}{\sqrt{\pi}}$, show that

$$f(t) = A\left[\frac{\Gamma'\left(\frac{1}{2}\right)}{\sqrt{\pi t}} - \frac{2\log t}{\sqrt{t}}\right],$$

where A is a constant, is the only solution of the integral equation as stated in Exercise 11.

13. Show that
    (a) $L_{k,t}\left[(x+a)^{-1}\right] = c_k (2k-1)!\, t^{k-1} a^k (t+a)^{-2k},\ (a>0,\ t>0,\ k=2,3,\ldots)$.

14. Prove that

$$\lim_{\varepsilon \to 0+} \frac{1}{\pi}\int_\varepsilon^\infty \left[\frac{f(x+u)-f(x-u)}{u}\right] du = \frac{1}{\pi}\oint_{-\infty}^\infty \frac{f(t)}{(t-x)}\, dt.$$

Chapter 8

Finite Fourier Sine and Cosine Transforms

8.1 Introduction

This chapter deals with the theory and applications of finite Fourier sine and cosine transforms. The basic operational properties including convolution theorem of these transforms are discussed in some detail. Special attention is given to the use of these transforms to the solutions of boundary value and initial-boundary value problems.

The finite Fourier sine transform was first introduced by Doetsch (1935). Subsequently, the method has been developed and generalized by several authors including Kneitz (1938), Koschmieder (1941), Roettinger (1947), and Brown (1944).

8.2 Definitions of the Finite Fourier Sine and Cosine Transforms and Examples

Both finite Fourier sine and cosine transforms are defined from the corresponding Fourier sine and Fourier cosine series.

Definition 8.2.1 (*The Finite Fourier Sine Transform*). If $f(x)$ is a continuous or piecewise continuous function on a finite interval $0 < x < a$, the *finite Fourier sine transform* of $f(x)$ is defined by

$$\mathcal{F}_s\{f(x)\} = \tilde{f}_s(n) = \int_0^a f(x) \sin\left(\frac{n\pi x}{a}\right) dx, \qquad (8.2.1)$$

where $n = 1, 2, 3, \ldots$.

It is a well-known result of the theory of Fourier series that the Fourier sine series for $f(x)$ in $0 < x < a$

$$\frac{2}{a} \sum_{n=1}^{\infty} \tilde{f}_s(n) \sin\left(\frac{n\pi x}{a}\right) \qquad (8.2.2)$$

converges to the value $f(x)$ at each point of continuity in the interval $0 < x < a$ and to the value $\frac{1}{2}[f(x+0) + f(x-0)]$ at each point x of the finite discontinuity in $0 < x < a$. In view of the definition (8.2.1), the *inverse Fourier sine transform* is given by

$$\mathcal{F}_s^{-1}\{\tilde{f}_s(n)\} = f(x) = \frac{2}{a}\sum_{n=1}^{\infty}\tilde{f}_s(n)\sin\left(\frac{n\pi x}{a}\right). \tag{8.2.3}$$

Clearly, both $\mathcal{F}_s$ and $\mathcal{F}_s^{-1}$ are linear transformations.

Definition 8.2.2 (*The Finite Fourier Cosine Transform*). If $f(x)$ is a continuous or piecewise continuous on a finite interval $0 < x < a$, the *finite Fourier cosine transform* of $f(x)$ is defined by

$$\mathcal{F}_c\{f(x)\} = \tilde{f}_c(n) = \int_0^a f(x)\cos\left(\frac{n\pi x}{a}\right)dx, \tag{8.2.4}$$

where $n = 0, 1, 2, \ldots$.

It is also a well known result of the theory of Fourier series that the Fourier cosine series for $f(x)$ in $0 < x < a$

$$f(x) = \frac{1}{a}\tilde{f}_c(0) + \frac{2}{a}\sum_{n=1}^{\infty}\tilde{f}_c(n)\cos\left(\frac{n\pi x}{a}\right) \tag{8.2.5}$$

converges to $f(x)$ at each point of continuity in $0 < x < a$, and to $\frac{1}{2}[f(x+0) + f(x-0)]$ at each point x of finite discontinuity in $0 < x < a$. By virtue of the definition (8.2.4), the *inverse Fourier cosine transform* is given by

$$\mathcal{F}_c^{-1}\{\tilde{f}_c(n)\} = f(x) = \frac{1}{a}\tilde{f}_c(0) + \frac{2}{a}\sum_{n=1}^{\infty}\tilde{f}_c(n)\cos\left(\frac{n\pi x}{a}\right). \tag{8.2.6}$$

Clearly, both $\mathcal{F}_c$ and $\mathcal{F}_c^{-1}$ are linear transformations.

When $a = \pi$, the finite Fourier sine and cosine transforms are defined respectively by (8.2.1) and (8.2.4) on the interval $0 < x < \pi$. The corresponding inverse transforms are given by the same results (8.2.3) and (8.2.6) with $a = \pi$. The transform of a function defined over an interval $0 < x < a$ can be written easily in terms of a transform on the standard interval $0 < x < \pi$. We substitute $\xi = \frac{\pi x}{a}$ to write (8.2.1) and (8.2.4) as follows:

$$\tilde{f}_s(n) = \int_0^a \sin\left(\frac{n\pi x}{a}\right)f(x)dx = \frac{a}{\pi}\int_0^\pi \sin(n\xi)f\left(\frac{a\xi}{\pi}\right)d\xi = \frac{a}{\pi}\mathcal{F}_s\left\{f\left(\frac{ax}{\pi}\right)\right\}$$

$$\tilde{f}_c(n) = \int_0^a \cos\left(\frac{n\pi x}{a}\right)f(x)dx = \frac{a}{\pi}\int_0^\pi \cos(n\xi)f\left(\frac{a\xi}{\pi}\right)d\xi = \frac{a}{\pi}\mathcal{F}_c\left\{f\left(\frac{ax}{\pi}\right)\right\}.$$

Example 8.2.1 Find the finite Fourier sine and cosine transforms of
 (a) $f(x) = 1$ and (b) $f(x) = x$.

(a) We have

$$\mathcal{F}_s(1) = \tilde{f}_s(n) = \int_0^a \sin\left(\frac{n\pi x}{a}\right)dx = \frac{a}{n\pi}\left[1-(-1)^n\right], \tag{8.2.7}$$

$$\mathcal{F}_c\{1\} = \tilde{f}_c(n) = \int_0^a \cos\left(\frac{n\pi x}{a}\right) dx = \begin{cases} a, & n=0 \\ 0, & n \neq 0 \end{cases}. \qquad (8.2.8)$$

(b)
$$\mathcal{F}_s\{x\} = \int_0^a x \sin\left(\frac{n\pi x}{a}\right) dx = \frac{(-1)^{n+1} a^2}{n\pi}. \qquad (8.2.9)$$

$$\mathcal{F}_c\{x\} = \int_0^a x \cos\left(\frac{n\pi x}{a}\right) dx = \begin{cases} \dfrac{a^2}{2}, & n=0 \\ \left(\dfrac{a}{n\pi}\right)^2 [(-1)^n - 1], & n \neq 0 \end{cases}. \qquad (8.2.10)$$

8.3. Basic Properties of Finite Fourier Sine and Cosine Transforms

As a preliminary to the solution of differential equations by the finite Fourier sine and cosine transforms, we now establish the transforms of derivatives of $f(x)$.

$$\mathcal{F}_s\{f'(x)\} = -\left(\frac{n\pi}{a}\right) \tilde{f}_c(n), \qquad (8.3.1)$$

$$\mathcal{F}_s\{f''(x)\} = -\left(\frac{n\pi}{a}\right)^2 \tilde{f}_s(n) + \left(\frac{n\pi}{a}\right)\left[f(0) + (-1)^{n+1} f(a)\right], \qquad (8.3.2)$$

$$\mathcal{F}_c\{f'(x)\} = \left(\frac{n\pi}{a}\right) \tilde{f}_s(n) + (-1)^n f(a) - f(0), \qquad (8.3.3)$$

$$\mathcal{F}_c\{f''(x)\} = -\left(\frac{n\pi}{a}\right)^2 \tilde{f}_c(n) + (-1)^n f'(a) - f'(0). \qquad (8.3.4)$$

Similar results can be obtained for the finite Fourier sine and cosine transforms of higher derivatives of $f(x)$.

Results (8.3.1)-(8.3.4) can be proved by integrating by parts. For example, we have

$$\mathcal{F}_s\{f'(x)\} = \int_0^a f'(x) \sin\left(\frac{n\pi x}{a}\right) dx,$$

which is, by integrating by parts,

$$= \left[f(x) \sin\left(\frac{n\pi x}{a}\right)\right]_0^a - \frac{n\pi}{a} \int_0^a f(x) \cos\left(\frac{n\pi x}{a}\right) dx$$

$$= -\left(\frac{n\pi}{a}\right) \tilde{f}_c(n).$$

Similarly, we find that
$$\mathscr{F}_c\{f'(x)\} = \int_0^a f'(x) \cos\left(\frac{n\pi x}{a}\right) dx$$
$$= \left[f(x)\cos\left(\frac{n\pi x}{a}\right)\right]_0^a + \frac{n\pi}{a}\int_0^a f(x)\sin\left(\frac{n\pi x}{a}\right) dx$$
$$= (-1)^n f(a) - f(0) + \frac{n\pi}{a}\tilde{f}_s(n).$$

This proves the result (8.3.3).

Results (8.3.2) and (8.3.4) and the results for higher derivatives can be obtained by the repeated application of the fundamental results (8.3.1) and (8.3.3).

Definition 8.3.1 (*Odd Periodic Extension*). A function $f_1(x)$ is said to be the *odd periodic extension* of the function $f(x)$, with period 2π if

$$f_1(x) = \begin{cases} f(x) & \text{for } 0 < x < \pi \\ -f(-x) & \text{for } -\pi < x < 0 \end{cases} \tag{8.3.5}$$

Or equivalently
$$f_1(x) = f(x) \quad \text{when } 0 < x < \pi,$$
$$f_1(-x) = -f_1(x), \quad f_1(x + 2\pi) = f_1(x) \quad \text{for } -\infty < x < \infty. \tag{8.3.6}$$

Similarly, the *even periodic extension* $f_2(x)$ of $f(x)$, with period 2π is defined in $-\pi < x < \pi$ by the equations

$$f_2(x) = \begin{cases} f(x) & \text{for } 0 < x < \pi \\ f(-x) & \text{for } -\pi < x < 0 \end{cases}. \tag{8.3.7}$$

Or equivalently
$$f_2(x) = f(x) \quad \text{when } 0 < x < \pi,$$
$$f_2(-x) = f_2(x), \quad f_2(x + 2\pi) = f_2(x) \quad \text{for } -\infty < x < \infty. \tag{8.3.8}$$

Theorem 8.3.1 If $f_1(x)$ is the odd periodic extension of $f(x)$ with period 2π, then, for any constant α,

$$\mathscr{F}_s\{f_1(x - \alpha) + f_1(x + \alpha)\} = 2\cos n\alpha\, \mathscr{F}_s\{f(x)\}. \tag{8.3.9}$$

In particular, when $\alpha = \pi$, and $n = 1, 2, 3, \ldots$,

$$\mathscr{F}_s\{f(x - \pi)\} = (-1)^n\, \mathscr{F}_s\{f(x)\}. \tag{8.3.10}$$

Similarly, we obtain
$$\mathscr{F}_c\{f_1(x + \alpha) - f_1(x - \alpha)\} = 2\sin(n\alpha)\, \mathscr{F}_s\{f(x)\}. \tag{8.3.11}$$

Proof. To prove (8.3.9), we follow Churchill (1972) and write the right hand side of (8.3.9) as

$$2 \cos n\alpha \, \tilde{f}_s(n) = 2 \cos (n\alpha) \int_0^\pi \sin(nx) f(x) \, dx$$

$$= 2 \int_0^\pi \cos n\alpha \, \sin nx \, f_1(x) \, dx$$

$$= \int_0^\pi [\sin n(x+\alpha) + \sin n(x-\alpha)] \, f_1(x) \, dx$$

which is, since the integrand is even function of x,

$$= \frac{1}{2} \int_{-\pi}^\pi [\sin n(x+\alpha) + \sin n(x-\alpha)] \, f_1(x) \, dx$$

which is, by putting $x + \alpha = t$ and $x - \alpha = t$,

$$= \frac{1}{2} \int_{-\pi+\alpha}^{\pi+\alpha} \sin nt \, f_1(t-\alpha) \, dt + \frac{1}{2} \int_{-(\pi+\alpha)}^{\pi-\alpha} \sin nt \, f_1(t+\alpha) \, dt$$

which is, since the integrands are periodic function of t with period 2π, and hence the limits of integration can be replaced with the limits $-\pi$ to π,

$$= \frac{1}{2} \int_{-\pi}^\pi \sin nt \, f_1(t-\alpha) \, dt + \frac{1}{2} \int_{-\pi}^\pi \sin nt \, f_1(t+\alpha) \, dt$$

$$= \frac{1}{2} \left[\int_{-\pi}^0 + \int_0^\pi \right] \{ \sin nt \, f_1(t-\alpha) \} \, dt$$

$$+ \frac{1}{2} \left[\int_{-\pi}^0 + \int_0^\pi \right] \{ \sin nt \, f_1(t+\alpha) \} \, dt. \quad (8.3.12)$$

Furthermore

$$\int_{-\pi}^0 \sin nt \, f_1(t-\alpha) \, dt = \int_0^\pi \sin nx \, f_1(x+\alpha) \, dx$$

in which $f_1(-x-\alpha) = -f_1(x+\alpha)$ is used.

Making a similar change in the third integral of (8.3.12), we obtain the formula

$$2 \cos n\alpha \, \tilde{f}_s(n) = \int_0^\pi \sin nt \, f_1(t-\alpha) \, dt + \int_0^\pi \sin nx \, f_1(x+\alpha) \, dx,$$

which gives the desired result (8.3.9).

Finally, $f_1(x+\pi) = f_1(2\pi + x - \pi) = f_1(x-\pi) = -f_1(\pi - x)$, and when $0 < x < \pi$, $f_1(\pi - x) = f(\pi - x)$. Thus when $\alpha = \pi$, result (8.3.9) becomes

$$f_s(n) \cos n\pi = \int_0^\pi \sin nx \, f(x-\pi) \, dx = \mathcal{F}_s \{ f(x-\pi) \},$$

which reduces to (8.3.10).

The proof of (8.3.11) is similar to that of (8.3.9), and hence is left to the reader.

Theorem 8.3.2 If $f_2(x)$ is the even periodic extension of $f(x)$ with period 2π, then, for any constant α,

$$\mathcal{F}_c\{f_2(x-\alpha)+f_2(x+\alpha)\} = 2\cos n\alpha\, \mathcal{F}_c\{f(x)\}, \qquad (8.3.13)$$

$$\mathcal{F}_s\{f_2(x-\alpha)-f_2(x+\alpha)\} = 2\sin n\alpha\, \mathcal{F}_c\{f(x)\}. \qquad (8.3.14)$$

This theorem is very much similar to that of Theorem 8.3.1, and hence the proof is left to the reader.

In the notation of Churchill (1972), we introduce the *convolution* of two sectionally continuous periodic functions $f(x)$ and $g(x)$ defined in $-\pi < x < \pi$ by

$$f(x) * g(x) = \int_{-\pi}^{\pi} f(x-u)\, g(u)\, du. \qquad (8.3.15)$$

Clearly, $f(x)*g(x)$ is continuous and periodic with period 2π. The convolution is symmetric, that is $f*g = g*f$. Furthermore, the convolution is an even function if $f(x)$ and $g(x)$ are both even or both odd. It is odd if either $f(x)$ or $g(x)$ is even or the other odd. We next prove the convolution theorem.

Theorem 8.3.3 (*Convolution*). If $f_1(x)$ and $g_1(x)$ are the odd periodic extensions of $f(x)$ and $g(x)$ respectively on $0 < x < \pi$, and if $f_2(x)$ and $g_2(x)$ are the even periodic extensions of $f(x)$ and $g(x)$ respectively on $0 < x < \pi$, then

$$\mathcal{F}_c\{f_1(x)*g_1(x)\} = -2\tilde{f}_s(n)\,\tilde{g}_s(n), \qquad (8.3.16)$$

$$\mathcal{F}_c\{f_2(x)*g_2(x)\} = 2\tilde{f}_c(n)\,\tilde{g}_c(n), \qquad (8.3.17)$$

$$\mathcal{F}_s\{f_1(x)*g_2(x)\} = 2\tilde{f}_s(n)\,\tilde{g}_c(n), \qquad (8.3.18)$$

$$\mathcal{F}_s\{f_2(x)*g_1(x)\} = 2\tilde{f}_c(n)\,\tilde{g}_s(n). \qquad (8.3.19)$$

Or equivalently,

$$\mathcal{F}_c^{-1}\{\tilde{f}_s(n)\,\tilde{g}_s(n)\} = -\frac{1}{2}\{f_1(x)*g_1(x)\}, \qquad (8.3.20)$$

$$\mathcal{F}_c^{-1}\{\tilde{f}_c(n)\,\tilde{g}_c(n)\} = \frac{1}{2}\{f_2(x)*g_2(x)\}, \qquad (8.3.21)$$

$$\mathcal{F}_s^{-1}\{\tilde{f}_s(n)\,\tilde{g}_c(n)\} = \frac{1}{2}\{f_1(x)*g_2(x)\}, \qquad (8.3.22)$$

$$\mathcal{F}_s^{-1}\{\tilde{f}_c(n)\,\tilde{g}_s(n)\} = \frac{1}{2}\{f_2(x)*g_1(x)\}. \qquad (8.3.23)$$

Proof. To prove (8.3.16), we consider the product

$$2\tilde{f}_s(n)\,\tilde{g}_s(n) = 2\int_0^\pi \tilde{f}_s(n)\,\sin nu\, g(u)\,du$$

which is, by using (8.3.11),

$$= \int_0^\pi g(u) \left[\mathscr{F}_c \{ f_1(x+u) - f_1(x-u) \} \right] du$$

$$= \int_0^\pi g(u) \left[\int_0^\pi \{ f_1(x+u) - f_1(x-u) \} \cos nx \right] du$$

which is, by interchanging the order of integration,

$$= \int_0^\pi \cos(nx) \left[\int_0^\pi \{ f_1(x+u) - f_1(x-u) \} g(u) \, du \right] dx. \tag{8.3.24}$$

Using the definition of convolution (8.3.15), introducing new variables of integration, and invoking the odd extension properties of $f_1(x)$ and $g_1(x)$, we obtain

$$f_1(x) * g_1(x) = \int_0^\pi [f_1(x-u) - f_1(x+u)] g(u) \, du \tag{8.3.25}$$

$$= I_1 - I_2 - I_3 + I_4, \tag{8.3.26}$$

where

$$I_1 = \int_0^x f(u) \, g(x+u) \, du, \quad I_2 = \int_x^\pi f(u) \, g(u-x) \, du, \tag{8.3.27ab}$$

$$I_3 = \int_0^{\pi-x} f(u) \, g(x+u) \, du, \quad I_4 = \int_x^\pi f(u) \, g(2\pi - x - u) \, du. \tag{8.3.28ab}$$

In view of (8.3.25), we thus obtain the desired result (8.3.16) from (8.3.24). This completes the proof.

The other results included in Theorem 8.3.3 can be proved by the above method of proof.

As an example of convolution theorem, we evaluate the inverse cosine Fourier transform of $(n^2 - a^2)^{-1}$. We write, for $n \neq 0$,

$$\frac{1}{(n^2 - a^2)} = \frac{n(-1)^{n+1}}{(n^2 - a^2)} \cdot \frac{(-1)^{n+1}}{n} = \tilde{f}_s(n) \, \tilde{g}_s(n),$$

where $\tilde{f}_s(n) = n(-1)^{n+1}(n^2 - a^2)^{-1}$ and $\tilde{g}_s(n) = \frac{(-1)^{n+1}}{n}$ so that

$$f(x) = \left(\frac{\sin ax}{\sin a\pi} \right) \text{ and } g(x) = \frac{x}{\pi}.$$

Evidently,

$$\frac{1}{(n^2 - a^2)} = \tilde{f}_s(n) \, \tilde{g}_s(n) = \mathscr{F}_s \left\{ \frac{\sin ax}{\sin a\pi} \right\} \mathscr{F}_s \left\{ \frac{x}{\pi} \right\}.$$

According to (8.3.20),

$$\mathcal{F}_c^{-1}\left\{\frac{1}{(n^2-a^2)}\right\} = \mathcal{F}_c^{-1}\left\{\tilde{f}_s(n)\,\tilde{g}_s(n)\right\} = -\frac{1}{2}f_1(x)*g_1(x), \quad (8.3.29)$$

where $f_1(x)$ is the periodic extension of the odd function $f(x)$ with period 2π and $g_1(x) = \frac{x}{\pi}$. Thus, it turns out that

$$\mathcal{F}_c^{-1}\left\{\frac{1}{(n^2-a^2)}\right\} = -\frac{1}{2\pi}\int_{-\pi}^{\pi} f_1(x-u)\,g_1(u)\,du = -\frac{1}{2\pi}\int_{-\pi}^{\pi} f_1(x-u)\,u\,du.$$

This integral can easily be evaluated by splitting up the interval of integration or by using (8.3.26), and hence

$$\mathcal{F}_c^{-1}\left\{\frac{1}{(n^2-a^2)}\right\} = -\frac{\cos\{a(\pi-x)\}}{a\sin a\pi}. \quad (8.3.30)$$

8.4 Applications of Finite Fourier Sine and Cosine Transforms

In this section we illustrate the use of finite Fourier sine and cosine transforms to the solutions of boundary value and initial-boundary value problems.

Example 8.4.1 (*Heat Conduction Problem in a Finite Domain with the Dirichlet Data at the Boundary*). We began by considering the solution of the temperature distribution $u(x,t)$ of the diffusion equation

$$u_t = \kappa\, u_{xx}, \quad 0 \le x \le a,\ t > 0, \quad (8.4.1)$$

with the boundary and initial conditions

$$u(0,t) = 0 = u(a,t), \quad (8.4.2ab)$$
$$u(x,0) = f(x) \quad \text{for } 0 \le x \le a. \quad (8.4.3)$$

Application of the finite Fourier sine transform (8.2.1) to this diffusion problem gives the initial value problem

$$\frac{d\tilde{u}_s}{dt} + \kappa\left(\frac{n\pi}{a}\right)^2 \tilde{u}_s = 0, \quad (8.4.4)$$

$$\tilde{u}_s(n,0) = \tilde{f}_s(n). \quad (8.4.5)$$

The solution of (8.4.4)-(8.4.5) is

$$\tilde{u}_s(n,t) = \tilde{f}_s(n)\,\exp\left\{-\kappa\left(\frac{n\pi}{a}\right)^2 t\right\}. \quad (8.4.6)$$

The inverse finite Fourier sine transform (8.2.3) leads to the solution

$$u(x,t) = \frac{2}{a}\sum_{n=1}^{\infty} \tilde{f}_s(n)\,\exp\left\{-\kappa\left(\frac{n\pi}{a}\right)^2 t\right\} \sin\left(\frac{n\pi x}{a}\right)$$

$$u(x,t) = \frac{2}{a}\sum_{n=1}^{\infty} \exp\left\{-\kappa\left(\frac{n\pi}{a}\right)^2 t\right\} \sin\left(\frac{n\pi x}{a}\right) \int_0^a f(\xi) \sin\left(\frac{n\pi\xi}{a}\right) d\xi. \tag{8.4.7}$$

If, in particular, $f(x) = T_0 =$ constant, then (8.4.7) becomes

$$u(x,t) = \frac{2T_0}{\pi} \sum_{n=1}^{\infty} \frac{1}{n}[1-(-1)^n] \exp\left\{-\kappa\left(\frac{n\pi}{a}\right)^2 t\right\} \sin\left(\frac{n\pi x}{a}\right). \tag{8.4.8}$$

This series solution can be evaluated numerically using the *Fast Fourier transform* which is an algorithm for the efficient calculation of the finite Fourier transform.

Example 8.4.2 (*Heat Conduction Problem in a Finite Domain with the Neumann Data at the Boundary*).

We consider the solution of the diffusion equation (8.4.1) with the prescribed heat flux at $x = 0$ and $x = a$, and the associated boundary and initial data are

$$u_x(0,t) = 0 = u_x(a,t) \quad \text{for} \quad t > 0, \tag{8.4.9}$$
$$u(x,0) = f(x) \quad \text{for} \quad 0 \leq x \leq a. \tag{8.4.10}$$

In this case, it is appropriate to use the finite Fourier cosine transform (8.2.4). So the application of this transform gives the initial value problem

$$\frac{d\tilde{u}_c}{dt} + \kappa\left(\frac{n\pi}{a}\right)^2 \tilde{u}_c = 0, \tag{8.4.11}$$

$$\tilde{u}_c(n,0) = \tilde{f}_c(n). \tag{8.4.12}$$

The solution of this problem is

$$\tilde{u}_c(n,t) = \tilde{f}_c(n) \exp\left\{-\kappa\left(\frac{n\pi}{a}\right)^2 t\right\}. \tag{8.4.13}$$

The inverse finite cosine transform (8.2.5) gives the formal solution

$$u(x,t) = \frac{1}{a}\tilde{f}_c(0) + \frac{2}{a}\sum_{n=1}^{\infty} \tilde{f}_c(n) \exp\left\{-\kappa\left(\frac{n\pi}{a}\right)^2 t\right\} \cos\left(\frac{n\pi x}{a}\right)$$

$$= \frac{1}{a}\int_0^a f(\xi) d\xi + \frac{2}{a} \sum_{n=1}^{\infty} \left[\int_0^a f(\xi) \cos\left(\frac{n\pi\xi}{a}\right) d\xi\right]$$

$$\times \exp\left\{-\kappa\left(\frac{n\pi}{a}\right)^2 t\right\} \cos\left(\frac{n\pi x}{a}\right). \tag{8.4.14}$$

Example 8.4.3 (*The Static Deflection of a Uniform Elastic Beam*). We consider the static deflection $y(x)$ of a uniform elastic beam of finite length ℓ which satisfies the equilibrium equation

$$\frac{d^4 y}{dx^4} = \frac{W(x)}{EI} = w(x), \quad 0 \leq x \leq \ell, \tag{8.4.15}$$

where $W(x)$ is the applied load per unit length of the beam, E is the Young's modulus of the beam and I is the moment of inertia of the cross section of the beam. If the beam is freely hinged at its ends, then

$$y(x) = y''(x) = 0 \quad \text{at} \quad x = 0 \quad \text{and} \quad x = \ell. \tag{8.4.16}$$

Application of the finite Fourier sine transform of $y(x)$ to (8.4.15) and (8.4.16) gives

$$\tilde{y}_s(n) = \left(\frac{\ell}{n\pi}\right)^4 \tilde{w}_s(n). \qquad (8.4.17)$$

Inverting this result, we find

$$y(x) = \frac{2\ell^3}{\pi^4} \sum_{n=1}^{\infty} \frac{1}{n^4} \sin\left(\frac{n\pi x}{\ell}\right) \tilde{w}_s(n)$$

$$= \frac{2\ell^3}{\pi^4} \sum_{n=1}^{\infty} \frac{1}{n^4} \sin\left(\frac{n\pi x}{\ell}\right) \int_0^{\ell} w(\xi) \sin\left(\frac{n\pi \xi}{\ell}\right) d\xi. \qquad (8.4.18)$$

In particular, if the applied load of magnitude W_0 is confined to the point $x = \alpha$, where $0 < \alpha < \ell$, then $w(x) = W_0\, \delta(x-\alpha)$ where W_0 is a constant. Consequently, the static deflection is

$$y(x) = \frac{2\ell^3 W_0}{\pi^4} \sum_{n=1}^{\infty} \frac{1}{n^4} \sin\left(\frac{n\pi x}{\ell}\right) \sin\left(\frac{n\pi \alpha}{\ell}\right). \qquad (8.4.19)$$

Example 8.4.4 (*Transverse Displacement of an Elastic Beam of Finite Length*). We consider the transverse displacement of an elastic beam at a point x in the downward direction where the equilibrium position of the beam is along the x-axis. With the applied load $W(x,t)$ per unit length of the beam, the displacement function $y(x,t)$ satisfies the equation of motion

$$\frac{\partial^4 y}{\partial x^4} + \frac{1}{a^2}\frac{\partial^2 y}{\partial t^2} = \frac{W(x,t)}{EI}, \quad 0 \leq x \leq \ell,\ t > 0, \qquad (8.4.20)$$

where $a^2 = EI/(\rho\alpha)$, α is the cross-sectional area and ρ is the line density of the beam.

If the beam is freely hinged at its ends, then

$$y(x,t) = \frac{\partial^2 y}{\partial x^2} = 0 \text{ at } x = 0 \text{ and } x = \ell. \qquad (8.4.21)$$

The initial conditions are

$$y(x,t) = f(x),\ \frac{\partial y}{\partial t} = g(x) \text{ at } t = 0 \text{ for } 0 < x < \ell. \qquad (8.4.22)$$

We use the joint Laplace transform with respect to t and the finite Fourier sine transform with respect to x defined by

$$\bar{\tilde{u}}_s(n,s) = \int_0^\infty e^{-st} dt \int_0^\ell u(x,t) \sin\left(\frac{n\pi x}{\ell}\right) dx. \qquad (8.4.23)$$

Application of the double transform to (8.4.20)-(8.4.22) gives the solution for $\bar{\tilde{y}}_s(n,s)$ as

$$\bar{\tilde{y}}_s(n,s) = \frac{s\,\tilde{f}_s(n) + \tilde{g}_s(n)}{(s^2 + c^2)} + \frac{a^2}{EI}\frac{\bar{W}_s(n,s)}{(s^2+c^2)}, \qquad (8.4.24)$$

where $c = a\left(\dfrac{n\pi}{\ell}\right)^2$.

The inverse Laplace transform gives
$$\tilde{y}_s(n,t) = \tilde{f}_s(n)\cos(ct) + \frac{\tilde{g}_s(n)}{c}\sin(ct)$$
$$+ \frac{a^2}{EIc}\int_0^t \sin c(t-\tau)\,\tilde{W}_s(n,\tau)\,d\tau. \tag{8.4.25}$$

Thus, the inverse finite Fourier sine transform yields the formal solution as
$$y(x,t) = \frac{2}{\ell}\sum_{n=1}^{\infty} y_s(n,t)\sin\left(\frac{n\pi x}{\ell}\right),$$
$$= \frac{2}{\ell}\sum_{n=1}^{\infty}\sin\left(\frac{n\pi x}{\ell}\right)\left[\left\{\tilde{f}_s(n)\cos(ct) + \frac{\tilde{g}_s(n)}{c}\sin(ct)\right\}\right.$$
$$\left. + \frac{a^2}{EIc}\int_0^t \sin c(t-\tau)\,\tilde{W}_s(n,\tau)\,d\tau\right], \tag{8.4.26}$$

where
$$\tilde{f}_s(n) = \int_0^\ell f(\xi)\sin\left(\frac{n\pi\xi}{\ell}\right)d\xi, \quad \tilde{g}_s(n) = \int_0^\ell g(\xi)\sin\left(\frac{n\pi\xi}{\ell}\right)d\xi. \tag{8.4.27ab}$$

The case of free vibrations is of interest. In this case, $W(x,t)\equiv 0$ and hence $\tilde{W}_s(n,t)\equiv 0$. Consequently, solution (8.4.26) reduces to a simple form
$$y(x,t) = \frac{2}{\ell}\sum_{n=1}^{\infty}\left[\tilde{f}_s(n)\cos ct + \frac{\tilde{g}_s(n)}{c}\sin ct\right]\sin\left(\frac{n\pi x}{\ell}\right), \tag{8.4.28}$$

where $\tilde{f}_s(n)$ and $\tilde{g}_s(n)$ are given by (8.4.27ab).

Example 8.4.5 (*Free Transverse Vibrations of an Elastic String of Finite Length*). We consider the free vibration of a string of length ℓ stretched to a constant tension T between two points $(0,0)$ and $(0,\ell)$ lying on the x-axis. The free transverse displacement function $u(x,t)$ satisfies the wave equation
$$\frac{\partial^2 u}{\partial t^2} = c^2 \frac{\partial^2 u}{\partial x^2}, \quad 0\leq x\leq \ell,\ t>0, \tag{8.4.29}$$

where $c^2 = \dfrac{T}{\rho}$ and ρ is the line density of the string.

The initial and boundary conditions are
$$u(x,t) = f(x), \quad \frac{\partial u}{\partial t} = g(x) \text{ at } t=0 \text{ for } 0\leq x\leq \ell, \tag{8.4.30ab}$$
$$u(x,t) = 0 \text{ at } x=0 \text{ and } x=\ell \text{ for } t>0. \tag{8.4.31ab}$$

Application of the joint Laplace transform with respect to t and the finite Fourier sine transform with respect to x defined by a result similar to (8.4.23) to (8.4.29)-(8.4.31ab) gives

$$\overline{\tilde{u}}_s(n,s) = \frac{s\,\tilde{f}_s(n)}{(s^2+a^2)} + \frac{\tilde{g}_s(n)}{(s^2+a^2)}, \qquad (8.4.32)$$

where $a^2 = \left(\dfrac{n\pi c}{\ell}\right)^2.$

The inverse Laplace transform gives

$$\tilde{u}_s(n,t) = \tilde{f}_s(n)\cos at + \frac{\tilde{g}_s(n)}{a}\sin at. \qquad (8.4.33)$$

The inverse finite Fourier sine transform leads to the solution for $u(x,t)$ as

$$u(x,t) = \frac{2}{\ell}\sum_{n=1}^{\infty}\left[\tilde{f}_s(n)\cos at + \frac{\tilde{g}_s(n)}{a}\sin at\right]\sin\left(\frac{n\pi x}{\ell}\right), \qquad (8.4.34)$$

where $\tilde{f}_s(n)$ and $\tilde{g}_s(n)$ are given by (8.4.27ab).

Example 8.4.6 (*Two-Dimensional Unsteady Couette Flow*). We consider two-dimensional unsteady viscous flow between the plate at $z = 0$ at rest and the plate $z = h$ in motion parallel to itself with a variable velocity $U(t)$ in the x direction. The fluid velocity $u(z,t)$ satisfies the equation of motion

$$\frac{\partial u}{\partial t} = -\frac{P(t)}{\rho} + v\frac{\partial^2 u}{\partial z^2}, \quad 0 \le z \le h,\ t > 0, \qquad (8.4.35)$$

and the boundary and initial conditions

$$u(z,t) = 0 \qquad \text{on } z = 0,\ t > 0; \qquad (8.4.36)$$
$$u(z,t) = U(t) \qquad \text{on } z = h,\ t > 0; \qquad (8.4.37)$$
$$u(z,t) = 0 \qquad \text{at } t \le 0 \text{ for } 0 \le z \le h; \qquad (8.4.38)$$

where the pressure gradient $p_x = P(t)$ and v is the kinematic viscosity of the fluid.

Application of the double transform defined by (8.4.23) to this initial-boundary value problem gives the solution for $\overline{\tilde{u}}_s(n,s)$ as

$$\left(s + \frac{vn^2\pi^2}{h^2}\right)\overline{\tilde{u}}_s(n,t) = -\frac{h\,\overline{P}(s)}{n\pi\rho}\left[1+(-1)^{n+1}\right] + \frac{vn\pi}{h}(-1)^{n+1}\overline{U}(s). \qquad (8.4.39)$$

The inverse Laplace transform yields

$$\tilde{u}_s(n,t) = -\frac{h}{n\pi\rho}\left[1+(-1)^{n+1}\right]\int_0^t P(t-\tau)\exp\left(-\frac{vn^2\pi^2\tau}{h^2}\right)d\tau$$
$$+ \frac{vn\pi}{h}(-1)^{n+1}\int_0^t U(t-\tau)\exp\left(-\frac{vn^2\pi^2\tau}{h^2}\right)d\tau. \qquad (8.4.40)$$

Finally, the inverse finite Fourier sine transform gives the formal solution

$$u(z,t) = \frac{2}{h}\sum_{n=1}^{\infty}\tilde{u}_s(n,t)\sin\left(\frac{n\pi z}{h}\right). \qquad (8.4.41)$$

If, in particular, $P(t) =$ constant P and $U(t) =$ constant $= U$, then (8.4.41) reduces to

$$u(z,t) = -\frac{2P}{\mu h}\sum_{n=1}^{\infty}\left(\frac{h}{n\pi}\right)^3\left[1+(-1)^{n+1}\right]\sin\left(\frac{n\pi z}{h}\right)\left[1-\exp\left(-\frac{vn^2\pi^2 t}{h^2}\right)\right]$$

$$+\frac{2U}{h}\sum_{n=1}^{\infty}(-1)^{n+1}\left(\frac{h}{n\pi}\right)\sin\left(\frac{n\pi z}{h}\right)\left[1-\exp\left(-\frac{vn^2\pi^2 t}{h^2}\right)\right]. \quad (8.4.42)$$

This solution for the velocity field consists of both steady-state and transient components. In the limit as $t \to \infty$, the transient component decays to zero, and the steady state is attained in the form

$$u(z,t) = -\frac{2P}{\mu h}\sum_{n=1}^{\infty}\left(\frac{h}{n\pi}\right)^3\left[1+(-1)^{n+1}\right]\sin\left(\frac{n\pi z}{\hbar}\right)$$

$$+\frac{2U}{h}\sum_{n=1}^{\infty}(-1)^{n+1}\left(\frac{h^2}{n\pi}\right)\sin\left(\frac{n\pi z}{h}\right). \quad (8.4.43)$$

In view of the inverse finite Fourier sine transforms

$$\mathcal{F}_s^{-1}\left\{2\left(\frac{h}{n\pi}\right)^3\left[1+(-1)^{n+1}\right]\right\} = z(h-z), \quad (8.4.44)$$

$$\mathcal{F}_s^{-1}\left\{(-1)^{n+1}\left(\frac{h^2}{n\pi}\right)\right\} = z, \quad (8.4.45)$$

solution (8.4.43) can be rewritten in the closed form

$$u(z,t) = \frac{Uz}{h} - \frac{h}{2\mu}\left(\frac{\partial p}{\partial x}\right)\left(1-\frac{z}{h}\right)z. \quad (8.4.46)$$

This is known as the *generalized Couette flow*. In the absence of the pressure gradient term, solution (8.4.46) reduces to the linear profile of simple *Couette flow*. On the other hand, if $U(t) \equiv 0$ and $P(t) \neq 0$, the solution (8.4.46) represents the *parabolic profile of Poiseuille flow* between parallel stationary plates due to an imposed pressure gradient.

8.5 Multiple Finite Fourier Transforms and Their Applications

The above analysis for the finite Fourier sine and cosine transforms of a function of one independent variable can readily be extended to a function of several independent variables. In particular, if $f(x,y)$ is a function of two independent variables x and y, defined in a region $0 \leq x \leq a$, $0 \leq y \leq b$, its *double finite Fourier sine transform* is defined by

$$\mathcal{F}_s\{f(x,y)\} = \tilde{f}_s(m,n) = \int_0^a\int_0^b \sin\left(\frac{m\pi x}{a}\right)\sin\left(\frac{n\pi y}{b}\right)dn\,dy. \quad (8.5.1)$$

The *inverse transform* is given by the double series

$$\mathscr{F}_s^{-1}\{\tilde{f}_s(m,n)\} = f(x,y) = \left(\frac{4}{ab}\right) \sum_{m=1}^{\infty} \sum_{n=1}^{\infty} \tilde{f}_s(m,n) \sin\left(\frac{m\pi x}{a}\right) \sin\left(\frac{n\pi y}{b}\right). \quad (8.5.2)$$

Similarly, we can define the double finite Fourier cosine transform and its inverse.

The double transforms of the partial derivatives of $f(x,y)$ can easily be obtained. If $f(x,y)$ vanishes on the boundary of the rectangular region $D\{0 \leq x \leq a, \ 0 \leq y \leq b\}$, then

$$\mathscr{F}_s\left[\frac{\partial^2 f}{\partial x^2} + \frac{\partial^2 f}{\partial y^2}\right] = -\pi^2 \left(\frac{m^2}{a^2} + \frac{n^2}{b^2}\right) \tilde{f}_s(m,n). \quad (8.5.3)$$

Example 8.5.1 (*Free Vibrations of a Rectangular Elastic Membrane*). The initial value problem for the transverse displacement field $u(x,y,t)$ satisfies the following equation and the boundary and initial data

$$c^2\left(\frac{\partial^2 u}{\partial x^2} + \frac{\partial^2 u}{\partial y^2}\right) = \frac{\partial^2 u}{\partial t^2}, \qquad \text{for all } (x,y) \text{ in } D, t > 0 \quad (8.5.4)$$

$$u(x,y,t) = 0 \qquad \text{on the boundary } \partial D \text{ for all } t > 0, \quad (8.5.5)$$

$$u(x,y,t) = f(x,y), \quad u_t(x,y,t) = g(x,y) \text{ at } t = 0, \text{ for all } (x,y) \text{ in } D. \quad (8.5.6\text{ab})$$

Application of the double finite Fourier sine transform defined by

$$\tilde{u}_s(m,n) = \int_0^a \int_0^b u(x,y) \sin\left(\frac{m\pi x}{a}\right) \sin\left(\frac{n\pi y}{b}\right) dx\,dy, \quad (8.5.7)$$

to the system (8.5.4)-(8.5.6ab) gives

$$\frac{d^2 \tilde{u}_s}{dt^2} + c^2\pi^2\left(\frac{m^2}{a^2} + \frac{n^2}{b^2}\right) \tilde{u}_s = 0, \qquad t > 0 \quad (8.5.8)$$

$$\tilde{u}_s(m,n,0) = \tilde{f}_s(m,n), \quad \left(\frac{du_s}{dt}\right)_{t=0} = \tilde{g}_s(m,n). \quad (8.5.9)$$

The solution of this transformed problem is

$$\tilde{u}_s(m,n,t) = \tilde{f}_s(m,n) \cos(c\pi\omega_{mn} t) + \tilde{g}_s(m,n) \sin(c\pi\omega_{mn} t), \quad (8.5.10)$$

where

$$\omega_{mn} = \left(\frac{m^2}{a^2} + \frac{n^2}{b^2}\right)^{\frac{1}{2}}. \quad (8.5.11)$$

The inverse transform gives the formal solution for $u(x,y,t)$ in the form

$$u(x,y,t) = \left(\frac{4}{ab}\right) \sum_{m=1}^{\infty} \sum_{n=1}^{\infty} \sin\left(\frac{m\pi x}{a}\right) \sin\left(\frac{n\pi y}{b}\right)$$

$$\times \left[\tilde{f}_s(m,n) \cos(c\pi\omega_{mn} t) + \tilde{g}_s(m,n) \sin(c\pi\omega_{mn} t)\right], \quad (8.5.12)$$

where

$$\tilde{f}_s(m,n) = \int_0^a \int_0^b f(\xi,\eta) \sin\left(\frac{m\pi\xi}{a}\right) \sin\left(\frac{n\pi\eta}{b}\right) d\xi d\eta, \qquad (8.5.13)$$

$$\tilde{g}_s(m,n) = \int_0^a \int_0^b g(\xi,\eta) \sin\left(\frac{m\pi\xi}{a}\right) \sin\left(\frac{n\pi\eta}{b}\right) d\xi d\eta. \qquad (8.5.14)$$

Example 8.5.2 (*Deflection of a Simply Supported Rectangular Elastic Plate*). The deflection $u(x,y)$ of the plate satisfies the *biharmonic equation*

$$\nabla^4 u \equiv \frac{\partial^4 u}{\partial x^4} + 2\frac{\partial^4 u}{\partial x^2 \partial y^2} + \frac{\partial^4 u}{\partial y^4} = \frac{w(x,y)}{\mathcal{D}}, \quad \text{in } D\{0 \le x \le a, \ 0 \le y \le b\}, (8.5.15)$$

where $w(x,y)$ represents the applied load at a point (x,y) and $\mathcal{D} = \dfrac{2Eh^3}{3(1-\sigma^2)}$ is the constant flexural rigidity of the plate.

On the edge of the simply supported plate the deflection and bending moments are zero; hence equation (8.5.15) has to be solved subject to the boundary conditions

$$\left.\begin{array}{ll} u(x,y) = 0 & \text{on } x = 0 \text{ and } x = a \\ u(x,y) = 0 & \text{on } y = 0 \text{ and } y = b \\ \dfrac{\partial^2 u}{\partial x^2} = 0 & \text{on } x = 0 \text{ and } x = a \\ \dfrac{\partial^2 u}{\partial y^2} = 0 & \text{on } y = 0 \text{ and } y = b \end{array}\right\} \qquad (8.5.16)$$

We first solve the problem due to a concentrated load W_0 at the point (ξ,η) inside D so that $w(x,y) = P\,\delta(x-\xi)\,\delta(y-\eta)$, where P is a constant.

Application of the double finite Fourier sine transform (8.5.7) to (8.5.15)-(8.5.16) gives

$$\pi^4 \left(\frac{m^2}{a^2} + \frac{n^2}{b^2}\right)^2 \tilde{u}_s(m,n) = \left(\frac{P}{\mathcal{D}}\right) \sin\left(\frac{m\pi\xi}{a}\right) \sin\left(\frac{n\pi\eta}{b}\right).$$

Or

$$\tilde{u}_s(m,n) = \left(\frac{P}{\mathcal{D}\pi^4 \omega_{mn}^4}\right) \sin\left(\frac{m\pi\xi}{a}\right) \sin\left(\frac{n\pi\eta}{b}\right), \qquad (8.5.17)$$

where ω_{mn} is defined by (8.5.1).

The inverse transform gives the formal solution

$$u(x,y) = \left(\frac{4P}{\pi^4 ab D}\right) \sum_{m=1}^{\infty} \sum_{n=1}^{\infty} \left[\frac{\sin\left(\frac{m\pi\xi}{a}\right)\sin\left(\frac{n\pi\eta}{b}\right)}{\omega_{mn}^4}\right]$$

$$\times \sin\left(\frac{m\pi x}{a}\right)\sin\left(\frac{n\pi y}{b}\right). \qquad (8.5.18)$$

For an arbitrary load $w(x,y)$ over the region $\alpha \le x \le \beta$, $\gamma \le y \le \delta$ inside the region D, we can replace P by $w(\xi,\eta)\,d\xi d\eta$ and integrate over the rectangle $\alpha \le \xi \le \beta$, $\gamma \le \eta \le \delta$.

Consequently, the formal solution is obtained from (8.5.18) and has the form

$$u(x,y) = \left(\frac{4}{\pi^4 ab}\right) \sum_{m=1}^{\infty} \sum_{n=1}^{\infty} \left\{ \int_{\alpha}^{\beta}\int_{\gamma}^{\delta} w(\xi,\eta)\sin\left(\frac{m\pi\xi}{a}\right)\sin\left(\frac{n\pi\eta}{b}\right) d\xi d\eta \right\}$$

$$\times \omega_{mn}^{-4} \sin\left(\frac{m\pi x}{a}\right)\sin\left(\frac{n\pi y}{b}\right). (8.5.19)$$

8.6 Exercises

1. Find the finite Fourier cosine transform of $f(x) = x^2$.

2. Use the result (8.3.2) to prove

    (a) $\mathcal{F}_s\{x^2\} = \frac{a^3}{n\pi}(-1)^{n+1} - 2\left(\frac{a}{n\pi}\right)^3 \left[1 + (-1)^{n+1}\right]$,

    (b) $\mathcal{F}_s\{x^3\} = (-1)^n \frac{a^4}{\pi^2}\left(\frac{6}{n^3\pi^3} - \frac{1}{n\pi}\right)$.

3. Solve the initial-boundary value problem in a finite domain

    $u_t = \kappa u_{xx}, \quad 0 \le x \le a, \; t > 0,$

    $u(x,0) = 0 \qquad$ for $0 \le x \le a,$

    $u(0,t) = f(t) \qquad$ for $t > 0,$

    $u(a,t) = 0 \qquad$ for $t > 0.$

4. Solve Exercise 3 above by replacing the only condition at $x = a$ with the radiation condition

    $u_x + hu = 0 \qquad$ at $x = a,$

    where h is a constant.

5. Solve the heat conduction problem

$$u_t = K u_{xx}, \quad 0 \le x \le a, \ t > 0,$$
$$\left. \begin{array}{l} u_x(0, t) = f(t) \\ u_x(a, t) + h u = 0 \end{array} \right\} \text{ for } t > 0,$$
$$u(x, 0) = 0 \quad \text{for } 0 \le x \le a.$$

6. Solve the diffusion equation (8.4.1) with the following boundary and initial data
$$u_x(0, t) = f(t), \quad u_x(a, t) = 0 \text{ for } t > 0$$
$$u(x, 0) = 0 \quad \text{for } 0 \le x \le a.$$

7. Solve the problem of free vibrations described in Example 8.4.4, when the beam is at rest in its equilibrium position at time $t = 0$, and an impulse I is applied at $x = \eta$, that is,
$$f(x) \equiv 0 \text{ and } g(x) = \frac{I}{\rho a} \delta(x - \eta).$$

8. Find the solution of the problem in Example 8.4.4 when
   (i) $W(x, t) = W_0 \phi(t) \delta(x - \eta), \quad 0 < \eta < \ell$;
   (ii) a concentrated applied load is moving along the beam with a constant speed U, that is, $W(x, t) = W_0 \phi(t) \delta(x - Ut) H(Ut - \ell)$ where W_0 is a constant.

9. Find the solution of the forced vibration of an elastic string of finite length ℓ which satisfies the forced wave equation
$$\frac{1}{c^2} \frac{\partial^2 y}{\partial t^2} = \frac{\partial^2 u}{\partial x^2} + F(x, t), \quad 0 \le x \le \ell, \ t > 0,$$
with the initial and boundary data
$$u(x, t) = f(x), \ u_t = g(x) \text{ at } t = 0 \quad \text{for } 0 \le x \le \ell,$$
$$u(0, t) = 0 = u(\ell, t) \quad \text{for } t > 0.$$
Derive the solution for special cases when $f(x) = 0 = g(x)$ with
(i) an arbitrary non-zero $F(x, t)$, and
(ii) $F(x, t) = \frac{P(t)}{T} \delta(x - a), \quad 0 \le a \le \ell$, where T is a constant.

10. For the finite Fourier sine transform defined over $(0, \pi)$, show that
    (a) $\mathcal{F}_s \left\{ \frac{x}{2} (\pi - x) \right\} = \frac{1}{n^3} \left[1 + (-1)^{n+1} \right]$
    (b) $\mathcal{F}_s \left\{ \frac{\sinh a(\pi - x)}{\sinh a\pi} \right\} = \frac{n}{(n^2 + a^2)}, \quad a \ne 0.$

11. For the finite Fourier cosine transform defined over $(0, \pi)$, show that
    (a) $\mathcal{F}_c \left\{ (\pi - x)^2 \right\} = \frac{2\pi}{n^2}$ for $n = 1, 2, 3, \ldots$; $\mathcal{F}_s \left\{ (\pi - x)^2 \right\} = \frac{\pi^3}{3}$ for $n = 0$.

(b) $\mathcal{F}_c\{\cosh a(\pi-x)\} = \dfrac{a \sinh(a\pi)}{(n^2+a^2)}$ for $a \neq 0$.

12. Use the finite Fourier sine transform to solve the problem of diffusion of electricity along a cable of length a. The potential $V(x,t)$ at any point x of the cable of resistance R and capacitance C per unit length satisfies the equation
$$V_t = \kappa V_{xx}, \qquad 0 \leq x \leq a, \ t > 0,$$
where $\kappa = (RC)^{-1}$ and the boundary conditions (the ends of the cable are earthed)
$$V(0,t) = 0 = V(a,t) \qquad \text{for } t > 0,$$
and the initial conditions
$$V(x,0) = \begin{cases} \left(\dfrac{2V_0}{a}\right)x, & 0 \leq x \leq \dfrac{a}{2} \\ \left(\dfrac{2V_0}{a}\right)(a-x), & \dfrac{a}{2} \leq x \leq a \end{cases},$$
where V_0 is a constant.

13. Establish the following results

(a) $\mathcal{F}_s\left[\dfrac{d}{dx}\{f_1(x) * g_1(x)\}\right] = 2n\, \tilde{f}_s(n)\, \tilde{g}_s(n)$,

(b) $\mathcal{F}_s\left[\int_0^x \{f_1(u) * g_1(u)\}\, du\right] = \dfrac{2}{n}\, \tilde{f}_s(n)\, \tilde{g}_s(n)$.

14. If p is not necessarily an integer, we write
$$\tilde{f}_c(p) = \int_0^\pi f(x) \cos px\, dx \quad \text{and} \quad \tilde{f}_s(p) = \int_0^\pi f(x) \sin px\, dx.$$
Show that, for any constant α,

(a) $\mathcal{F}_c\{2 f(x) \cos \alpha x\} = \tilde{f}_c(n-\alpha) + \tilde{f}_c(n+\alpha)$,

(b) $\mathcal{F}_c\{2 f(x) \sin \alpha x\} = \tilde{f}_s(n+\alpha) - \tilde{f}_s(n-\alpha)$,

(c) $\mathcal{F}_s\{2 f(x) \cos \alpha x\} = \tilde{f}_s(n+\alpha) + \tilde{f}_s(n-\alpha)$,

(d) $\mathcal{F}_s\{2 f(x) \sin \alpha x\} = \tilde{f}_c(n-\alpha) - \tilde{f}_c(n+\alpha)$.

15. Solve the problem in Example 8.5.2 for uniform load W_0 over the region $\alpha \leq x \leq \beta$ and $\gamma \leq y \leq \delta$.

16. Solve the problem of free oscillations of a rectangular elastic plate of density ρ bounded by the two parallel planes $z = \pm h$. The deflection $u(x,y,t)$ satisfies the equation
$$\mathcal{D}\nabla^4 u + \rho h\, u_{tt} = 0, \quad 0 \leq x \leq a, \ 0 \leq y \leq b, \ t > 0$$
where the deflection and the bending moments are all zero at the edges.

Chapter 9

Finite Laplace Transforms

9.1 Introduction

The Laplace transform method is normally used to find the response of a linear system at any time t to the initial data at $t = 0$ and the disturbance $f(t)$ acting for $t \geq 0$. If the disturbance or input function is $f(t) = \exp(at^2)$, $a > 0$, the usual Laplace transform cannot be used to find the solution of an initial value problem because the Laplace transform of $f(t)$ does not exist. From a physical point of view, there seems to be no reason at all why the function $f(t)$ cannot be used as an acceptable disturbance for a system. It is often true that the solution at times later than t would not affect the state at time t. This leads to the idea of introducing the finite Laplace transform in $0 \leq t \leq T$ in order to extend the power and usefulness of the usual Laplace transform in $0 \leq t < \infty$.

This chapter deals with the definition and basic operational properties of the finite Laplace transform. In Section 9.4, the method of the finite Laplace transform is used to solve the initial value problems and the boundary value problems. This chapter is essentially based on papers by Debnath and Thomas (1976) and Dunn (1967).

9.2 Definition of the Finite Laplace Transform and Examples

The *finite Laplace transform* of a continuous (or an almost piecewise continuous) function $f(t)$ in $(0, T)$ is denoted by $\mathscr{L}_T\{f(t)\} = \bar{f}(s, T)$, and defined by

$$\mathscr{L}_T\{f(t)\} = \bar{f}(s, T) = \int_0^T f(t)\, e^{-st}\, dt, \qquad (9.2.1)$$

where s is a real or complex number and T is a finite number that may be positive or negative so that (9.2.1) can be defined in any interval $(-T_1, T_2)$. Clearly, $\mathscr{L}_T$ is a linear integral transformation.

The *inverse transform* is defined by the complex integral

$$f(t) = \mathscr{L}_T^{-1}\{\bar{f}(s, T)\} = \frac{1}{2\pi i} \int_{c-i\infty}^{c+i\infty} \bar{f}(s, T)\, e^{st}\, ds, \qquad (9.2.2)$$

where the integral is taken over any open contour Γ joining any two points $c - iR$ and $c + iR$ in the finite complex s plane as $R \to \infty$.

If $f(t)$ is almost piecewise continuous, that is, it has at most a finite number of simple discontinuities in $0 \leq t \leq T$. Moreover, in the intervals where $f(t)$ is continuous, it satisfies a Lipschitz condition of order $\alpha > 0$. Under these conditions, it can be shown that the inversion integral (9.2.2) is equal to

$$\frac{1}{2\pi i} \int_\Gamma \bar{f}(s, T) e^{st} ds = \frac{1}{2}[f(t-0) + f(t+0)], \tag{9.2.3}$$

where Γ is an arbitrary open contour which terminates with finite constant C as $R \to \infty$. This is due to the fact that $\bar{f}(s, T)$ is an entire function of s.

It follows from (9.2.1) that if

$$\int f(t) e^{-st} dt = -F(s, t) e^{-st}, \tag{9.2.4}$$

then

$$\bar{f}(s, T) = F(s, 0) - F(s, T) e^{-sT} \tag{9.2.5}$$
$$= \tilde{f}(s) - F(s, T) e^{-sT}, \tag{9.2.6}$$

where $\tilde{f}(s)$ is the usual Laplace transform defined by (3.2.1) and hence

$$\tilde{f}(s) = F(s, 0) = \int_0^\infty e^{-st} f(t) \, dt. \tag{9.2.7}$$

Furthermore, using (9.2.2) and (9.2.6), the inversion formula can be written as

$$f(t) = \frac{1}{2\pi i} \int_\Gamma F(s, 0) e^{st} ds - \frac{1}{2\pi i} \int_\Gamma F(s, T) e^{s(t-T)} ds. \tag{9.2.8}$$

It is noted that the first integral may be closed in the left half of the complex plane. On the other hand, for $t < T$, the contour of the second integral must be closed in the right half-plane. We select Γ so that all poles of $F(s, 0)$ lie to the left of Γ. Thus, the first integral represents the solution of the initial value problem, and for $t < T$, the second integral vanishes. When $t > T$, the second integral may be closed in the left half of the complex plane so that $f(t) = 0$ for $t > T$. Thus, for the solution of the initial value problem, there is no need to consider the second integral, and this case is identical with the usual Laplace transform.

So unlike the usual Laplace transform of a function $f(t)$, there is no restriction needed on the transform variable s for the existence of the finite Laplace transform $\mathscr{L}_T\{f(t)\} = \bar{f}(s, T)$. Further, the existence of (9.2.1) does not require the exponential order property of $f(t)$. If a function $f(t)$ has the usual Laplace transform, then it also has the finite Laplace transform. In other words, if $\tilde{f}(s) = \mathscr{L}\{f(t)\}$ exists, then $\mathscr{L}_T\{f(t)\} = \bar{f}(s, T)$ exists as shown below.

$$\tilde{f}(s) = \int_0^T e^{-st} f(t) dt + \int_T^\infty e^{-st} f(t) dt. \tag{9.2.9}$$

Since $\bar{f}(s)$ exists, both the integrals on the right of (9.2.9) exist. Hence the first integral in (9.2.9) exists and defines $\bar{f}(s, T)$.

However, the converse of this result is not necessarily true. This can be shown by an example. It is well known that the usual Laplace transform of $f(t) = \exp(at^2)$, $a > 0$, does not exist. But the finite Laplace transform of this function exists as shown below.

$$\bar{f}(s, T) = \mathscr{L}_T\{\exp(at^2)\} = \int_0^T \exp(-st + at^2) dt$$

$$= \exp\left(-\frac{s^2}{4a}\right) \int_0^T \exp\left[-\left(\sqrt{a}\, t - \frac{s}{2\sqrt{a}}\right)i\right]^2 dt$$

$$= \frac{1}{2i}\left(\frac{\pi}{a}\right)^{\frac{1}{2}} \exp\left(-\frac{s^2}{4a}\right)\left[\operatorname{erf}\left(\sqrt{a}\, T - \frac{s}{2\sqrt{a}}\right)i + \operatorname{erf}\left(\frac{si}{2\sqrt{a}}\right)\right]. \quad (9.2.10)$$

In the limit as $T \to \infty$, (9.2.10) does not exist as seen below.

We use the result (see Carslaw and Jeager, 1953, p 48), to obtain

$$\operatorname{erf}(\dot{z}) = \exp(-z^2)\left[1 + \frac{2i}{\pi}\int_0^z e^{x^2} dx\right] \to \infty \quad \text{as } T \to \infty, \quad (9.2.11)$$

where

$$z = \left[T\sqrt{a} - \frac{s}{2\sqrt{a}}\right]i.$$

This ensures that the right hand side of (9.2.10) tends to infinity as $T \to \infty$. Thus the usual Laplace transform of $\exp(at^2)$ does not exist as expected.

The solution of the final value problem is denoted by f_{fi} and defined by

$$f_{fi}(t) = \frac{1}{2\pi i} \int_\Gamma F(s, T)\, e^{s(t-T)} ds, \quad (9.2.12)$$

where the contour Γ lies to the left of the singularities of $F(s, t)$ or $F(s, 0)$.

Theorem 9.2.1 The solution of an initial value problem is identical with that of the final value problem.

Proof. Suppose f_{in} is the solution of the initial value problem, and it is given by

$$f_{in}(t) = \frac{1}{2\pi i} \int_{Br} F(s, 0)\, e^{st} ds \quad (9.2.13)$$

where Br is the Bromwich contour extending from $c - iR$ to $c + iR$ as $R \to \infty$. We next reverse the direction of Γ in (9.2.12) and then subtract (9.2.13) from (9.2.12) to obtain

$$f_{in}(t) - f_{fi}(t) = \frac{1}{2\pi i} \int_C \{F(s, 0) - F(s, T)\, e^{-sT}\} e^{st} ds$$

$$= \frac{1}{2\pi i} \int_C \bar{f}(s, T) e^{st} ds, \tag{9.2.14}$$

where C is a closed contour which contains all the singularities of $F(s,0)$ or $F(s, T)$. Thus the integrand of (9.2.14) is an entire function of s and hence the integral around a contour C must vanish by Cauchy's Fundamental Theorem. Hence

$$f_{in}(t) = f_{fi}(t) = f(t). \tag{9.2.15}$$

This completes the proof.

We next calculate the finite Laplace transform of several elementary functions:

Example 9.2.1 If $f(t) = 1$, then

$$\mathscr{L}_T\{1\} = \bar{f}(s, T) = \int_0^T e^{-st} dt = \frac{1}{s}(1 - e^{-sT}). \tag{9.2.16}$$

Example 9.2.2 If $f(t) = e^{at}$, then

$$\mathscr{L}_T\{e^{at}\} = \bar{f}(s, T) = \int_0^T e^{-(s-a)t} dt = \frac{1 - e^{-(s-a)T}}{(s - a)}. \tag{9.2.17}$$

Example 9.2.3 If $f(t) = \sin at$ or $\cos at$, then

$$\mathscr{L}_T\{\sin at\} = \int_0^T \sin at \, e^{-st} dt = \frac{a}{s^2 + a^2} - \frac{e^{-sT}}{(s^2 + a^2)}(s \sin aT + a \cos aT). \tag{9.2.18}$$

$$\mathscr{L}_T\{\cos at\} = \frac{s}{s^2 + a^2} + \frac{e^{-sT}}{s^2 + a^2}(a \sin aT - s \cos aT). \tag{9.2.19}$$

Example 9.2.4 If $f(t) = t$ then

$$\mathscr{L}_T\{t\} = \int_0^T t e^{-st} dt = \frac{1}{s^2} - \frac{e^{-sT}}{s}\left(\frac{1}{s} + T\right). \tag{9.2.20}$$

Example 9.2.5 If $f(t) = t^2$, then

$$\mathscr{L}_T\{t^2\} = \int_0^T t^2 e^{-st} dt = \frac{2}{s^3} - \frac{e^{-sT}}{s}\left(T^2 + \frac{2T}{s} + \frac{2}{s^2}\right). \tag{9.2.21}$$

More generally, if $f(t) = t^n$, then

$$\mathscr{L}_T\{t^n\} = \int_0^T t^n e^{-st} dt$$

$$= \frac{n!}{s^{n+1}} - \frac{e^{-sT}}{s}\left\{T^n + \frac{n}{s}T^{n-1} + \frac{n(n-1)}{s^2}T^{n-2} + \cdots + \frac{n!\,T}{s^{n-1}} + \frac{n!}{s^n}\right\}. \tag{9.2.22}$$

Example 9.2.6 If $f(t) = t^a$, $a(> -1)$ is a real number, then

$$\mathscr{L}_T\{t^n\} = s^{-(a+1)} \gamma(a+1, sT), \tag{9.2.23}$$

where $\gamma(\alpha, x)$ is called the *incomplete gamma function* and is defined by

$$\gamma(\alpha, x) = \int_0^x e^{-u} u^{\alpha-1} du. \tag{9.2.24}$$

We have

$$\mathscr{L}_T\{t^a\} = \int_0^T t^a e^{-st} dt, \quad (u = st)$$

$$= s^{-(a+1)} \int_0^{sT} u^a e^{-u} du$$

$$= s^{-(a+1)} \gamma(a+1, sT).$$

Example 9.2.7 If $0 < a < T$, then

$$= e^{-as} \int_0^{T-a} e^{-\tau s} f(\tau) d\tau = e^{-as} \bar{f}(s, T-a). \tag{9.2.25}$$

In particular,

$$\mathscr{L}_T\{H(t-a)\} = \int_a^T e^{-st} dt = \frac{1}{s}\left(e^{-as} - e^{-sT}\right). \tag{9.2.26}$$

Example 9.2.8

$$\mathscr{L}_T\{erf\ at\} = s^{-1} \exp\left(\frac{s^2}{4a^2}\right)\left[erf\left(aT + \frac{s}{2a}\right) - erf\left(\frac{s}{2a}\right)\right]$$

$$- \frac{e^{-sT}}{s} erf(aT). \tag{9.2.27}$$

We have, by definition,

$$\mathscr{L}_T\{erf(at)\} = \int_0^T erf(at) e^{-st} dt$$

which is, by integrating by parts and using the definition of $erf(at)$,

$$= -\left[s^{-1} e^{-st} erf(at)\right]_0^T + \frac{1}{s} \frac{2}{\sqrt{\pi}} \int_0^T \exp\left[-\left(st + a^2 t^2\right)\right] dt$$

$$= -\frac{e^{-sT}}{s} erf(aT) + \frac{1}{s} \exp\left(\frac{s^2}{4a^2}\right) \frac{2}{\sqrt{\pi}} \int_{\frac{s}{2a}}^{aT+\frac{s}{2a}} e^{-u^2} du$$

$$= -\frac{e^{-sT}}{s} erf(aT) + \frac{1}{s} \exp\left(\frac{s^2}{4a^2}\right)\left[erf\left(aT + \frac{s}{2a}\right) - erf\left(\frac{s}{2a}\right)\right].$$

Example 9.2.9 If $f(t)$ is a periodic function with period ω, then

$$\bar{f}(s, T) = \mathscr{L}_T\{f(t)\} = \frac{\left(1 - e^{-sT}\right)}{1 - e^{-s\omega}} \bar{f}(s, \omega), \tag{9.2.28}$$

where $T = n\omega$, and n is a finite positive integer.

By definition, we have
$$\mathcal{L}_T\{f(t)\} = \int_0^T f(t) e^{-st} dt$$
$$= \int_0^\omega f(t) e^{-st} dt + \int_\omega^{2\omega} f(t) e^{-st} dt + \cdots + \int_{(n-1)\omega}^{n\omega} f(t) e^{-st} dt$$

which is, substituting $t = u + \omega$, $t = u + 2\omega, \ldots, t = u + (n-1)\omega$ in the second, third, and the last integral respectively,

$$= \int_0^\omega f(u) e^{-su} du + e^{-s\omega} \int_0^\omega f(u) e^{-su} du + \cdots + e^{-s(n-1)\omega} \int_0^\omega f(u) e^{-su} du$$

$$= \left[1 + e^{-s\omega} + \cdots + e^{-s\omega(n-1)}\right] \int_0^\omega e^{-su} f(u) du$$

$$= \frac{\left(1 - e^{-ns\omega}\right)}{1 - e^{-s\omega}} \bar{f}(s, \omega).$$

In the limit as $n \to \infty$ ($T \to \infty$), (9.2.28) reduces to the known result

$$\bar{f}(s) = \left(1 - e^{-s\omega}\right)^{-1} \int_0^\omega e^{-su} f(u) du. \tag{9.2.29}$$

9.3 Basic Operational Properties of the Finite Laplace Transform

Theorem 9.3.1 If $\mathcal{L}_T\{f(t)\} = \bar{f}(s, T)$, then

(a) (Shifting) $\quad \mathcal{L}_T\{e^{-at} f(t)\} = \bar{f}(s + a, T),$ (9.3.1)

(b) (Scaling) $\quad \mathcal{L}_T\{f(at)\} = \dfrac{1}{a} \bar{f}\left(\dfrac{s}{a}, aT\right).$ (9.3.2)

The proofs are easy exercises for the reader.

Theorem 9.3.2 (*Finite Laplace Transforms of Derivatives*). If $\mathcal{L}_T\{f(t)\} = \bar{f}(s, T)$, then

$$\mathcal{L}_T\{f'(t)\} = s\bar{f}(s, T) - f(0) + e^{-sT} f(T), \tag{9.3.3}$$

$$\mathcal{L}_T\{f''(t)\} = s^2 \bar{f}(s, T) - sf(0) - f'(0) + sf(T) e^{-sT} + f'(T) e^{sT}, \tag{9.3.4}$$

More generally,

$$\mathcal{L}_T\{f^{(n)}(t)\} = s^n \bar{f}(s, T) - \sum_{\kappa=1}^n s^{n-\kappa} f^{(\kappa-1)}(0) + e^{-sT} \sum_{\kappa=1}^n s^{n-\kappa} f^{\kappa-1}(T). \tag{9.3.5}$$

Proof. We have, by integrating by parts,

$$\mathcal{L}_T\{f'(t)\} = \int_0^T f'(t) e^{-st} dt = \left[f(t) e^{-st}\right]_0^T + s \int_0^T f(t) e^{-st} dt$$

$$= s\bar{f}(s, T) - f(0) + f(T) e^{-sT}.$$

Repeating this process gives (9.3.4). By induction, we can prove (9.3.5).

Theorem 9.3.3 (*Finite Laplace Transform of Integrals*). If

$$F(t) = \int_0^t f(u)\,du \tag{9.3.6}$$

so that $F'(t) = f(t)$ for all t, then

$$\mathscr{L}_T\left\{\int_0^t f(u)\,du\right\} = \frac{1}{s}\left\{\bar{f}(s, T) - e^{-sT} F(T)\right\}. \tag{9.3.7}$$

Proof. We have from (9.3.3)
$$\mathscr{L}_T\{F'(t)\} = s\,\mathscr{L}_T\{F(t)\} - F(0) + e^{-sT} F(T).$$

Or
$$\bar{f}(s, T) = s\,\mathscr{L}_T\left\{\int_0^t f(u)\,du\right\} + e^{-sT} F(T).$$

Hence,
$$\mathscr{L}_T\left\{\int_0^t f(u)\,du\right\} = \frac{1}{s}\left[\bar{f}(s, T) - e^{-sT} F(T)\right].$$

Theorem 9.3.4 If $\mathscr{L}_T\{f(t)\} = \bar{f}(s, T)$, then

$$\frac{d}{ds}\bar{f}(s, T) = \mathscr{L}_T\{(-t)f(t)\}, \tag{9.3.8}$$

$$\frac{d^2}{ds^2}\bar{f}(s, T) = \mathscr{L}_T\{(-t)^2 f(t)\}. \tag{9.3.9}$$

More generally,

$$\frac{d^n}{ds^n}\bar{f}(s, T) = \mathscr{L}_T\{(-t)^n f(t)\}. \tag{9.3.10}$$

The proofs of these results are easy exercises.

These results can be used to find the finite Laplace transform of the product t^n and any derivatives of $f(t)$ in terms of $\bar{f}(s, T)$. In other words,

$$\mathscr{L}_T\{t^n f'(t)\} = (-1)^n \frac{d^n}{ds^n}\left[\mathscr{L}_T\{f'(t)\}\right]$$

$$= (-1)^n \frac{d^n}{ds^n}\left[s\,\bar{f}(s, T) - f(0) + f(T)\,e^{-sT}\right].$$

Similarly, we obtain a more general result

$$\mathscr{L}_T\{t^n f^{(m)}(t)\} = (-1)^n \frac{d^n}{ds^n}\left[\mathscr{L}_T\{f^{(m)}(t)\}\right],$$

which is, by (9.3.5),

$$= (-1)^n \frac{d^n}{ds^n}\left[s^m \bar{f}(s, T) - \sum_{\kappa=1}^{m} s^{m-\kappa} f^{(\kappa-1)}(0)\right.$$

$$\left. + e^{-sT} \sum_{\kappa=1}^{m} s^{m-\kappa} f^{(\kappa-1)}(T)\right].$$

Finally, we can find

$$\int_s^T \bar{f}(s, T) \, ds = \int_s^T ds \int_0^T f(t) \, e^{-st} dt$$

$$= \int_0^T f(t) \, dt \int_s^T e^{-st} ds$$

$$= \int_0^T \frac{f(t)}{t} e^{-st} dt - \int_0^T \frac{f(t)}{t} e^{-Tt} dt$$

$$= \bar{g}(s, T) - \bar{g}(T, T),$$

where $g(t) = \dfrac{f(t)}{t}$ and the existence of $\bar{g}(s, T)$ is assumed.

9.4 Applications of Finite Laplace Transforms

Example 9.4.1 Use the finite Laplace transform to solve the initial value problem

$$\frac{dx}{dt} + \alpha x = At, \quad 0 \le t \le T, \tag{9.4.1}$$

$$x(t = 0) = a. \tag{9.4.2}$$

Application of the finite Laplace transform gives

$$s\bar{x}(s, T) - x(0) + e^{-sT} x(T) + \alpha \bar{x}(s, T) = A \left[\frac{1}{s^2} - \frac{1}{s} e^{-sT} \left(\frac{1}{s} + T \right) \right].$$

Or

$$\bar{x}(s, T) = \frac{a}{s + \alpha} - \frac{e^{-sT} x(T)}{s + \alpha} + A \left[\frac{1}{s^2} - \frac{1}{s} e^{-sT} \left(\frac{1}{s} + T \right) \right]. \tag{9.4.3}$$

This is not an entire function. But it becomes an entire function by setting

$$x(T) = \frac{AT}{\alpha} - \frac{A}{\alpha^2} + \frac{1}{\alpha^2} e^{-\alpha T} + a e^{-\alpha T} \tag{9.4.4}$$

so that

$$\bar{x}(s, T) = \left(a + \frac{A}{\alpha^2} \right) \left[\frac{1 - e^{-(s+\alpha)T}}{s + \alpha} \right] + \frac{1}{\alpha} \left[\frac{A}{s^2} (1 - e^{-sT}) - \frac{AT}{s} e^{-sT} \right]$$

$$- \frac{A}{\alpha^2} \left[\frac{1}{s} (1 - e^{-sT}) \right]. \tag{9.4.5}$$

Using Table B-11 of finite Laplace transforms gives the final solution

$$x(t) = \left(a + \frac{A}{\alpha^2} \right) e^{-\alpha t} + \frac{At}{\alpha} - \frac{A}{\alpha^2}. \tag{9.4.6}$$

Example 9.4.2 Solve the simple harmonic oscillator governed by

$$\frac{d^2x}{dt^2} + \omega^2 x = F, \tag{9.4.7}$$

$$x(t=0) = a, \quad \dot{x}(t=0) = u, \tag{9.4.8ab}$$

where F, a, and u are constants.

Application of the finite Laplace transform gives the solution as

$$\bar{x}(s,T) = \frac{as}{s^2 + \omega^2} + \frac{u}{s^2 + \omega^2} - \frac{se^{-sT}x(T)}{s^2 + \omega^2}$$
$$- \frac{e^{-sT}\dot{x}(T)}{s^2 + \omega^2} + \frac{F}{s(s^2 + \omega^2)}\{1 - e^{-sT}\}, \tag{9.4.9}$$

Since $\bar{x}(s,T)$ is not an entire function, we choose $x(T)$ such that

$$x(T) = \left(a - \frac{F}{\omega^2}\right)\cos \omega T + \frac{u}{\omega}\sin \omega T + \frac{F}{\omega^2} \tag{9.4.10}$$

in order to make $\bar{x}(s,T)$ an entire function. Consequently, (9.4.9) becomes

$$\bar{x}(s,T) = \left(a - \frac{F}{\omega^2}\right)\left[\frac{s}{s^2+\omega^2} + \frac{e^{-sT}}{s^2+\omega^2}\{\omega \sin \omega T - s \cos \omega T\}\right]$$
$$+ \frac{u}{\omega}\left[\frac{\omega}{s^2+\omega^2} - \frac{e^{-sT}}{s^2+\omega^2}\{s \sin \omega T + \omega \cos \omega T\}\right]$$
$$+ \frac{F}{\omega^2}\left\{\frac{1 - e^{-sT}}{s}\right\}. \tag{9.4.11}$$

Using Table B-11 of finite Laplace transforms, we invert (9.4.11) so that the solution becomes

$$x(t) = \left(a - \frac{F}{\omega^2}\right)\cos \omega t + \frac{u}{\omega}\sin \omega t + \frac{F}{\omega^2}. \tag{9.4.12}$$

Example 9.4.3 (*Boundary Value Problem*). The equation for the upward displacement of a taut string caused by a concentrated or distributed load $W(x)$ normalized with respect to the tension of the string of length L is

$$\frac{d^2y}{dx^2} = W(x), \quad 0 \le x \le L \tag{9.4.13}$$

and the associated boundary conditions are

$$y(0) = y(L) = 0. \tag{9.4.14}$$

We solve this boundary value problem due to a concentrated load of unit magnitude at a point a where

$$W(x) = \delta(x - a), \quad 0 < a < L.$$

The use of the finite Laplace transform defined by

$$\bar{y}(s,L) = \int_0^L y(x) e^{-sx} dx \tag{9.4.15}$$

leads to the solution of (9.4.13)-(9.4.14) in the form

$$\bar{y}(s,L) = \frac{1}{s^2}\left[e^{-sa} + y'(0) - e^{-sL}y'(L)\right], \tag{9.4.16}$$

where $y'(x)$ denotes the derivative of $y(x)$ with respect to x. The function $\bar{y}(s, L)$ is not an entire function of s unless the condition $y'(0) = y'(L) - 1$ is satisfied. Using this condition, solution (9.4.16) can be put in the form

$$\bar{y}(s, L) = \frac{y'(0)}{s^2}\left[1 - e^{-sL} - sLe^{-sL}\right] + \frac{e^{-sa}}{s^2}\left[1 - e^{-s(L-a)} + y'(0)\, sLe^{-s(L-a)}\right]. \quad (9.4.17)$$

In order to complete the inversion of (9.4.17), we set $Ly'(0) = a - L$ so that the inversion gives the solution

$$y(x) = x y'(0) + (x - a)\, H(x - a). \quad (9.4.18)$$

Example 9.4.4 (*Transient Current in a Simple Circuit*). The current $I(t)$ in a simple circuit (see Figure 4.4 on page 137) containing a resistance R, and an inductance L with an oscillating voltage $E(t) = E_0 \cos \omega t$ is given by

$$L\frac{dI}{dt} + RI = E_0 \cos \omega t, \quad 0 \leq t \leq T, \quad (9.4.19)$$

$$I(t) = 0 \quad \text{at} \quad t = 0. \quad (9.4.20)$$

Application of the finite Laplace transform to (9.4.19)-(9.4.20) gives

$$s\bar{I}(s, T) + e^{-sT}I(T) + \frac{R}{L}\bar{I}(s, T)$$

$$= \frac{E_0}{L}\left[\frac{s}{s^2 + \omega^2} + \frac{e^{-sT}}{s^2 + \omega^2}(\omega \sin \omega T - s \cos \omega T)\right].$$

Or

$$\bar{I}(s, T) = -\frac{e^{-sT}I(T)}{\left(s + \dfrac{R}{L}\right)} + \frac{E_0}{L\left(s + \dfrac{R}{L}\right)}$$

$$\times \left[\frac{s}{s^2 + \omega^2} + \frac{e^{-sT}}{s^2 + \omega^2}(\omega \sin \omega T - s \cos \omega T)\right]. \quad (9.4.21)$$

Since $\bar{I}(s, T)$ is not an entire function, we make it entire by setting

$$I(T) = \frac{E_0}{L}\frac{\omega}{\left(\omega^2 + \dfrac{R^2}{L^2}\right)}\left(\frac{R}{\omega L}\cos \omega T + \sin \omega T - \frac{R}{\omega L}e^{-\frac{RT}{L}}\right). \quad (9.4.22)$$

Putting this into (9.4.21) gives

$$\bar{I}(s, T) = \frac{\omega E_0}{L\left(\omega^2 + \dfrac{R^2}{L^2}\right)}\left[\frac{R}{\omega L}\left\{\frac{s}{s^2 + \omega^2} + \frac{e^{-sT}}{s^2 + \omega^2}(\omega \sin \omega T - s \cos \omega T)\right\}\right]$$

$$+ \frac{\omega E_0}{L\left(\omega^2 + \dfrac{R^2}{L^2}\right)}\left[\left\{\frac{\omega}{s^2 + \omega^2} - \frac{e^{-sT}}{s^2 + \omega^2}(s \sin \omega T + \omega \cos \omega T)\right\}\right]$$

$$-\frac{\omega E_0}{L\left(\omega^2+\dfrac{R^2}{L^2}\right)}\left[\frac{R}{\omega L\left(s+\dfrac{R}{L}\right)}\left\{1-e^{-\left(s+\frac{R}{L}\right)T}\right\}\right]. \qquad (9.4.23)$$

Using the table of the finite Laplace transform, we can invert (9.4.23) to obtain the solution

$$I(t)=\frac{\omega E_0}{L\left(\omega^2+\dfrac{R^2}{L^2}\right)}\left\{\frac{R}{\omega L}\cos\omega t+\sin\omega t-\frac{R}{\omega L}\exp\left(-\frac{Rt}{L}\right)\right\}. \qquad (9.4.24)$$

Obviously, the first two terms in the curly bracket represent the steady-state current field, and the last term represents the transient current. In the limit $t \to \infty$, the transient term decays and the steady state is attained.

Example 9.4.5 (*Moments of a Random Variable*). Find the nth order moments of a random variable X with the density function $f(x)$ in $0 \le x \le T$.

It follows from definition (9.2.1) that the finite Laplace transform $\bar{f}(s, T)$ of the density function $f(x)$ can be interpreted as the mathematical expectation of $\exp(sX)$. In other words,

$$\bar{f}(s,T)=\mathscr{L}_T\{f(x)\}=E\{\exp(sX)\}=\int_0^T e^{sx}f(x)dx, \qquad (9.4.25)$$

where s is real parameter.

Consequently,

$$\frac{d}{ds}\bar{f}(s,T)=\int_0^T x\,e^{sx}f(x)dx.$$

This result gives the definition of the expectation of X as

$$m_1=\int_0^T x\,f(x)\,dx=\left[\frac{d}{ds}\bar{f}(s,T)\right]_{s=0}. \qquad (9.4.26)$$

This implies that the mean of X is expressed in terms of the derivative of the finite Laplace transform of the density function $f(x)$.

Similarly, differentiating (9.4.25) n times with respect to s, we obtain

$$m_n=\int_0^T x^n f(x)\,dx=\left[\frac{d^n}{ds^n}\bar{f}(s,T)\right]_{s=0}. \qquad (9.4.27)$$

In view of the result, the standard deviation and the variance of X can be obtained in terms of the derivatives of the finite Laplace transform of the density function.

9.5 Tauberian Theorems

Theorem 9.5.1 If $\mathcal{L}_T\{f(t)\} = \bar{f}(s, T)$ exists, then
$$\lim_{s \to \infty} \bar{f}(s, T) = 0. \tag{9.5.1}$$
If, in addition, $\mathcal{L}_T\{f'(t)\}$ exists, then
$$\lim_{s \to \infty} \left[s\bar{f}(s, T)\right] = \lim_{t \to 0} f(t). \tag{9.5.2}$$
Theorem 9.5.2 If $\mathcal{L}_T\{f(t)\} = \bar{f}(s, T)$ exists, then
$$\lim_{s \to 0} \bar{f}(s, T) = \int_0^T f(t)\,dt. \tag{9.5.3}$$
If, in addition, $\mathcal{L}_T\{f'(t)\}$ exists, then
$$\lim_{s \to 0} s\,\bar{f}(s, T) = 0. \tag{9.5.4}$$

The proofs of these theorems are similar to those for the usual Laplace transforms discussed in Section 3.8.

9.6 Exercises

1. Find the finite Laplace transform of each of the following functions:
   (a) $\cosh at$, (b) $\sinh at$,
   (c) $\exp(-at^2)$, $a > 0$, (d) $H(t)$,
   (e) $t^n e^{-at}$, $a > 0$.

2. If $f(t)$ has a finite discontinuity at $t = a$, where $0 < a < T$, show that
$$\mathcal{L}_T\{f'(t)\} = s\,\bar{f}(s, T) + f(T)\,e^{-sT} - f(0) - e^{-sa}[f]_a,$$
where $[f]_a = f(a+0) - f(a-0)$.

   Generalize this result if $f(t)$ has a finite number of finite discontinuities at $t = a_1, a_2, \ldots, a_n$ in $[0, T]$.

3. Verify the result (9.3.7) when $f(t) = \sin at$.

4. Verify the Tauberian theorems for the function $f(t) = \exp(-t)$.

5. Solve the initial value problem
$$\frac{d^2x}{dt^2} + \omega^2 x = A\,\exp(at^2), \quad (a > 0), \quad 0 \le t \le T$$
$$x(0) = a, \quad \dot{x}(0) = u,$$
where a and u are constants.

Chapter 10
Z Transforms

10.1 Introduction

We begin this chapter with a brief introduction to the input-output characteristics of a linear dynamic system. Some special features of linear dynamic systems are briefly discussed. Analogous to the Fourier and Laplace transforms applied to the continuous linear systems, the Z transform applicable to linear time-invariant discrete-time systems is studied in this chapter. The basic operational properties including the convolution theorem, initial and final value theorems, the Z transform of partial derivatives, and the inverse Z transform are presented in some detail. Applications of the Z transform to difference equations and to the summation of infinite series are discussed with examples.

10.2 Dynamic Linear Systems and Impulse Response

In physical and engineering problems, a *system* is referred to as a physical device which can transform a *forcing* or *input function (input signal* or simply *signal) $f(t)$* into an *output function (output signal* or *response) $g(t)$* where t is an independent time variable. In other words, the output is simply the response of the input due to the action of the physical device. Both input and output signals are functions of the continuous time variable. These may include steps or impulses. However, the input or the output or both may be sequences in the sense that they can assume values defined only for discrete values of time t. One of the essential features of a system is that the output $g(t)$ is completely determined by the given input function $f(t)$, and the characteristics of the system, and in some instances by the initial data. Usually, the action of the system is mathematically represented by

$$g(t) = L\, f(t), \qquad (10.2.1)$$

where L is a transformation (or operator) that transforms the input signal $f(t)$ to the output signal $g(t)$. The system is called *linear* if its operator L is linear, that is, L satisfies the *principle of superposition*. Obvious examples of linear operators are integral transformations.

Another fundamental characteristic of linear systems is that the response to an arbitrary input can be found by analyzing the input components of standard type and adding the responses to the individual components. The very nature of the delta function, $\delta(t)$, suggests that it can be used to represent the *unit impulse function*. In Section 2.3, it was shown that $\delta(t)$ satisfies the following fundamental property

$$f(t)\,\delta(t-t_n) = f(t_n)\,\delta(t-t_n), \tag{10.2.2}$$

where t_n (n is an integer) is any particular value of t and $f(t)$ is a continuous function in any interval containing the point $t = t_n$. Result (10.2.2) is very important in the theory of sampling systems. Sampling of signals is very common in communication and digital systems. It is also used in pulse modulation systems and in all kinds of feedback systems where a digital computer is one of the common elements.

Summing (10.2.2) over all integral n gives the *sampled function* $f^*(t)$ as

$$f^*(t) = f(t) \sum_{n=-\infty}^{\infty} \delta(t-t_n) = \sum_{n=-\infty}^{\infty} f(t_n)\,\delta(t-t_n). \tag{10.2.3}$$

Thus the sampled function is approximately represented by a train of impulse functions, each having an area equal to the function at the sampling instant.

With $t_n = nT$, the series $\sum_{n=-\infty}^{\infty} \delta(t - nT)$ is called the *impulse train* as shown in Figure 10.1.

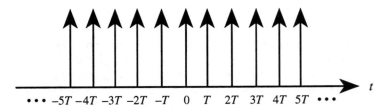

Figure 10.1 The impulse train $\sum_{n=-\infty}^{\infty} \delta(t - nT)$.

As the sampling period T assumes a small value $d\tau$, the function $f(t)$ involved in (10.2.3) can be written in the form

$$f(t) = \frac{\sum_{n=-\infty}^{\infty} f(n\,d\tau)\,\delta(t - n\,d\tau)}{\sum_{n=-\infty}^{\infty} \delta(t - n\,d\tau)}. \tag{10.2.4}$$

Multiplying the numerator and the denominator of (10.2.4) by $d\tau$ and replacing $n\,d\tau$ by $d\tau$, we obtain

$$f(t) = \frac{\int_{-\infty}^{\infty} f(\tau)\, \delta(t-\tau)\, d\tau}{\int_{-\infty}^{\infty} \delta(t-\tau)\, d\tau} = \int_{-\infty}^{\infty} f(t)\, \delta(t-\tau)\, d\tau. \qquad (10.2.5)$$

Denoting the impulse response of the system to the input $\delta(t)$ by $h(t)$, the output is mathematically represented by the Fourier convolution as

$$g(t) = \int_{-\infty}^{\infty} f(\tau)\, h(t-\tau)\, d\tau = f(t) * h(t). \qquad (10.2.6)$$

The convolution of $f(t)$ and the impulse train is $g(t)$ so that

$$g(t) = f(t) * \sum_{n=-\infty}^{\infty} \delta(t-nT)$$

$$= \sum_{n=-\infty}^{\infty} \int_{-\infty}^{\infty} f(\tau)\, \delta(t-nT-\tau)\, d\tau$$

$$= \sum_{n=-\infty}^{\infty} f(t-nT). \qquad (10.2.7)$$

This represents the superposition of all translations of $f(t)$ by nT.

If the input function $f(\tau)$ is the impulse $\delta(\tau-\tau_n)$ located at τ_n, then $h(t-\tau_n)$ is the response (output) of the system to the above impulse. This follows from (10.2.6) as

$$g(t) = \int_{-\infty}^{\infty} h(t-\tau)\, \delta(\tau-\tau_n)\, d\tau = h(t-\tau_n).$$

If the impulse is located at $\tau_0 = 0$, then $g(t) = h(t)$. This explains why the term impulse response of $h(t)$ was coined in the system's analysis, for the system's response to this particular case of an impulse located at $\tau_0 = 0$. The most important result in this section is (10.2.6), which gives the output $g(t)$ as the Fourier convolution product of the input signal $f(t)$ and the impulse response of the system $h(t)$. This shows an application of Fourier integral analysis to the analysis of linear dynamic systems.

Usually, the input is applied only for $t \geq 0$, and $h(t) = 0$ for $t < 0$. Hence the output represented by (10.2.6) reduces to the Laplace convolution as

$$g(t) = \int_0^t f(\tau)\, h(t-\tau)\, d\tau = f(t) * h(t). \qquad (10.2.8)$$

Physically, this represents the response for any input when the impulse response of any linear time invariant system is known.

We consider a certain waveform $f(t)$ shown in Figure 10.2 which is sampled periodically by a switch.

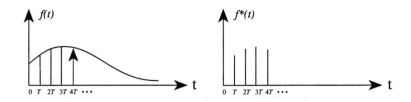

Figure 10.2 Input and sampled functions.

We have seen earlier that the delta function takes on the value of the function at the instant at which it is applied, the sampled function $f^*(t)$ can be expressed as

$$f^*(t) = f(t) \sum_{n=0}^{\infty} \delta(t - nT)$$

$$= \sum_{n=0}^{\infty} f(nT) \, \delta(t - nT). \qquad (10.2.9)$$

Result (10.2.9) can be considered as the amplitude modulation of unit impulses by the waveform $f(t)$. Evidently, this result is very useful for analyzing the systems where signals are sampled at a time interval T. Thus the above discussion enables us to introduce the Z transform in the next section.

10.3 Definition of the Z Transform and Examples

We take the Laplace transform of the sampled function given by (10.2.9) so that

$$\mathcal{L}\{f^*(t)\} = \bar{f}^*(s) = \sum_{n=0}^{\infty} f(nT) \exp(-nsT). \qquad (10.3.1)$$

It is convenient to make a change of variable $z = \exp(sT)$ so that (10.3.1) becomes

$$\mathcal{L}\{f^*(t)\} = F(z) = \sum_{n=0}^{\infty} f(nT) \, z^{-n}. \qquad (10.3.2)$$

Thus $F(z)$ is called the *Z transform* of $f(nT)$. Since the interval T between the samples has no effect on the properties and the use of the Z transform, it is convenient to set $T = 1$. We now define the *Z transform* of $f(n)$ as the function $F(z)$ of a complex variable z defined by

$$Z\{f(n)\} = F(z) = \sum_{n=0}^{\infty} f(n) \, z^{-n}. \qquad (10.3.3)$$

Thus Z is a linear transformation and can be considered as an operator mapping sequences of scalars into functions of the complex variable z. It is assumed in this chapter that there exists an R such that (10.3.3) converges for $|z| > R$. Since $|z| = |\exp(sT)| = |\exp(\sigma + i\mu)T| = |\exp(\sigma T)|$, it follows that, when $\sigma < 0$ (that is, when in the left half of the complex s plane), $|z| < 1$, and thus the left half of the s plane corresponds to the interior of the unit circle in the complex z plane. Similarly, the right half of the s plane corresponds to the exterior $(|z| > 1)$ of the unit circle in the z plane. And $\sigma = 0$ in the s plane corresponds to the unit circle in the z plane.

The inverse Z transform is given by the complex integral

$$Z^{-1}\{F(z)\} = f(n) = \frac{1}{2\pi i} \oint_C F(z) z^{n-1} \, dz, \tag{10.3.4}$$

where C is a simple closed contour enclosing the origin and lying outside the circle $|z| = R$. The existence of the inverse imposes restrictions on $f(n)$ for uniqueness. We require that $f(n) = 0$ for $n < 0$.

To obtain the inversion integral, we consider

$$F(z) = \sum_{n=0}^{\infty} f(n) z^{-n}$$
$$= f(0) + f(1) z^{-1} + f(2) z^{-2} + \cdots + f(n) z^{-n} + f(n+1) z^{-(n+1)} + \cdots.$$

Multiplying both sides by z^{n-1} and integrating along the closed contour C, which usually encloses all singularities of $F(z)$, we obtain

$$\frac{1}{2\pi i} \oint_C F(z) z^{n-1} \, dz = \frac{1}{2\pi i} \left[\oint_C f(0) z^{n-1} \, dz + \oint_C f(1) z^{n-2} \, dz \right.$$

$$\left. + \cdots + \oint_C f(n) z^{-1} \, dz + \oint_C f(n+1) z^{-2} \, dz + \cdots \right].$$

By Cauchy's Fundamental Theorem all integrals on the right vanish except

$$\frac{1}{2\pi i} \oint_C f(n) \frac{dz}{z} = f(n).$$

This leads to the inversion integral for the Z transform in the form

$$Z^{-1}\{F(z)\} = f(n) = \frac{1}{2\pi i} \oint_C F(z) z^{n-1} \, dz.$$

Similarly, we can define the so called *bilateral Z transform* by

$$Z\{f(n)\} = F(z) = \sum_{n=-\infty}^{\infty} f(n) z^{-n}. \tag{10.3.5}$$

This reduces to the unilateral Z transform (10.3.3) if $f(n) = 0$ for $n < 0$. The inverse Z transform is given by a complex integral similar to (10.3.4).

Example 10.3.1 If $f(n) = a^n$, then

$$Z\{a^n\} = \sum_{n=0}^{\infty} \left(\frac{a}{z}\right)^n = \frac{1}{1 - \frac{a}{z}} = \frac{z}{z-a}, \qquad |z| > a. \tag{10.3.6}$$

When $a = 1$, we obtain

$$Z\{1\} = \sum_{n=0}^{\infty} z^{-n} = \frac{z}{z-1}, \qquad |z| > 1. \tag{10.3.7}$$

Example 10.3.2 If $f(n) = \exp(inx)$, then

$$Z\{\exp(inx)\} = \frac{z}{z - \exp(ix)}. \tag{10.3.8}$$

This follows immediately from (10.3.6).
Furthermore,

$$Z\{\cos nx\} = \frac{z(z - \cos x)}{z^2 - 2z \cos x + 1}. \tag{10.3.9}$$

$$Z\{\sin nx\} = \frac{z \sin x}{z^2 - 2z \cos x + 1}. \tag{10.3.10}$$

These follow readily from (10.3.8) by writing $\exp(inx) = \cos nx + i \sin nx$.

Example 10.3.3 If $f(n) = n$, then

$$Z\{n\} = \sum_{n=0}^{\infty} n z^{-n} = z \sum_{n=0}^{\infty} n z^{-(n+1)}$$

$$= -z \frac{d}{dz} \left(\sum_{n=0}^{\infty} z^{-n} \right) = \frac{z}{(z-1)^2}, \qquad |z| > 1. \tag{10.3.11}$$

Example 10.3.4 If $f(n) = \frac{1}{n!}$, then

$$Z\left\{\frac{1}{n!}\right\} = \sum_{n=0}^{\infty} \frac{1}{n!} z^{-n} = \exp\left(\frac{1}{z}\right) \qquad \text{for all } z. \tag{10.3.12}$$

Example 10.3.5 If $f(n) = \cosh nx$, then

$$Z\{\cosh nx\} = \frac{z(z - \cosh x)}{z^2 - 2z \cosh x + 1}. \tag{10.3.13}$$

We have

$$Z\{\cosh nx\} = \frac{1}{2} Z\{e^{nx} + e^{-nx}\} = \frac{1}{2} \left[\frac{z}{z - e^x} + \frac{z}{z - e^{-x}}\right]$$

$$= \frac{z(z - \cosh x)}{z^2 - 2z \cosh x + 1}.$$

Example 10.3.6 Show that

$$Z\{n^2\} = \frac{z(z+1)}{(z-1)^3}. \tag{10.3.14}$$

We have, from (10.4.13) in section 10.4,

$$Z\{n \cdot n\} = -z \frac{d}{dz} Z\{n\} = -z \frac{d}{dz} \frac{z}{(z-1)^2} = \frac{z(z+1)}{(z-1)^3}.$$

Example 10.3.7 If $f(n)$ is a periodic sequence of integral period N, then
$$F(z) = Z\{f(n)\} = \frac{z^N}{z^N - 1} F_1(z), \qquad (10.3.15)$$
where
$$F_1(z) = \sum_{k=0}^{N-1} f(k) z^{-k}.$$

We have, by definition,
$$F(z) = Z\{f(n)\} = \sum_{n=0}^{\infty} f(n) z^{-n}$$
$$= z^N \sum_{n=0}^{\infty} f(n+N) z^{-(n+N)}$$
$$= z^N \sum_{k=N}^{\infty} f(k) z^{-k}, \qquad (n+N=k)$$
$$= z^N \left[\sum_{k=0}^{\infty} f(k) z^{-k} - \sum_{k=0}^{N-1} f(k) z^{-k} \right]$$
$$= z^N F(z) - z^N F_1(z)$$

Thus
$$F(z) = \frac{z^N}{(z^N - 1)} F_1(z).$$

10.4 Basic Operational Properties

Theorem 10.4.1 (*Translation*). If $Z\{f(n)\} = F(z)$ and $m \geq 0$, then
$$Z\{f(n-m)\} = z^{-m} \left[F(z) + \sum_{r=-m}^{-1} f(r) z^{-r} \right], \qquad (10.4.1)$$
$$Z\{f(n+m)\} = z^{m} \left[F(z) - \sum_{r=0}^{m-1} f(r) z^{-r} \right]. \qquad (10.4.2)$$

In particular, if $m = 1, 2, 3, \ldots$, then
$$Z\{f(n-1)\} = z^{-1} F(z). \qquad (10.4.3)$$
$$Z\{f(n-2)\} = z^{-2} F(z) + \sum_{r=-2}^{-1} f(r) z^{-r}. \qquad (10.4.4)$$

and so on.

Similarly, it follows from (10.4.2) that
$$Z\{f(n+1)\} = z\{F(z) - f(0)\}, \qquad (10.4.5)$$
$$Z\{f(n+2)\} = z^2 \{F(z) - f(0)\} - z f(1), \qquad (10.4.6)$$

$$Z\{f(n+3)\} = z^3\{F(z) - f(0)\} - z^2 f(1) - z f(2). \qquad (10.4.7)$$

More generally, for $m > 0$,

$$Z\{f(n+m)\} = z^m\{F(z) - f(0)\} - z^{m-1} f(1) - \cdots - z f(m-1) \qquad (10.4.8)$$

All these results are widely used for the solution of initial value problems involving difference equations. Result (10.4.8) is somewhat similar to (3.4.12) for this Laplace transform, and has been used to solve initial value problems involving differential equations.

Proof. We have, by definition,

$$Z\{f(n-m)\} = \sum_{m=0}^{\infty} f(n-m) z^{-n}$$

which is, by letting $n - m = r$,

$$= z^{-m} \sum_{r=-m}^{\infty} f(r) z^{-r}$$

$$= z^{-m} \sum_{r=0}^{\infty} f(r) z^{-r} + z^{-m} \sum_{r=-m}^{-1} f(r) z^{-r}.$$

If $f(r) = 0$ for all $r < 0$, then

$$Z\{f(n-m)\} = z^{-m} \sum_{r=0}^{\infty} f(r) z^{-r}. \qquad (10.4.9)$$

When $m = 1$, this result gives

$$Z\{f(n-1)\} = z^{-1} F(z). \qquad (10.4.10)$$

Similarly, we prove (10.4.2) by writing

$$Z\{f(n+m)\} = \sum_{n=0}^{\infty} f(n+m) z^{-n}$$

which is, by putting $n + m = r$,

$$= z^m \sum_{r=m}^{\infty} f(r) z^{-r}$$

$$= z^m \sum_{r=0}^{\infty} f(r) z^{-r} - z^m \sum_{r=0}^{m-1} f(r) z^{-r}$$

$$= z^m \left[F(z) - \sum_{r=0}^{m-1} f(r) z^{-m} \right].$$

When $m = 1, 2, 3, \ldots$, results (10.4.5)-(10.4.7) follow immediately.

Theorem 10.4.2 (*Multiplication*). If $Z\{f(n)\} = F(z)$, then

$$Z\{a^n f(n)\} = F\left(\frac{z}{a}\right), \qquad |z| > |a|. \qquad (10.4.11)$$

$$Z\{e^{-nb} f(n)\} = F(z e^b), \qquad |z| > |e^{-b}|. \qquad (10.4.12)$$

$$Z\{n f(n)\} = -z \frac{d}{dz} F(z). \qquad (10.4.13)$$

Proof. Result (10.4.11) follows immediately from the definition (10.3.3). And (10.4.12) follows from (10.4.11) by writing $a = e^{-b}$.

If $f(n) = 1$ so that $Z\{f(n)\} = \dfrac{z}{z-1}$, and $a = e^b$, then

$$Z\{(e^b)^n\} = \frac{ze^{-b}}{ze^{-b}-1} = \frac{z}{z-e^b}, \qquad |z| > |e^b|. \tag{10.4.14}$$

Putting $b = ix$ also gives (10.3.8)

To prove (10.4.13), we use the definition (10.3.3) to obtain

$$Z\{n\,f(n)\} = \sum_{n=0}^{\infty} n\,f(n)\,z^{-n}$$

$$= z\sum_{n=0}^{\infty} n\,f(n)\,z^{-(n+1)}$$

$$= z\sum_{n=0}^{\infty} f(n)\left\{-\frac{d}{dz}z^{-n}\right\}$$

$$= -z\frac{d}{dz}\left\{\sum_{n=0}^{\infty} f(n)\,z^{-n}\right\} = -z\frac{d}{dz}F(z).$$

Theorem 10.4.3 (*Division*). $Z\left\{\dfrac{f(n)}{n+m}\right\} = -z^m \displaystyle\int_0^z \dfrac{F(\xi)\,d\xi}{\xi^{m+1}}.$ (10.4.15)

Proof. We have

$$Z\left\{\frac{f(n)}{n+m}\right\} = \sum_{n=0}^{\infty} \frac{f(n)}{n+m}\,z^{-n}, \quad (m \geq 0),$$

$$= \sum_{n=0}^{\infty} f(n)\left[-\int_0^z \xi^{-(n+m+1)}\,d\xi\right]$$

$$= -z^m \int_0^z \xi^{-(m+1)}\left[\sum_{n=0}^{\infty} f(n)\,\xi^{-n}\right]$$

$$= -z^m \int_0^z \xi^{-(m+1)} F(\xi)\,d\xi$$

When $m = 0, 1, 2, \ldots$, several particular results follow from (10.4.15).

Theorem 10.4.4 (*Convolution*). If $Z\{f(n)\} = F(z)$ and $Z\{g(n)\} = G(z)$, then the Z transform of the convolution $f(n) * g(n)$ is given by

$$Z\{f(n) * g(n)\} = Z\{f(n)\}\,Z\{g(n)\}, \tag{10.4.16}$$

where the convolution is defined as

$$f(n) * g(n) = \sum_{m=0}^{\infty} f(n-m)\,g(m). \tag{10.4.17}$$

Or equivalently
$$Z^{-1}\{F(z)G(z)\} = \sum_{m=0}^{\infty} f(n-m)g(m). \tag{10.4.18}$$

Proof. We proceed formally to obtain
$$Z\{f(n) * g(n)\} = \sum_{n=0}^{\infty} z^{-n} \sum_{m=0}^{\infty} f(n-m)g(m)$$
which is, interchanging the order of summation,
$$= \sum_{m=0}^{\infty} g(m) \sum_{n=0}^{\infty} f(n-m) z^{-n}$$
Substituting $n - m = r$, we obtain
$$Z\{f(n) * g(n)\} = \sum_{m=0}^{\infty} g(m) z^{-m} \sum_{r=-m}^{\infty} f(r) z^{-r}$$
which is, in view of $f(r) = 0$ for $r < 0$,
$$= \sum_{m=0}^{\infty} g(m) z^{-m} \sum_{r=0}^{\infty} f(r) z^{-r}$$
$$= Z\{f(n)\} Z\{g(n)\}.$$
This proves the theorem.

More generally, the convolution $f(n) * g(n)$ is defined by
$$f(n) * g(n) = \sum_{m=-\infty}^{\infty} f(n-m)g(m). \tag{10.4.19}$$
If we assume $f(n) = 0 = g(n)$ for $n < 0$, then (10.4.19) becomes (10.4.17).

However, the Z transform of (10.4.19) gives
$$Z\{f(n) * g(n)\} = \sum_{n=-\infty}^{\infty} z^{-n} \sum_{m=-\infty}^{\infty} f(n-m) g(m)$$
which is, interchanging the order of summation,
$$= \sum_{m=-\infty}^{\infty} g(m) \sum_{n=-\infty}^{\infty} f(n-m) z^{-n}$$
$$= \sum_{m=-\infty}^{\infty} z^{-m} g(m) \sum_{n=-\infty}^{\infty} f(n-m) z^{-(n-m)}$$
$$= \sum_{m=-\infty}^{\infty} z^{-m} g(m) \sum_{r=-\infty}^{\infty} f(r) z^{-r}, \; (r = n - m)$$
$$= f(n) g(n). \tag{10.4.20}$$
This is the convolution theorem for the bilateral Z transform.

Theorem 10.4.5 (*Initial Value Theorem*). If $Z\{f(n)\} = F(z)$, then
$$f(0) = \lim_{z \to \infty} F(z). \tag{10.4.21}$$

Also, if $f(0) = 0$, then
$$f(1) = \lim_{z \to \infty} z\, F(z). \tag{10.4.22}$$

Proof. We have, by definition,
$$F(z) = \sum_{n=0}^{\infty} f(n)\, z^{-n} = f(0) + \frac{f(1)}{z} + \frac{f(2)}{z^2} + \cdots. \tag{10.4.23}$$

The initial value of $f(n)$ at $n = 0$ is obtained from (10.4.23) by letting $z \to \infty$, and hence
$$f(0) = \lim_{z \to \infty} F(z).$$

If $f(0) = 0$, then (10.4.23) gives
$$f(1) = \lim_{z \to \infty} z\, F(z).$$

This proves the theorem.

Theorem 10.4.6 (*Final Value Theorem*). If $Z\{f(n)\} = F(z)$, then
$$\lim_{n \to \infty} f(n) = \lim_{z \to 1} \{(z-1)\, F(z)\}. \tag{10.4.24}$$
provided the limits exist.

Proof. We have, from (10.3.3) and (10.4.5),
$$Z\{f(n+1) - f(n)\} = z\{F(z) - f(0)\} - F(z).$$

Or equivalently,
$$\sum_{n=0}^{\infty} [f(n+1) - f(n)]\, z^{-n} = (z-1)\, F(z) - z\, f(0).$$

In the limit as $z \to 1$, we obtain
$$\lim_{z \to 1} \sum_{n=0}^{\infty} [f(n+1) - f(n)]\, z^{-n} = \lim_{z \to 1} (z-1)\, F(z) - f(0)$$

Or
$$f(\infty) - f(0) = \lim_{z \to 1} (z-1)\, F(z) - f(0)$$

Thus
$$\lim_{n \to \infty} f(n) = \lim_{z \to 1} (z-1)\, F(z),$$

provided the limits exist.

This proves the theorem. The reader is referred to Zadeh and Desoer (1963) for a rigorous proof.

Example 10.4.1 Verify the initial value theorem for the function
$$F(z) = \frac{z}{(z-a)(z-b)}.$$

We have
$$f(0) = \lim_{z \to \infty} \frac{z}{(z-a)(z-b)} = 0.$$
$$f(1) = \lim_{z \to \infty} z\, F(z) = 1.$$

Theorem 10.4.7 (*The Z Transform of Partial Derivatives*).
$$Z\left\{\frac{\partial}{\partial a}f(n,a)\right\} = \frac{\partial}{\partial a}[Z\{f(n,a)\}] \qquad (10.4.25)$$

Proof.
$$Z\left\{\frac{\partial}{\partial a}f(n,a)\right\} = \sum_{n=0}^{\infty}\left[\frac{\partial}{\partial a}f(n,a)\right]z^{-n}$$
$$= \frac{\partial}{\partial a}\left[\sum_{n=0}^{\infty}f(n,a)z^{-n}\right]$$
$$= \frac{\partial}{\partial a}[Z\{f(n,a)\}].$$

As an example of this result, we show
$$Z\{ne^{an}\} = Z\left\{\frac{\partial}{\partial a}e^{na}\right\} = \frac{\partial}{\partial a}Z\{e^{na}\}$$
$$= \frac{\partial}{\partial a}\left(\frac{z}{z-e^a}\right) = \frac{ze^a}{(z-e^a)^2}.$$

10.5 The Inverse Z Transform and Examples

The inverse Z transform is given by the complex integral (10.3.4) which can be evaluated by using the Cauchy residue theorem. However, we discuss other simple way of finding the inverse transform of a given $F(z)$. This includes a method from the definition (10.3.3) which leads to the expansion of $F(z)$ as a series of inverse powers of z in the form
$$F(z) = f(0) + f(1)z^{-1} + f(2)z^{-2} + \cdots + f(n)z^{-n} + \cdots. \qquad (10.5.1)$$
The coefficient of z^{-n} in this expansion is
$$f(n) = Z^{-1}\{F(z)\}. \qquad (10.5.2)$$

Example 10.5.1 Find the inverse Z transform of $F(z) = z(z-a)^{-1}$.
We have
$$F(z) = \frac{z}{z-a} = \left(1 - \frac{a}{z}\right)^{-1}$$
$$= 1 + az^{-1} + a^2 z^{-2} + \cdots + a^n z^{-n} + \cdots.$$
so that
$$f(0) = 1, \ f(1) = a, \ f(2) = a^2, \ \ldots, \ f(n) = a^n, \ \ldots.$$
Obviously
$$f(n) = Z^{-1}\left\{\frac{z}{(z-a)}\right\} = a^n.$$

Example 10.5.2 Find $Z^{-1}\left\{\exp\left(\dfrac{1}{z}\right)\right\}$.

Obviously

$$\exp\left(\dfrac{1}{z}\right) = 1 + 1\cdot z^{-1} + \dfrac{1}{2!}z^{-2} + \cdots + \dfrac{1}{n!}z^{-n} + \cdots.$$

This gives

$$f(n) = \dfrac{1}{n!} = Z^{-1}\left\{\exp\left(\dfrac{1}{z}\right)\right\}.$$

Other methods for inversion use partial fractions and the Convolution Theorem 10.4.4. We illustrate these methods by the following examples.

Example 10.5.3 Find the inverse Z transform of

$$F(z) = \dfrac{z}{z^2 - 6z + 8}.$$

We write

$$F(z) = \dfrac{z}{(z-2)(z-4)} = \dfrac{1}{2}\left(\dfrac{z}{z-4} - \dfrac{z}{z-2}\right).$$

It follows from the table of Z transforms that

$$f(n) = Z^{-1}\{F(z)\} = \dfrac{1}{2}\left[Z^{-1}\left\{\dfrac{z}{z-4}\right\} - Z^{-1}\left\{\dfrac{z}{z-2}\right\}\right]$$

$$= \dfrac{1}{2}(4^n - 2^n).$$

Example 10.5.4 Use the Convolution Theorem 10.4.4 to find the inverse of $\dfrac{z^2}{(z-a)(z-b)}$.

We set

$$F(z) = \dfrac{z}{z-a},\quad G(z) = \dfrac{z}{z-b}$$

so that

$$f(n) = Z^{-1}\{F(z)\} = a^n,\quad g(n) = Z^{-1}\{G(z)\} = b^n.$$

Thus the Convolution Theorem gives

$$Z^{-1}\{F(z)\,G(z)\} = \sum_{m=0}^{n} a^{n-m}\,b^m = a^n \sum_{m=0}^{n}\left(\dfrac{b}{a}\right)^m$$

$$= a^n \cdot \left\{\dfrac{1 - \left(\dfrac{b}{a}\right)^{n+1}}{1 - \dfrac{b}{a}}\right\}$$

$$= \dfrac{a^{n+1}}{(a-b)}\left\{1 - \left(\dfrac{b}{a}\right)^{n+1}\right\}.$$

Example 10.5.5 Find the inverse Z transform of
$$F(z) = \frac{3z^2 - z}{(z-1)(z-2)^2}.$$
We write $F(z)$ as partial fractions
$$F(z) = \frac{3z^2 - z}{(z-1)(z-2)^2} = 2 \cdot \frac{z}{(z-1)} - 2 \cdot \frac{z}{(z-2)} + \frac{5}{2} \cdot \frac{2z}{(z-2)^2}$$
so that the inverse is
$$f(n) = Z^{-1}\left\{\frac{2z}{(z-1)}\right\} - Z^{-1}\left\{2 \cdot \frac{z}{(z-2)}\right\} + \frac{5}{2} Z^{-1}\left\{\frac{2z}{(z-2)^2}\right\}$$
which is, by (10.3.6) and (10.4.11),
$$= 2 - 2^{n+1} + \frac{5}{2} \cdot n\, 2^n = 2 - 2^{n+1} + 5 \cdot n\, 2^{n-1}.$$

Example 10.5.6 Use the Convolution Theorem to show that
$$Z^{-1}\left\{\frac{z(z+1)}{(z-1)^3}\right\} = n^2.$$
We write
$$\frac{z(z+1)}{(z-1)^3} = \frac{z}{(z-1)^2}\left(\frac{z+1}{z-1}\right) = \frac{z}{(z-1)^2}\left[\frac{z}{z-1} + \frac{1}{z-1}\right].$$
Letting
$$F(z) = \frac{z}{(z-1)^2} \quad \text{and} \quad G(z) = \frac{z}{z-1} + \frac{1}{z-1},$$
we obtain
$$f(n) = n \quad \text{and} \quad g(n) = H(n) + H(n-1)$$
Thus
$$Z^{-1}\left\{\frac{z(z+1)}{(z-1)^3}\right\} = f(n) * g(n) = \sum_{m=0}^{n} m[H(n-m) + H(n-m-1)] = n^2.$$

10.6 Applications of Z Transforms to Finite Difference Equations

Example 10.6.1 (*First Order Difference Equation*). Solve the initial value problem for the difference equation
$$f(n+1) - f(n) = 1, \qquad f(0) = 0. \tag{10.6.1}$$
Application of the Z transform to (10.6.1) combined with (10.4.5) gives
$$z[F(z) - f(0)] - F(z) = \frac{z}{z-1}$$
Or
$$F(z) = \frac{z}{(z-1)^2}.$$

Integral Transforms and Their Applications 309

The inverse Z transform (see result (10.3.11)) gives the solution
$$f(n) = Z^{-1}\left\{\frac{z}{(z-1)^2}\right\} = n. \tag{10.6.2}$$

Example 10.6.2 (*First Order Difference Equation*). Solve the equation
$$f(n+1) + 2f(n) = n, \quad f(0) = 1. \tag{10.6.3}$$
The use of the Z transform to this problem gives
$$z\{F(z) - f(0)\} + 2F(z) = \frac{z}{(z-1)^2}$$
Or
$$F(z) = \frac{z}{z+2} + \frac{z}{(z+2)(z-1)^2}$$
$$= \frac{z}{z+2} + \frac{1}{9} \cdot \frac{z}{z+2} + \frac{3}{9} \cdot \frac{z}{(z-1)^2} - \frac{1}{9} \cdot \frac{z}{z-1}$$
$$= \left(\frac{10}{9}\right)\frac{z}{(z+2)} + \frac{3}{9} \cdot \frac{z}{(z-1)^2} - \frac{1}{9}\frac{z}{(z-1)}.$$
The inverse Z transform yields the solution
$$f(n) = \frac{1}{9}\left[10(-2)^n + 3n - 1\right]. \tag{10.6.4}$$

Example 10.6.3 (*The Fibonacci Sequence*). The Fibonacci sequence is defined as a sequence in which every term is the sum of the two proceeding terms. So it satisfies the difference equation
$$u_{n+1} = u_n + u_{n-1}, \quad u_1 = u(0) = 1. \tag{10.6.5}$$
Application of the Z transform gives
$$U(z) = \frac{z^2}{z^2 - z - 1}, \quad \text{where } U(z) = Z\{u_n\}.$$
Thus the inverse transform leads to the solution
$$u_n = Z^{-1}\left\{\frac{z^2}{z^2 - z - 1}\right\} = Z^{-1}\left\{\frac{z^2}{(z-a)(z-b)}\right\},$$
where $a = \frac{1}{2}(1+\sqrt{5})$ and $b = \frac{1}{2}(1-\sqrt{5})$.

Using Example 10.5.4, the Fibonacci sequence is
$$u_n = \frac{a^{n+1} - b^{n+1}}{(a-b)}, \quad n = 0, 1, 2, \ldots \tag{10.6.6}$$
More explicitly, the Fibonacci sequence is given by 1, 1, 2, 3, 5,... .

Example 10.6.4 (*Second Order Difference Equation*). Solve the initial value problem
$$f(n+2) - 3f(n+1) + 2f(n) = 0, \quad f(0) = 1, \quad f(1) = 2. \tag{10.6.7}$$
Application of the Z transform gives
$$z^2\{F(z) - f(0)\} - z f(1) - 3\left[z\{F(z) - f(0)\}\right] + 2 F(z) = 0$$

Or
$$(z^2 - 3z + 2) F(z) = (z^2 - z).$$
Hence
$$F(z) = \frac{z}{(z-2)}.$$
Thus the inversion gives the solution
$$f(n) = Z^{-1}\left\{\frac{z}{(z-2)}\right\} = 2^n. \tag{10.6.8}$$

Example 10.6.5 (*Periodic Solution*). Find the solution of the initial value problem
$$u(n+2) - u(n+1) + u(n) = 0, \tag{10.6.9}$$
$$u(0) = 1 \text{ and } u(1) = 2. \tag{10.6.10}$$
The Z transform of (10.6.9)-(10.6.10) gives
$$\{z^2 U(z) - z^2 - 2z\} - \{zU(z) - z\} + U(z) = 0.$$
Or
$$U(z) = \frac{z^2 + z}{(z^2 - z + 1)} = \frac{\left(z^2 - \frac{1}{2}z\right)}{z^2 - z + 1} + \frac{\sqrt{3}\left(\frac{\sqrt{3}}{2}z\right)}{z^2 - z + 1}. \tag{10.6.11}$$

Writing $x = \frac{\pi}{3}$ in (10.3.9) and (10.3.10), the inverse Z transform of (10.6.11) gives the periodic solution
$$u(n) = \cos\left(\frac{n\pi}{3}\right) + \sqrt{3}\sin\left(\frac{n\pi}{3}\right). \tag{10.6.12}$$

Example 10.6.6 (*Second Order Nonhomogeneous Difference Equation*). Solve the initial value problem
$$u(n+2) - 5u(n+1) + 6u(n) = 2^n, \ u(0) = 1, \ u(1) = 0. \tag{10.6.13}$$

The Z transform of (10.6.13) yields
$$(z^2 - 5z + 6) U(z) = z^2 - 5z + \frac{z}{z-2}. \tag{10.6.14}$$
Or
$$U(z) = z\left[\frac{z-5}{(z-2)(z-3)} + \frac{1}{(z-2)^2(z-3)}\right]$$
$$= z\left[\left(\frac{3}{z-2} - \frac{2}{z-3}\right) + \left(\frac{1}{z-3} - \frac{1}{z-2} - \frac{1}{(z-2)^2}\right)\right]$$
$$= z\left[\frac{2}{z-2} - \frac{1}{z-3} - \frac{1}{(z-2)^2}\right]. \tag{10.6.15}$$

The inverse Z transform of (10.6.15) gives the solution
$$u(n) = 2^{n+1} - 3^n - n\,2^{n-1}. \tag{10.6.16}$$

Example 10.6.7 (*Chebyshev Polynomials*). Solve the second order difference equation
$$u_{n+2} - 2x\,u_{n+1} + u_n = 0, \quad |x| \le 1, \tag{10.6.17}$$
$$u(0) = u_0 \quad \text{and} \quad u(1) = u_1, \tag{10.6.18}$$
where u_0 and u_1 are constants.

The Z transform of equation (10.6.17) with (10.6.18) gives
$$z^2 U(z) - z^2 u_0 - z u_1 - 2x\left[z U(z) - z u_0\right] + U(z) = 0.$$

Or
$$U(z) = u_0 \left[\frac{z^2 - zx}{z^2 - 2xz + 1}\right] + (u_1 - 2x\,u_0)\left[\frac{z}{z^2 - 2xz + 1}\right]. \tag{10.6.19}$$

$$= u_0 \left[\frac{z^2 - zx}{z^2 - 2xz + 1}\right] + \frac{(u_1 - 2x\,u_0)}{\sqrt{1 - x^2}}\left[\frac{z\sqrt{1-x^2}}{z^2 - 2xz + 1}\right]$$

$$= u_0\left[\frac{z^2 - zx}{z^2 - 2xz + 1}\right] + v_0\left[\frac{z\sqrt{1-x^2}}{z^2 - 2xz + 1}\right], \tag{10.6.20}$$

where $v_0 = (u_1 - 2x\,u_0)(1 - x^2)^{-\frac{1}{2}}$ is independent of z.

Since $|x| \le 1$, we may write $x = \cos t$ and then take the inverse Z transform with the aid of (10.3.9)–(10.3.10) to obtain the solution
$$u_n = u_0 \cos nt + v_0 \sin nt \tag{10.6.21}$$
$$= u_0 \cos(n\cos^{-1} x) + v_0 \sin(n\cos^{-1} x). \tag{10.6.22}$$

Usually, the function
$$T_n(x) = \cos(n\cos^{-1} x), \tag{10.6.23}$$
is called a *Chebyshev polynomial of the first kind of degree n.*

The properties of this polynomial are presented in Appendix A-4. This polynomial plays an important role in the theory of special functions, and is found to be extremely useful in approximation theory and modern numerical analysis.

10.7 Summation of Infinite Series

Theorem 10.7.1 If $Z\{f(n)\} = F(z)$, then

(i)
$$\sum_{k=1}^{n} f(k) = Z^{-1}\left\{\frac{z}{z-1} F(z)\right\}, \tag{10.7.1}$$

and

(ii)
$$\sum_{k=1}^{\infty} f(k) = \lim_{z \to 1} F(z) = F(1). \tag{10.7.2}$$

Proof. We write
$$g(n) = \sum_{k=0}^{n} f(k)$$
so that
$$g(n) = f(n) + g(n-1)$$
Application of the Z transform gives
$$G(z) = F(z) + z^{-1} G(z)$$
so that
$$G(z) = \frac{z}{(z-1)} F(z).$$
Or
$$Z\{g(n)\} = Z\left\{\sum_{k=0}^{n} f(k)\right\} = \frac{z}{(z-1)} F(z).$$
In the limit as $z \to 1$ together with the Final Value Theorem 10.4.6 gives
$$\lim_{n \to \infty} \sum_{k=0}^{n} f(k) = \lim_{z \to 1} (z-1) \cdot \frac{z}{z-1} F(z) = F(1).$$
This proves the theorem.

Example 10.7.1 Use the Z transform to show that
$$\sum_{n=0}^{\infty} \frac{x^n}{n!} = e^x. \tag{10.7.3}$$

We have, from (10.4.11),
$$Z\{x^n f(n)\} = F\left(\frac{z}{x}\right).$$
Setting $f(n) = \frac{1}{n!}$ so that $F(z) = \exp\left(\frac{1}{z}\right)$, we find
$$Z\left\{\frac{x^n}{n!}\right\} = \exp\left(\frac{x}{z}\right).$$
The use of Theorem 10.7.1(ii) gives
$$\sum_{n=0}^{\infty} \frac{x^n}{n!} = \lim_{z \to 1} \exp\left(\frac{x}{z}\right) = e^x.$$

Example 10.7.2 Show that
$$\sum_{n=0}^{\infty} (-1)^n \frac{x^{n+1}}{n+1} = \log(1+x). \tag{10.7.4}$$

Using (10.3.6), we find
$$Z\{x^{n+1}\} = \frac{zx}{z-x}$$
whence, in view of exercise 4(b) in 10.8,

$$Z\left\{\frac{x^{n+1}}{n+1}\right\} = z\int_z^\infty \frac{zx}{(z-x)} \cdot \frac{dz}{z^2}$$

$$= xz\int_z^\infty \frac{dz}{z(z-x)} = xz\left[\frac{1}{x}\log\left(\frac{z-x}{z}\right)\right]_z^\infty$$

$$= -z\log\left(\frac{z-x}{z}\right).$$

Replacing x by $(-x)$ in this result, we obtain

$$Z\left\{(-1)^n \frac{x^{n+1}}{n+1}\right\} = z\log\left(\frac{z+x}{z}\right).$$

Application of Theorem 10.7.1(ii) gives

$$\sum_{n=0}^\infty (-1)^n \cdot \frac{x^{n+1}}{n+1} = \lim_{z\to 1} z\log\left(\frac{z+x}{z}\right) = \log(1+x).$$

Example 10.7.3 Find the sum of the series

$$\sum_{n=0}^\infty a^n \sin nx.$$

We know from (10.3.10) and (10.4.11) that

$$Z\{f(n)\} = Z\{\sin nx\} = \frac{z\sin x}{z^2 - 2z\cos x + 1},$$

$$Z\{a^n \sin nx\} = F\left(\frac{z}{a}\right) = \frac{az\sin x}{a^2 - 2az\cos x + z^2}.$$

Hence Theorem 10.7.1(ii) gives

$$\sum_{n=0}^\infty a^n \sin nx = \lim_{z\to 1} F\left(\frac{z}{a}\right) = \frac{a\sin x}{a^2 - 2a\cos x + 1}. \quad (10.7.5)$$

10.8 Exercises

1. Find the Z transform of the following functions:

   (a) n^3,   (b) $\frac{a^n}{n!}$,   (c) $n\exp\{(n-1)\alpha\}$,

   (d) $H(n) - H(n-2)$,   (e) $n^2 a^n$.

2. Show that

   (a) $Z\{\sinh na\} = \dfrac{z(\sinh a)}{z^2 - 2z\cosh a + 1}$,

   (b) $Z\{\exp(-an)\cos bn\} = \dfrac{z(z - z e^{-a}\cos b)}{z^2 - 2z e^{-a}\cos b + e^{-2a}}.$

3. Show that
$$Z\{n a^n f(n)\} = -z \frac{d}{dz}\left\{F\left(\frac{z}{a}\right)\right\}.$$

4. Prove that

(a) $Z\left\{\dfrac{f(n)}{n}\right\} = \int_z^\infty \dfrac{F(z)}{z} dz.$

(b) $Z\left\{\dfrac{f(n)}{n+m}\right\} = z^m \int_z^\infty \dfrac{F(z)\, dz}{z^{m+1}}.$

Hence deduce that
$$Z\left\{\frac{1}{n+1}\right\} = z \log\left(\frac{z}{z-1}\right).$$

5. Show that

(a) $Z\{n a^{n-1}\} = \dfrac{z}{(z-a)^2},$ (c) $|z| > |a|,$

(b) $Z\left\{\dfrac{n(n-1)\cdots(n-m+1)}{m!} a^{n-m}\right\} = \dfrac{z}{(z-a)^{m+1}},$ (d) $|z| > |a|.$

6. Find the inverse Z transform of the following functions:

(a) $\dfrac{z^2}{(z-2)(z-3)},$ (b) $\dfrac{z^2-1}{z^2+1},$ (c) $\dfrac{z}{(z-1)^2},$

(d) $\dfrac{z}{(z-a)^2},$ (e) $\dfrac{1}{(z-a)^2},$ (f) $\dfrac{1}{(z-1)^2(z-2)},$

(g) $\dfrac{z+3}{(z+1)(z+2)},$ (h) $\dfrac{z^3}{(z^2-1)(z-2)},$ (i) $\dfrac{z^2}{(z-1)\left(z-\dfrac{1}{2}\right)}.$

7. Solve the following difference equations:

(a) $f(n+1) + 3 f(n) = n, \quad f(0) = 1.$

(b) $f(n+1) - 5 f(n) = \sin n, \quad f(0) = 0.$

(c) $f(n+1) - a f(n) = a^n, \quad f(0) = x_0.$

(d) $f(n+1) - f(n) = a[1 - f(n)], \quad f(0) = x_0.$

(e) $f(n+2) - f(n+1) - 6 f(n) = 0, \quad f(0) = 0, \quad f(1) = 3.$

(f) $f(n+2) + 4 f(n+1) + 3 f(n) = 0, \quad f(0) = 1, \quad f(1) = 1.$

(g) $f(n+2) - f(n+1) - 6 f(n) = \sin\left(\dfrac{n\pi}{2}\right) \quad (n \geq 2),$

$f(0) = 0, \quad f(1) = 3.$

(h) $f(n+2) - 2 f(n+1) + f(n) = 0, \quad f(0) = 2, \quad f(1) = 0.$

(i) $f(n+2) - 2a f(n+1) + a^2 f(n) = 0, \quad f(0) = 0, \quad f(1) = a.$

(j) $f(n+3) - f(n+2) - f(n+1) + f(n) = 0, \; f(0) = 1, \; f(1) = f(2) = 0.$

8. Show that the solution of the resistive ladder network governed by the difference equation for the current field $i(n)$

$$i(n+2) - 3i(n+1) + i(n) = 0, \quad i(0) = 1, \quad i(1) = 2i(0) - \frac{V}{R}.$$

is

$$i(n) = \cosh(xn) + \frac{2}{\sqrt{5}}\left(\frac{1}{2} - \frac{V}{R}\right)\sinh(nx),$$

where $\cosh x = \frac{3}{2}$ and $\sinh x = \frac{\sqrt{5}}{2}$.

9. Use the Initial Value Theorem to find $f(0)$ for $F(z)$ given by

(a) $\dfrac{z}{z-\alpha}$,

(b) $\dfrac{z}{(z-\alpha)(z-\beta)}$,

(c) $\dfrac{z(z-\cos x)}{z^2 - 2z\cos x + 1}$,

(d) $\dfrac{1}{(z-a)^m}$.

10. Use the Final Value Theorem to find $\lim\limits_{n\to\infty} f(n)$ for the following $F(z)$:

(a) $\dfrac{z}{z-a}$,

(b) $\dfrac{z^2 - z\cos a}{(z^2 - 2z\cos a + 1)}$.

11. Find the sum of the following series using the Z transform:

(a) $\sum\limits_{n=0}^{\infty} a^n e^{inx}$,

(b) $\sum\limits_{n=0}^{\infty} (-1)^n \dfrac{e^{-n}}{n+1}$,

(c) $\sum\limits_{n=0}^{\infty} \exp[-x(2n+1)]$.

12. Solve the second order difference equation
$$3f(n+2) - 2f(n+1) - f(n) = 0, \quad f(0) = 1, \quad f(1) = 2$$
and show that $f(n) \to \dfrac{7}{4}$ as $n \to \infty$.

13. Solve the simultaneous difference equations
$$u(n+1) = 2v(n) + 2$$
$$v(n+1) = 2u(n) - 1$$
with the initial data $u(0) = v(0) = 0$.

14. Show that the solution of the third order difference equation
$$u(n+3) - 3u(n+2) + 3u(n+1) - u(n) = 0,$$
$$u(0) = 1, \quad u(1) = 0, \quad u(2) = 1,$$
is
$$u(n) = (n-1)^2.$$

15. Show that the solution of the initial value problem
$$u(n+2) - 4u(n+1) + 3u(n) = 0, \quad u(0) = u_0 \text{ and } u(1) = u_1$$
is
$$u_n = \frac{1}{2}(3u_0 - u_1) + \frac{1}{2}(u_1 - u_0)3^n,$$

16. Find the solution of the following initial value problems:
    (a) $u_{n+2} + 2u_{n+1} - 3u_n = 0, \quad u_0 = 1, \quad u_1 = 0,$
    (b) $3u_{n+2} - 5u_{n+1} + 2u_n = 0, \quad u_0 = 1, \quad u_1 = 0,$
    (c) $u_{n+2} - 4u_{n+1} + 5u_n = 0, \quad u_0 = \frac{1}{2}, \quad u_1 = 3.$

Chapter 11

Finite Hankel Transforms

11.1 Introduction

This chapter is devoted to the study of the finite Hankel transform and its basic operational properties. The usefulness of this transform is shown by solving several initial-boundary problems of physical interest. The method of finite Hankel transforms was first introduced by Sneddon (1946).

11.2 Definition of the Finite Hankel Transform and Examples

Just as problems on finite invervals $-a < x < a$ lead to Fourier series, problems on finite intervals $0 < r < a$, where r is the cylindrical polar coordinate, lead to the Fourier-Bessel series representation of a function $f(r)$ which can be stated in the following theorem:

Theorem 11.2.1 If $f(r)$ is defined in $0 \leq r \leq a$ and

$$\tilde{f}_n(k_i) = \int_0^a r\, f(r)\, J_n(rk_i)\, dr, \tag{11.2.1}$$

then $f(r)$ can be represented by the Fourier-Bessel series as

$$f(r) = \frac{2}{a^2} \sum_{i=1}^{\infty} \tilde{f}_n(k_i) \frac{J_n(rk_i)}{J_{n+1}^2(ak_i)}, \tag{11.2.2}$$

where $k_i\,(0 < k_1 < k_2 < \cdots)$ are the roots of the equation $J_n(ak_i) = 0$, that means

$$J_n'(ak_i) = J_{n-1}(ak_i) = -J_{n+1}(ak_i), \tag{11.2.3}$$

due to the standard recurrence relations among $J_n'(x)$, $J_{n-1}(x)$, and $J_{n+1}(x)$.

Proof. We write formally the Bessel series expansion of $f(r)$ as

$$f(r) = \sum_{i=1}^{\infty} c_i\, J_n(rk_i), \tag{11.2.4}$$

where the summation is taken over all the positive zeros $k_1, k_2, \ldots$ of the Bessel function $J_n(ak_i)$. Multiplying (11.2.4) by $rJ_n(rk_i)$, integrating the both sides of the result from 0 to a, and then using the orthogonal property of the Bessel functions, we obtain

$$\int_0^a r\, f(r) J_n(rk_i)\, dr = c_i \int_0^a r J_n^2(rk_i)\, dr.$$

Or
$$\tilde{f}_n(k_i) = \frac{2c_i}{a^2} J_{n+1}^2(ak_i),$$
whence we obtain
$$c_i = \frac{a^2}{2} \frac{\tilde{f}_n(k_i)}{J_{n+1}^2(ak_i)}. \tag{11.2.5}$$
Substituting the value of c_i into (11.2.4) gives (11.2.2).

Definition 11.2.1 The *finite Hankel transform* of order n of a function $f(r)$ is denoted by $\mathcal{H}_n\{f(r)\} = \tilde{f}_n(k_i)$ and is defined by
$$\mathcal{H}_n\{f(r)\} = \tilde{f}_n(k_i) = \int_0^a r\, f(r)\, J_n(rk_i)\, dr. \tag{11.2.6}$$
The *inverse finite Hankel transform* is then defined by
$$\mathcal{H}_n^{-1}\{\tilde{f}_n(k_i)\} = f(r) = \frac{2}{a^2} \sum_{i=1}^{\infty} \tilde{f}_n(k_i) \frac{J_n(rk_i)}{J_{n+1}^2(ak_i)}, \tag{11.2.7}$$
where the summation is taken over all positive roots of $J_n(ak) = 0$.

The *zero-order finite Hankel transform* and its *inverse* are defined by
$$\mathcal{H}_0\{f(r)\} = \tilde{f}_0(k_i) = \int_0^a r\, f(r) J_0(rk_i)\, dr, \tag{11.2.8}$$
$$\mathcal{H}_0^{-1}\{\tilde{f}_0(k_i)\} = f(r) = \frac{2}{a^2} \sum_{i=1}^{\infty} \tilde{f}_0(k_i) \frac{J_0(rk_i)}{J_1^2(ak_i)}, \tag{11.2.9}$$
where the summation is taken over the positive roots of $J_0(ak) = 0$.

Similarly, the *first-order finite Hankel transform* and its *inverse* are defined by
$$\mathcal{H}_1\{f(r)\} = \tilde{f}_1(k_i) = \int_0^a r\, f(r)\, J_1(rk_i)\, dr, \tag{11.2.10}$$
$$\mathcal{H}_1^{-1}\{\tilde{f}_1(k_i)\} = f(r) = \frac{2}{a^2} \sum_{i=1}^{\infty} \tilde{f}_1(k_i) \frac{J_1(rk_i)}{J_2^2(ak_i)}, \tag{11.2.11}$$
where k_i is chosen as a positive root of $J_1(ak) = 0$.

We now give examples of *finite Hankel transforms* of some elementary functions.

Example 11.2.1 If $f(r) = r^n$, then
$$\mathcal{H}_n\{r^n\} = \int_0^a r^{n+1} J_n(rk_i)\, dr = \frac{a^{n+1}}{k_i} J_{n+1}(ak_i). \tag{11.2.12}$$
When $n = 0$,
$$\mathcal{H}_0\{1\} = \frac{a}{k_i} J_1(ak_i). \tag{11.2.13}$$

Example 11.2.2 If $f(r) = (a^2 - r^2)$, then
$$\mathcal{H}_0\{(a^2 - r^2)\} = \int_0^a r(a^2 - r^2) J_0(ak_i)\, dr = \frac{4a}{k_i^3} J_1(ak_i) - \frac{2a^2}{k_i^2} J_0(ak_i).$$
Since k_i are the roots of $J_0(ak) = 0$, we find

$$\mathcal{H}_0\{(a^2 - r^2)\} = \frac{4a}{k_i^3} J_1(ak_i). \qquad (11.2.14)$$

11.3 Basic Operational Properties

We state the following operational properties of the *finite Hankel transforms*:

$$\mathcal{H}_n\{f'(r)\} = \frac{k_i}{2n}\left[(n-1)\mathcal{H}_{n+1}\{f(r)\} - (n+1)\mathcal{H}_{n-1}\{f(r)\}\right], \quad n \geq 1, \qquad (11.3.1)$$

provided $f(r)$ is finite at $r = 0$.

When $n = 1$, we obtain

$$\mathcal{H}_1\{f'(r)\} = -k_i \mathcal{H}_0\{f(r)\} = -k_i \tilde{f}_0(k_i). \qquad (11.3.2)$$

$$\mathcal{H}_n\left[\frac{1}{r}\frac{d}{dr}\{rf'(r)\} - \frac{n^2}{r^2}f(r)\right] = -k_i^2 \tilde{f}_n(k_i) - a f(a) k_i J'_n(ak_i), \qquad (11.3.3)$$

When $n = 0$

$$\mathcal{H}_0\left[f''(r) + \frac{1}{r}f'(r)\right] = -k_i^2 \tilde{f}_0(k_i) + a f(a) k_i J_1(ak_i). \qquad (11.3.4)$$

If $n = 1$, (11.3.3) becomes

$$\mathcal{H}_1\left[f''(r) + \frac{1}{r}f'(r) - \frac{1}{r^2}f(r)\right] = -k_i^2 \tilde{f}_1(k_i) - a f(a) k_i J'_1(ak_i) \qquad (11.3.5)$$

Results (11.3.4) and (11.3.5) are very useful for finding solution of differential equations in cylindrical polar coordinates.

The proofs of the above results are elementary exercises for the reader.

11.4 Applications of Finite Hankel Transforms

Example 11.4.1 (*Temperature Distribution in a Long Circular Cylinder*). Find the solution of the axisymmetric heat conduction equation

$$\frac{\partial u}{\partial t} = \kappa\left(\frac{\partial^2 u}{\partial r^2} + \frac{1}{r}\frac{\partial u}{\partial r}\right), \quad 0 \leq r \leq a, \; t > 0 \qquad (11.4.1)$$

with the boundary and initial conditions

$$u(r,t) = f(t) \quad \text{on } r = a, t > 0 \qquad (11.4.2)$$
$$u(r,0) = 0, \quad 0 \leq r \leq a. \qquad (11.4.3)$$

Application of the *finite Hankel transform* defined by

$$\tilde{u}(k_i, t) = \mathcal{H}_0\{u(r,t)\} = \int_0^a r J_o(rk_i) u(r,t) dr, \qquad (11.4.4)$$

yields the given system with the boundary condition

$$\tilde{u}_t + \kappa k_i^2 \tilde{u} = \kappa a k_i J_1(ak_i) f(t),$$
$$\tilde{u}(k_i, 0) = 0. \qquad (11.4.5ab)$$

The solution of the first order system is
$$\tilde{u}(k_i, t) = \kappa\, a\, k_i\, J_1(ak_i)\int_0^t f(\tau)\exp\{-\kappa k_i^2(t-\tau)\}d\tau. \tag{11.4.6}$$
The inverse transform gives the formal solution
$$u(r,t) = \frac{2\kappa}{a}\sum_{i=1}^{\infty}\frac{k_i J_0(rk_i)}{J_1(ak_i)}\int_0^t f(\tau)\exp\{-\kappa k_i^2(t-\tau)\}d\tau. \tag{11.4.7}$$
In particular, if $f(t) = T_0 =$ constant,
$$u(r,t) = \frac{2T_0}{a}\sum_{i=1}^{\infty}\frac{J_0(rk_i)}{k_i J_1(ak_i)}\left[1-\exp\{-\kappa k_i^2 t\}\right]. \tag{11.4.8}$$
Using the inverse version of (11.2.7) gives the final solution
$$u(r,t) = T_0 - \frac{2T_0}{a}\sum_{i=1}^{\infty}\frac{J_0(rk_i)}{k_i J_1(ak_i)}\exp\{-\kappa k_i^2 t\}. \tag{11.4.9}$$
This solution representing the temperature distribution consists of the steady-state term and the transient term, which decays to zero as $t\to\infty$. Consequently, the steady temperature is attained in the limit as $t\to\infty$.

Example 11.4.2 (*Unsteady Viscous Flow in a Rotating Long Circular Cylinder*). The axisymmetric unsteady motion of a viscous fluid in an infinitely long circular cylinder of radius a is governed by
$$u_t = v\left(u_{rr} + \frac{1}{r}u_r - \frac{u}{r^2}\right), \quad 0\le r\le a, \quad t>0, \tag{11.4.10}$$
where $u = u(r,t)$ is the tangential fluid velocity and v is the constant kinematic viscosity of the fluid.

The cylinder is initially at rest at $t = 0$, and it is then allowed to rotate with constant angular velocity Ω. Thus the boundary and initial conditions are
$$u(r,t) = a\Omega \quad \text{on } r = a, t > 0, \tag{11.4.11}$$
$$u(r,t) = 0 \quad \text{at } t = 0 \text{ for } 0 < r < a. \tag{11.4.12}$$

We solve the problem by using the joint Laplace and the finite Hankel transform of order one defined by
$$\bar{\tilde{u}}(k_i, s) = \int_0^{\infty} e^{-st}dt \int_0^a r\, J_1(k_i r)u(r,t)\,dr, \tag{11.4.13}$$
where k_i are the positive roots of $J_1(ak_i) = 0$.

Application of the joint transform gives
$$s\bar{\tilde{u}}(k_i, s) = -v k_i^2 \bar{\tilde{u}}(k_i, s) - \frac{va^2\Omega k_i}{s}J_1'(ak_i)$$
Or
$$\bar{\tilde{u}}(k_i, s) = -\frac{va^2\Omega\, k_i\, J_1'(ak_i)}{s(s+vk_i^2)}. \tag{11.4.14}$$
The inverse Laplace transform gives
$$\tilde{u}(k_i, t) = -\frac{a^2\Omega}{k_i}J_1'(ak_i)\left[1-\exp\{-vk_i^2 t\}\right]. \tag{11.4.15}$$

Thus the final solution is found from (11.4.15) by using the *inverse Hankel transform* with $J_1'(ak_i) = -J_2(ak_i)$ in the form

$$u(r,t) = 2\Omega \sum_{i=1}^{\infty} \frac{J_1(rk_i)}{k_i J_2(ak_i)} \left[1 - \exp\{-vtk_i^2\}\right]. \tag{11.4.16}$$

This solution is the sum of the steady-state and the transient fluid velocities.

In view of (11.2.12) for $n = 1$, we can write

$$r = \mathcal{H}_1^{-1}\left\{\frac{a^2}{k_i} J_2(ak_i)\right\} = 2\sum_{i=1}^{\infty} \frac{J_1(rk_i)}{k_i J_2(ak_i)}. \tag{11.4.17}$$

This result is used to simplify (11.4.16) so that the final solution for $u(r,t)$ takes the form

$$u(r,t) = r\Omega - 2\Omega \sum_{i=1}^{\infty} \frac{J_1(rk_i)}{k_i J_2(ak_i)} \exp\left(-vtk_i^2\right). \tag{11.4.18}$$

In the limit as $t \to \infty$, the transient velocity component decays to zero, and the ultimate steady state flow is attained in the form

$$u(r,t) = r\Omega. \tag{11.4.19}$$

Physically, this represents the rigid body rotation of the fluid inside the cylinder.

Example 11.4.3 (*Vibrations of a Circular Membrane*). The free symmetric vibration of a thin circular membrane of radius a is governed by the wave equation

$$u_{tt} = c^2\left(u_{rr} + \frac{1}{r}u_r\right), \qquad 0 < r < a, \quad t > 0 \tag{11.4.20}$$

with the initial and boundary data

$$u(r,t) = f(r), \quad \frac{\partial u}{\partial t} = g(r) \quad \text{at } t = 0 \text{ for } 0 < r < a, \tag{11.4.21 ab}$$

$$u(a,t) = 0 \qquad \text{for all } t > 0. \tag{11.4.22}$$

Application of the zero-order *finite Hankel transform* of $u(r,t)$ defined by (11.4.4) to (11.4.20)-(11.4.22) gives

$$\frac{d^2\tilde{u}}{dt^2} + c^2 k_i^2 \tilde{u} = 0, \tag{11.4.23}$$

$$\tilde{u} = \tilde{f}(k_i) \text{ and } \left(\frac{d\tilde{u}}{dt}\right)_{t=0} = \tilde{g}(k_i). \tag{11.4.24ab}$$

The solution of this system is

$$\tilde{u}(k_i, t) = \tilde{f}(k_i)\cos(ctk_i) + \frac{\tilde{g}(k_i)}{ck_i}\sin(ctk_i). \tag{11.4.25}$$

The inverse transform yields the formal solution

$$u(r,t) = \frac{2}{a^2}\sum_{i=1}^{\infty} \tilde{f}(k_i)\cos(ctk_i)\frac{J_0(rk_i)}{J_1^2(ak_i)}$$

$$+ \frac{2}{ca^2}\sum_{i=1}^{\infty} \tilde{g}(k_i)\sin(ctk_i)\frac{J_0(rk_i)}{k_i J_1^2(ak_i)}. \tag{11.4.26}$$

where the summation is taken over all positive roots of $J_0(ak_i) = 0$.

We consider a more general form of the *finite Hankel transform* associated with a more general boundary condition

$$f'(r) + h f(r) = 0 \quad \text{at } r = a, \tag{11.4.27}$$

where h is a constant.

We define the *finite Hankel transform* of $f(r)$ by

$$\mathcal{H}_n\{f(r)\} = \tilde{f}_n(k_i) = \int_0^a r J_n(rk_i) f(r) dr, \tag{11.4.28}$$

where k_i are the roots of the equation

$$k_i J_n'(ak_i) + h J_n(ak_i) = 0. \tag{11.4.29}$$

The corresponding inverse transform is given by

$$f(r) = \mathcal{H}_n^{-1}\{\tilde{f}_n(k_i)\} = 2 \sum_{i=1}^{\infty} \frac{k_i^2 \tilde{f}_n(k_i) J_n(rk_i)}{\{(k_i^2 + h^2)a^2 - n^2\} J_n^2(ak_i)}. \tag{11.4.30}$$

This *finite Hankel transform* has the following operational property

$$\mathcal{H}_n\left[\frac{1}{r}\frac{d}{dr}\{rf'(r)\} - \frac{n^2}{r^2} f(r)\right] = -k_i^2 \tilde{f}_n(k_i)$$

$$+ a[f'(a) + h f(a)] J_n(ak_i), \tag{11.4.31}$$

which is, by (11.4.29)

$$= -k_i^2 \tilde{f}_n(k_i) - \frac{ak_i}{h}[f'(a) + h f(a)] J_n'(ak_i). \tag{11.4.32}$$

Thus result (11.4.32) involves $f'(a) + h f(a)$ as the boundary condition.

We apply this more general *finite Hankel transform* pairs (11.4.28) and (11.4.30) to solve the following axisymmetric initial-boundary value problem.

Example 11.4.4 (*Temperature Distribution of Cooling of a Circular Cylinder*). Solve the axisymmetric heat conduction problem for an infinitely long circular cylinder of radius $r = a$ with the initial constant temperature T_0, and the cylinder is cooling by radiation of heat from its boundary surface at $r = a$ to the outside medium at zero temperature according to Newton's law of cooling, which satisfies the boundary condition

$$\frac{\partial u}{\partial r} + hu = 0 \quad \text{at } r = a, \quad t > 0, \tag{11.4.33}$$

where h is a constant.

The problem is governed by the axisymmetric heat conduction equation

$$u_t = \kappa(u_{rr} + \frac{1}{r}u_r), \quad 0 \le r \le a, \quad t > 0, \tag{11.4.34}$$

with the boundary condition (11.4.33) and the initial condition

$$u(r, 0) = T_0 \quad \text{at } t = 0, \quad \text{for } 0 < r < a. \tag{11.4.35}$$

Application of the zero-order Hankel transform (11.4.28) with (11.4.29) to the system (11.4.33)-(11.4.35) gives

$$\frac{d\tilde{u}}{dt} + \kappa k_i^2 \tilde{u} = 0, \quad t > 0 \tag{11.4.36}$$

$$\tilde{u}(k_i,0) = aT_0 \int_0^a r J_0(rk_i)\,dr = \frac{aT_0}{k_i} J_1(ak_i). \qquad (11.4.37)$$

The solution of (11.4.36)-(11.4.37) is

$$\tilde{u}(k_i,t) = \left(\frac{aT_0}{k_i}\right) J_1(ak_i) \exp(-\kappa t k_i^2). \qquad (11.4.38)$$

The inverse transform (11.4.30) with $n = 0$ and $k_i J_0'(ak_i) + h J_0(ak_i) = 0$, that is, $k_i J_1(ak_i) = h J_0(ak_i)$, leads to the formal solution

$$u(r,t) = \left(\frac{2hT_0}{a}\right) \sum_{i=1}^{\infty} \frac{J_0(rk_i)\exp(-\kappa t k_i^2)}{(k_i^2 + h^2) J_0(ak_i)}, \qquad (11.4.39)$$

where the summation is taken over all the positive roots of $k_i J_1(ak_i) = h J_0(ak_i)$.

11.5 Exercises

1. Find the zero-order *finite Hankel transform* of
    (a) $f(r) = r^2$, (b) $f(r) = J_0(\alpha r)$.

2. Show that
$$\mathcal{H}_n\left\{\frac{J_n(\alpha r)}{J_n(\alpha a)}\right\} = \frac{ak_i}{(\alpha^2 - k_i^2)} J_n'(ak_i)$$

3. If $\mathcal{H}_n\{f(r)\}$ is the finite Hankel transform of $f(r)$ defined by (11.2.6), and if $n > 0$, show that
    (a) $\mathcal{H}_n\{r^{-1} f'(r)\} = \frac{1}{2} k_i \left[\mathcal{H}_{n+1}\{r^{-1} f(r)\} - \mathcal{H}_{n-1}\{r^{-1} f(r)\}\right]$,
    (b) $\mathcal{H}_0\{r^{-1} f'(r)\} = k_i \mathcal{H}_1\{r^{-1} f(r)\} - f(a)$.

4. Solve the initial-boundary value problem
$$c^2\left(u_{rr} + \frac{1}{r} u_r\right) = u_{tt}, \quad 0 < r < a,\ t > 0,$$
$$u(r,0) = 0,\ u_t(r,0) = u_0,$$
$$u(a,t) = 0,$$
where u_0 is a constant.

5. Obtain a solution of the initial-boundary problem
$$\kappa\left(u_{rr} + \frac{1}{r} u_r\right) = u_t,\ 0 < r < a,\ t > 0,$$
$$u(r,0) = f(r),\ \text{for}\ 0 < r < a$$
$$u(a,t) = 0.$$

6. If we define the *finite Hankel transform* of $f(r)$ by
$$\mathcal{H}_n\{f(r)\} = \tilde{f}_n(k_i) = \int_a^b r f(r) A_n(rk_i)\,dr, \qquad b > a,$$
where

$$A_n(rk_i) = J_n(rk_i)Y_n(ak_i) - Y_n(rk_i)J_n(ak_i),$$

and $Y_n(x)$ is the Bessel function of the second kind of order n, show that the inverse transform is

$$\mathcal{H}_n^{-1}\{\tilde{f}_n(k_i)\} = f(r) = \frac{\pi^2}{2} \sum_{i=1}^{\infty} \frac{k_i^2 \tilde{f}_n(k_i) A_n(rk_i) J_n^2(bk_i)}{J_n^2(ak_i) - J_n^2(bk_i)},$$

where k_i are the positive roots of $A_n(bk_i) = 0$.

7. For the transform defined in exercise 6, show that

$$\mathcal{H}_n\left[f''(r) + \frac{1}{r}f'(r) - \frac{n^2 f(r)}{r^2}\right] = -k_i^2 \tilde{f}_n(k_i) + \frac{2}{\pi}\left[f(b)\frac{J_n(ak_i)}{J_n(bk_i)} - f(a)\right].$$

8. Viscous fluid of kinematic viscosity v is bounded between two infinitely long concentric circular cylinders of radii a and b. The inner cylinder is stationery and the outer cylinder begins to rotate with uniform angular velocity Ω at $t = 0$. The axisymmetric flow is governed by (11.4.10) with $v(a,0) = 0$ and $v(b,0) = \Omega b$. show that

$$v(r,t) = \pi b \Omega \sum_{i=1}^{\infty} \frac{J_1(ak_i) J_1(bk_i) A_1(rk_i)\left[1 - \exp(-vtk_i^2)\right]}{J_1^2(ak_i) - J_1^2(bk_i)},$$

where

$$A_1(rk_i) = J_n(rk_i)Y_n(ak_i) - Y_n(rk_i)J_n(ak_i),$$

and k_i are the positive roots of $A_1(bk_i) = 0$.

9. Find the solution of the forced symmetric vibrations of a thin elastic membrane that satisfy the initial-boundary value problem

$$u_{rr} + \frac{1}{r}u_r - \frac{1}{c^2}u_{tt} = -\frac{p(r,t)}{T_0},$$

where $p(r,t)$ is the applied pressure which produces vibrations, and the membrane is stretched by a constant tension T_0. The membrane is set into motion from rest in its equilibrium position so that

$$u(r,t) = 0 = \left(\frac{\partial u}{\partial t}\right) \quad \text{at } t = 0.$$

10. Use the joint Hankel and Laplace transform method to the axisymmetric diffusion problem in an infinitely long circular cylinder of radius a:

$$u_t = \kappa\left(u_{rr} + \frac{1}{r}u_r\right) + Q(r,t), \quad 0 < r < a, \quad t > 0,$$

$$u(a,t) = 0 \quad \text{for } t > 0,$$

$$u(r,0) = 0 \quad \text{for } 0 < r \leq a,$$

where $Q(r,t)$ represents a heat source inside the cylinder. Find the explicit solution for two special cases:

(a) $Q(r,t) = \dfrac{\kappa Q_0}{k}$, (b) $Q(r,t) = Q_0 \dfrac{\delta(r)}{r} f(t)$,

where Q_0, κ, and k are constants.

Chapter 12

Legendre Transforms

12.1 Introduction

We consider in this chapter the Legendre transform with a Legendre polynomial as kernel and discuss basic operational properties including the Convolution Theorem. Legendre transforms are then used to solve boundary value problems in potential theory. This chapter is based on papers by Churchill (1954) and Churchill and Dolph (1954) listed in the bibliography.

12.2 Defintion of the Legendre Transform and Examples

Churchill (1954) defined the *Legendre transform* of a function $f(x)$ defined in $-1 < x < 1$ by the integral

$$\mathcal{T}_n\{f(x)\} = \tilde{f}(n) = \int_{-1}^{1} P_n(x) f(x) dx, \qquad (12.2.1)$$

provided the integral exists and where $P_n(x)$ is the *Legendre polynomial* of degree n (≥ 0). Obviously $\mathcal{T}_n$ is a linear integral transformation.

When $x = \cos\theta$, (12.2.1) becomes

$$\mathcal{T}_n\{f(\cos\theta)\} = \tilde{f}(n) = \int_0^{\pi} P_n(\cos\theta) f(\cos\theta) \sin\theta \, d\theta. \qquad (12.2.2)$$

The *inverse Legendre transform* is given by

$$f(x) = \mathcal{T}_n^{-1}\{\tilde{f}(n)\} = \sum_{n=0}^{\infty} \left(\frac{2n+1}{2}\right) \tilde{f}(n) P_n(x). \qquad (12.2.3)$$

This follows from the expansion of any function $f(x)$ in the form

$$f(x) = \sum_{n=0}^{\infty} a_n P_n(x), \qquad (12.2.4)$$

where the coefficient a_n can be determined from the orthogonal property of $P_n(x)$. It turns out that

$$a_n = \left(\frac{2n+1}{2}\right) \int_{-1}^{1} P_n(x) f(x) dx = \left(\frac{2n+1}{2}\right) \tilde{f}(n), \qquad (12.2.5)$$

and hence (12.2.3) follows.

Example 12.2.1

$$\mathscr{T}_n\{\exp(i\alpha x)\} = \left(\frac{2\pi}{\alpha}\right)^{1/2} i^n J_{n+1/2}(\alpha), \qquad (12.2.6)$$

where $J_\nu(x)$ is the Bessel function.

We have, by definition,

$$\mathscr{T}_n\{\exp(i\alpha x)\} = \int_{-1}^{1} \exp(i\alpha x) P_n(x) dx,$$

which is, by a result in Copson (1935, p 341),

$$= \sqrt{\frac{2\pi}{\alpha}} \, i^n J_{n+1/2}(\alpha).$$

Similarly,

$$\mathscr{T}_n\{\exp(\alpha x)\} = \sqrt{\frac{2\pi}{\alpha}} \, I_{n+1/2}(\alpha), \qquad (12.2.7)$$

where $I_\nu(x)$ is the modified Bessel function of the first kind.

Example 12.2.2

(a) $$\mathscr{T}_n\left\{(1-x^2)^{-1/2}\right\} = \pi P_n^2(0). \qquad (12.2.8)$$

(b) $$\mathscr{T}_n\left\{\frac{1}{2(t-x)}\right\} = Q_n(t), \qquad |t| > 1, \qquad (12.2.9)$$

where $Q_n(t)$ is the Legendre function of the second kind given by

$$Q_n(t) = \frac{1}{2} \int_{-1}^{1} (t-x)^{-1} P_n(x) dx.$$

These results are easy to verify with the aid of results given in Copson (1935, p 292 and p 310).

Example 12.2.3 If $|r| \leq 1$, then

(a) $$\mathscr{T}_n\left\{(1-2rx+r^2)^{-1/2}\right\} = \frac{2r^n}{(2n+1)}. \qquad (12.2.10)$$

(b) $$\mathscr{T}_n\left\{(1-2rx+r^2)^{-3/2}\right\} = \frac{2r^n}{(1-r^2)}. \qquad (12.2.11)$$

We have, from the generating function of $P_n(x)$,

$$(1-2rx+r^2)^{-1/2} = \sum_{n=0}^{\infty} r^n P_n(x), \qquad |r| < 1.$$

Multiplying this result by $P_n(x)$ and using the orthogonality condition of the Legendre polynomial gives

$$\int_{-1}^{1} (1-2rx+r^2)^{-1/2} P_n(x) dx = \frac{2r^n}{(2n+1)}. \qquad (12.2.12)$$

In particular, when $r = 1$, we obtain

$$\mathcal{T}_n\{(1-x)^{-1/2}\} = \frac{2\sqrt{2}}{2n+1}. \tag{12.2.13}$$

Differentiating (12.2.12) with respect to r gives

$$\frac{1}{2}\int_{-1}^{1}(1-2rx+r^2)^{-3/2}(2rx-2r^2)\,P_n(x)\,dx = \frac{2nr^n}{(2n+1)},$$

so that

$$-\mathcal{T}_n\{(1-2rx+r^2)^{-1/2}\} + (1-r^2)\mathcal{T}_n\{(1-2rx+r^2)^{-3/2}\} = \frac{2nr^n}{(2n+1)}.$$

Using (12.2.10), we obtain (12.2.11).

Example 12.2.4 If $|r| < 1$ and $\alpha > 0$, then

$$\mathcal{T}_n\left\{\int_0^r \frac{t^{\alpha-1}\,dt}{(1-2xt+t^2)^{1/2}}\right\} = \frac{2r^{n+\alpha}}{(2n+1)(n+\alpha)}. \tag{12.2.14}$$

We replace r by t in (12.2.10) and multiply the result by $t^{\alpha-1}$ to obtain

$$\mathcal{T}_n\{t^{\alpha-1}(1-2xt+t^2)^{-1/2}\} = \frac{2t^{n+\alpha-1}}{(2n+1)}.$$

Integrating this result on $(0,r)$ we find (12.2.14).

Example 12.2.5 If $H(x)$ is a Heaviside unit step function, then

$$\mathcal{T}_n\{H(x)\} = \begin{cases} 1, & n = 0 \\ \dfrac{P_{n-1}(0) - P_{n+1}(0)}{(2n+1)}, & n \geq 1 \end{cases}. \tag{12.2.15}$$

Obviously,

$$\mathcal{T}_n\{H(x)\} = \int_0^1 P_n(x)\,dx = 1 \quad \text{when } n = 0.$$

However, for $n > 1$, we use the recurrence relation for $P_n(x)$ as

$$(2n+1)P_n(x) = P'_{n+1}(x) - P'_{n-1}(x) \tag{12.2.16}$$

to derive

$$\mathcal{T}_n\{H(x)\} = \frac{1}{(2n+1)}\int_0^1 [P'_{n+1}(x) - P'_{n-1}(x)]\,dx$$

$$= \frac{1}{2n+1}[P_{n-1}(0) - P_{n+1}(0)].$$

Debnath and Harrel (1976) introduced the *associated Legendre transform* defined by

$$\mathcal{T}_{n,m}\{f(x)\} = \tilde{f}(n,m) = \int_{-1}^{1}(1-x^2)^{-m/2}P_n^m(x)f(x)\,dx, \tag{12.2.17}$$

where $P_n^m(x)$ is the *associated Legendre function* of the first kind.

The inverse transform is given by

$$f(x) = \mathcal{T}_{n,m}^{-1}\{\tilde{f}(n,m)\} = \sum_{n=0}^{\infty} \frac{(2n+1)}{2} \frac{(n-m)!}{(n+m)!} \tilde{f}(n,m)(1-x^2)^{m/2} P_n^m(x). \quad (12.2.18)$$

The reader is referred to Debnath and Harrel (1976) for a detailed discussion of this transform.

12.3 Basic Operational Properties of Legendre Transforms

Theorem 12.3.1 If $f'(x)$ is continuous and $f''(x)$ is bounded and integrable in each subinterval of $-1 \leq x \leq 1$, and if $\mathcal{T}_n\{f(x)\}$ exists and

$$\lim_{|x|\to 1}(1-x^2)f(x) = \lim_{|x|\to 1}(1-x^2)f'(x) = 0, \quad (12.3.1)$$

then

$$\mathcal{T}_n\{R[f(x)]\} = -n(n+1)\tilde{f}(n), \quad (12.3.2)$$

where $R[f(x)]$ is a differential form given by

$$R[f(x)] = \frac{d}{dx}\left[(1-x^2)\frac{d}{dx}f(x)\right], \quad n > 0. \quad (12.3.3)$$

Proof. We have, by definition,

$$\mathcal{T}_n\{R[f(x)]\} = \int_{-1}^{1} \frac{d}{dx}\left[(1-x^2)\frac{d}{dx}f(x)\right] P_n(x)\, dx$$

which is, by integrating by parts together with (12.3.1),

$$= -\int_{-1}^{1}(1-x^2) P_n'(x) \frac{d}{dx} f(x)\, dx.$$

Integrating this result by parts again, we obtain

$$\mathcal{T}_n\{R[f(x)]\} = -\left[(1-x^2) P_n'(x) f(x)\right]_{-1}^{1} + \int_{-1}^{1} \frac{d}{dx}\left[(1-x^2) P_n'(x)\right] f(x)\, dx.$$

Using (12.3.1) and the differential equation for the Legendre polynomial

$$\frac{d}{dx}\left[(1-x^2)\frac{dy}{dx}\right] + n(n+1)y = 0, \quad (12.3.4)$$

we obtain the desired result

$$\mathcal{T}_n\{R[f(x)]\} = -n(n+1)\tilde{f}(n).$$

We may extend this result to evaluate the Legendre transforms of the differential forms $R^2[f(x)]$, $R^3[f(x)]$, ..., $R^k[f(x)]$.

Clearly
$$\mathcal{T}_n\{R^2[f(x)]\} = \mathcal{T}_n\{R[R[f(x)]]\}$$
$$= -n(n+1)\mathcal{T}_n\{R[f(x)]\} = n^2(n+1)^2 \tilde{f}(n) \quad (12.3.5)$$
provided $f'(x)$ and $f''(x)$ satisfy the conditions of Theorem 12.3.1.
Similarly,
$$\mathcal{T}_n\{R^3[f(x)]\} = (-1)^3 n^3 (n+1)^3 \tilde{f}(n). \quad (12.3.6)$$
More generally, for a positive integer k,
$$\mathcal{T}_n\{R^k[f(x)]\} = (-1)^k n^k (n+1)^k \tilde{f}(n). \quad (12.3.7)$$

Corollary 12.3.1 If $\mathcal{T}_n\{R[f(x)]\} = -n(n+1)\tilde{f}(n)$, then
$$\mathcal{T}_n\left\{\frac{1}{4}f(x) - R[f(x)]\right\} = \left(n+\frac{1}{2}\right)^2 \tilde{f}(n). \quad (12.3.8)$$

Proof. We replace $n(n+1)$ by $\left(n+\frac{1}{2}\right)^2 - \frac{1}{4}$ in (12.3.2) to obtain
$$\mathcal{T}_n\{R[f(x)]\} = -\left[\left(n+\frac{1}{2}\right)^2 - \frac{1}{4}\right]\tilde{f}(n). \quad (12.3.9)$$
Rearranging the terms in (12.3.9) gives
$$\mathcal{T}_n\left\{\frac{1}{4}f(x) - R[f(x)]\right\} = \left(n+\frac{1}{2}\right)^2 \tilde{f}(n).$$
In general, this result can be written as
$$(-1)^k \mathcal{T}_n\{R^k[f(x)] - 4^{-k} f(x)\} = \sum_{r=0}^{k-1} (-1)^r \binom{k}{r} \left[4^{-r}\left(n+\frac{1}{2}\right)^{2k-2r}\right] \tilde{f}(n). \quad (12.3.10)$$
The proof of (12.3.10) follows from (12.3.7) by replacing $n(n+1)$ with $\left(n+\frac{1}{2}\right)^2 - \frac{1}{4}$ and using the binomial expansion.

Example 12.3.1
$$\mathcal{T}_n\{\log(1-x)\} = \begin{cases} 2(\log 2 - 1), & n=0 \\ -\dfrac{2}{n(n+1)}, & n>0 \end{cases}. \quad (12.3.11)$$

Clearly,
$$R[\log(1-x)] = \frac{d}{dx}\left[(1-x^2)\frac{d}{dx}\log(1-x)\right] = -1.$$
Although $\dfrac{d}{dx}\log(1-x)$ does not satisfy the conditions of Theorem 12.3.1, we integrate by parts to obtain

$$\mathcal{T}_n\{R[\log(1-x)]\} = \int_{-1}^{1} R[\log(x)] P_n(x) \, dx$$

$$= [-(1+x) P_n(x)]_{-1}^{1} + \int_{-1}^{1} (1+x) P_n'(x) \, dx$$

which is, since $(1+x) = -(1-x^2)\dfrac{d}{dx}\log(1-x)$, and by integrating by parts,

$$= -2 + \int_{-1}^{1} \log(1-x) \frac{d}{dx}\left[(1-x^2) P_n'(x)\right] dx. \quad (12.3.12)$$

By integrating by parts twice, result (12.3.12) gives

$$\mathcal{T}_n\{R[\log(1-x)]\} = -2 + \int_{-1}^{1} \frac{d}{dx}\left[(1-x^2)\frac{d}{dx}\log(1-x)\right] P_n(x) \, dx$$

which is, by (12.3.2),

$$= -2 - n(n+1)\, \tilde{f}(n), \quad (12.3.13)$$

where $\tilde{f}(n) = \mathcal{T}_n\{\log(1-x)\}$.

However, $R[\log(1-x)] = -1$ so that $\mathcal{T}_n\{R[\log(1-x)]\} = 0$ for all $n > 0$ and hence (12.3.13) gives

$$\mathcal{T}_n[\log(1-x)] = \tilde{f}(n) = -\frac{2}{n(n+1)}.$$

On the other hand, since $P_0(x) = 1$, we have

$$\mathcal{T}_0\{[\log(1-x)]\} = \int_{-1}^{1} \log(1-x) \, dx$$

which is, by direct integration,

$$= -[(1-x)\{\log(1-x) - x\}]_{-1}^{1} = 2(\log 2 - 1).$$

Theorem 12.3.2 If $f(x)$ and $f'(x)$ are piecewise continuous in $-1 < x < 1$, $R^{-1}[f(x)] = h(x)$, and $f(0) = \int_{-1}^{1} f(x) \, dx = 0$, then

$$\mathcal{T}_n^{-1}\left\{\frac{\tilde{f}(n)}{n(n+1)}\right\} = A - \int_{0}^{x} \frac{ds}{(1-s^2)} \int_{-1}^{s} f(t) \, dt, \quad (12.3.14)$$

where A is an arbitrary constant of integration.

Proof. We have

$$R[h(x)] = f(x)$$

or

$$\frac{d}{dx}\left[(1-x^2)\frac{d}{dx}h(x)\right] = f(x).$$

Integral Transforms and Their Applications 331

Integrating over $(-1, x)$ gives

$$\int_{-1}^{x} f(t)\,dt = (1-x^2)\frac{d}{dx}h(x), \qquad (12.3.15)$$

which is a continuous function of x in $|x|<1$ with limit zero as $|x|\to 1$.

Integration of (12.3.15) gives

$$h(x) = \int_{0}^{x} \frac{ds}{(1-s^2)} \int_{-1}^{s} f(t)\,dt - A$$

where A is an arbitrary constant. Clearly, $h(x)$ satisfies the conditions of Theorem 12.3.1, and there exists a positive real constant $m < 1$ such that

$$|h(x)| = O\{(1-x^2)^{-m}\} \qquad \text{as } |x|\to 1.$$

Hence $\mathcal{T}_n\{R[h(x)]\}$ exists, and by Theorem 12.3.1, it follows that

$$\mathcal{T}_n\{R[h(x)]\} = -n(n+1)\,\mathcal{T}_n\{h(x)\} = -n(n+1)\,\mathcal{T}_n\{R^{-1}[f(x)]\}, \qquad (12.3.16)$$

whence it turns out that

$$\mathcal{T}_n\{R^{-1}\{f(x)\}\} = -\frac{\tilde{f}(n)}{n(n+1)}. \qquad (12.3.17)$$

Inversion leads to the result

$$\mathcal{T}_n^{-1}\left\{\frac{\tilde{f}(n)}{n(n+1)}\right\} = -R^{-1}\{f(x)\} = -h(x)$$

$$= A - \int_{0}^{x} \frac{ds}{1-s^2}\int_{-1}^{s} f(t)\,dt. \qquad (12.3.18)$$

This proves the theorem.

Theorem 12.3.3 If $f(x)$ is continuous in each subinterval of $(-1,1)$ and a continuous function $g(x)$ is defined by

$$g(x) = \int_{-1}^{x} f(t)\,dt, \qquad (12.3.19)$$

then

$$\mathcal{T}_n\{g'(x)\} = \tilde{f}(n) = g(1) - \int_{-1}^{1} g(x)\,P'_n(x)\,dx. \qquad (12.3.20)$$

Proof. We have, by definition,

$$\mathcal{T}_n\{g'(x)\} = \int_{-1}^{1} g'(x)\,P_n(x)\,dx,$$

which is, by integrating by parts,

$$= [P_n(x)\,g(x)]_{-1}^{1} - \int_{-1}^{1} g(x)\,P'_n(x)\,dx.$$

Since $P_n(1) = 1$ and $g(-1) = 0$, the preceding result becomes (12.3.20).

Corollary 12.3.2 If result (12.3.20) is true and $g(x)$ is given by (12.3.19), then

$$\mathcal{T}_n\{g(x)\} = f(0) - f(1) \quad \text{when} \quad n = 0$$
$$= \frac{\tilde{f}(n-1) - \tilde{f}(n+1)}{(2n+1)} \quad \text{when} \quad n > 1 \quad . \quad (12.3.21)$$

Proof. We write $\tilde{f}(n-1)$ and $\tilde{f}(n+1)$ using (12.3.20) and then subtract so that the resulting expression gives (12.3.21) with the help of (12.2.16).

Corollary 12.3.3 If $g'(x)$ is a sectionally continuous function and $g(x)$ is the continuous function given by (12.3.19), then

$$\mathcal{T}_n\{g'(x)\} = g(1), \quad \text{when} \quad n = 0$$
$$= g(1) - (2n-1)\tilde{g}(n-1) - (2n-5)\tilde{g}(n-3) - \cdots - g(0)$$
$$\quad \text{when} \quad n = 1, 2, 3, \ldots \quad . \quad (12.3.22)$$
$$= g(1) - 2(2n-1)\tilde{g}(n-1) - (2n-5)\tilde{g}(n-3) - \cdots - 3g(1)$$
$$\quad \text{when} \quad n = 2, 4, 6, \ldots$$

These results can readily be verified using (12.3.20) and (12.2.16).

Theorem 12.3.4 (*Convolution*). If $\mathcal{T}_n\{f(x)\} = \tilde{f}(n)$ and $\mathcal{T}_n\{g(x)\} = \tilde{g}(n)$, then

$$\mathcal{T}_n\{f(x) * g(x)\} = \tilde{f}(n)\,\tilde{g}(n), \quad (12.3.23)$$

where the convolution $f(x) * g(x)$ is given by

$$f(x) * g(x) = h(x) = \frac{1}{\pi} \int_0^\pi f(\cos\mu) \sin\mu\, d\mu \int_0^\pi g(\cos\lambda)\, d\beta, \quad (12.3.24)$$

with

$$x = \cos\nu \text{ and } \cos\lambda = \cos\mu\,\cos\nu + \sin\mu\,\sin\nu\,\cos\beta. \quad (12.3.25)$$

Proof. We have, by definition (12.2.2),

$$\tilde{f}(n)\tilde{g}(n) = \int_0^\pi f(\cos\mu) P_n(\cos\mu) \sin\mu\, d\mu \int_0^\pi g(\cos\lambda) P_n(\cos\lambda) \sin\lambda\, d\lambda$$

$$= \int_0^\pi f(\cos\mu) \sin\mu \left[\int_0^\pi g(\cos\lambda) P_n(\cos\lambda) P_n(\cos\mu) \sin\lambda\, d\lambda \right] d\mu,$$

$$(12.3.26)$$

where $f(x) = f(\cos\mu)$ and $g(x) = g(\cos\lambda)$.

With the aid of an addition formula [Sansone, 1959, p. 169] given as

$$P_n(\cos\lambda) P_n(\cos\mu) = \frac{1}{\pi} \int_0^\pi P_n(\cos\nu)\, d\alpha, \quad (12.3.27)$$

where $\cos v = \cos \lambda \cos \mu + \sin \lambda \sin \mu \cos \alpha$, the product can be rewritten in the form

$$\tilde{f}(n)\tilde{g}(n) = \frac{1}{\pi} \int_0^\pi f(\cos\mu)\sin\mu \left[\int_0^\pi \int_0^\pi g(\cos\mu) P_n(\cos\mu)\sin\lambda\, d\alpha\, d\lambda \right] d\mu.$$
(12.3.28)

We next use Churchill and Dolph's (1954, pp 94-96) geometrical arguments to replace the double integral inside the square bracket by

$$\int_0^\pi \int_0^\pi g(\cos\mu \cos v + \sin\mu \sin v \cos\beta) P_n(\cos v)\sin v\, dv. \quad (12.3.29)$$

Substituting this result in (12.3.26) and changing the order of integration, we obtain

$$\tilde{f}(n)\tilde{g}(n) = \frac{1}{\pi} \int_0^\pi P_n(\cos v)\sin v \left[\int_0^\pi \int_0^\pi f(\cos\mu)\sin\mu\; g(\cos\lambda)\, d\mu\, d\beta \right] dv$$

$$= \int_0^\pi h(\cos v) P_n(\cos v)\sin v\, dv, \quad (12.3.30)$$

where
$$\cos\lambda = \cos\mu \cos v + \sin\mu \sin v \cos\beta, \quad (12.3.31)$$

and
$$h(\cos v) = \frac{1}{\pi} \int_0^\pi f(\cos\mu)\sin\mu\, d\mu \int_0^\pi g(\cos\lambda)\, d\beta.$$

This proves the theorem.

In particular, when $v = 0$, (12.3.24) becomes

$$h(1) = \int_{-1}^1 f(t) g(t)\, dt, \quad (12.3.32)$$

and when $v = \pi$, (12.3.24) gives

$$h(-1) = \int_{-1}^1 f(t) g(-t)\, dt. \quad (12.3.33)$$

12.4 Applications of Legendre Transforms to Boundary Value Problems

We solve the Dirichlet problem for the potential $u(r,\theta)$ inside a unit sphere $r = 1$, which satisfies the Laplace equation

$$\frac{\partial}{\partial r}\left[r^2 \frac{\partial u}{\partial r}\right] + \frac{\partial}{\partial x}\left[(1-x^2)\frac{\partial u}{\partial x}\right] = 0, \quad 0 < r < 1, \quad (12.4.1)$$

with the boundary condition $(x = \cos\theta)$
$$u(1, x) = f(x), \quad -1 < x < 1. \tag{12.4.2}$$

We introduce the Legendre transform $\tilde{u}(r,n) = \mathcal{T}_n\{u(r,\theta)\}$ defined by (12.2.1). Application of this transform to (12.4.1)-(12.4.2) gives
$$r^2 \frac{d^2\tilde{u}(r,n)}{dr^2} + 2r\frac{d\tilde{u}}{dr} - n(n+1)\tilde{u}(r,n) = 0, \tag{12.4.3}$$
$$\tilde{u}(1,n) = \tilde{f}(n), \tag{12.4.4}$$
where $\tilde{u}(r,n)$ is to be continuous function of r for $0 \le r < 1$.

The bounded solution of (12.4.3)-(12.4.4) is
$$\tilde{u}(r,n) = \tilde{f}(n) r^n, \quad 0 \le r < 1, \quad \text{for } n = 0, 1, 2, 3, \ldots. \tag{12.4.5}$$

Thus the solution for $u(r, x)$ can be found by the inverse transform so that
$$u(r, x) = \sum_{n=0}^{\infty}\left(n + \frac{1}{2}\right)\tilde{f}(n) r^n P_n(x) \quad \text{for } 0 < r \le 1, \ |x| < 1. \tag{12.4.6}$$

The Convolution Theorem allows us to give another representation of the solution. In view of (12.2.11), we find
$$\mathcal{T}_n^{-1}\{r^n\} = \frac{1}{2}(1-r^2)(1-2rx+r^2)^{-3/2}.$$

Thus it follows from (12.4.5) that
$$u(r, \cos\theta) = \mathcal{T}_n^{-1}\{\tilde{f}(n) r^n\}$$
$$= \frac{1}{2\pi}\int_0^\pi f(\cos\mu)\sin\mu \, d\mu \int_0^\pi \frac{(1-r^2)d\lambda}{(1-2r\cos v + r^2)^{3/2}}, \tag{12.4.7}$$

where
$$\cos v = \cos\mu \cos\theta + \sin\mu \sin\theta \cos\lambda. \tag{12.4.8}$$

Integral (12.4.7) is called the *Poisson integral formula* for the potential inside the unit sphere for the Dirichlet problem.

On the other hand, for the Dirichlet exterior problem, the potential $w(r, \cos\theta)$ outside the unit sphere $(r > 1)$ can be obtained with the boundary condition $w(1, \cos\theta) = f(\cos\theta)$. The solution of the Legendre transformed problem is
$$\tilde{w}(r,n) = \frac{1}{r}\tilde{f}(n) r^{-n}, \quad n = 0, 1, 2, \ldots, \tag{12.4.9}$$

which is, in terms of w,
$$w(r, \cos\theta) = \frac{1}{r} w\left(\frac{1}{r}, \cos\theta\right), \quad r > 1 \tag{12.4.10}$$
$$= \frac{1}{2\pi}\int_0^\pi f(\cos\mu)\sin\mu \, d\mu \int_0^\pi \frac{(r^2-1) d\lambda}{(1-2r\cos v + r^2)^{3/2}}, \tag{12.4.11}$$

where $\cos v$ is given by (12.4.8).

12.5 Exercises

1. Show that, if $|r|<1$,

   (a) $\mathcal{T}_n\{x^n\} = \dfrac{2^{n+1}(n!)^2}{(2n+1)!}$.

   (b) $\mathcal{T}_n\left[\log\left\{\dfrac{r-x+\left(1-2rx+r^2\right)^{1/2}}{1-x}\right\}\right] = \dfrac{2r^{n+1}}{(n+1)(2n+1)}$.

   (c) $\mathcal{T}_n\left[\left\{2r\left(1-rx+r^2\right)^{-1/2}\right\} - \log\left\{\dfrac{r-x+\left(1-2rx+r^2\right)^{1/2}}{1-x}\right\}\right] = \dfrac{2r^{n+1}}{n+1}$.

   (d) $\mathcal{T}_n\left[-\log\dfrac{1}{2}\left\{1-rx+\left(1-2rx+r^2\right)^{1/2}\right\}\right] = \begin{cases} 0, & n=0 \\ \dfrac{2r^n}{n(2n+1)}, & n>0 \end{cases}$.

   (e) $\mathcal{T}_n\left[\left(1-2rx+r^2\right)^{-1/2} - \dfrac{1}{2}\log\left\{\dfrac{1-rx+\left(1-2rx+r^2\right)^{1/2}}{2}\right\}\right] = \dfrac{r^n}{n}$.

2. Using the recurrence relation for the Legendre polynomials, show that
   $$\mathcal{T}_n[\,x f(x)] = (2n+1)^{-1}\left[(n+1)\tilde{f}(n+1) + n\tilde{f}(n-1)\right].$$
   Hence find $\mathcal{T}_n\{x^2 f(x)\}$.

3. Use the definition of the even Legendre-transform pairs (Tranter, 1966)
   $$\mathcal{T}_{2n}\{f(x)\} = \tilde{f}(2n) = \int_0^1 f(x) P_{2n}(x)\,dx, \qquad n=0,1,2,\ldots$$
   $$f(x) = \mathcal{T}_{2n}^{-1}\{\tilde{f}(2n)\} = \sum_{n=0}^{\infty}(4n+1)\tilde{f}(2n) P_{2n}(x), \qquad 0<x<1,$$
   to show that
   $$\mathcal{T}_{2n}\left[\dfrac{d}{dx}\left\{(1-x^2)f'(x)\right\}\right] = -2n(2n+1)\tilde{f}(2n) - f'(0) P_{2n}(0).$$
   Hence deduce
   $$\mathcal{T}_{2n}\{x\} = -\dfrac{P_{2n}(0)}{(2n-1)(2n+2)}.$$

4. Use the definition of the odd Legendre-transform pairs (Tranter, 1966)

$$\mathscr{T}_{2n+1} = \tilde{f}(2n+1) = \int_0^1 P_{2n+1}(x) f(x) dx, \quad n = 0, 1, 2, \ldots.$$

$$f(x) = \mathscr{T}_{2n+1}^{-1}\{\tilde{f}(2n+1)\} = \sum_{n=0}^{\infty} (4n+3) P_{2n+1}(x) \tilde{f}(2n+1),$$

to prove the result

$$\mathscr{T}_{2n+1}\left[\frac{d}{dx}\{(1-x^2) f'(x)\}\right] = -(2n+1)(2n+2) \tilde{f}(2n+1)$$
$$+ f(0) P'_{2n+1}(0).$$

Hence derive

$$\mathscr{T}_{2n+1}\{1\} = \frac{P'_{2n+1}(0)}{(2n+1)(2n+2)}.$$

5. From the definition of the even Legendre transform, show that

$$\mathscr{T}_{2n}\{x^{2r}\} = \frac{2^{2n}(2r)!(r+n)!}{(2r+2n+1)!(r-n)!}.$$

6. Show that the Legendre transform solution of the Dirichlet boundary value problem for $u(r, \theta)$

$$u_{rr} + \frac{1}{r} u_r + (1-x^2) u_{xx} - 2x u_x = 0, \quad 0 \le r \le a, \quad 0 \le \theta \le \pi$$
$$u(a, \theta) = f(x), \quad 0 \le \theta \le \pi,$$

where $x = \cos\theta$, is

$$\tilde{u}(r, n) = \left(\frac{r}{a}\right)^n \tilde{f}(n).$$

Obtain the solution for $u(r, \theta)$ with the help of (12.2.11) and the Convolution Theorem 12.3.4.

7. Solve the problem of the electrified disk for the potential $u(\xi, \eta)$ which satisfies the equation (see Tranter, 1966, p 99)

$$\frac{\partial}{\partial \xi}\left[(1-\xi^2)\frac{\partial u}{\partial \xi}\right] + \frac{\partial}{\partial \eta}\left[(1-\eta^2)\frac{\partial u}{\partial \eta}\right] = 0,$$

and the boundary data

$$u(\xi, \eta) = 0 \quad \text{on} \quad \eta = 0, \quad \text{and} \quad \frac{\partial u}{\partial \xi} = 0 \quad \text{on} \quad \xi = 0,$$

where (ξ, η) are the oblate spheroidal coordinates related to the cylindrical polar coordinates (r, z) by $r = (1-\xi^2)^{1/2}(1-\eta^2)^{1/2}$ and $z = \xi\eta$.

Chapter 13

Jacobi and Gegenbauer Transforms

13.1 Introduction

This chapter deals with Jacobi and Gegenbauer transforms and their basic operational properties. The former is a fairly general finite integral transform in the sense that both Gegenbauer and Legendre transforms follow as special cases of the Jacobi transform. Some applications of both Jacobi and Gegenbauer transforms are discussed. This chapter is based on papers by Debnath (1963, 1967), Scott (1953), Conte (1955) and Lakshmanarao (1954). In Chapters 12-15, we discussed several special transforms with orthogonal polynomials as kernels. All these special transforms have been unified by Eringen (1954) in his paper on the finite Sturm-Liouville transform.

13.2 Definition of the Jacobi Transform and Examples

Debnath (1963) introduced the *Jacobi transform* of a function $F(x)$ defined in $-1 < x < 1$ by the integral

$$J\{F(x)\} = f^{(\alpha,\beta)}(n) = \int_{-1}^{1} (1-x)^\alpha (1+x)^\beta P_n^{(\alpha,\beta)}(x) F(x)\, dx, \quad (13.2.1)$$

where $P_n^{(\alpha,\beta)}(x)$ is the Jacobi polynomial of degree n and orders $\alpha(>-1)$ and $\beta(>-1)$.

We assume that $F(x)$ admits the following series expansion

$$F(x) = \sum_{n=1}^{\infty} a_n\, P_n^{(\alpha,\beta)}(x). \quad (13.2.2)$$

In view of the orthogonal relation

$$\int_{-1}^{1} (1-x)^\alpha (1+x)^\beta P_n^{(\alpha,\beta)}(x) P_m^{(\alpha,\beta)}(x)\, dx = \delta_n\, \delta_{mn}, \quad (13.2.3)$$

where δ_{nm} is the Kronecker delta symbol,

$$\delta_n = \frac{2^{\alpha+\beta+1}\, \Gamma(n+\alpha+1)\, \Gamma(n+\beta+1)}{n!\, (\alpha+\beta+2n+1)\, \Gamma(n+\alpha+\beta+)}, \quad (13.2.4)$$

and the coefficients a_n in (13.2.2) are given by

$$a_n = \frac{1}{\delta_n} \int_{-1}^{1} F(x) \, P_n^{(\alpha,\beta)}(x) \, dx = \frac{f^{(\alpha,\beta)}(n)}{\delta_n}. \tag{13.2.5}$$

Thus the *inverse Jacobi transform* is given by

$$J^{-1}\left\{f^{(\alpha,\beta)}(n)\right\} = F(x) = \sum_{n=0}^{\infty} (\delta_n)^{-1} f^{(\alpha,\beta)}(n) \, P_n^{(\alpha,\beta)}(x). \tag{13.2.6}$$

Note that both J and J^{-1} are linear transformations.

Example 13.2.1 If $F(x)$ is a polynomial of degree $m < n$, then

$$J\{F(x)\} = 0. \tag{13.2.7}$$

Example 13.2.2

$$J\left\{P_m^{(\alpha,\beta)}(x)\right\} = \delta_{mn}. \tag{13.2.8}$$

Example 13.2.3 From the uniformly convergent expansion of the generating function for $|z| < 1$

$$2^{\alpha+\beta} \, Q^{-1} (1-z+Q)^{-\alpha} (1+z+Q)^{-\beta} = \sum_{n=0}^{\infty} z^n \, P_n^{(\alpha,\beta)}(x), \tag{13.2.9}$$

it turns out that

$$J\left\{2^{\alpha+\beta} \, Q^{-1} (1-z+Q)^{-\alpha} (1+z+Q)^{-\beta}\right\}$$

$$= \sum_{n=0}^{\infty} z^n \int_{-1}^{1} (1-x)^\alpha (1+x)^\beta \, P_n^{(\alpha,\beta)}(x) \, P_n^{(\alpha,\beta)}(x) \, dx$$

$$= \sum_{n=0}^{\infty} (\delta_n) z^n. \tag{13.2.10}$$

Example 13.2.4

$$J\{x^n\} = \int_{-1}^{1} (1-x)^\alpha (1+x)^\beta \, P_n^{(\alpha,\beta)}(x) \, x^n \, dx$$

$$= 2^{n+\alpha+\beta+1} \, \frac{\Gamma(n+\alpha+1) \, \Gamma(n+\beta+1)}{\Gamma(n+\alpha+\beta+1)}. \tag{13.2.11}$$

Example 13.2.5 If $p > \beta - 1$, then

$$J\left\{(1+x)^{p-\beta}\right\} = \int_{-1}^{1} (1-x)^\alpha (1+x)^p \, P_n^{(\alpha,\beta)}(x) \, dx$$

$$= \binom{n+\alpha}{n} 2^{\alpha+p+1} \, \frac{\Gamma(p+1)\Gamma(\alpha+1)\Gamma(p-\beta+1)}{\Gamma(\alpha+p+n+2)\Gamma(p-\beta+n+1)}. \tag{13.2.12}$$

In particular, when $\alpha = \beta = 0$, the above results reduce to the corresponding results for the Legendre transform defined by (12.2.1) so that

$$\mathcal{J}_n\{(1+x)^p\} = \int_{-1}^{1} (1+x)^p \, P_n(x) \, dx$$

$$= \frac{2^{p+1}\{\Gamma(1+p)\}^2}{\Gamma(p+n+2)\,\Gamma(p+n+1)}, \quad (p > -1) \quad (13.2.13)$$

Example 13.2.6 If Re $\sigma > -1$, then

$$J\{(1-x)^{\sigma-\alpha}\} = \int_{-1}^{1} (1-x)^\sigma \,(1+x)^\beta \, P_n^{(\alpha,\beta)}(x) \, dx$$

$$= \frac{2^{\sigma+\beta+1}}{n!\,\Gamma(\alpha-\sigma)} \cdot \frac{\Gamma(\sigma+1)\,\Gamma(n+\beta+1)\,\Gamma(\alpha-\sigma+n)}{\Gamma(\beta+\sigma+n+2)}, \quad \text{Re } \sigma > -1, \quad (13.2.14)$$

Example 13.2.7 If Re $\sigma > -1$, then

$$J\{(1+x)^{\sigma-\beta} \, P_m^{(\alpha,\sigma)}(x)\}$$

$$= \int_{-1}^{1} (1-x)^\alpha \, (1+x)^\sigma \, P_n^{(\alpha,\beta)}(x) \, P_m^{(\alpha,\sigma)}(x) \, dx$$

$$= \frac{2^{\alpha+\sigma+1} \, \Gamma(n+\alpha+1)\,\Gamma(\alpha+\beta+m+n+1)\,\Gamma(\sigma+m+1)\,\Gamma(\sigma-\beta+1)}{m!\,(n-m)!\,\Gamma(\alpha+\beta+n+1)\,\Gamma(\alpha+\sigma+m+n+2)\,\Gamma(\alpha-\beta+m+1)}. \quad (13.2.15)$$

13.3 Basic Operational Properties

Theorem 13.3.1 If $J\{F(x)\} = f^{(\alpha,\beta)}(n)$,

$$\lim_{|x| \to 1} (1-x)^{\alpha+1} (1+x)^{\beta+1} F(x) = 0 = \lim_{|x| \to 1} (1-x)^{\alpha+1} (1+x)^{\beta+1} F'(x) = 0, \quad (13.3.1\text{ab})$$

and if

$$R[F(x)] = (1-x)^{-\alpha} (1+x)^{-\beta} \frac{d}{dx}\left[(1-x)^{\alpha+1} (1+x)^{\beta+1} \frac{d}{dx} F(x)\right], \quad (13.3.2)$$

then $J\{R[F(x)]\}$ exists and is given by

$$J\{R[F(x)]\} = -n(n+\alpha+\beta+1)\, f^{(\alpha,\beta)}(n), \quad (13.3.3)$$

where $n = 0, 1, 2, 3, \ldots$.

Proof. We have, by definition,

$$J\{R[F(x)]\} = \int_{-1}^{1} \frac{d}{dx}\left[(1-x)^{\alpha+1} (1+x)^{\beta+1} \frac{dF}{dx}\right] P_n^{(\alpha,\beta)}(x) \, dx$$

which is, by integrating by parts and using the orthogonal relation (13.2.3),

$$= -n(n+\alpha+\beta+1) \int_{-1}^{1} (1-x)^\alpha (1+x)^\beta \, P_n^{(\alpha,\beta)}(x) \, F(x) \, dx$$

$$= -n(n+\alpha+\beta+1)\, f^{(\alpha,\beta)}(n).$$

This completes the proof.

If $F(x)$ and $R[F(x)]$ satisfy the conditions of Theorem 13.3.1, then $J\{R[R[F(x)]]\}$ exists and is given by

$$J\{R^2[F(x)]\} = J\{R[R[F(x)]]\} = (-1)^2 \, n^2 (n+\alpha+\beta+1)^2 \, f^{(\alpha,\beta)}(n). \quad (13.3.4)$$

More generally, if $F(x)$ and $R^k[F(x)]$ satisfy the conditions of Theorem 13.3.1, where $k = 1, 2, \ldots, m-1$, then

$$J\{R^m[F(x)]\} = (-1)^m \, n^m \, (n+\alpha+\beta+1)^m \, f^{(\alpha,\beta)}(n), \quad (13.3.5)$$

where m is a positive integer.

When $\alpha = \beta = 0$, $P_n^{(0,0)}(x)$ becomes the Legendre polynomial $P_n(x)$ and the Jacobi transform pairs (13.2.1) and (13.2.5) reduce to the Legendre transform pairs (12.2.1) and (12.2.3). All results for the Jacobi transform also reduce to those given in Chapter 12.

13.4 Applications of Jacobi Transforms to the Generalized Heat Conduction Problem

The one-dimensional generalized heat equation for temperature $u(x,t)$ is given by

$$\frac{\partial}{\partial x}\left[\kappa \frac{\partial u}{\partial x}\right] + Q(x,t) = \rho c \frac{\partial u}{\partial t}, \quad (13.4.1)$$

where κ is the thermal conductivity, $Q(x,t)$ is a continuous heat source within the medium, ρ and c are density and specific heat respectively. If the thermal conductivity is $\kappa = a(1-x^2)$, where a is a real constant, and the source is $Q(x,t) = (\mu x + v)\frac{\partial u}{\partial x}$, then the heat equation (13.4.1) reduces to

$$\frac{\partial}{\partial x}\left[(1-x^2)\frac{\partial u}{\partial x}\right] + \left(\frac{\mu x + v}{a}\right)\frac{\partial u}{\partial x} = \left(\frac{\rho c}{a}\right)\frac{\partial u}{\partial t}. \quad (13.4.2)$$

We consider a non-homogeneous beam with ends at $x = \pm 1$ whose lateral surface is insulated. Since $\kappa = 0$ at the ends, the ends of the beam are also insulated. We assume the initial conditions as

$$u(x,0) = G(x) \quad \text{for all } -1 < x < 1, \quad (13.4.3)$$

where $G(x)$ is a suitable function so that $J\{G(x)\}$ exists.

If we write $\frac{\mu}{a} = -(\alpha+\beta)$ and $\frac{v}{a} = \beta - \alpha$ so that $(\alpha,\beta) = -\left(\frac{\mu+v}{2a}, \frac{\mu-v}{2a}\right)$, the left hand side of (13.4.2) becomes

$$\frac{\partial}{\partial x}\left[(1-x^2)\frac{\partial u}{\partial x}\right] + [(\beta-\alpha)-(\beta+\alpha)x]\frac{\partial u}{\partial x}$$

$$= \frac{\partial}{\partial x}\left[(1-x^2)\frac{\partial u}{\partial x}\right] + [(1-x)\beta - (1+x)\alpha]\frac{\partial u}{\partial x}$$

$$= (1-x)^{-\alpha}(1+x)^{-\beta}\left\{(1-x)^{\alpha}(1+x)^{\beta}\frac{\partial}{\partial x}\left[(1-x^2)\frac{\partial u}{\partial x}\right]\right.$$

$$\left. + \left[\beta(1+x)^{\beta}(1-x)^{\alpha+1} - \alpha(1-x)^{\alpha}(1+x)^{\beta+1}\right]\frac{\partial u}{\partial x}\right\}$$

$$= (1-x)^{-\alpha}(1+x)^{-\beta}\left\{\frac{\partial}{\partial x}\left[(1-x)^{\alpha+1}(1+x)^{\beta+1}\frac{\partial u}{\partial x}\right]\right\}$$

$$= R[u(x,t)].$$

Thus, equation (13.4.2) reduces to

$$R[u(x,t)] = \left(\frac{1}{d}\right)\frac{\partial u}{\partial t}, \quad d = \left(\frac{a}{\rho c}\right). \tag{13.4.4}$$

Application of the Jacobi transform to (13.4.4) and (13.4.3) gives

$$\frac{d}{dt}u^{(\alpha,\beta)}(n,t) = -d\,n\,(n+\alpha+\beta+1)\,u^{(\alpha,\beta)}(n,t), \tag{13.4.5}$$

$$u^{(\alpha,\beta)}(n,0) = g^{(\alpha,\beta)}(n). \tag{13.4.6}$$

The solution of this system is

$$u^{(\alpha,\beta)}(n,t) = g^{(\alpha,\beta)}(n)\exp\left[-n(n+\alpha+\beta+1)td\right]. \tag{13.4.7}$$

The inverse Jacobi transform gives the formal solution

$$u(x,t) = \sum_{n=0}^{\infty}\delta_n^{-1}\,g^{(\alpha,\beta)}(n)\,P_n^{(\alpha,\beta)}(x)\exp\left[-n(n+\alpha+\beta+1)td\right], \tag{13.4.8}$$

where $\alpha = -\frac{1}{2a}(\mu+\nu)$ and $\beta = \frac{1}{2a}(\mu-\nu)$.

13.5 The Gegenbauer Transform and its Basic Operational Properties

When $\alpha = \beta = \nu - \frac{1}{2}$, the Jacobi polynomial $P_n^{(\alpha,\beta)}(x)$ becomes the Gegenbauer polynomial $C_n^\nu(x)$ which satisfies the self-adjoint differential form

$$\frac{d}{dx}\left[(1-x^2)^{\nu+\frac{1}{2}}\frac{dy}{dx}\right] + n(n+2\nu)(1-x^2)^{\nu-\frac{1}{2}}y = 0, \tag{13.5.1}$$

and the orthogonal relation

$$\int_{-1}^{1}(1-x^2)^{\nu-\frac{1}{2}}C_m^\nu(x)\,C_n^\nu(x)\,dx = \delta_n\,\delta_{mn}, \tag{13.5.2}$$

where

$$\delta_n = \frac{2^{1-2\nu} \pi \Gamma(n+2\nu)}{n!(n+\nu)[\Gamma(\nu)]^2}. \tag{13.5.3}$$

Thus, when $\alpha = \beta = \nu - \frac{1}{2}$, the Jacobi transform pairs (13.2.1) and (13.2.6) reduce to the *Gegenbauer transform* pairs, in the form

$$G\{F(x)\} = f^{(\nu)}(n) = \int_{-1}^{1} (1-x^2)^{\nu-\frac{1}{2}} C_n^{\nu}(x) F(x) dx, \tag{13.5.4}$$

$$G^{-1}\{f^{(\nu)}(n)\} = F(x) = \sum_{n=0}^{\infty} \delta_n^{-1} C_n^{\nu}(x) f^{(\nu)}(n), \qquad -1 < x < 1. \tag{13.5.5}$$

Obviously, G and G^{-1} stand for the Gegenbauer transformation and its inverse respectively. They are linear integral transformations.

When $\alpha = \beta = \nu - \frac{1}{2}$, the differential form (13.3.2) becomes

$$R[F(x)] = (1-x^2) \frac{d^2 F}{dx^2} - (2\nu+1) x \frac{dF}{dx}, \tag{13.5.6}$$

which can be expressed as

$$R[F(x)] = (1-x^2)^{\frac{1}{2}-\nu} \frac{d}{dx}\left[(1-x^2)^{\nu+\frac{1}{2}} \frac{dF}{dx}\right]. \tag{13.5.7}$$

Under the Gegenbauer transformation G, the differential form (13.5.6) is reduced to the algebraic form

$$G\{R[F(x)]\} = -n(n+2\nu) f^{(\nu)}(n). \tag{13.5.8}$$

This follows directly from (13.3.3).

Similarly, we obtain

$$G\{R^2[F(x)]\} = (-1)^2 n^2 (n+2\nu)^2 f^{(\nu)}(n). \tag{13.5.9}$$

More generally,

$$G\{R^k[F(x)]\} = (-1)^k n^k (n+2\nu)^k f^{(\nu)}(n), \tag{13.5.10}$$

where $k = 1, 2, \dots$.

Convolution Theorem 13.5.1 If $G\{F(x)\} = f^{(\nu)}(n)$ and $G\{G(x)\} = g^{(\nu)}(n)$, then

$$f^{(\nu)}(n) g^{(\nu)}(n) = G\{H(x)\} = h^{(\nu)}(n), \tag{13.5.11}$$

where

$$H(x) = G^{-1}\{h^{(\nu)}(n)\} = G^{-1}\{f^{(\nu)}(n) g^{(\nu)}(n)\} = F(x) * G(x), \tag{13.5.12}$$

and $H(x)$ is given by

$$H(\cos \psi) = A (\sin \psi)^{1-2\nu} \int_0^{\pi}\int_0^{\pi} F(\cos \theta) G(\cos \phi)(\sin \theta)^{2\nu}$$
$$\times (\sin \phi)^{2\nu-1} (\sin \lambda)^{2\nu-1} d\theta\, d\alpha. \tag{13.5.13}$$

Proof. We have, by definition,

$$f^{(v)}(n) g^{(v)}(n) = \int_{-1}^{1} F(x)(1-x^2)^{v-\frac{1}{2}} C_n^v(x) dx$$

$$\times \int_{-1}^{1} G(x)(1-x^2)^{v-\frac{1}{2}} C_n^v(x) dx$$

$$= \int_0^{\pi} F(\cos \theta)(\sin \theta)^{2v} C_n^v(\cos \theta) d\theta$$

$$\times \int_0^{\pi} G(\cos \phi)(\sin \phi)^{2v} C_n^v(\cos \phi) d\phi$$

$$= \int_0^{\pi} F(\cos \theta)(\sin \theta)^{2v} \left[\int_0^{\pi} G(\cos \phi) C_n^v(\cos \theta) C_n^v(\cos \phi) \right.$$

$$\left. \times (\sin \phi)^{2v} d\phi \right] d\theta. \quad (13.5.14)$$

The addition formula for the Gegenbauer polynomial (see Erdélyi, 1953, p 177) is

$$C_n^v(\cos \theta) C_n^v(\cos \phi) = A \int_0^{\pi} C_n^v(\cos \psi)(\sin \lambda)^{2v-1} d\lambda, \quad (13.5.15)$$

where

$$A = \{\Gamma(n+2v)/n! \, 2^{2v-1} \, \Gamma^2(v)\}, \quad (13.5.16)$$

and

$$\cos \psi = \cos \theta \cos \phi + \sin \theta \sin \phi \cos \lambda. \quad (13.5.17)$$

In view of this formula, result (13.5.14) assumes the form

$$f^{(v)}(n) g^{(v)}(n)$$

$$= A \int_0^{\pi} F(\cos \theta)(\sin \theta)^{2v} \left[\int_0^{\pi}\int_0^{\pi} G(\cos \phi) C_n^v(\cos \psi) \right.$$

$$\left. \times (\sin \phi)^{2v} (\sin \lambda)^{2v-1} d\lambda \, d\phi \right] d\theta. \quad (13.5.18)$$

We next introduce a new variable α defined by the relation

$$\cos \phi = \cos \theta \cos \psi + \sin \theta \sin \psi \cos \alpha. \quad (13.5.19)$$

Thus, under the transformation of coordinates defined by (13.5.17) and (13.5.19), the elementary area $d\lambda \, d\phi = (\sin \psi / \sin \phi) d\psi \, d\alpha$, where $(\sin \psi / \sin \phi)$ is the Jacobian of the transformation. In view of this transformation, the square region of the $\phi - \lambda$ plane given by $(0 \le \phi \le \pi, \, 0 \le \lambda \le \pi)$ transforms into a square region of the same dimension in the $\psi - \alpha$ plane. Consequently, the double integral inside the square bracket in (13.5.18) reduces to

$$\int_0^{\pi}\int_0^{\pi} G(\cos \phi) C_n^v(\cos \psi)(\sin \phi)^{2v-1} (\sin \lambda)^{2v-1} \sin \psi \, d\psi \, d\alpha, \quad (13.5.20)$$

where $\cos \psi$ is defined by (13.5.17) and $\cos \phi$ is defined by (13.5.19). If the double integral (13.5.20) is substituted into (13.5.18), and if the order of integration is interchanged, (13.5.18) becomes

$$f^{(v)}(n) g^{(v)}(n) = \int_0^\pi (\sin \psi)^{2v} C_n^v(\cos \psi) H(\cos \psi) d\psi$$

$$= G\{H(\cos \psi)\}, \qquad (13.5.21)$$

where

$$H(\cos \psi) = A(\sin \psi)^{1-2v} \int_0^\pi \int_0^\pi F(\cos \theta) G(\cos \phi) (\sin \theta)^{2v}$$

$$\times (\sin \phi)^{2v-1} (\sin \lambda)^{2v-1} d\theta \, d\alpha. \qquad (13.5.22)$$

When $v = \frac{1}{2}$, $C_n^{\frac{1}{2}}(x)$ becomes the Legendre polynomial, the Gegenbauer transform pairs (13.5.4) and (13.5.5) reduce to the Legendre transform pairs (12.2.1) and (12.2.3), and the Convolution Theorem 13.5.1 reduces to the corresponding Convolution Theorem 12.3.4 for the Legendre transform.

13.6 Application of the Gegenbauer Transform

The generalized one-dimensional heat equation in a non-homogeneous solid beam for the temperature $u(x,t)$ is

$$\frac{\partial}{\partial x}\left[(1-x^2) \frac{\partial u}{\partial x}\right] - (2v+1) x \frac{\partial u}{\partial x} = \frac{1}{d} \frac{\partial u}{\partial t}, \qquad (13.6.1)$$

where $\kappa = (1-x^2)$ is the thermal conductivity, $d = \frac{a}{\rho c}$, and the second term on the left hand side represents the continuous source of heat within the solid beam. We assume that the beam is bounded by the planes at $x = \pm 1$ and its lateral surfaces are insulated. The initial condition is

$$u(x,0) = G(x) \quad \text{for } -1 < x < 1, \qquad (13.6.2)$$

where $G(x)$ is a given function so that its Gegenbauer transform exists.

Application of the Gegenbauer transform to (13.6.1) and (13.6.2) and the use of (13.5.8) gives

$$\frac{d}{dt} u^{(v)}(n,t) = - dn(n+2v) u^{(v)}(n,t), \qquad (13.6.3)$$

$$u^{(v)}(n,0) = g^{(v)}(n). \qquad (13.6.4)$$

The solution of this system is

$$u^{(v)}(n,t) = g^{(v)}(n) \exp[-n(n+2v)td]. \qquad (13.6.5)$$

The inverse transform gives the formal solution

$$u(x,t) = \sum_{n=0}^\infty \delta_n^{-1} C_n^v(x) g^{(v)}(n) \exp[-n(n+2v)td]. \qquad (13.6.6)$$

Chapter 14

Laguerre Transforms

14.1 Introduction

This chapter is devoted to the study of the Laguerre transform and its basic operational properties. It is shown that the Laguerre transform can be used effectively to solve the heat conduction problem in a semi-infinite medium with variable thermal conductivity in the presence of a heat source within the medium.

This chapter is based on a series of papers by Debnath (1960-65) and McCully (1960) listed in the Bibliography.

14.2 Definition of the Laguerre Transform and Examples

Debnath (1960) introduced the *Laguerre transform* of a function $f(x)$ defined in $0 \leq x < \infty$ by means of the integral

$$L\{f(x)\} = \tilde{f}_\alpha(n) = \int_0^\infty e^{-x} x^\alpha L_n^\alpha(x) f(x) dx, \qquad (14.2.1)$$

where $L_n^\alpha(x)$ is the Laguerre polynomial of degree $n (\geq 0)$ and order $\alpha (> -1)$ which satisfies the ordinary differential equation expressed in the self-adjoint form

$$\frac{d}{dx}\left[e^{-x} x^{\alpha+1} \frac{d}{dx} L_n^\alpha(x)\right] + n e^{-x} x^\alpha L_n^\alpha(x) = 0. \qquad (14.2.2)$$

In view of the orthogonal property of the Laguerre polynomials

$$\int_0^\infty e^{-x} x^\alpha L_n^\alpha(x) L_m^\alpha(x) dx = \binom{n+\alpha}{n} \Gamma(\alpha+1) \delta_{mn} = \delta_n \delta_{nm}, \qquad (14.2.3)$$

where δ_{mn} is the Kronecker delta symbol, and δ_n is given by

$$\delta_n = \binom{n+\alpha}{n} \Gamma(\alpha+1). \qquad (14.2.4)$$

The *inverse Laguerre transform* is given by

$$f(x) = L^{-1}\{\tilde{f}_\alpha(n)\} = \sum_{n=0}^\infty (\delta_n)^{-1} \tilde{f}_\alpha(n) L_n^\alpha(x). \qquad (14.2.5)$$

When $\alpha = 0$, the Laguerre transform pairs due to McCully (1960) follow from (14.2.1) and (14.2.5) in the form

$$L\{f(x)\} = \tilde{f}_0(n) = \int_0^\infty e^{-x} L_n(x) f(x) dx, \qquad (14.2.6)$$

$$L^{-1}\{\tilde{f}_0(n)\} = f(x) = \sum_{n=0}^\infty \tilde{f}_0(n) L_n(x), \qquad (14.2.7)$$

where $L_n(x)$ is the Laguerre polynomial of degree n and order zero.

Obviously, L and L^{-1} are linear integral transformations. The following examples (Debnath, 1960) illustrate the Laguerre transform of some simple functions.

Example 14.2.1 If $f(x) = L_m^\alpha(x)$ then $L\{L_m^\alpha(x)\} = \delta_n \delta_{nm}$. (14.2.8)

This follows directly from the definitions, (14.2.1) and (14.2.3).

Example 14.2.2 If $f(x) = x^{s-1}$ and s is a positive real number, then

$$L\{x^{s-1}\} = \int_0^\infty e^{-x} x^{\alpha+s-1} L_n^\alpha(x) dx = \frac{\Gamma(s+\alpha)\Gamma(n-s+1)}{n!\,\Gamma(1-s)}, \qquad (14.2.9)$$

in which a result due to Howell (1938) is used.

Example 14.2.3 If $a > -1$, and $f(x) = e^{-ax}$ then

$$L\{e^{-ax}\} = \int_0^\infty e^{-x(1+a)} x^\alpha L_n^\alpha(x) dx$$

$$= \frac{\Gamma(n+\alpha+1)\,a^n}{n!\,(a+1)^{n+\alpha+1}}, \qquad (14.2.10)$$

where Erdélyi et al.'s (1954, vol. 2, p 191) result is used.

Example 14.2.4 If $f(x) = e^{-ax} L_m^\alpha(x)$, then

$$L\{e^{-ax} L_m^\alpha(x)\} = \int_0^\infty e^{-x(a+1)} x^\alpha L_n^\alpha(x) L_m^\alpha(x) dx,$$

which is, due to Howell (1938),

$$= \frac{1}{n!\,m!} \frac{\Gamma(n+\alpha+1)\Gamma(m+\alpha+1)}{\Gamma(1+\alpha)} \cdot \frac{(a-1)^{n-m+\alpha+1}}{a^{n+m+2\alpha+2}}$$

$$\times {}_2F_1\left(n+\alpha+1,\,\frac{m+a+1}{a+1};\,\frac{1}{a^2}\right), \qquad (14.2.11)$$

where ${}_2F_1(x, \alpha, \beta)$ is the hypergeometric function.

Example 14.2.5

$$L\{f(x) x^{\beta-\alpha}\} = \int_0^\infty e^{-x} x^\beta L_n^\alpha(x) f(x) dx.$$

We use a result from Erdélyi (1953, vol. 2, p 192) as

$$L_n^\alpha(x) = \sum_{m=0}^n (m!)^{-1} (\alpha-\beta)_m L_{n-m}^\beta(x) \qquad (14.2.12)$$

to obtain the following result:

$$L\{f(x) x^{\beta-\alpha}\} = \sum_{m=0}^{n} (m!)^{-1} (\alpha-\beta)_m \tilde{f}_\beta(n-m). \quad (14.2.13)$$

In particular, when $\beta = \alpha - 1$, we obtain

$$L\left\{\frac{f(x)}{x}\right\} = \sum_{m=0}^{n} (m!)^{-1} \tilde{f}_{\alpha-1}(n-m).$$

Example 14.2.6

$$L\{e^x x^{-\alpha} \Gamma(\alpha, x)\} = \sum_{n=0}^{\infty} \frac{\delta_n}{(n+1)}, \quad -1 < \alpha < 0. \quad (14.2.14)$$

We use a result from Erdélyi (1953, vol. 2, p 215) as

$$e^x x^{-\alpha} \Gamma(\alpha, x) = \sum_{n=0}^{\infty} (n+1)^{-1} L_n^\alpha(x), \quad (\alpha > -1, x > 0),$$

in the definition (14.2.1) to derive (14.2.14).

Example 14.2.7 If $\beta > 0$, then

$$L\{x^\beta\} = \Gamma(\alpha+\beta+1) \sum_{n=0}^{\infty} \frac{(-\beta)_n \delta_n}{\Gamma(n+\alpha+1)}. \quad (14.2.15)$$

Using the result from Erdélyi (1953, vol. 2, p 214)

$$x^\beta = \Gamma(\alpha+\beta+1) \sum_{n=0}^{\infty} \frac{(-\beta)_n}{\Gamma(n+\alpha+1)} L_n^\alpha(x),$$

where

$$-\beta < 1 + \min\left(\alpha, \frac{\alpha}{2} - \frac{1}{4}\right), \quad x > 0, \alpha > -1,$$

we can easily obtain (14.2.15).

Example 14.2.8 If $|z| < 1$ and $\alpha \geq 0$, then

(a) $$L\left\{(1-z)^{-(\alpha+1)} \exp\left(\frac{xz}{z-1}\right)\right\} = \sum_{n=0}^{\infty} \delta_n z^n, \quad (14.2.16)$$

(b) $$L\left\{(xz)^{-\frac{\alpha}{2}} e^z J_\alpha\left[2(xz)^{\frac{1}{2}}\right]\right\} = \sum_{n=0}^{\infty} \frac{\delta_n z^n}{\Gamma(n+\alpha+1)}. \quad (14.2.17)$$

We have the following generating functions (Erdélyi, 1953, vol. 2, p 189)

$$(1-z)^{-(\alpha+1)} \exp\left(\frac{xz}{z-1}\right) = \sum_{n=0}^{\infty} L_n^\alpha(x) z^n, \quad |z| < 1,$$

$$(xz)^{-\alpha/2} e^z J_\alpha\left[2\sqrt{xz}\right] = \sum_{n=0}^{\infty} \frac{z^n L_n^\alpha(x)}{\Gamma(n+\alpha+1)}, \quad |z| < 1.$$

In view of these results combined with the orthogonality relation (14.2.3), we obtain (14.2.16) and (14.2.17).

Example 14.2.9 (*Recurrence Relations*).

(a) $$\tilde{f}_{\alpha+1}(n) = (n+\alpha+1) \tilde{f}_\alpha(n) - (n+1) \tilde{f}_\alpha(n+1), \quad (14.2.18)$$

(b) $$n! \tilde{f}_{m-n}(n) = (-1)^{n-m} m! \sum_{k=0}^{m} (k!)^{-1} (2n-2m)_k \tilde{f}_{m-n}(m-k). \quad (14.2.19)$$

We have

$$\tilde{f}_{\alpha+1}(n) = \int_0^\infty e^{-x} x^{\alpha+1} L_n^{\alpha+1}(x) f(x) dx,$$

which is, by using the recurrence relation for the Laguerre polynomial,

$$= \int_0^\infty e^{-x} x^\alpha \left[(n+\alpha+1) L_n^\alpha(x) - (n+1) L_{n+1}^\alpha(x) \right] f(x) dx$$

$$= (n+\alpha+1) \tilde{f}_\alpha(n) - (n+1) \tilde{f}_\alpha(n+1).$$

Similarly, we find

$$n! \tilde{f}_{m-n}(n) = \int_0^\infty e^{-x} x^{m-n} n! L_n^{m-n}(x) f(x) dx.$$

We next use a result due to Howell (1938), as

$$n! L_n^{m-n}(x) = (-1)^{n-m} m! L_m^{n-m}(x)$$

to obtain

$$n! \tilde{f}_{m-n}(n) = (-1)^{n-m} m! \int_0^\infty e^{-x} x^{m-n} L_m^{n-m}(x) f(x) dx$$

$$= (-1)^{n-m} m! \sum_{k=0}^m (k!)^{-1} (2n-2m)_k \tilde{f}_{m-n}(m-k).$$

14.3 Basic Operational Properties

We obtain the Laguerre transform of derivatives of $f(x)$ as

$$L\{f'(x)\} = \tilde{f}_\alpha(n) - \alpha \sum_{k=0}^n f_{\alpha-1}(k) + \sum_{k=0}^{n-1} f_\alpha(k), \qquad (14.3.1)$$

$$L\{f''(x)\} = \tilde{f}_\alpha(n) - 2\alpha \sum_{m=0}^n \tilde{f}_{\alpha-1}(n-m) + 2 \sum_{m=0}^{n-1} \tilde{f}_\alpha(n-m-1)$$

$$- 2\alpha \sum_{m=0}^{n-1} (m+1) \tilde{f}_{\alpha+1}(n-m-1)$$

$$+ \alpha(\alpha-1) \sum_{m=0}^n (m+1) f_{\alpha-2}(n-m)$$

$$+ \sum_{m=0}^{n-2} (m+1) \tilde{f}_\alpha(n-m-2), \qquad (14.3.2)$$

and so on for the Laguerre transforms of higher derivatives.

We have, by definition,

$$L\{f'(x)\} = \int_0^\infty e^{-x} x^\alpha L_n^\alpha(x) f'(x) dx$$

$$= \left[e^{-x} n^\alpha L_n^\alpha(x) f(x) \right]_0^\infty + \int_0^\infty e^{-x} x^\alpha L_n^\alpha(x) f(x) dx$$

$$-\alpha \int_0^\infty e^{-x} x^{\alpha-1} L_n^\alpha(x) f(x) dx - \int_0^\infty e^{-x} x^\alpha \left[\frac{d}{dx} L_n^\alpha(x) \right] f(x) dx,$$

which is, due to Erdélyi (1953, vol. 2, p 192),

$$= \tilde{f}_\alpha(n) - \alpha \sum_{k=0}^n \tilde{f}_{\alpha-1}(k) + \sum_{k=0}^{n-1} f_\alpha(k).$$

Similarly, we can derive (14.3.2).

Theorem 14.3.1 If $g(x) = \int_0^x f(t) dt$ so that $g(x)$ is absolutely continuous and $g'(x)$ exists, and if $g'(x)$ is bounded and integrable, then

$$\tilde{f}_\alpha(n) - \tilde{f}_\alpha(n-1) = \tilde{g}_\alpha(n) - \alpha \tilde{g}_{\alpha-1}(n), \tag{14.3.3}$$

and

$$L\left\{ \int_0^x f(t) dt \right\} = \tilde{f}_0(n) - \tilde{f}_0(n-1), \tag{14.3.4}$$

where L stands for the zero-order Laguerre transform defined by (14.2.6).

Proof. We have

$$\tilde{f}_\alpha(n) = \int_0^\infty e^{-x} x^\alpha L_n^\alpha(x) g'(x) dx,$$

which is, by integrating by parts,

$$= \int_0^\infty e^{-x} x^\alpha L_n^\alpha(x) g(x) dx - \alpha \int_0^\alpha e^{-x} x^{\alpha-1} L_n^\alpha(x) g(x) dx$$

$$- \int_0^\infty e^{-x} x^\alpha \left[\frac{d}{dx} L_n^\alpha(x) \right] g(x) dx.$$

Thus,

$$\tilde{f}_\alpha(n) - \tilde{f}_\alpha(n+1)$$

$$= \int_0^\infty e^{-x} x^\alpha \left[L_n^\alpha(x) - L_{n+1}^\alpha(x) \right] g(x) dx + \alpha \int_0^\infty e^{-x} x^\alpha \left[L_{n+1}^\alpha(x) - L_n^\alpha(x) \right] g(x) dx$$

$$- \int_0^\infty e^{-x} x^\alpha \frac{d}{dx} \left[L_n^\alpha(x) - L_{n+1}^\alpha(x) \right] g(x) dx.$$

Thus,

$$\tilde{f}_\alpha(n) - \tilde{f}_\alpha(n+1) = \int_0^\infty e^{-x} x^\alpha \left[L_n^\alpha(x) - L_{n+1}^\alpha(x)\right] g(x) dx$$

$$+ \alpha \int_0^\infty e^{-x} x^\alpha L_{n+1}^{\alpha-1}(x) g(x) dx - \int_0^\infty e^{-x} x^\alpha L_n^\alpha(x) g(x) dx$$

$$= -\tilde{g}_\alpha(n+1) + \alpha \tilde{g}_{\alpha-1}(n+1).$$

This proves (14.3.3).

Putting $\alpha = 0$, and replacing n by $n-1$ gives

$$\tilde{g}_0(n) = \tilde{f}_0(n) - \tilde{f}_0(n-1).$$

Or

$$L\left\{\int_0^x f(t) dt\right\} = \tilde{f}_0(n) - \tilde{f}_0(n-1).$$

Theorem 14.3.2 If $L\{f(x)\} = \tilde{f}_\alpha(n)$ exists, then

$$L\{R[f(x)]\} = -n \tilde{f}_\alpha(n), \qquad (14.3.5)$$

where $R[f(x)]$ is the differential operator given by

$$R[f(x)] = e^x x^{-\alpha} \frac{d}{dx}\left[e^{-x} x^{\alpha+1} \frac{d}{dx} f(x)\right]. \qquad (14.3.6)$$

Proof. We have, by definition,

$$L\{R[f(x)]\} = \int_0^\infty L_n^\alpha(x) \frac{d}{dx}\left[e^{-x} x^{\alpha+1} \frac{df}{dx}\right] dx,$$

which is, by integrating by parts and using (14.2.2),

$$= -n \int_0^\infty e^{-x} x^\alpha L_n^\alpha(x) f(x) dx$$

$$= -n \tilde{f}_\alpha(n).$$

This completes the proof of the basic operational property. This result can easily be extended as follows:

$$L\{R^2[f(x)]\} = L\{R[R[f(x)]]\} = (-1)^2 n^2 \tilde{f}_\alpha(n). \qquad (14.3.7)$$

More generally,

$$L\{R^m[f(x)]\} = (-1)^m n^m \tilde{f}_\alpha(n), \qquad (14.3.8)$$

where m is a non-negative integer.

The Convolution Theorem for the Laguerre transform can be stated as follows:

Theorem 14.3.3 (Convolution Theorem). If $L\{f(x)\} = \tilde{f}_\alpha(n)$ and $L\{g(x)\} = \tilde{g}_\alpha(n)$, then

$$L^{-1}\{\tilde{f}_\alpha(n) \tilde{g}_\alpha(n)\} = h(x), \qquad (14.3.9)$$

where $h(x)$ is given by the following repeated integral

$$h(x) = \frac{\Gamma(n+\alpha+1)}{\sqrt{\pi}\,\Gamma(n+1)} \int_0^\infty e^{-t} t^\alpha f(t)dt \int_0^\pi \exp\left(-\sqrt{xt}\,\cos\phi\right)$$

$$\times \sin^{2\alpha}\phi\; g\!\left(x+t+2\sqrt{xt}\,\cos\phi\right) \frac{J_{\alpha-\frac{1}{2}}\!\left(\sqrt{xt}\,\sin\phi\right)d\phi}{\left[\frac{1}{2}\!\left(\sqrt{xt}\,\sin\phi\right)\right]^{\alpha-\frac{1}{2}}}. \qquad (14.3.10)$$

In order to avoid long proof of this Convolution theorem 14.3.3, we will not present the proof here, but refer the reader to the article of Debnath (1969). However, when $\alpha = 0$ and ϕ is replaced by $(\pi - \theta)$, and the standard result

$$J_{-\frac{1}{2}}(x) = \sqrt{\frac{2}{\pi x}}\,\cos x \qquad (14.3.11)$$

is used, the Convolution theorem 14.3.3 reduces to that of McCully's (1960). We now state and prove McCully's Theorem as follows:

Theorem 14.3.4 (Convolution Theorem). If $L\{f(x)\} = \tilde{f}_0(n)$ and $L\{g(x)\} = \tilde{g}_0(n)$, then

$$L^{-1}\{\tilde{f}_0(n)\tilde{g}_0(n)\} = h(x), \qquad (14.3.12)$$

where $h(x)$ is given by the formula

$$h(x) = \frac{1}{\pi}\int_0^\infty e^{-t} f(t)dt \int_0^\pi \exp\left(\sqrt{xt}\,\cos\theta\right)\cos\left(\sqrt{xt}\,\sin\theta\right)$$

$$\times g\!\left(x+t-2\sqrt{xt}\,\cos\theta\right)d\theta. \qquad (14.3.13)$$

Proof. We have, by definition,

$$\tilde{f}_0(n)\,\tilde{g}_0(n) = \int_0^\infty e^{-x} L_n(x) f(x)dx \int_0^\infty e^{-y} L_n(y) g(y)dy$$

$$= \int_0^\infty e^{-x} f(x)dx \int_0^\infty e^{-y} L_n(x) L_n(y) g(y)dy. \qquad (14.3.14)$$

This can be written in the form

$$\tilde{f}_0(n)\,\tilde{g}_0(n) = L\{h(t)\} = \int_0^\infty e^{-t} L_n(t) h(t)dt.$$

This shows that h is the convolution of f and g and has the representation

$$h(x) = f(x) * g(x). \qquad (14.3.15)$$

It follows from a formula of Bateman (1944, p 457) that

$$L_n(x) L_n(y) = \frac{1}{\pi}\int_0^\pi e^{\sqrt{xy}\,\cos\theta}\cos\left(\sqrt{xy}\,\sin\theta\right) L_n\!\left(x+y-2\sqrt{xy}\,\cos\theta\right)d\theta. \qquad (14.3.16)$$

In view of this result, (14.3.14) becomes

$$\pi \tilde{f}_0(n)\, \tilde{g}_0(n) = \int_0^\infty e^{-x} f(x)dx \left[\int_0^\infty e^{-y} g(y) \int_0^\pi \exp\left(\sqrt{xy}\, \cos\theta\right) \right.$$
$$\left. \times \cos\left(\sqrt{xy}\, \sin\theta\right) L_n\left(x+y-2\sqrt{xy}\,\cos\theta\right) d\theta\, dy \right]. \quad (14.3.17)$$

Using $\sqrt{y}$ as the variable of integration combined with polar coordinates, the integral inside the square bracket in (14.3.17) can be reduced to the form

$$\int_0^\infty e^{-t} L_n(t)dt \int_0^\pi \exp\left(\sqrt{xt}\,\cos\phi\right) \cos\left(\sqrt{xt}\,\sin\phi\right)$$
$$\times g\left(x+t-2\sqrt{xt}\,\cos\phi\right)d\phi, \quad (14.3.18)$$

so that (14.3.17) becomes

$$\tilde{f}_0(n)\, \tilde{g}_0(n) = L\{h(t)\} = \int_0^\infty e^{-t} L_n(t)\, h(t)dt,$$

where $h(x)$ is given by

$$h(x) = \frac{1}{\pi} \int_0^\infty e^{-t} f(t)dt \int_0^\pi \exp\left(\sqrt{xt}\,\cos\theta\right) \cos\left(\sqrt{xt}\,\sin\theta\right)$$
$$\times g\left(x+t-2\sqrt{xt}\,\cos\theta\right)d\theta. \quad (14.3.19)$$

This proves the Convolution Theorem for the Laguerre transform (14.2.6).

14.4 Applications of Laguerre Transforms

Example 14.4.1 *(Heat Conduction Problem).* The diffusion equation for one-dimensional linear flow of heat in a semi-infinite medium $0 \le x < \infty$ with a source $Q(x,t)$ in the medium is

$$\frac{\partial}{\partial x}\left[\kappa \frac{\partial u}{\partial x}\right] + Q(x,t) = \rho c \frac{\partial u}{\partial x}, \quad t > 0, \quad (14.4.1)$$

where $\kappa = \kappa(x) = \lambda\, e^{-x}\, x^\beta$ is the variable thermal conductivity; $Q(x,t) = \mu\, e^{-x}\, x^{\beta'}\, \frac{\partial u}{\partial x}$; $\rho = v\, e^{-x}\, x^{\beta'}$; λ, μ, v, and c are constants; and $\beta \ge 1$ and $\beta - \beta' = 1$. Thus, the above equation reduces to

$$\frac{\partial}{\partial x}\left[e^{-x} x^\beta \frac{\partial u}{\partial x}\right] + \frac{\mu}{\lambda} e^{-x} x^{\beta'} \frac{\partial u}{\partial x} = \frac{vc}{\lambda} e^{-x} x^{\beta'} \frac{\partial u}{\partial t}. \quad (14.4.2)$$

The initial condition is

$$u(x,0) = g(x), \quad 0 \le x < \infty. \quad (14.4.3)$$

Clearly, equation (14.4.2) assumes the form

$$e^x x^{-\alpha} \frac{\partial}{\partial x}\left(e^{-x} x^{\alpha+1} \frac{\partial u}{\partial x}\right) = \gamma \frac{\partial u}{\partial t}, \quad (14.4.4)$$

Integral Transforms and Their Applications　　　　　　　　　　　353

where $\alpha = \dfrac{\mu}{\lambda} + \beta - 1$ and $\gamma = \dfrac{vc}{\lambda}$.

Application of the Laguerre transform to (14.4.4) gives
$$\frac{d}{dt} u_\alpha(n,t) = -\frac{n}{\gamma} u_\alpha(n,t)$$
$$u_\alpha(n,0) = g_\alpha(n)$$

Thus the solution of this system is
$$u_\alpha(n,t) = g_\alpha(n) \exp\left(-\frac{nt}{\gamma}\right). \tag{14.4.5}$$

The inverse transform (14.2.5) gives the formal solution
$$u(x,t) = \sum_{n=0}^{\infty} (\delta_n)^{-1} g_\alpha(n) L_n^\alpha(x) \exp\left(-\frac{nt}{\gamma}\right). \tag{14.4.6}$$

Example 14.4.2 (*Diffusion Equation*). Solve equation (14.4.1) with $\kappa = x\,e^{-x}$, $Q(x,t) = e^{-x} f(t)$, and $\rho c = e^{-x}$.

In this case, the diffusion equation
$$\frac{\partial u}{\partial t} = e^x \frac{\partial}{\partial x}\left(xe^{-x} \frac{\partial u}{\partial x}\right) + f(t), \quad 0 \le x < \infty,\ t > 0, \tag{14.4.7}$$

has to be solved with the initial-boundary data
$$\left.\begin{aligned} u(x,0) &= g(x), & 0 \le x < \infty \\ \frac{\partial}{\partial t} u(x,t) &= f(x), \quad \text{at } t = 0, \quad \text{for } x > 0 \end{aligned}\right\}. \tag{14.4.8}$$

Application of the Laguerre transform $L\{u(x,t)\} = \tilde{u}_0(n,t)$ to (14.4.7)-(14.4.8) gives
$$\tilde{u}_0(n,t) = g_0(n)\,e^{-nt}, \quad n = 1, 2, 3,\ldots \tag{14.4.9}$$
$$\tilde{u}_0(0,t) = g_0(0) + \int_0^t f(\tau)d\tau. \tag{14.4.10}$$

The inverse Laguerre transform (14.2.5) leads to the formal solution
$$u(x,t) = g_0(0) + \int_0^t f(\tau)d\tau + \sum_{n=1}^{\infty} g_0(n)\,e^{-nt} L_n(x)$$
$$= \int_0^t f(\tau)d\tau + \sum_{n=0}^{\infty} g_0(n)\,e^{-nt} L_n(x). \tag{14.4.11}$$

In view of the Convolution Theorem 14.3.4, this result takes the form
$$u(x,t) = \int_0^t f(\tau)d\tau + \frac{1}{\pi}\int_0^\infty e^{-\tau}(e^\tau - 1)^{-1} \exp\left(\frac{-\tau}{e^\tau - 1}\right)$$
$$\times \int_0^\pi \exp\left(\sqrt{x\tau}\cos\theta\right)\cos\left(\sqrt{x\tau}\sin\theta\right) g\left(x + \tau - 2\sqrt{x\tau}\cos\theta\right) d\theta d\tau. \tag{14.4.12}$$

This result is obtained by McCully (1960).

Another application of the Laguerre transform to the problem of oscillations of a very long and heavy chain with variable tension was discussed by Debnath (1961).

14.5 Exercises

1. Find the zero-order Laguerre transform of each of the following functions:
   (a) $H(x-a)$ for constant $a \geq 0$,   (b) e^{-ax} $(a > -1)$,
   (c) $A L_m(x)$,   (d) x^m.   (e) $L_n(x)$.

2. If $L\{f(x)\} = f_0(n) = \int_0^\infty e^{-x} L_n(x) f(x) dx$, and $a > 0$, show that

   (a) $L\{\sin ax\} = \dfrac{a^n}{(1+a^2)^{\frac{n+1}{2}}} \sin\left[n \tan^{-1}\left(\dfrac{1}{a}\right) + \tan^{-1}(-a)\right]$,

   (b) $L\{\cos ax\} = \dfrac{a^n}{(1+a^2)^{\frac{n+1}{2}}} \cos\left[n \tan^{-1}\left(\dfrac{1}{a}\right) + \tan^{-1}(-a)\right]$.

3. If $L\{f(x)\} = \tilde{f}_0(n) = \int_0^\infty e^{-x} L_n(x) f(x) dx$, prove the following properties:

   (a) $L\{x f'(x)\} = -(n+1) \tilde{f}_0(n+1) + n \tilde{f}_0(n)$,

   (b) $L\left[e^x \dfrac{d}{dx}\{x e^{-x} f'(x)\}\right] = -n \tilde{f}_0(n)$,

   (c) $L\left[e^{-x} \dfrac{d}{dx}\{x e^x f'(x)\}\right] = n \tilde{f}_0(n) - 2(n+1) \tilde{f}_0(n+1)$,

   (d) $L\left[\dfrac{d}{dx}\{x f'(x)\}\right] = -(n+1) \tilde{f}_0(n+1)$.

4. Show that

   (a) $\tilde{f}_\alpha(n) = L\{L_n^\alpha(x)\} = \dfrac{\Gamma(n+\alpha+1)}{n!}$ for $\alpha > -1$.

   (b) $\tilde{f}_\alpha(n) = L\{x L_n^\alpha(x)\} = \dfrac{\Gamma(n+\alpha+1)}{n!}(2n+\alpha+1)$ for $\alpha > -1$.

Chapter 15

Hermite Transforms

15.1 Introduction

In this chapter we introduce the Hermite transform with a kernel involving a Hermite polynomial and discuss its basic operational properties, including the Convolution Theorem. Debnath (1964) first introduced this transform and proved some of its basic operational properties. This chapter is based on papers by Debnath (1964, 1968) and Dimovski and Kalla (1988).

15.2 Definition of the Hermite Transform and Examples

Debnath (1964) defined the *Hermite transform* of a function $F(x)$ defined in $-\infty < x < \infty$ by the integral

$$H\{F(x)\} = f_H(n) = \int_{-\infty}^{\infty} \exp(-x^2) H_n(x) F(x) \, dx, \qquad (15.2.1)$$

where $H_n(x)$ is the well-known Hermite polynomial of degree n.

The inverse Hermite transform is given by

$$H^{-1}\{f_H(n)\} = F(x) = \sum_{n=0}^{\infty} \delta_n^{-1} f_H(n) H_n(x), \qquad (15.2.2)$$

where δ_n is given by

$$\delta_n = \sqrt{\pi}\, n!\, 2^n. \qquad (15.2.3)$$

This follows from the expansion of any function $F(x)$ in the form

$$F(x) = \sum_{n=0}^{\infty} a_n H_n(x), \qquad (15.2.4)$$

where the coefficients a_n can be determined from the orthogonal relation of the Hermite polynomial $H_n(x)$ as

$$\int_{-\infty}^{\infty} \exp(-x^2) H_n(x) H_n(x) \, dx = \delta_{nm}\, \delta_n. \qquad (15.2.5)$$

Multiplying (15.2.4) by $\exp(-x^2) H_m(x)$ and integrating over $(-\infty, \infty)$ and using (15.2.4), we obtain

$$a_n = \delta_n^{-1} f_H(n) \qquad (15.2.6)$$

so that (15.2.2) follows immediately.

Example 15.2.1 If $F(x)$ is a polynomial of degree m, then
$$f_H(n) = 0 \qquad \text{for } n > m. \tag{15.2.7}$$

Example 15.2.2 If $F(x) = H_m(x)$, then
$$H\{H_m(x)\} = \int_{-\infty}^{\infty} \exp(-x^2) H_n(x) H_m(x) \, dx = \delta_n \, \delta_{nm}. \tag{15.2.8}$$

Example 15.2.3 If
$$\exp(2xt - t^2) = \sum_{n=0}^{\infty} \frac{t^n}{n!} H_n(x) \tag{15.2.9}$$

is the generating function of $H_n(x)$, then
$$H\{\exp(2xt - t^2)\} = \sqrt{\pi} \sum_{n=0}^{\infty} (2t)^n, \qquad |t| < \frac{1}{2}. \tag{15.2.10}$$

We have, by definition,
$$H\{\exp(2xt - t^2)\} = \sum_{n=0}^{\infty} \frac{t^n}{n!} \int_{-\infty}^{\infty} \exp(-x^2) H_n^2(x) \, dx$$

$$= \sum_{n=0}^{\infty} \delta_n \frac{t^n}{n!}$$

$$= \sqrt{\pi} \sum_{n=0}^{\infty} (2t)^n, \qquad |t| < \frac{1}{2}.$$

Example 15.2.4 If $F(x) = H_m(x) H_p(x)$, then
$$H\{H_m(x) H_p(x)\} = \begin{cases} \dfrac{\sqrt{\pi} \, 2^k \, m! \, n! \, p!}{(k-m)!(k-n)!(k-p)!}, & m+n+p = 2k, \\ & k \geq m, n, p \\ 0, & \text{otherwise} \end{cases}. \tag{15.2.11}$$

This follows from a result proved by Bailey (1939).

Example 15.2.5 If $F(x) = H_m^2(x) H_n(x)$, then
$$H\{H_m^2(x) H_n(x)\} = 2^m \, \delta_n \sum_{k=0}^{n} \binom{m}{k}\binom{n}{k}\binom{2k}{k}, \qquad \text{if } m > n. \tag{15.2.12}$$

Using a result proved by Feldheim (1938), (15.2.12) follows immediately.

Example 15.2.6 If $F(x) = H_{n+p+q}(x) H_p(x) H_q(x)$, then
$$H\{F(x)\} = \delta_{n+p+q}. \tag{15.2.13}$$

We have, by definition,

$$H\{F(x)\} = \int_{-\infty}^{\infty} \exp(-x^2) H_{n+p+q}(x) H_p(x) H_q(x) dx$$

which is, by a result due to Bailey (1939),
$$= \delta_{n+p+q}.$$
where δ_n is given by (15.2.3).

Example 15.2.7 If $F(x) = \exp(ax)$, then
$$H\{\exp(ax)\} = \sqrt{\pi} \sum a^n \exp\left(\frac{1}{4}a^2\right). \tag{15.2.14}$$

This result follows from the standard result
$$\int_{-\infty}^{\infty} \exp(-x^2 + 2bx) H_n(x) dx = \sqrt{\pi} (2b)^n \exp(b^2).$$

Example 15.2.8 If $|2z| < 1$, show that
$$H\{\exp(z^2) \sin(\sqrt{2}\, xz)\} = \begin{cases} 0, & n \ne 2m+1 \\ \sqrt{\pi} \sum_{m=0}^{\infty} (-1)^m (2z)^{2m+1}, & n = 2m+1 \end{cases}. \tag{15.2.15}$$

We have, by definition,
$$H\{\exp(z^2) \sin(\sqrt{2}\, xz)\} = \int_{-\infty}^{\infty} \exp(z^2 - x^2) H_n(x) \sin(\sqrt{2}\, xz) dx.$$

We use a result (see, Erdélyi et al. (1954), vol. 2, p 194)
$$\exp(z^2) \sin(\sqrt{2}\, xz) = \sum_{m=0}^{\infty} (-1)^m H_{2m+1}(x) \frac{z^{2m+1}}{(2m+1)!}, \tag{15.2.16}$$

to derive
$$H\{\exp(z^2) \sin(\sqrt{2}\, xz)\}$$
$$= \sum_{m=0}^{\infty} (-1)^m \frac{z^{2m+1}}{(2m+1)!} \int_{-\infty}^{\infty} \exp(-x^2) H_n(x) H_{2m+1}(x) dx$$
$$= \begin{cases} \sqrt{\pi} \sum_{m=0}^{\infty} (-1)^m (2z)^{2m+1}, & n = 2m+1 \\ 0, & n \ne 2m+1 \end{cases}.$$

Example 15.2.9
$$H\left[(1-z^2)^{-\frac{1}{2}} \exp\left\{\frac{2xyz - (x^2+y^2)z^2}{(1-z^2)}\right\}\right] = \sqrt{\pi} \sum_{m=0}^{\infty} z^m H_m(y) \delta_{mn}. \tag{15.2.17}$$

We use a result (see Erdélyi et al. (1954), vol. 2, p 194)

to derive

$$(1-z^2)^{-\frac{1}{2}} \exp\left\{\frac{2xyz - (x^2+y^2)z^2}{(1-z^2)}\right\} = \sum_{m=0}^{\infty} \left(\frac{1}{2}z\right)^m \frac{1}{m!} H_m(x) H_m(y)$$

$$H\left[(1-z^2)^{-\frac{1}{2}} \exp\left\{\frac{2xyz - (x^2+y^2)z^2}{(1-z^2)}\right\}\right]$$

$$= \sum_{m=0}^{\infty} \left(\frac{1}{2}z\right)^m \frac{1}{m!} H_m(y) \int_{-\infty}^{\infty} \exp(-x^2) H_n(x) H_m(x) dx$$

$$= \sum_{m=0}^{\infty} \left(\frac{1}{2}z\right)^m \frac{1}{m!} H_m(y) \delta_m \delta_{mn}$$

$$= \sqrt{\pi} \sum_{m=0}^{\infty} z^m H_m(y) \delta_{mn}.$$

15.3 Basic Operational Properties

Theorem 15.3.1 If $F'(x)$ is continuous and $F''(x)$ is bounded and locally integrable in the interval $-\infty < x < \infty$, and if $H\{F(x)\} = f_H(n)$, then

$$H\{R[F(x)]\} = -2n\, f_H(n), \tag{15.3.1}$$

where $R[F(x)]$ is the differential form given by

$$R[F(x)] = \exp(x^2) \frac{d}{dx}\left[\exp(-x^2) \frac{dF}{dx}\right]. \tag{15.3.2}$$

Proof. We have, by definition,

$$H\{R[F(x)]\} = \int_{-\infty}^{\infty} \frac{d}{dx}\left[\exp(-x^2) \frac{dF}{dx}\right] H_n(x)\, dx$$

which is, by integrating by parts and using the orthogonal relation (15.2.8),

$$= -2n \int_{-\infty}^{\infty} \exp(-x^2) H_n(x) F(x)\, dx$$

$$= -2n\, f_H(n).$$

Thus the theorem is proved.

If $F(x)$ and $R[F(x)]$ satisfy the conditions of Theorem 15.3.1, then

$$H\{R^2[F(x)]\} = H\{R[R[F(x)]]\} = (-1)^2 (2n)^2 f_H(n). \tag{15.3.3}$$

$$H\{R^3[F(x)]\} = (-1)^3 (2n)^3 f_H(n). \tag{15.3.4}$$

More generally,

$$H\{R^m[F(x)]\} = (-1)^m (2n)^m f_H(n), \tag{15.3.5}$$

where $m = 1, 2, \ldots, m-1$.

Theorem 15.3.2 If $F(x)$ is bounded and locally integrable in $-\infty < x < \infty$, and $f_H(0) = 0$, then $H\{F(x)\} = f_H(n)$ exists and for each constant C,

$$H^{-1}\left\{-\frac{f_H(n)}{2n}\right\} = R^{-1}[F(x)] = \int_0^x \exp(s^2) \int_{-\infty}^s \exp(-t^2) F(t) \, dt \, ds + C, \quad (15.3.6)$$

where R^{-1} is the inverse of the differential operator R and n is a positive integer.

Proof. We write

$$R^{-1}[F(x)] = Y(x)$$

so that $Y(x)$ is a solution of the differential equation

$$R[Y(x)] = F(x). \quad (15.3.7)$$

Since $f_H(0) = 0$, and $H_0(x) = 1$, then

$$\int_{-\infty}^{\infty} \exp(-x^2) F(x) \, dx = 0.$$

The first integral of (15.3.7) is

$$\exp(-x^2) Y'(x) = \int_{-\infty}^x \exp(-t^2) F(t) \, dt$$

which is a continuous function of x and tends to zero as $|x| \to \infty$. The second integral

$$Y(x) = \int_0^x \exp(s^2) \int_{-\infty}^s \exp(-t^2) F(t) \, dt \, ds + C,$$

where C is an arbitrary constant, is also continuous. Evidently,

$$\lim_{|x| \to \infty} \exp(-x^2) Y(x) = 0$$

provided $Y(x)$ is bounded.

Then $H\{Y(x)\}$ exists and

$$H\{R[Y(x)]\} = -2n \, H\{Y(x)\}.$$

Or

$$H[F(x)] = -2n \, H\{Y(x)\}.$$

Hence

$$f_H(n) = -2n \, H\{R^{-1}[F(x)]\}.$$

Thus, for any positive integer n,

$$H\{R^{-1}[F(x)]\} = -\frac{f_H(n)}{2n}.$$

Theorem 15.3.3 If $F(x)$ has bounded derivatives of order m and if $H\{F(x)\} = f_H(n)$ exists, then

$$H\{F^{(m)}(x)\} = f_H(n+m). \tag{15.3.8}$$

Proof. We have, by definition,

$$H\{F'(x)\} = \int_{-\infty}^{\infty} \exp(-x^2) H_n(x) F'(x) dx$$

which is, by integrating by parts,

$$= \left[\exp(-x^2) F(x) H_n(x)\right]_{-\infty}^{\infty} - \int_{-\infty}^{\infty} F(x) \frac{d}{dx}\left[e^{-x^2} H_n(x)\right] dx$$

$$= 2 \int_{-\infty}^{\infty} x \exp(-x^2) H_n(x) F(x) - \int_{-\infty}^{\infty} F(x) \exp(-x^2) H_n'(x) dx. \tag{15.3.9}$$

We use the following recurrence relations for the Hermite polynomial:

$$H_{n+1}(x) = 2x H_n(x) - 2n H_{n-1}(x), \tag{15.3.10}$$

and

$$H_n'(x) = 2n H_{n-1}(x), \tag{15.3.11}$$

to rewrite (15.3.9) in the form

$$H\{F'(x)\} = \int_{-\infty}^{\infty} \exp(-x^2) \left[H_{n+1}(x) + 2n H_{n-1}(x)\right] F(x) dx$$

$$- 2n \int_{-\infty}^{\infty} \exp(-x^2) H_{n-1}(x) F(x) dx$$

$$= \int_{-\infty}^{\infty} \exp(-x^2) H_{n+1}(x) F(x) dx = f_H(n+1).$$

Proceeding in a similar manner, we can prove

$$H\{F^{(m)}(x)\} = f_H(n+m).$$

Thus the theorem is proved.

Theorem 15.3.4 If the Hermite transforms of $F(x)$ and $x F^{(m-1)}(x)$ exist, then

$$H\{x F^{(m)}(x)\} = n f_H(m+n-1) + \frac{1}{2} f_H(m+n+1). \tag{15.3.12}$$

Proof. We have, by definition,

$$H\{x F^{(m)}(x)\} = \int_{-\infty}^{\infty} \exp(-x^2) H_n(x) \left\{x \frac{d^m F(x)}{dx^m}\right\} dx$$

$$= \left[x \exp(-x^2) H_n(x) F^{(m-1)}(x)\right]_{-\infty}^{\infty}$$

$$- \int_{-\infty}^{\infty} \frac{d}{dx}\left[x \exp(-x^2) H_n(x)\right] F^{(m-1)}(x) dx.$$

Thus

$$H\{x\, F^{(m)}(x)\} = \int_{-\infty}^{\infty} 2x^2 \exp(-x^2) H_n(x) F^{(m-1)}(x)\, dx$$

$$- \int_{-\infty}^{\infty} \exp(-x^2) H_n(x) F^{(m-1)}(x)\, dx$$

$$- n \int_{-\infty}^{\infty} 2x \exp(-x^2) H_{n-1}(x) F^{(m-1)}\, dx$$

which is, by the recurrence relations (15.3.10)-(15.3.11), and (15.3.8),

$$= \int_{-\infty}^{\infty} x \exp(-x^2) [H_{n+1}(x) + 2n\, H_{n-1}(x)] F^{(m-1)}(x)\, dx$$

$$- n \int_{-\infty}^{\infty} \exp(-x^2) [H_n(x) + 2(n-1) H_{n-2}(x)] F^{(m-1)}(x)\, dx - f_H(n+m+1)$$

$$= \frac{1}{2} \int_{-\infty}^{\infty} \exp(-x^2) [H_{n+2}(x) + 2(n+1) H_n(x)] F^{(m-1)}(x)\, dx$$

$$+ n \int_{-\infty}^{\infty} \exp(-x^2) [H_n(x) + 2(n-1) H_{n-2}(x)] F^{(m-1)}(x)\, dx$$

$$- n f_H(n+m-1) - 2n(n-1) f_H(n+m-3) - f_H(n+m+1)$$

$$= \frac{1}{2} f_H(n+m+1) + (n+1) f_H(n+m-1)$$

$$+ n[f_H(n+m-1) + 2(n-1) f_H(n+m-3)]$$

$$- n f_H(n+m-1) - 2n(n-1) f_H(n+m-3) - f_H(n+m+1)$$

$$= n f_H(n+m-1) + \frac{1}{2} f_H(n+m+1).$$

In particular, when $m = 1$ and $m = 2$, we obtain

$$H\{x\, F'(x)\} = n f_H(n) + \frac{1}{2} f_H(n+2), \qquad (15.3.13)$$

$$H\{x\, F''(x)\} = n f_H(n-1) + \frac{1}{2} f_H(n+3). \qquad (15.3.14)$$

The reader is referred to a paper by Debnath (1968) for other results similar to those of (15.3.13)-(15.3.14).

Definition 15.3.1 (Generalized Convolution). The generalized convolution of $F(x)$ and $G(x)$ for the Hermite transform defined by

$$H\{F(x) * G(x)\} = \mu_n\, H\{F(x)\} H\{G(x)\}$$
$$= \mu_n\, f_H(n) g_H(n), \qquad (15.3.15)$$

where μ_n is a non-zero quantity given by

$$\mu_n = \sqrt{\pi}\,(-1)^n \left\{ 2^{2n+1} \Gamma\left(n + \frac{3}{2}\right) \right\}^{-1}. \tag{15.3.16}$$

Debnath (1968) first proved the convolution theorem of the Hermite transform for odd functions. However, Dimovski and Kalla (1988) extended the theorem for both odd and even functions. We follow Dimovski and Kalla to state and prove the convolution theorem of the Hermite transform. Before we discuss the theorem, it is observed that, if $F(x)$ is an odd function, then

$$H\{F(x);\ 2n\} = f_H(2n) = \int_{-\infty}^{\infty} \exp(-x^2)\, H_{2n}(x)\, F(x)\, dx = 0, \tag{15.3.17}$$

but

$$H\{F(x);\ 2n+1\} = f_H(2n+1) \neq 0. \tag{15.3.18}$$

On the other hand, if $F(x)$ is an even function, then

$$H\{F(x);\ 2n+1\} = f_H(2n+1) = 0, \tag{15.3.19}$$

but

$$H\{F(x);\ 2n\} = f_H(2n) \neq 0. \tag{15.3.20}$$

Theorem 15.3.5 If $F(x)$ and $G(x)$ are odd functions and n is an odd positive integer, then

$$H\left\{F(x) \overset{\circ}{*} G(x);\ 2n+1\right\} = \mu_n\, f_H(2n+1)\, g_H(2n+1), \tag{15.3.21}$$

where $\overset{\circ}{*}$ denotes the convolution operation for odd functions and is given by

$$F(x) \overset{\circ}{*} G(x) = \frac{x}{\pi} \int_{-\infty}^{\infty} \exp(-t^2)\, t\, F(t)\, dt \int_0^{\pi} \exp(-xt\cos\phi) \sin\phi$$

$$\times \int_0^{\pi} \frac{G\left[(x^2 + t^2 + 2xt\cos\phi)^{\frac{1}{2}}\right]}{(x^2 + t^2 + 2xt\cos\phi)^{\frac{1}{2}}}\, J_0\left(xt \sin\phi\right) d\phi, \tag{15.3.22}$$

and $J_0(z)$ is the Bessel function of the first kind of order zero.

Proof. We have, by definition,

$$f_H(2n+1) = \int_{-\infty}^{\infty} \exp(-x^2)\, H_{2n+1}(x)\, F(x)\, dx. \tag{15.3.23}$$

We replace $H_{2n+1}(x)$ by using a result from Erdélyi (1953, vol. 2, p 193)

$$H_{2n+1}(x) = (-1)^n\, 2^{2n+1}\, n!\, x\, L_n^{\frac{1}{2}}(x^2), \tag{15.3.24}$$

where $L_n^\alpha(x)$ is the Laguerre polynomial of degree n and order α so that (15.3.23) reduces to the form

$$f_H(2n+1) = (-1)^n\, 2^{2n+2}\, n! \int_0^{\infty} x \exp(-x^2)\, L_n^{\frac{1}{2}}(x^2)\, F(x)\, dx. \tag{15.3.25}$$

Invoking the change of variable $x^2 = t$, we obtain

$$H\{F(x);\ 2n+1\} = (-1)^n\ 2^{2n+1}\ n!\int_0^\infty \sqrt{t}\ \exp(-t)\ L_n^{\frac{1}{2}}(t)\frac{F(\sqrt{t})}{\sqrt{t}}\ dt. \quad (15.3.26)$$

It is convenient to introduce the transformation T by

$$(T\ F)(t) = \frac{F(\sqrt{t})}{\sqrt{t}},\quad 0 \le t < \infty \quad (15.3.27)$$

so that the inverse of T is given by

$$T^{-1}(\Phi)(x) = x\ \Phi(x^2). \quad (15.3.28)$$

Consequently, (15.3.26) takes the form

$$H\{F(x);\ 2n+1\} = (-1)^n\ 2^{2n+1}\ n!\ L\{T\ F(x)\}, \quad (15.3.29)$$

where L is the Laguerre transformation of degree n and order $\alpha = \frac{1}{2}$ defined by (14.2.1) in Chapter 14.

The use of (15.3.29) allows us to write the product of two Hermite transforms as the product of two Laguerre transforms as

$$f_H(2n+1)\ g_H(2n+1) = 2^{4n+2}(n!)^2\ L\{T\ F(x)\}\ L\{T\ G(x)\}. \quad (15.3.30)$$

We now apply the Convolution Theorem for the Laguerre transform (when $\alpha = 0$) proved by Debnath (1969) in the form

$$L\{F \tilde{*} G(x)\} = \frac{n!\sqrt{\pi}}{\Gamma\left(n + \frac{3}{2}\right)}\ L\{F(x)\}\ L\{G(x)\}, \quad (15.3.31)$$

where $F \tilde{*} G$ is given by

$$F \tilde{*} G(x) = \int_0^\infty \exp(-\tau)\sqrt{\tau}\ F(\tau)\ d\tau \int_0^\pi \exp\left(-\sqrt{t\tau}\cos\phi\right)\sin\phi$$
$$\times G\left(t + \tau + 2\sqrt{t\tau}\cos\phi\right) J_0\left(\sqrt{t\tau}\sin\phi\right) d\phi. \quad (15.3.32)$$

Substituting (5.3.31) into (5.3.30), we obtain

$$f_H(2n+1)\ g_H(2n+1) = \pi^{-\frac{1}{2}}\ 2^{4n+2}\ n!\ \Gamma\left(n + \frac{3}{2}\right) L\{TF \tilde{*} TG\},$$

which is, by (15.3.29),

$$= \frac{2^{2n+1}\ \Gamma\left(n + \frac{3}{2}\right)}{(-1)^n\ \sqrt{\pi}}\ H\{T^{-1}\left(TF \tilde{*} TG\right)\}. \quad (15.3.33)$$

Or equivalently,

$$H\{F \overset{\circ}{*} G(x);\ 2n+1\} = \mu_n\ H\{F(x)\}\ H\{G(x)\}, \quad (15.3.34)$$

where

$$F \overset{\circ}{*} G(x) = T^{-1}\left\{TF \overset{\circ}{*} TG(x)\right\} \quad (15.3.35)$$

This coincides with (15.3.22). Thus the proof is complete.

Theorem 15.3.6 (*Convolution of the Hermite Transform for Even Functions*). If $F(x)$ and $G(x)$ are even functions and n is an even positive integer, then

$$H\{F(x) \overset{e}{*} G(x);\ 2n\} = \mu_n\, H\{F(x);\ 2n\}\, H\{G(x);\ 2n\}. \quad (15.3.36)$$

Proof. We use result (15.3.8), that is,
$$H\{F'(x);\ n\} = H\{F(x);\ n+1\}$$
so that
$$H\{I\, F(x);\ 2n+1\} = H\{F(x),\ 2n\}, \quad (15.3.37)$$
where
$$I\, F(x) = \int_0^x F(t)\, dt \quad \text{and} \quad [I\, F(x)]' = F(x).$$

Obviously,
$$H\{F(x) \overset{e}{*} G(x);\ 2n\} = H\left\{\left[I\, F(x) \overset{e}{*} I\, G(x)\right]';\ 2n\right\}$$
$$= H\{I\, F(x) \overset{o}{*} I\, G(x);\ 2n+1\}$$
$$= \mu_n\, H\{I\, F(x);\ 2n+1\}\, H\{I\, G(x);\ 2n+1\}$$
$$= \mu_n\, H\{F(x);\ 2n\}\, H\{G(x);\ 2n\}.$$

This proves the theorem.

Theorem 15.3.7 If $F(x)$ and $G(x)$ are two arbitrary functions such that their Hermite transforms exist, then
$$H\{F(x) * G(x);\ n\} = \mu_{[n/2]}\, H\{F(x);\ n\}\, H\{G(x);\ n\}, \quad (15.3.38)$$
where
$$F(x) * G(x) = F_0(x) \overset{o}{*} G_0(x) + F_e(x) \overset{e}{*} G_e(x), \quad (15.3.39)$$
and
$$F_0(x) = \frac{1}{2}[F(x) - F(-x)] \quad \text{and} \quad F_e(x) = \frac{1}{2}[F(x) + F(-x)]. \quad (15.3.40)$$

Proof. We first note that arbitrary functions $F(x)$ and $G(x)$ can be expressed as sums of even and odd functions, that is, $F(x) = F_0(x) + F_e(x)$ and $G(x) = G_0(x) + G_e(x)$ so that result (15.3.40) follows.

Suppose n is odd. Then
$$H\{F(x);\ n\} = H\{F_0(x);\ n\},\quad H\{G(x);\ n\} = H\{G_0(x);\ n\},$$
and
$$H\{F(x) + G(x);\ n\} = H\{F_0(x) + G_0(x);\ n\}.$$

Clearly,

$$H\{F(x) * G(x); \ 2n+1\}$$
$$= H\left\{F_0(x) \overset{\circ}{\underset{*}{}} G_0(x); \ 2n+1\right\} + H\left\{F_e(x) \overset{e}{\underset{*}{}} G_e(x); \ 2n+1\right\}$$
$$= \mu_n \, H\{F_0(x)\} \, H\{G_0(x)\}$$
$$= \mu_n \, H\{F(x)\} \, H\{G(x)\}.$$

Similarly, the case for even n can be handled without any difficulty.

15.4 Exercises

1. Find the Hermite transform of the following functions:
   (a) $\exp(-x^2) H_n(x)$,   (b) x^m,   (c) $x^2 H_n(x)$.

2. Show that
$$H\{x^n\} = \sqrt{\pi} \, n! \, P_n(1),$$
where $P_n(x)$ is the Legendre polynomial.

3. Show that
$$H\{H_n^2(x)\} = \sqrt{\pi} \sum_{r=0}^{n} \binom{n}{r} 2^{r+n} \, (2r)! \, n!$$

Appendix A

Some Special Functions and Their Properties

The main purpose of this appendix is to introduce several special functions and to state their basic properties that are most frequently used in the theory and applications of integral transforms. The subject is, of course, too vast to be treated adequately in so short a space, so that only the more important results can be stated. For a fuller discussion of these topics and of further properties of these functions the reader is referred to the standard treatises on the subject.

A-1 Gamma, Beta, and Error Functions

The *Gamma function* (also called the *factorial function*) is defined by a definite integral in which a variable appears as a parameter

$$\Gamma(x) = \int_0^\infty e^{-t} t^{x-1} dt, \quad x > 0. \tag{A-1.1}$$

In view of the fact that the integral (A-1.1) is uniformly convergent for all x in [a, b] where $0 < a \le b < \infty$, $\Gamma(x)$ is a continuous function for all $x > 0$.

Integrating (A-1.1) by parts, we obtain the the fundamental property of $\Gamma(x)$

$$\Gamma(x) = \left[-e^{-t} t^{x-1}\right]_0^\infty + (x-1)\int_0^\infty e^{-t} t^{x-2} dt$$

$$= (x-1)\Gamma(x-1), \quad \text{for} \quad x - 1 > 0.$$

Then we replace x by $x+1$ to obtain the fundamental result

$$\Gamma(x+1) = x\,\Gamma(x). \tag{A-1.2}$$

In particular, when $x = n$ is a positive integer, we make repeated use of (A-1.2) to obtain

$$\Gamma(n+1) = n\,\Gamma(n) = n(n-1)\,\Gamma(n-1) = \cdots$$

$$= n(n-1)(n-2)\cdots 3\cdot 2\cdot 1\,\Gamma(1) = n!, \tag{A-1.3}$$

where $\Gamma(1) = 1$.

We put $t = u^2$ in (A-1.1) to obtain

$$\Gamma(x) = 2\int_0^\infty \exp(-u^2)\, u^{2x-1} du, \quad x > 0. \tag{A-1.4}$$

Letting $x = \frac{1}{2}$, we find

$$\Gamma\left(\frac{1}{2}\right) = 2\int_0^\infty \exp(-u^2)\, du = 2\frac{\sqrt{\pi}}{2} = \sqrt{\pi}. \qquad (A-1.5)$$

Using (A-1.2), we deduce

$$\Gamma\left(\frac{3}{2}\right) = \frac{1}{2}\Gamma\left(\frac{1}{2}\right) = \frac{\sqrt{\pi}}{2}. \qquad (A-1.6)$$

Similarly, we can obtain the values of $\Gamma\left(\frac{5}{2}\right)$, $\Gamma\left(\frac{7}{2}\right)$, ..., $\Gamma\left(\frac{2n+1}{2}\right)$.

The gamma function can also be defined for negative values of x by the rewritten form of (A-1.2) as

$$\Gamma(x) = \frac{\Gamma(x+1)}{x}, \quad x \neq 0, -1, -2, \ldots \qquad (A-1.7)$$

For example

$$\Gamma\left(-\frac{1}{2}\right) = \frac{\Gamma\left(\frac{1}{2}\right)}{-\frac{1}{2}} = -2\,\Gamma\left(\frac{1}{2}\right) = -2\sqrt{\pi}, \qquad (A-1.8)$$

$$\Gamma\left(-\frac{3}{2}\right) = \frac{\Gamma\left(-\frac{1}{2}\right)}{-\frac{3}{2}} = \frac{4}{3}\sqrt{\pi}. \qquad (A-1.9)$$

We differentiate (A-1.1) with respect to x to obtain

$$\frac{d}{dx}\Gamma(x) = \Gamma'(x) = \int_0^\infty \frac{d}{dx}(t^x)\frac{e^{-t}}{t}\, dt$$

$$= \int_0^\infty \frac{d}{dx}[\exp(x \log t)]\frac{e^{-t}}{t}\, dt$$

$$= \int_0^\infty t^{x-1}(\log t)e^{-t}\, dt. \qquad (A-1.10)$$

At $x = 1$, this gives

$$\Gamma'(1) = \int_0^\infty e^{-t} \log t\, dt = -\gamma, \qquad (A-1.11)$$

where γ is called the *Euler constant* and has the value 0.5772.

The graph of the gamma function is shown in Figure A.1.

Integral Transforms and Their Applications

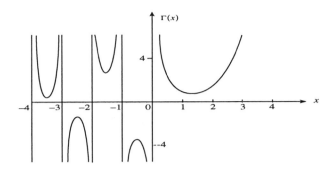

Figure A.1 The gamma function.

Several useful properties of the gamma function are recorded below for reference without proof.

Legendre Duplication Formula

$$2^{2x-1}\,\Gamma(x)\,\Gamma\!\left(x+\frac{1}{2}\right) = \sqrt{\pi}\,\Gamma(2x), \qquad (A\text{-}1.12)$$

In particular, when $x = n\ (n = 0, 1, 2, \ldots)$

$$\Gamma\!\left(n+\frac{1}{2}\right) = \frac{\sqrt{\pi}\,(2n)!}{2^{2n}\,n!}, \qquad (A\text{-}1.13)$$

The following properties also hold for $\Gamma(x)$:

$$\Gamma(x)\,\Gamma(1-x) = \pi\,\operatorname{cosec}\pi x, \qquad x \text{ is a noninteger}, \qquad (A\text{-}1.14)$$

$$\Gamma(x) = p^x \int_0^\infty \exp(-pt)\, t^{x-1}\, dt, \qquad (A\text{-}1.15)$$

$$\Gamma(x) = \int_{-\infty}^{\infty} \exp\!\left(xt - e^t\right) dt. \qquad (A\text{-}1.16)$$

$$\Gamma(x+1) \sim \sqrt{2\pi}\,\exp(-x)\, x^{x+\frac{1}{2}} \qquad \text{for large } x, \qquad (A\text{-}1.17)$$

$$n! \sim \sqrt{2\pi}\,\exp(-n)\, n^{n+\frac{1}{2}} \qquad \text{for large } n. \qquad (A\text{-}1.18)$$

The *incomplete gamma function*, $\gamma(x, a)$, is defined by the integral

$$\gamma(a, x) = \int_0^x e^{-t} t^{a-1}\, dt, \qquad a > 0. \qquad (A\text{-}1.19)$$

The *complementary incomplete gamma* function, $\Gamma(a, x)$, is defined by the integral

$$\Gamma(a, x) = \int_x^\infty e^{-t}\, t^{a-1}\, dt, \qquad a > 0. \qquad (A\text{-}1.20)$$

Thus it follows that
$$\gamma(a,x)+\Gamma(a,x)=\Gamma(a). \quad (A\text{-}1.21)$$

The *beta function*, denoted by $B(x, y)$ is defined by the integral
$$B(x,y)=\int_0^1 t^{x-1}(1-t)^{y-1}\,dt, \quad x>0,\ y>0. \quad (A\text{-}1.22)$$

The beta function $B(x, y)$ is *symmetric* with respect to its arguments x and y, that is,
$$B(x,y)=B(y,x). \quad (A\text{-}1.23)$$

This follows from (A-1.22) by the change of variable $1-t=u$, that is,
$$B(x,y)=\int_0^1 u^{y-1}(1-u)^{x-1}\,du=B(y,x).$$

If we make the change of variable $t=u/(1+u)$ in (A-1.22), we obtain another integral representation of the beta function
$$B(x,y)=\int_0^\infty u^{x-1}(1+u)^{-(x+y)}\,du=\int_0^\infty u^{y-1}(1+u)^{-(x+y)}\,du, \quad (A\text{-}1.24)$$

Putting $t=\cos^2\theta$ in (A-1.22), we derive
$$B(x,y)=2\int_0^{\pi/2}\cos^{2x-1}\theta\ \sin^{2y-1}\theta\,d\theta. \quad (A\text{-}1.25)$$

Several important results are recorded below for ready reference without proof.
$$B(1,1)=1, \quad B\left(\frac{1}{2},\frac{1}{2}\right)=\pi, \quad (A\text{-}1.26)$$

$$B(x,y)=\left(\frac{x-1}{x+y-1}\right)B(x-1,y), \quad (A\text{-}1.27)$$

$$B(x,y)=\frac{\Gamma(x)\,\Gamma(y)}{\Gamma(x+y)}, \quad (A\text{-}1.28)$$

$$B\left(\frac{1+x}{2},\frac{1-x}{2}\right)=\pi\sec\left(\frac{\pi x}{2}\right), \quad 0<x<1. \quad (A\text{-}1.29)$$

The *error function*, $erf(x)$ is defined by the integral
$$erf(x)=\frac{2}{\sqrt{\pi}}\int_0^x \exp(-t^2)\,dt, \quad -\infty<x<\infty. \quad (A\text{-}1.30)$$

Clearly it follows from (A-1.30) that
$$erf(-x)=-erf(x), \quad (A\text{-}1.31)$$

$$\frac{d}{dx}[erf(x)]=\frac{2}{\sqrt{\pi}}\exp(-x^2), \quad (A\text{-}1.32)$$

$$erf(0)=0,\quad erf(\infty)=1. \quad (A\text{-}1.33)$$

The *complementary error function*, $erfc(x)$ is defined by the integral

$$erfc(x) = \frac{2}{\sqrt{\pi}} \int_x^\infty \exp(-t^2) dt. \qquad (A\text{-}1.34)$$

Clearly it follows that
$$erfc(x) = 1 - erf(x), \qquad (A\text{-}1.35)$$
$$erfc(0) = 1, \quad erfc(\infty) = 0. \qquad (A\text{-}1.36)$$

$$erfc(x) \sim \frac{1}{x\sqrt{\pi}} \exp(-x^2) \qquad \text{for large } x. \qquad (A\text{-}1.37)$$

The graphs of $erf(x)$ and $erfc(x)$ are shown in Figure A.2.

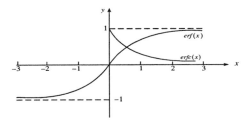

Figure A.2. The error function and the complementary error function.

Closely associated with the error function are the Fresnel integrals, which are defined by

$$C(x) = \int_0^x \cos\left(\frac{\pi t^2}{2}\right) dt \quad \text{and} \quad S(x) = \int_0^x \sin\left(\frac{\pi t^2}{2}\right) dt. \qquad (A\text{-}1.38)$$

These integrals arises in diffraction problems in optics, in water waves and in elasticity and elsewhere.

Clearly it follows from (A-1.38) that
$$C(0) = 0 = S(0) \qquad (A\text{-}1.39)$$

$$C(\infty) = S(\infty) = \frac{\pi}{2}, \qquad (A\text{-}1.40)$$

$$\frac{d}{dx} C(x) = \cos\left(\frac{\pi x^2}{2}\right), \quad \frac{d}{dx} S(x) = \sin\left(\frac{\pi x^2}{2}\right). \qquad (A\text{-}1.41)$$

It also follows from (A-1.38) that $C(x)$ has extrema at the points where $x^2 = (2n+1)$, $n = 0, 1, 2, 3, \ldots$, and $S(x)$ has extrema at the points where $x^2 = 2n$, $n = 1, 2, 3, \ldots$. The largest maxima occur first and are found to be $C(1) = 0.7799$ and $S(\sqrt{2}) = 0.7139$. We also infer that both $C(x)$ and $S(x)$ are oscillatory about the line $y = 0.5$. The graphs of $C(x)$ and $S(x)$ for non-negative real x are shown in Figure A.3.

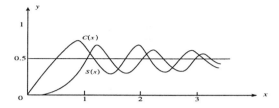

Figure A.3 The Fresnel integrals $C(x)$ and $S(x)$.

A.2 Bessel and Airy Functions

The Bessel function of the first kind of order v (non-negative real number) is denoted by $J_v(x)$, and defined by

$$J_v(x) = x^v \sum_{r=0}^{\infty} \frac{(-1)^r x^{2r}}{2^{2r+v}\, r!\, \Gamma(r+v+1)}. \tag{A-2.1}$$

This series is convergent for all x.

The Bessel function $y = J_v(x)$ satisfies the *Bessel equation*

$$x^2 y'' + x y' + (x^2 - v^2) y = 0. \tag{A-2.2}$$

When v is *not* a positive integer or zero, $J_v(x)$ and $J_{-v}(x)$ are two linearly independent solutions so that

$$y = A\, J_v(x) + B\, J_{-v}(x) \tag{A-2.3}$$

is the general solution of (A-2.2), where A and B are arbitrary constants of integration.

However, when $v = n$, where n is a *positive integer* or *zero*, $J_n(x)$ and $J_{-n}(x)$ are no longer independent, but are related by the equation

$$J_{-n}(x) = (-1)^n J_n(x). \tag{A-2.4}$$

Thus, when n is a positive integer or zero, equation (A-2.2) has only *one* solution given by

$$J_n(x) = \sum_{r=0}^{\infty} \frac{(-1)^r}{r!(n+r)!} \left(\frac{x}{2}\right)^{n+2r}. \tag{A-2.5}$$

A second solution, known as *Neumann's or Weber's solution*, $Y_n(x)$ is given by

$$Y_n(x) = \lim_{v \to n} Y_v(x), \tag{A-2.6}$$

where

$$Y_v(x) = \frac{(\cos v\pi) J_v(x) - J_{-v}(x)}{\sin v\pi}. \tag{A-2.7}$$

Thus the general solution of (A-2.2) is

$$y(x) = A\, J_n(x) + B\, Y_n(x), \tag{A-2.8}$$

where A and B are arbitrary constants.

In particular, from (A-2.5),

$$J_0(x) = \sum_{r=0}^{\infty} \frac{(-1)^r}{(r!)^2} \left(\frac{x}{2}\right)^{2r}, \qquad (A\text{-}2.9)$$

$$J_1(x) = \sum_{r=0}^{\infty} \frac{(-1)^r}{r!(r+1)!} \left(\frac{x}{2}\right)^{2r+1}. \qquad (A\text{-}2.10)$$

Clearly, it follows from (A-2.9) and (A-2.10) that

$$J_0'(x) = -J_1(x). \qquad (A\text{-}2.11)$$

Bessel's equation may not always arise in the standard form given in (A-2.2), but more frequently as

$$x^2 y'' + xy' + (k^2 x^2 - v^2) y = 0 \qquad (A\text{-}2.12)$$

with the general solution

$$y(x) = A\, J_v(kx) + B\, Y_v(kx). \qquad (A\text{-}2.13)$$

The *recurrence relations* are recorded below for easy reference without proof.

$$J_{v+1}(x) = \left(\frac{v}{x}\right) J_v(x) - J_v'(x), \qquad (A\text{-}2.14)$$

$$J_{v-1}(x) = \left(\frac{v}{x}\right) J_v(x) + J_v'(x), \qquad (A\text{-}2.15)$$

$$J_{v-1}(x) + J_{v+1}(x) = \left(\frac{2v}{x}\right) J_v(x), \qquad (A\text{-}2.16)$$

$$J_{v-1}(x) - J_{v+1}(x) = 2 J_v'(x). \qquad (A\text{-}2.17)$$

We have, from (A-2.5),

$$x^n J_n(x) = \sum_{r=0}^{\infty} \frac{(-1)^r 2^{-(n+2r)}}{r!(n+r)!} x^{2n+2r}.$$

Differentiating both sides of this result with respect to x and using the fact that $2(n+r)/(n+r)! = 2/(n+r-1)!$, it turns out that

$$\frac{d}{dx}\left[x^n J_n(x)\right] = \sum_{r=0}^{\infty} \frac{(-1)^r 2^{-(n+2r+1)}}{r!(n+r-1)!} x^{2n+2r-1} = x^n J_{n-1}(x) \qquad (A\text{-}2.18)$$

Similarly, we can show

$$\frac{d}{dx}\left[x^{-n} J_n(x)\right] = -x^{-n} J_{n+1}(x). \qquad (A\text{-}2.19)$$

The generating function for the Bessel function is

$$\exp\left[\frac{1}{2} x\left(t - \frac{1}{t}\right)\right] = \sum_{n=-\infty}^{\infty} t^n J_n(x). \qquad (A\text{-}2.20)$$

The integral representation is

$$J_n(x) = \frac{1}{\pi} \int_0^{\pi} \cos(n\theta - x \sin\theta)\, d\theta. \qquad (A\text{-}2.21)$$

The following are known as the *Lommel integrals*:

$$\int_0^a x J_n(px) J_n(qx)\, dx$$

$$= \frac{a}{(q^2 - p^2)} \left[p J_n(qa) J'_n(pa) - q J_n(pa) J'_n(qa) \right], \quad p \neq q. \quad \text{(A-2.22)}$$

and

$$\int_0^a x J_n^2(px)\, dx = \frac{a^2}{2} \left[J'^2_n(pa) + \left(1 - \frac{n^2}{p^2 a^2}\right) J_n^2(pa) \right]. \quad \text{(A-2.23)}$$

When $v = \pm \frac{1}{2}$,

$$J_{\frac{1}{2}}(x) = \sqrt{\frac{2}{\pi x}} \sin x, \quad J_{-\frac{1}{2}}(x) = \sqrt{\frac{2}{\pi x}} \cos x \quad \text{(A-2.24)}$$

A rough idea of the shape of the Bessel functions when x is large may be obtained from equation (A-2.2). Substitution of $y = x^{-\frac{1}{2}} u(x)$ eliminates the first derivative, and hence gives the equation

$$u'' + \left(1 - \frac{4n^2 - 1}{4x^2}\right) u = 0 \quad \text{(A-2.25)}$$

For large x, this equation approximately becomes

$$u'' + u = 0. \quad \text{(A-2.26)}$$

This equation admits the solution $u(x) = A \cos(x + \varepsilon)$ that is,

$$y = \frac{A}{\sqrt{x}} \cos(n + \varepsilon). \quad \text{(A-2.27)}$$

This suggests that $J_n(x)$ is oscillatory and has an infinite number of zeros. It also tends to zero as $x \to \infty$. The graphs of $J_n(x)$ for $n = 0, 1, 2$ and for $n = \pm \frac{1}{2}$ are shown in Figure A.4 and Figure A.5 respectively.

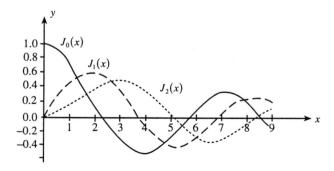

Figure A.4 Graphs of $y = J_0(x)$, $J_1(x)$ and $J_2(x)$.

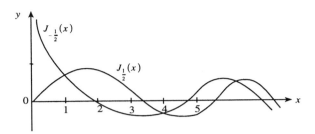

Figure A.5. Graphs of $J_{\frac{1}{2}}(x)$ and $J_{-\frac{1}{2}}(x)$.

An important special case arises in particular physical problems when $k^2 = -1$ in equation (A-2.12). We then have the *modified Bessel equation*

$$x^2 y'' + xy' - (x^2 + v^2) y = 0, \qquad \text{(A-2.28)}$$

with the general solution

$$y = A\, J_v(ix) + B\, Y_v(ix). \qquad \text{(A-2.29)}$$

We now define a new function

$$I_v(x) = i^{-v} J_v(ix), \qquad \text{(A-2.30)}$$

and then use the series (A-2.1) for $J_v(x)$ so that

$$I_v(x) = i^{-v} \sum_{r=0}^{\infty} \frac{(-1)^r}{r!\,\Gamma(r+v+1)} \left(\frac{ix}{2}\right)^{v+2r}$$

$$= \sum_{r=0}^{\infty} \frac{1}{r!\,\Gamma(r+v+1)} \left(\frac{x}{2}\right)^{v+2r}. \qquad \text{(A-2.31)}$$

Similarly, we can find the second solution, $K_v(x)$ of the modified Bessel equation (A-2.28). Usually, $I_v(x)$ and $K_v(x)$ are called *modified Bessel functions* and their properties can be obtained in a similar way to those of $J_v(x)$ and $Y_v(x)$. The graphs of $Y_0(x)$ and $Y_1(x)$ are shown in Figure A-6.

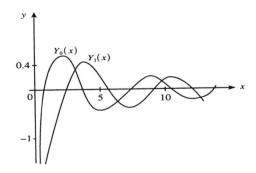

Figure A-6 Graphs of $y = Y_0(x),\ Y_1(x)$.

We state a few important infinite integrals involving Bessel functions which arise frequently in the application of Hankel transforms.

$$\int_0^\infty \exp(-at)\, J_\nu(bt)\, t^\nu dt = \frac{(2b)^\nu \Gamma\left(\nu+\frac{1}{2}\right)}{\sqrt{\pi}\,(a^2+b^2)^{\nu+\frac{1}{2}}}, \quad \nu > -\frac{1}{2}, \quad \text{(A-2.32)}$$

$$\int_0^\infty \exp(-at)\, J_\nu(bt)\, t^{\nu+1} dt = \frac{2a(2b)^\nu \Gamma\left(\nu+\frac{3}{2}\right)}{\sqrt{\pi}\,(a^2+b^2)^{\nu+\frac{3}{2}}}, \quad \nu > -1, \quad \text{(A-2.33)}$$

$$\int_0^\infty \exp(-a^2 t^2)\, J_\nu(bt)\, t^{\nu+1} dt = \frac{b^\nu}{(2a^2)^{\nu+1}} \exp\left(-\frac{b^2}{4a^2}\right), \quad \nu > -1, \quad \text{(A-2.34)}$$

$$\int_0^\infty \exp(-a^2 t^2)\, J_\nu(bt)\, J_\nu(ct)\, t\, dt = \frac{1}{2a^2} \exp\left(-\frac{b^2+c^2}{4a^2}\right) I_\nu\left(\frac{bc}{2a^2}\right), \quad \nu > -1, \quad \text{(A-2.35)}$$

$$\int_0^\infty t^{2\mu-\nu-1} J_\nu(t)\, dt = \frac{2^{2\mu-\nu-1}\Gamma(\mu)}{\Gamma(\nu-\mu+1)}, \quad 0 < \mu < \frac{1}{2},\ \nu > -\frac{1}{2}. \quad \text{(A-2.36)}$$

The *Airy function*, $y = Ai(x)$ is the first solution of the differential equation
$$y'' - xy = 0. \quad \text{(A-2.37)}$$
The second solution is denoted by $Bi(x)$. Then these functions are given by

$$Ai(x) = \sqrt{\frac{x}{3}} \left[I_{-\frac{1}{3}}\left(\frac{2}{3} x^{3/2}\right) - I_{\frac{1}{3}}\left(\frac{2}{3} x^{3/2}\right) \right], \quad \text{(A-2.38)}$$

$$Bi(x) = \sqrt{\frac{x}{3}} \left[I_{-\frac{1}{3}}\left(\frac{2}{3} x^{3/2}\right) + I_{\frac{1}{3}}\left(\frac{2}{3} x^{3/2}\right) \right]. \quad \text{(A-2.39)}$$

The integral representation for $Ai(x)$ is

$$Ai(x) = \frac{1}{\pi} \int_0^\infty \cos\left(\frac{1}{3} t^3 + xt\right) dt. \quad \text{(A-2.40)}$$

The graph of $y = Ai(x)$ is shown in Figure A-7.

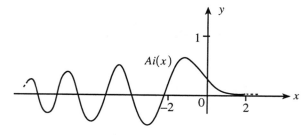

Figure A-7. The Airy function.

A-3 Legendre and Associated Legendre Functions

The Legendre polynomials $P_n(x)$ are defined by the *Rodrigues formula*

$$P_n(x) = \frac{1}{2^n n!} \frac{d^n}{dx^n} (x^2 - 1)^n. \qquad (A\text{-}3.1)$$

The seven Legendre polynomials are

$$P_0(x) = 1$$

$$P_1(x) = x$$

$$P_2(x) = \frac{1}{2}(3x^2 - 1)$$

$$P_3(x) = \frac{1}{2}(5x^3 - 3x)$$

$$P_4(x) = \frac{1}{8}(35x^4 - 30x^2 + 3)$$

$$P_5(x) = \frac{1}{8}(63x^5 - 70x^3 + 15x)$$

$$P_6(x) = \frac{1}{16}(231 x^6 - 315 x^4 + 105 x^2 - 5).$$

The generating function for the Legendre polynomial is

$$(1 - 2xt + t^2)^{-\frac{1}{2}} = \sum_{n=0}^{\infty} t^n P_n(x). \qquad (A\text{-}3.2)$$

This function provides more information about the Legendre polynomials. For example,

$$P_n(1) = 1, \quad P_n(-1) = (-1)^n, \qquad (A\text{-}3.3)$$

$$P_{2n}(0) = (-1)^n \frac{1 \cdot 3 \cdot 5 \cdots (2n-1)}{2^n n!} = (-1)^n \frac{(2n-1)!!}{(2n)!!}, \qquad (A\text{-}3.4)$$

$$P_{2n+1}(0) = 0, \ n = 0, 1, 2, \ldots, \qquad (A\text{-}3.5)$$

$$P_n(-x) = (-1)^n P_n(x), \qquad \frac{d^n}{dx^n} P_n(x) = \frac{(2n)!}{2^n n!}, \qquad (A\text{-}3.6)$$

where the double factorial is defined by

$$(2n-1)!! = 1 \cdot 3 \cdot 5 \cdots (2n-1) \text{ and } (2n)!! = 2 \cdot 4 \cdot 6 \cdots (2n).$$

The recurrence relations for the Legendre polynomials are

$$(n+1) P_{n+1}(x) = (2n+1)x P_n(x) - n P_{n-1}(x), \qquad (A\text{-}3.7)$$

$$P'_{n+1}(x) - P'_{n-1}(x) = (2n+1) P_n(x), \qquad (A\text{-}3.8)$$

$$(1 - x^2) P'_n(x) = n P_{n-1}(x) - n x P_n(x), \qquad (A\text{-}3.9)$$

$$(1 - x^2) P'_n(x) = (n+1) x P_n(x) - (n+1) P_{n+1}(x), \qquad (A\text{-}3.10)$$

The Legendre polynomials $y = L_n(x)$ satisfy the Legendre differential equation

$$(1 - x^2) y'' - 2x y' + n(n+1) y = 0. \qquad (A\text{-}3.11)$$

If n is *not* an integer, both solutions of (A-3.11) diverge at $x = \pm 1$.
The orthogonal relation is

$$\int_{-1}^{1} P_n(x) P_m(x) \, dx = \frac{2}{(2n+1)} \delta_{nm}. \qquad (A\text{-}3.12)$$

The graphs of the first three Legendre polynomials are shown in Figure A-8.

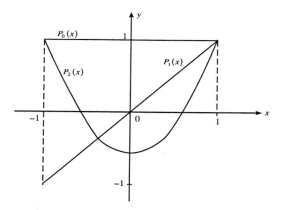

Figure A-8. Graphs of $y = P_0(x)$, $P_1(x)$, and $P_2(x)$.

The *associated Legendre functions* are defined by

$$P_n^m(x) = \left(1-x^2\right)^{\frac{m}{2}} \frac{d^m}{dx^m} P_n(x) = \frac{1}{2^n n!} \left(1-x^2\right)^{\frac{m}{2}} \frac{d^{m+n}}{dx^{m+n}} \left(x^2-1\right)^n, \qquad (A\text{-}3.13)$$

where $0 \le m \le n$.
Clearly, it follows that

$$P_n^0(x) = P_n(x), \qquad (A\text{-}3.14)$$

$$P_n^m(-x) = (-1)^{n+m} P_n^m(x), \qquad P_n^{-m}(x) = (-1)^m \frac{(n-m)!}{(n+m)!} P_n^m(x). \qquad (A\text{-}3.15)$$

The generating function for $P_n^m(x)$ is

$$\frac{(2m)! \left(1-x^2\right)^{\frac{m}{2}}}{2^m m! \left(1-2tx+t^2\right)^{m+\frac{1}{2}}} = \sum_{r=0}^{\infty} P_{r+m}^m(x) t^r. \qquad (A\text{-}3.16)$$

The recurrence relations are

$$(2n+1) x \, P_n^m(x) = (n+m) P_{n-1}^m(x) + (n-m+1) P_{n+1}^m(x), \qquad (A\text{-}3.17)$$

$$2\left(1-x^2\right)^{\frac{1}{2}} \frac{d}{dx} P_n^m(x) = P_n^{m+1}(x) - (n+m)(n-m+1) P_n^{m-1}(x). \qquad (A\text{-}3.18)$$

The associated Legendre functions $P_n^m(x)$ are solutions of the differential equation

$$(1-x^2)y'' - 2xy' + \left[n(n+1) - \frac{m^2}{(1-x^2)}\right]y = 0. \qquad \text{(A-3.19)}$$

This reduces to the Legendre equation when $m = 0$.

Listed below are few associated Legendre functions with $x = \cos\theta$:

$$P_1^1(x) = (1-x^2)^{\frac{1}{2}} = \sin\theta,$$

$$P_2^1(x) = 3x(1-x^2)^{\frac{1}{2}} = 3\cos\theta \sin\theta$$

$$P_2^2(x) = 3(1-x^2) = 3\sin^2\theta$$

$$P_3^1(x) = \frac{3}{2}(5x^2 - 1)(1-x^2)^{\frac{1}{2}} = \frac{3}{2}(5\cos^2\theta - 1)\sin\theta$$

$$P_3^2(x) = 15x(1-x^2) = 15\cos\theta \sin^2\theta$$

$$P_3^3(x) = 15(1-x^2)^{3/2} = 15\sin^3\theta.$$

The orthogonal relations are

$$\int_{-1}^{1} P_n^m(x) P_\ell^m(x)\, dx = \frac{2}{(2\ell+1)} \cdot \frac{(\ell+m)!}{(\ell-m)!}\, \delta_{n\ell}, \qquad \text{(A-3.20)}$$

$$\int_{-1}^{1} (1-x^2)^{-1} P_n^m(x) P_n^\ell(x)\, dx = \frac{(n+m)!}{m(n-m)!}\, \delta_{m\ell}. \qquad \text{(A-3.21)}$$

A-4 Jacobi and Gegenbauer Polynomials

The *Jacobi polynomials* $P_n^{(\alpha,\beta)}(x)$ of degree n are defined by the Rodrigues formula

$$P_n^{(\alpha,\beta)}(x) = \frac{(-1)^n}{2^n n!} (1-x)^{-\alpha}(1+x)^{-\beta} \frac{d^n}{dx^n}\left[(1-x)^{\alpha+n}(1+x)^{\beta+n}\right], \qquad \text{(A-4.1)}$$

where $\alpha > -1$ and $\beta > -1$.

When $\alpha = \beta = 0$, the Jacobi polynomials become Legendre polynomials, that is,

$$P_n(x) = P_n^{(0,0)}(x), \qquad n = 0,1,2,\ldots \qquad \text{(A-4.2)}$$

On the other hand, the associated Laguerre functions arise as the limit

$$L_n^\alpha(x) = \lim_{\beta\to\infty} P_n^{(\alpha,\beta)}\left(1 - \frac{2x}{\beta}\right). \qquad \text{(A-4.3)}$$

The recurrence relations for $P_n^{(\alpha,\beta)}(x)$ are

$$2(n+1)(\alpha+\beta+n+1)(\alpha+\beta+2n)\, P_{n+1}^{(\alpha,\beta)}(x)$$
$$=(\alpha+\beta+2n+1)\left[(\alpha^2-\beta^2)+x(\alpha+\beta+2n+2)(\alpha+\beta+2n)\right]P_n^{(\alpha,\beta)}(x)$$
$$-2(\alpha+n)(\beta+n)(\alpha+\beta+2n+2)\, P_{n-1}^{(\alpha,\beta)}(x), \qquad (A\text{-}4.4)$$

where $n = 1, 2, 3, \ldots$, and

$$P_n^{(\alpha,\beta-1)}(x) - P_n^{(\alpha-1,\beta)}(x) = P_{n-1}^{(\alpha,\beta)}(x). \qquad (A\text{-}4.5)$$

The generating function for Jacobi polynomials is

$$2^{(\alpha+\beta)}\, R^{-1}\,(1-t+R)^{-\alpha}\,(1+t+R)^{-\beta} = \sum_{n=0}^{\infty} P_n^{(\alpha,\beta)}(x)\, t^n, \qquad (A\text{-}4.6)$$

where $R = \left(1 - 2xt + t^2\right)^{\frac{1}{2}}$.

The Jacobi polynomials, $y = P_n^{(\alpha,\beta)}(x)$, satisfy the differential equation

$$(1-x^2)y'' + \left[(\beta-\alpha)-(\alpha+\beta+2)x\right]y' + n(n+\alpha+\beta+1)y = 0. \qquad (A\text{-}4.7)$$

The orthogonal relation is

$$\int_{-1}^{1} (1-x)^{\alpha}(1+x)^{\beta}\, P_n^{(\alpha,\beta)}(x)\, P_m^{(\alpha,\beta)}(x)\, dx = \begin{cases} 0, & n \neq m \\ \delta_n, & n = m \end{cases}, \qquad (A\text{-}4.8)$$

where

$$\delta_n = \frac{2^{\alpha+\beta+1}\,\Gamma(n+\alpha+1)\,\Gamma(n+\beta+1)}{n!\,(\alpha+\beta+2n+1)\,\Gamma(\alpha+\beta+n+1)}. \qquad (A\text{-}4.9)$$

When $\alpha = \beta = v - \frac{1}{2}$, the Jacobi polynomials reduce to the *Gegenbauer polynomials* $C_n^v(x)$, which are defined by the Rodrigues formula

$$C_n^v(x) = \frac{(-1)^n}{2^n n!}\,(1-x^2)^{v-\frac{1}{2}}\,\frac{d^n}{dx^n}\left[(1-x^2)^{v+n-\frac{1}{2}}\right], \qquad (A\text{-}4.10)$$

The generating function for $C_n^v(x)$ of degree n is

$$(1 - 2xt + t^2)^{-v} = \sum_{n=0}^{\infty} C_n^v(x)\, t^n, \qquad |t| < 1, \quad |x| \leq 1, \qquad (A\text{-}4.11)$$

where $v > -\frac{1}{2}$.

The recurrence relations are

$$(n+1)\, C_{n+1}^v(x) - 2(v+n)x\, C_n^v(x) + (2v+n-1)C_{n-1}^v(x) = 0, \qquad (A\text{-}4.12)$$
$$(n+1)\, C_{n+1}^v(x) - 2v\, C_n^{v+1}(x) + 2v\, C_{n-1}^{v+1}(x) = 0, \qquad (A\text{-}4.13)$$
$$\frac{d}{dx}\left[C_n^v(x)\right] = 2v\, C_{n+1}^{v+1}(x). \qquad (A\text{-}4.14)$$

The differential equation satisfied by $y = C_n^v(x)$ is

$$(1-x^2)\, y'' - (2v+1)\, xy' + n(n+2v)\, y = 0. \qquad (A\text{-}4.15)$$

The orthogonal property is

$$\int_{-1}^{1} (1-x^2)^{v-\frac{1}{2}} C_n^v(x) C_m^v(x) \, dx = \delta_n \delta_{nm}, \qquad (A\text{-}4.16)$$

where

$$\delta_n = \frac{2^{1-2v} n \Gamma(n+2v)}{n!(n+v)[\Gamma(v)]^2}. \qquad (A\text{-}4.17)$$

When $v = \frac{1}{2}$, the Gegenbauer polynomials reduce to Legendre polynomials, that is

$$C_n^{\frac{1}{2}}(x) = P_n(x). \qquad (A\text{-}4.18)$$

The Hermite polynomials can also be obtained from the Gegenbauer polynomials as the limit

$$H_n(x) = n! \lim_{v \to \infty} v^{-n/2} C_n^v\left(\frac{x}{\sqrt{v}}\right). \qquad (A\text{-}4.19)$$

Finally, when $\alpha = \beta = \frac{1}{2}$, the Gegenbauer polynomials reduce to the well-known *Chebyshev polynomials*, $T_n(x)$, which are defined by a solution of the second order difference equation (see Example 10.6.7)

$$u_{n+2} - 2x\, u_{n+1} + u_n = 0, \quad |x| \leq 1 \qquad (A\text{-}4.20)$$

$$u(0) = u_0 \text{ and } u(1) = u_1. \qquad (A\text{-}4.21)$$

The generating function for $T_n(x)$ is

$$\frac{(1-t^2)}{(1-2xt+t^2)} = T_0(x) + 2\sum_{n=1}^{\infty} T_n(x) t^n, \quad |x| \leq 1 \quad t < 1. \qquad (A\text{-}4.22)$$

The first seven Chebyshev polynomials of degree n of the first kind are

$$T_0(x) = 1$$
$$T_1(x) = x$$
$$T_2(x) = 2x^2 - 1$$
$$T_3(x) = 4x^3 - 3x$$
$$T_4(x) = 8x^4 - 8x^2 + 1$$
$$T_5(x) = 16x^5 - 20x^3 + 5x$$
$$T_6(x) = 32x^6 - 48x^4 + 18x^2 - 1.$$

The graphs of the first three Chebyshev polynomials are shown in Figure A-9.

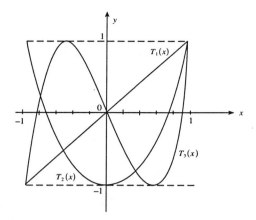

Figure A-9. Chebyshev polynomials $y = T_n(x)$.

The Chebyshev polynomials $y = T_n(x)$ satisfy the differential equation
$$(1-x^2)y'' - xy' + n^2 y = 0. \quad (A\text{-}4.23)$$
It follows from (A-4.22) that $T_n(x)$ satisfies the recurrence relations
$$T_{n+1}(x) - 2x\, T_n(x) + T_{n-1}(x) = 0, \quad (A\text{-}4.24)$$
$$T_{n+m}(x) - 2\, T_n(x)\, T_m(x) + T_{n-m}(x) = 0, \quad (A\text{-}4.25)$$
$$(1-x^2)\, T_n'(x) + nx\, T_n(x) - n\, T_{n-1}(x) = 0. \quad (A\text{-}4.26)$$
The parity relation for $T_n(x)$ is
$$T_n(-x) = (-1)^n\, T_n(x). \quad (A\text{-}4.27)$$
The Rodrigues relation is
$$T_n(x) = \frac{\sqrt{\pi}\,(-1)^n\,(1-x^2)^{\frac{1}{2}}}{2^n\left(n-\frac{1}{2}\right)!} \cdot \frac{d^n}{dx^n}\left[(1-x^2)^{n-\frac{1}{2}}\right]. \quad (A\text{-}4.28)$$
The orthogonal relation for $T_n(x)$ is
$$\int_{-1}^{1} (1-x^2)^{-\frac{1}{2}} T_m(x)\, T_n(x)\, dx = \begin{cases} 0, & m \neq n \\ \dfrac{\pi}{2}, & m = n \\ \pi, & m = n = 0 \end{cases}. \quad (A\text{-}4.29)$$
The *Chebyshev polynomials of the second kind*, $U_n(x)$, are defined by
$$U_n(x) = (1-x^2)^{-\frac{1}{2}} \sin\!\left[(n+1)\cos^{-1} x\right], \quad -1 \leq x \leq 1. \quad (A\text{-}4.30)$$
The generating function for $U_n(x)$ is
$$(1 - 2xt + t^2)^{-1} = \sum_{n=0}^{\infty} U_n(x)\, t^n, \quad |x| < 1, \quad |t| < 1. \quad (A\text{-}4.31)$$

The first seven Chebyshev polynomials $U_n(x)$ are given by

$$U_0(x) = 1$$
$$U_1(x) = 2x$$
$$U_2(x) = 4x^2 - 1$$
$$U_3(x) = 8x^3 - 4x$$
$$U_4(x) = 16x^4 - 12x^2 + 1$$
$$U_5(x) = 32x^5 - 32x^3 + 6x$$
$$U_6(x) = 64x^6 - 80x^4 + 24x^2 - 1.$$

The differential equation for $y = U_n(x)$ is

$$(1-x^2)y'' - 3xy' + n(n+2)y = 0. \tag{A-4.32}$$

The recurrence relations are

$$U_{n+1}(x) - 2x\, U_n(x) + U_{n-1}(x) = 0. \tag{A-4.33}$$
$$(1-x^2)U_n'(x) + nx\, U_n(x) - (n+1)\, U_{n-1}(x) = 0 \tag{A-4.34}$$

The parity relation is

$$U_n(-x) = (-1)^n U_n(x) \tag{A-4.35}$$

The Rodrigues formula is

$$U_n(x) = \frac{\sqrt{\pi}\,(-1)^n (n+1)}{2^{n+1}\left(n+\frac{1}{2}\right)!(1-x^2)^{\frac{1}{2}}} \frac{d^n}{dx^n}\left[(1-x^2)^{n+\frac{1}{2}}\right]. \tag{A-4.36}$$

The orthogonal relation for $U_n(x)$ is

$$\int_{-1}^{1} (1-x^2)^{\frac{1}{2}} U_m(x) U_n(x)\, dx = \frac{\pi}{2} \delta_{mn}. \tag{A-4.37}$$

A-5 Laguerre and Associated Laguerre Functions

The Laguerre polynomials $L_n(x)$ are defined by the *Rodrigues formula*

$$L_n(x) = e^x \frac{d^n}{dx^n}(x^n e^{-x}), \tag{A-5.1}$$

where $n = 0, 1, 2, 3, \ldots$.

The first seven Laguerre polynomials are

$$L_0(x) = 1$$
$$L_1(x) = 1 - x$$
$$L_2(x) = 2 - 4x + x^2$$
$$L_3(x) = 6 - 18x + 9x^2 - x^3$$
$$L_4(x) = 24 - 96x + 72x^2 - 16x^3 + x^4$$

$$L_5(x) = 120 - 600x + 600x^2 - 200x^3 + 25x^4 - x^5$$
$$L_6(x) = 720 - 4320x + 5400x^2 - 2400x^3 + 450x^4 - 36x^5 + x^6$$

The generating function is

$$(1-t)^{-1} \exp\left(\frac{xt}{1-t}\right) = \sum_{n=0}^{\infty} t^n L_n(x). \quad \text{(A-5.2)}$$

In particular

$$L_n(0) = 1. \quad \text{(A-5.3)}$$

The orthogonal relation for the Laguerre polynomial is

$$\int_0^{\infty} e^{-x} L_m(x) L_n(x) dx = (n!)^2 \delta_{nm}. \quad \text{(A-5.4)}$$

The recurrence relations are

$$(n+1) L_{n+1}(x) = (2n+1-x) L_n(x) - n L_{n-1}(x), \quad \text{(A-5.5)}$$
$$x L_n'(x) = n L_n(x) - n L_{n-1}(x). \quad \text{(A-5.6)}$$

The Laguerre polynomials $y = L_n(x)$ satisfy the *Laguerre differential equation*

$$xy'' + (1-x)y' + ny = 0. \quad \text{(A-5.7)}$$

The *associated Laguerre polynomials* are defined by

$$L_n^m(x) = \frac{d^m}{dx^m} L_n(x) \quad \text{for } n \geq m. \quad \text{(A-5.8)}$$

These polynomials $y = L_n^m(x)$ satisfy the differential equation

$$xy'' + (m+1-x)y' + (n-m)y = 0. \quad \text{(A-5.9)}$$

The generating function for $L_n^m(x)$ is

$$(1-z)^{-(m+1)} \exp\left[-\frac{xz}{1-z}\right] = \sum_{n=0}^{\infty} L_n^m(x) z^n, \quad |z| < 1. \quad \text{(A-5.10)}$$

It follows from this that

$$L_n^m(0) = \frac{(n+m)!}{n! m!} \quad \text{(A-5.11)}$$

The associated Laguerre function satisfies the *recurrence relation*

$$(n+1)L_{n+1}^m(x) = (2n+m+1-x)L_n^m(x) - (n+m)L_{n-1}^m(x), \quad \text{(A-5.12)}$$
$$x \frac{d}{dx} L_n^m(x) = n L_n^m(x) - (n+m)L_{n-1}^m(x). \quad \text{(A-5.13)}$$

The associated Laguerre function $y = L_n^m(x)$ satisfies the associated Laguerre differential equation

$$x y'' + (m+1-x)y' + n y = 0. \quad \text{(A-5.14)}$$

The Rodrigues formula for $L_n^m(x)$ is

$$L_n^m(x) = \frac{e^x x^{-m}}{n!} \frac{d^n}{dx^n}\left(e^{-x} x^{n+m}\right). \quad \text{(A-5.15)}$$

The *orthogonal* relation for $L_n^m(x)$ is

$$\int_0^\infty e^{-x} x^m L_n^m(x) L_l^m(x)\, dx = \frac{(n+m)!}{n!} \delta_{nl}. \tag{A-5.16}$$

A-6 Hermite and Weber-Hermite Functions

The Hermite polynomials $H_n(x)$ are defined by the Rodrigues formula

$$H_n(x) = (-1)^n \exp(x^2) \frac{d^n}{dx^n}\left[\exp(-x^2)\right], \tag{A-6.1}$$

where $n = 0, 1, 2, 3, \ldots$.

The first seven Hermite polynomials are

$H_0(x) = 1$
$H_1(x) = 2x$
$H_2(x) = 4x^2 - 2$
$H_3(x) = 8x^3 - 12x$
$H_4(x) = 16x^4 - 48x^2 + 12$
$H_5(x) = 32x^5 - 16x^3 + 120x$
$H_6(x) = 64x^6 - 480x^4 + 720x^2 - 120.$

The *generating function* is

$$\exp(2xt - t^2) = \sum_{n=0}^\infty \frac{t^n}{n!} H_n(x). \tag{A-6.2}$$

It follows from (A-6.2) that $H_n(x)$ satisfies the *parity relation*

$$H_n(-x) = (-1)^n H_n(x). \tag{A-6.3}$$

Also, it follows from (A-6.2) that

$$H_{2n+1}(0) = 0, \quad H_{2n}(0) = (-1)^n \frac{(2n)!}{n!} \tag{A-6.4}$$

The *recurrence relations* for Hermite polynomials are

$$H_{n+1}(x) - 2x H_n(x) + 2n H_{n-1}(x) = 0, \tag{A-6.5}$$
$$H_n'(x) = 2x H_{n-1}(x). \tag{A-6.6}$$

The Hermite polynomials, $y = H_n(x)$, are solutions of the Hermite differential equation

$$y'' - 2xy' + 2ny = 0. \tag{A-6.7}$$

The orthogonal property of Hermite polynomials is

$$\int_{-\infty}^\infty \exp(-x^2) H_n(x) H_m(x)\, dx = 2^n n! \sqrt{\pi}\, \delta_{mn}. \tag{A-6.8}$$

With repeated use of integration by parts, it follows from (A-6.1) that

$$\int_{-\infty}^{\infty} \exp(-x^2) H_n(x) x^m \, dx = 0, \quad m = 0,1,\ldots,(n-1), \qquad (A\text{-}6.9)$$

$$\int_{-\infty}^{\infty} \exp(-x^2) H_n(x) x^n dx = \sqrt{\pi}\, n!. \qquad (A\text{-}6.10)$$

The *Weber-Hermite function*

$$y = \exp\left(-\frac{x^2}{2}\right) H_n(x) \qquad (A\text{-}6.11)$$

satisfies the differential equation

$$y'' + (\lambda - x^2) y = 0, \qquad (A\text{-}6.12)$$

where $\lambda = 2n+1$. If $\lambda \neq 2n+1$, then y is not finite as $|x| \to \infty$.

The normalized Weber-Hermite functions are given by

$$\psi_n(x) = 2^{-n/2} \pi^{-\frac{1}{4}} (n!)^{-\frac{1}{2}} \exp\left(-\frac{x^2}{2}\right) H_n(x) \qquad (A\text{-}6.13)$$

Physically, they represent quantum mechanical oscillator wave functions. The graphs of these functions are shown in Figure A-10.

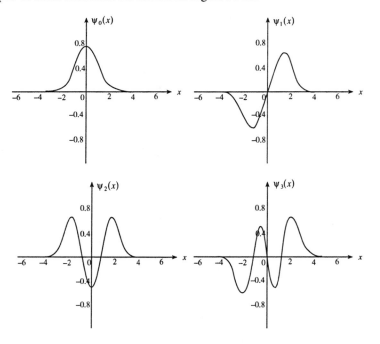

Figure A.10. The normalized Weber-Hermite functions.

Appendix B

Tables of Integral Transforms

In this Appendix we provide a set of *short* tables of integral transforms of the functions that are either cited in the text or in most common use in mathematical, physical and engineering applications. In these tables no attempt is made to give complete lists of transforms. For exhaustive lists of integral transforms, the reader is referred to Erdélyi et al. (1954), Campbell and Foster (1948), Ditkin and Prudnikov (1965), Doetsch (1950-56, 1970), Marichev (1983), and Oberhettinger (1972, 1974).

Table B-1. Fourier Transforms

	$f(x)$	$F(k) = \dfrac{1}{\sqrt{2\pi}} \int\limits_{-\infty}^{\infty} \exp(-ikx) f(x) dx$
1	$\exp(-a\lvert x \rvert), \quad a > 0$	$\left(\sqrt{\dfrac{2}{\pi}}\right) a\left(a^2 + k^2\right)^{-1}$
2	$x \exp(-a\lvert x \rvert)$	$\left(\sqrt{\dfrac{2}{\pi}}\right)(-2aik)\left(a^2 + k^2\right)^{-2}$
3	$\exp(-ax^2), \quad a > 0$	$\dfrac{1}{\sqrt{2a}} \exp\left(-\dfrac{k^2}{4a}\right)$
4	$\left(x^2 + a^2\right)^{-1}, \quad a > 0$	$\sqrt{\dfrac{\pi}{2}} \, \dfrac{\exp(-a\lvert k \rvert)}{a}$
5	$x\left(x^2 + a^2\right)^{-1}$	$\sqrt{\dfrac{\pi}{2}} \left(\dfrac{ik}{2a}\right) \exp(-a\lvert k \rvert)$
6	$\begin{cases} c, & a \le x \le b \\ 0, & \text{outside} \end{cases}$	$\dfrac{ic}{\sqrt{2\pi}} \dfrac{1}{k}\left(e^{-ibk} - e^{-iak}\right)$

	$f(x)$	$F(k) = \dfrac{1}{\sqrt{2\pi}} \displaystyle\int_{-\infty}^{\infty} \exp(-ikx) f(x)\,dx$
7	$\lvert x \rvert \exp(-a\lvert x \rvert),\quad a>0$	$\sqrt{\dfrac{2}{\pi}}\,(a^2-k^2)(a^2+k^2)^{-2}$
8	$\dfrac{\sin ax}{x}$	$\sqrt{\dfrac{\pi}{2}}\,H(a-\lvert k \rvert)$
9	$\exp\{-x(a-i\omega)\}\,H(x)$	$\dfrac{1}{\sqrt{2\pi}}\,\dfrac{i}{(\omega-k+ia)}$
10	$(a^2-x^2)^{-\tfrac{1}{2}}\,H(a-\lvert x \rvert)$	$\sqrt{\dfrac{\pi}{2}}\,J_0(ak)$
11	$\dfrac{\sin\!\left[b(x^2+a^2)^{\tfrac{1}{2}}\right]}{(x^2+a^2)^{\tfrac{1}{2}}}$	$\sqrt{\dfrac{\pi}{2}}\,J_0\!\left(a\sqrt{b^2-k^2}\right) H(b-\lvert k \rvert)$
12	$\dfrac{\cos\!\left(b\sqrt{a^2-x^2}\right)}{(a^2-x^2)^{\tfrac{1}{2}}}\,H(a-\lvert x \rvert)$	$\sqrt{\dfrac{\pi}{2}}\,J_0\!\left(a\sqrt{b^2+k^2}\right)$
13	$e^{-ax}H(x),\quad a>0$	$\dfrac{1}{\sqrt{2\pi}}\,(a-ik)(a^2+k^2)^{-1}$
14	$\dfrac{1}{\sqrt{\lvert x \rvert}}\exp(-a\lvert x \rvert)$	$(a^2+k^2)^{-\tfrac{1}{2}}\!\left[a+(a^2+k^2)^{\tfrac{1}{2}}\right]^{\tfrac{1}{2}}$
15	$\delta(x)$	$\dfrac{1}{\sqrt{2\pi}}$
16	$\delta^{(n)}(x)$	$\dfrac{1}{\sqrt{2\pi}}\,(ik)^n$
17	$\delta(x-a)$	$\dfrac{1}{\sqrt{2\pi}}\,\exp(-iak)$
18	$\delta^{(n)}(x-a)$	$\dfrac{1}{\sqrt{2\pi}}\,(ik)^n \exp(-iak)$
19	$\exp(iax)$	$\sqrt{2\pi}\,\delta(k-a)$

	$f(x)$	$F(k) = \dfrac{1}{\sqrt{2\pi}} \displaystyle\int_{-\infty}^{\infty} \exp(-ikx) f(x)\,dx$
20	1	$\sqrt{2\pi}\,\delta(k)$
21	x	$\sqrt{2\pi}\,i\,\delta'(k)$
22	x^n	$\sqrt{2\pi}\,i^n\,\delta^{(n)}(k)$
23	$H(x)$	$\sqrt{\dfrac{\pi}{2}} \left[\dfrac{1}{i\pi k} + \delta(k) \right]$
24	$H(x-a)$	$\sqrt{\dfrac{\pi}{2}} \left[\dfrac{\exp(-ika)}{\pi i k} + \delta(k) \right]$
25	$H(x) - H(-x)$	$\sqrt{\dfrac{2}{\pi}} \left(-\dfrac{i}{k} \right)$
26	$x^n \exp(iax)$	$\sqrt{2\pi}\,i^n\,\delta^{(n)}(k-a)$
27	$|x|^{-1}$	$\dfrac{1}{\sqrt{2\pi}} (A - 2\log|k|)$, A is a constant
28	$\log(|x|)$	$-\sqrt{\dfrac{\pi}{2}}\,\dfrac{1}{|k|}$
29	$H(a - |x|)$	$\sqrt{\dfrac{2}{\pi}} \left(\dfrac{\sin ak}{k} \right)$
30	$|x|^\alpha$ (α not integer)	$\sqrt{\dfrac{2}{\pi}}\,\Gamma(\alpha+1)\,|k|^{-(1+\alpha)}$ $\times \cos\left[\dfrac{\pi}{2}(\alpha+1) \right]$
31	$\operatorname{sgn} x$	$\sqrt{\dfrac{2}{\pi}}\,\dfrac{1}{(ik)}$
32	$x^{-n-1} \operatorname{sgn} x$	$\dfrac{1}{\sqrt{2\pi}}\,\dfrac{(-ik)^n}{n!}(A - 2\log|k|)$

	$f(x)$	$F(k) = \dfrac{1}{\sqrt{2\pi}} \displaystyle\int_{-\infty}^{\infty} \exp(-ikx) f(x)\, dx$
33	$\dfrac{1}{x}$	$-i\sqrt{\dfrac{\pi}{2}}\, \text{sgn}\, k$
34	$\dfrac{1}{x^n}$	$-i\sqrt{\dfrac{\pi}{2}}\left[\dfrac{(-ik)^{n-1}}{(n-1)!}\, \text{sgn}\, k\right]$
35	$x^n \exp(iax)$	$\sqrt{2\pi}\, i^n \delta^{(n)}(k-a)$
36	$x^\alpha H(x)$, α is not an integer	$\dfrac{\Gamma(\alpha+1)}{\sqrt{2\pi}}\, |k|^{-(\alpha+1)}$ $\times \exp\left[-\left(\dfrac{\pi i}{2}\right)(\alpha+1)\, \text{sgn}\, k\right]$
37	$x^n \exp(iax)\, H(x)$	$\sqrt{\dfrac{\pi}{2}}\left[\dfrac{n!}{i\pi(k-a)^{n+1}} + i^n\, \delta^{(n)}(k-a)\right]$
38	$\exp(iax)\, H(x-b)$	$\sqrt{\dfrac{\pi}{2}}\left[\dfrac{\exp[-ib(k-a)]}{i\pi(k-a)} + \delta(k-a)\right]$
39	$\dfrac{1}{x-a}$	$-i\sqrt{\dfrac{\pi}{2}}\, \exp(-iak)\, \text{sgn}\, k$
40	$\dfrac{1}{(x-a)^n}$	$-i\sqrt{\dfrac{\pi}{2}}\, \exp(-iak)\, \dfrac{(-ik)^{n-1}}{(n-1)!}\, \text{sgn}\, k$
41	$\dfrac{e^{iax}}{(x-b)}$	$i\sqrt{\dfrac{\pi}{2}}\, \exp[ib(a-k)][1-2H(k-a)]$
42	$\dfrac{e^{iax}}{(x-b)^n}$	$i\sqrt{\dfrac{\pi}{2}}\, [1-2H(k-a)]$ $\times \dfrac{\exp\{ib(a-k)\}}{(n-1)!}\, [-i(k-a)]^{n-1}$
43	$|x|^\alpha\, \text{sgn}\, x$ (α not integer)	$\sqrt{\dfrac{2}{\pi}}\, \dfrac{(-i)\Gamma(\alpha+1)}{|k|^{\alpha+1}}\, \cos\left(\dfrac{\pi\alpha}{2}\right)\, \text{sgn}\, k$

Table B-2. Fourier Cosine Transforms

	$f(x)$	$F_c(k) = \sqrt{\dfrac{2}{\pi}} \int_0^\infty \cos(kx) f(x)\,dx$
1	$\exp(-ax),\quad a>0$	$\left(\sqrt{\dfrac{2}{\pi}}\right) a\left(a^2+k^2\right)^{-1}$
2	$x\exp(-ax)$	$\left(\sqrt{\dfrac{2}{\pi}}\right)\left(a^2-k^2\right)\left(a^2+k^2\right)^{-2}$
3	$\exp\left(-a^2 x^2\right)$	$\dfrac{1}{a\sqrt{2}}\exp\left(-\dfrac{k^2}{4a^2}\right)$
4	$H(a-x)$	$\sqrt{\dfrac{2}{\pi}}\left(\dfrac{\sin ak}{k}\right)$
5	$x^{a-1},\quad 0<a<1$	$\sqrt{\dfrac{2}{\pi}}\,\Gamma(a)\,k^{-a}\cos\left(\dfrac{a\pi}{2}\right)$
6	$\cos\left(ax^2\right)$	$\dfrac{1}{2\sqrt{a}}\left[\cos\left(\dfrac{k^2}{4a}\right)+\sin\left(\dfrac{k^2}{4a}\right)\right]$
7	$\sin\left(ax^2\right),\ a>0$	$\dfrac{1}{2\sqrt{a}}\left[\cos\left(\dfrac{k^2}{4a}\right)-\sin\left(\dfrac{k^2}{4a}\right)\right]$
8	$\left(a^2-x^2\right)^{v-\frac{1}{2}}H(a-x),\ v>-\dfrac{1}{2}$	$2^{v-\frac{1}{2}}\Gamma\left(v+\dfrac{1}{2}\right)\left(\dfrac{a}{k}\right)^v J_v(ak)$
9	$\left(a^2+x^2\right)^{-1}J_0(bx),\ a,b>0$	$\sqrt{\dfrac{\pi}{2}}\,a^{-1}\exp(-ak)I_0(ab),\ b<k<\infty$
10	$x^{-v}J_v(ax),\quad v>-\dfrac{1}{2}$	$\dfrac{\left(a^2-k^2\right)^{v-\frac{1}{2}}H(a-k)}{2^{v-\frac{1}{2}}a^v\,\Gamma\left(v+\dfrac{1}{2}\right)}$
11	$\left(x^2+a^2\right)^{-\frac{1}{2}}\exp\left[-b\left(x^2+a^2\right)^{\frac{1}{2}}\right]$	$K_0\left[a\left(k^2+b^2\right)^{\frac{1}{2}}\right],\ a>0,\ b>0$

	$f(x)$	$F_c(k) = \sqrt{\dfrac{2}{\pi}} \displaystyle\int_0^\infty \cos(kx) f(x)\,dx$
12	$(2ax-x^2)^{v-\frac{1}{2}} H(2a-x),\ v>-\dfrac{1}{2}$	$\sqrt{2}\,\Gamma\!\left(v+\dfrac{1}{2}\right)\!\left(\dfrac{2a}{k}\right)^{v} \times \cos(ak)\,J_v(ak)$
13	$x^{v-1} e^{-ax},\qquad v>0,\ a>0$	$\sqrt{\dfrac{2}{\pi}}\,\Gamma(v)\,r^{-v}\cos n\theta,$ where $r=(a^2+k^2)^{\frac{1}{2}},\ \theta=\tan^{-1}\!\left(\dfrac{k}{a}\right)$
14	$\dfrac{2}{x} e^{-x}\sin x$	$\sqrt{\dfrac{2}{\pi}}\,\tan^{-1}\!\left(\dfrac{2}{k^2}\right)$
15	$\sin\!\left[a(b^2-x^2)^{\frac{1}{2}}\right] H(b-x)$	$\sqrt{\dfrac{\pi}{2}}\,(ab)(a^2+k^2)^{-\frac{1}{2}} \times J_1\!\left[b(a^2+k^2)^{\frac{1}{2}}\right]$
16	$\dfrac{(1-x^2)}{(1+x^2)^2}$	$\sqrt{\dfrac{\pi}{2}}\,k\,\exp(-k)$
17	$x^{-\alpha},\qquad 0<\alpha<1$	$\sqrt{\dfrac{\pi}{2}}\,\dfrac{k^{\alpha-1}}{\Gamma(\alpha)}\,\sec\!\left(\dfrac{\pi\alpha}{2}\right)$
18	$\left(\dfrac{1}{a}+x\right)e^{-ax}$	$\sqrt{\dfrac{2}{\pi}}\,\dfrac{2a^2}{(a^2+k^2)^2}$
19	$\log\!\left(1+\dfrac{a^2}{x^2}\right),\ a>0$	$\sqrt{2\pi}\,\dfrac{(1-e^{-ak})}{k}$
20	$\log\!\left(\dfrac{a^2+x^2}{b^2+x^2}\right),\qquad a,b>0$	$\sqrt{2\pi}\,\dfrac{(e^{-bk}-e^{-ak})}{k}$
21	$a(x^2+a^2)^{-1},\qquad a>0$	$\sqrt{\dfrac{\pi}{2}}\,\exp(-ak),\qquad k>0$

Table B-3. Fourier Sine Transforms

	$f(x)$	$F_s(k) = \sqrt{\dfrac{2}{\pi}} \displaystyle\int_0^\infty \sin(kx)\, f(x)\, dx$
1	$\exp(-ax), \quad a>0$	$\sqrt{\dfrac{2}{\pi}}\, k\left(a^2+k^2\right)^{-1}$
2	$x\exp(-ax), \quad a>0$	$\sqrt{\dfrac{2}{\pi}}\,(2ak)\left(a^2+k^2\right)^{-2}$
3	$x^{\alpha-1}, \quad 0<\alpha<1$	$\sqrt{\dfrac{2}{\pi}}\, k^{-\alpha}\,\Gamma(\alpha)\,\sin\!\left(\dfrac{\pi\alpha}{2}\right)$
4	$\dfrac{1}{\sqrt{x}}$	$\dfrac{1}{\sqrt{k}}, \quad k>0$
5	$x^{\alpha-1}e^{-ax}, \quad \alpha>-1,\ a>0$	$\sqrt{\dfrac{2}{\pi}}\,\Gamma(\alpha)\, r^{-\alpha}\sin(\alpha\theta),\ \text{where}$ $r=\left(a^2+k^2\right)^{\frac{1}{2}},\ \theta=\tan^{-1}\!\left(\dfrac{k}{a}\right)$
6	$x^{-1}e^{-ax}, \quad a>0$	$\sqrt{\dfrac{2}{\pi}}\,\tan^{-1}\!\left(\dfrac{k}{a}\right), \quad k>0$
7	$x\exp\left(-a^2x^2\right)$	$2^{-3/2}\left(\dfrac{k}{a^3}\right)\exp\!\left(-\dfrac{k^2}{4a^2}\right)$
8	$\operatorname{erfc}(ax)$	$\sqrt{\dfrac{2}{\pi}}\,\dfrac{1}{k}\left[1-\exp\!\left(-\dfrac{k^2}{4a^2}\right)\right]$
9	$x\left(a^2+x^2\right)^{-1}$	$\sqrt{\dfrac{\pi}{2}}\,\exp(-ak), \quad a>0$
10	$x\left(a^2+x^2\right)^{-2}$	$\dfrac{1}{\sqrt{2\pi}}\left(\dfrac{k}{a}\right)\exp(-ak), \quad (a>0)$
11	$x\left(a^2-x^2\right)^{\nu-\frac{1}{2}}H(a-x),\ \nu>-\dfrac{1}{2}$	$2^{\nu-\frac{1}{2}} a^{\nu+1} k^{-\nu}\,\Gamma\!\left(\nu+\dfrac{1}{2}\right) J_{\nu+1}(ak)$

	$f(x)$	$F_s(k) = \sqrt{\dfrac{2}{\pi}} \displaystyle\int_0^\infty \sin(kx) f(x)\, dx$
12	$\tan^{-1}\left(\dfrac{x}{a}\right)$	$\sqrt{\dfrac{\pi}{2}}\, k^{-1} \exp(-ak)$
13	$x^{-\nu} J_{\nu+1}(ax), \quad \nu > -\dfrac{1}{2}$	$\dfrac{k(a^2 - k^2)^{\nu - \frac{1}{2}}}{2^{\nu - \frac{1}{2}}\, a^{\nu+1}\, \Gamma\left(\nu + \dfrac{1}{2}\right)}\, H(a-k)$
14	$x^{-1} J_0(ax)$	$\begin{cases} \sqrt{\dfrac{2}{\pi}}\, \sin^{-1}\left(\dfrac{k}{a}\right), & 0 < k < a \\ \sqrt{\dfrac{\pi}{2}}, & a < k < \infty \end{cases}$
15	$x(a^2 + x^2)^{-1} J_0(bx), \quad a>0,\ b>0$	$\sqrt{\dfrac{\pi}{2}}\, e^{-ak} I_0(ab), \quad a < k < \infty$
16	$J_0(a\sqrt{x}), \quad a > 0$	$\sqrt{\dfrac{2}{\pi}}\, \dfrac{1}{k} \cos\left(\dfrac{a^2}{4k}\right)$
17	$(x^2 - a^2)^{\nu - \frac{1}{2}} H(x-a), \quad |\nu| < \dfrac{1}{2}$	$2^{\nu - \frac{1}{2}} \left(\dfrac{a}{k}\right)^\nu \Gamma\left(\nu + \dfrac{1}{2}\right) J_{-\nu}(ak)$
18	$x^{1-\nu}(x^2 + a^2)^{-1} J_\nu(ax),$ $\nu > -\dfrac{3}{2}, \quad a,b>0$	$\sqrt{\dfrac{\pi}{2}}\, a^{-\nu} \exp(-ak)\, I_\nu(ab),$ $a < k < \infty$
19	$H(a-x), \quad a>0$	$\sqrt{\dfrac{2}{\pi}}\, \dfrac{1}{k}(1 - \cos ak)$
20	$\mathrm{erfc}(ax)$	$\sqrt{\dfrac{2}{\pi}}\, \dfrac{1}{k}\left[1 - \exp\left(-\dfrac{k^2}{4a^2}\right)\right]$
21	$x^{-\alpha}, \quad 0 < \alpha < 2$	$\Gamma(1-\alpha)\, k^{\alpha-1} \cos\left(\dfrac{\alpha\pi}{2}\right)$
22	$(ax - x^2)^{\alpha - \frac{1}{2}} H(a-x), \quad \alpha > -\dfrac{1}{2}$	$\sqrt{2}\, \Gamma\left(\alpha + \dfrac{1}{2}\right)\left(\dfrac{a}{k}\right)^\alpha$ $\times \sin\left(\dfrac{ak}{2}\right) J_\alpha\left(\dfrac{ak}{2}\right)$

Table B-4. Laplace Transforms

	$f(t)$	$\bar{f}(s) = \int_0^\infty \exp(-st)\, f(t)\, dt$
1	$t^n \quad (n = 0, 1, 2, 3, \ldots)$	$\dfrac{n!}{s^{n+1}}$
2	e^{at}	$\dfrac{1}{s-a}$
3	$\cos at$	$\dfrac{s}{s^2 + a^2}$
4	$\sin at$	$\dfrac{a}{s^2 + a^2}$
5	$\cosh at$	$\dfrac{s}{s^2 - a^2}$
6	$\sinh at$	$\dfrac{a}{s^2 - a^2}$
7	$t^n e^{-at}$	$\dfrac{\Gamma(n+1)}{(s+a)^{n+1}}$
8	$t^a \quad (a > -1)$	$\dfrac{\Gamma(a+1)}{s^{a+1}}$
9	$e^{at} \cos bt$	$\dfrac{s-a}{(s-a)^2 + b^2}$
10	$e^{at} \sin bt$	$\dfrac{b}{(s-a)^2 + b^2}$
11	$(e^{at} - e^{bt})$	$\dfrac{a-b}{(s-a)(s-b)}$
12	$\dfrac{1}{(a-b)}(a\, e^{at} - b\, e^{bt})$	$\dfrac{s}{(s-a)(s-b)}$
13	$t \sin at$	$\dfrac{2as}{(s^2 + a^2)^2}$
14	$t \cos at$	$\dfrac{s^2 - a^2}{(s^2 + a^2)^2}$
15	$\sin at \sinh at$	$\dfrac{2sa^2}{(s^4 + 4a^4)}$
16	$(\sinh at - \sin at)$	$\dfrac{2a^3}{(s^4 - a^4)}$

	$f(t)$	$\bar{f}(s) = \int_0^\infty \exp(-st)\, f(t)\, dt$
17	$(\cosh at - \cos at)$	$\dfrac{2a^2 s}{(s^4 - a^4)}$
18	$\dfrac{\cos at - \cos bt}{(b^2 - a^2)} \quad (a^2 \neq b^2)$	$\dfrac{s}{(s^2 + a^2)(s^2 + b^2)}$
19	$\dfrac{1}{\sqrt{t}}$	$\sqrt{\dfrac{\pi}{s}}$
20	$2\sqrt{t}$	$\dfrac{1}{s}\sqrt{\dfrac{\pi}{s}}$
21	$t \cosh at$	$(s^2 + a^2)(s^2 - a^2)^{-2}$
22	$t \sinh at$	$2as(s^2 - a^2)^{-2}$
23	$\dfrac{\sin(at)}{t}$	$\tan^{-1}\left(\dfrac{a}{s}\right)$
24	$t^{-1/2} \exp\left(-\dfrac{a}{t}\right)$	$\sqrt{\dfrac{\pi}{s}} \exp(-2\sqrt{as})$
25	$t^{-3/2} \exp\left(-\dfrac{a}{t}\right)$	$\sqrt{\dfrac{\pi}{a}} \exp(-2\sqrt{as})$
26	$\dfrac{1}{\sqrt{\pi t}} (1 + 2at) e^{at}$	$\dfrac{s}{(s-a)\sqrt{s-a}}$
27	$(1 + at)\, e^{at}$	$\dfrac{s}{(s-a)^2}$
28	$\dfrac{1}{2\sqrt{\pi t^3}} (e^{bt} - e^{at})$	$\sqrt{s-a} - \sqrt{s-b}$
29	$\exp(a^2 t)\, \text{erf}(a\sqrt{t})$	$\dfrac{a}{\sqrt{s}\,(s - a^2)}$
30	$\exp(a^2 t)\, \text{erfc}(a\sqrt{t})$	$\dfrac{1}{\sqrt{s}\,(\sqrt{s} + a)}$
31	$\dfrac{1}{\sqrt{\pi t}} + a\exp(a^2 t)\, \text{erf}(a\sqrt{t})$	$\dfrac{\sqrt{s}}{(s - a^2)}$

	$f(t)$	$\bar{f}(s) = \int_0^\infty \exp(-st)\, f(t)\, dt$
32	$\dfrac{1}{\sqrt{\pi t}} - a\exp(a^2 t)\,\mathrm{erfc}(a\sqrt{t})$	$\dfrac{1}{\sqrt{s}+a}$
33	$\dfrac{\exp(-at)}{\sqrt{b-a}}\,\mathrm{erf}\left(\sqrt{(b-a)t}\right)$	$\dfrac{1}{(s+a)\sqrt{s+b}}$
34	$\dfrac{1}{2}e^{i\omega t}\Big[\exp(-\lambda z)\,\mathrm{erfc}\left(\zeta - \sqrt{i\omega t}\right)$ $\quad + \exp(\lambda z)\,\mathrm{erfc}\left(\zeta + \sqrt{i\omega t}\right)\Big]$, where $\zeta = z/2\sqrt{vt}$.	$(s-i\omega)^{-1}\exp\left(-z\sqrt{\dfrac{s}{v}}\right)$
35	$\dfrac{1}{2}\Bigg[\exp(-ab)\,\mathrm{erfc}\left(\dfrac{b-2at}{2\sqrt{t}}\right)$ $\quad + \exp(ab)\,\mathrm{erfc}\left(\dfrac{b+2at}{2\sqrt{t}}\right)\Bigg]$	$\exp\left[-b(s+a^2)^{\frac{1}{2}}\right]$
36	$Si(t) = \int_0^t \dfrac{\sin x}{x}\,dx$	$\dfrac{1}{s}\cot^{-1}(s)$
37	$Ci(t) = -\int_t^\infty \dfrac{\cos x}{x}\,dx$	$-\dfrac{1}{2s}\log(1+s^2)$
38	$-Ei(-t) = \int_t^\infty \dfrac{e^{-x}}{x}\,dx$	$\dfrac{1}{s}\log(1+s)$
39	$J_0(at)$	$(s^2+a^2)^{-\frac{1}{2}}$
40	$I_0(at)$	$(s^2-a^2)^{-\frac{1}{2}}$
41	$t^{\alpha-1}\exp(-at),\quad \alpha>0$	$\Gamma(\alpha)(s+a)^{-\alpha}$
42	$\dfrac{\sqrt{\pi}}{\Gamma\left(v+\dfrac{1}{2}\right)}\left(\dfrac{t}{2a}\right)^v J_v(at)$	$(s^2+a^2)^{-\left(v+\frac{1}{2}\right)},\quad \mathrm{Re}\,v > -\dfrac{1}{2}$
43	$t^{-1}J_v(at)$	$v^{-1}a^v\left(\sqrt{s^2+a^2}+s\right)^{-v},\quad \mathrm{Re}\,v > -\dfrac{1}{2}$
44	$J_0(a\sqrt{t})$	$\dfrac{1}{s}\exp\left(-\dfrac{a^2}{4s}\right)$

	$f(t)$	$\bar{f}(s) = \int_0^\infty \exp(-st)\, f(t)\,dt$
45	$\left(\dfrac{2}{a}\right)^{\nu} t^{\nu/2} J_{\nu}\left(a\sqrt{t}\right)$	$s^{-(\nu+1)} \exp\left(-\dfrac{a^2}{4s}\right)$, $\quad \operatorname{Re} \nu > -\dfrac{1}{2}$
46	$\dfrac{a}{2t\sqrt{\pi t}} \exp\left(-\dfrac{a^2}{4t}\right)$	$\exp(-a\sqrt{s})$, $\quad a > 0$
47	$\dfrac{1}{\sqrt{\pi t}} \exp\left(-\dfrac{a^2}{4t}\right)$	$\dfrac{1}{\sqrt{s}} \exp(-a\sqrt{s})$, $\quad a \geq 0$
48	$\exp\left(-\dfrac{a^2 t^2}{4}\right)$	$\dfrac{\sqrt{\pi}}{a} \exp\left(\dfrac{s^2}{a^2}\right) \operatorname{erfc}\left(\dfrac{s}{a}\right)$, $\quad a > 0$
49	$\left(t^2 - a^2\right)^{-\frac{1}{2}} H(t - a)$	$K_0(as)$, $\quad a > 0$
50	$\delta(t - a)$	$\exp(-as)$, $\quad a > 0$
51	$H(t - a)$	$\dfrac{1}{s} \exp(-as)$, $\quad a \geq 0$
52	$\delta^{(n)}(t)$	s^n
53	$\delta^{(n)}(t - a)$	$s^n \exp(-as)$
54	$\lvert \sin at \rvert$, $\quad (a > 0)$	$\dfrac{a}{(s^2 + a^2)} \coth\left(\dfrac{\pi s}{2a}\right)$
55	$\dfrac{1}{\sqrt{\pi t}} \cos(2\sqrt{at})$	$\dfrac{1}{\sqrt{s}} \exp\left(-\dfrac{a}{s}\right)$
56	$\dfrac{1}{\sqrt{\pi t}} \sin(2\sqrt{at})$	$\dfrac{1}{s\sqrt{s}} \exp\left(-\dfrac{a}{s}\right)$
57	$\dfrac{1}{\sqrt{\pi a}} \cosh(2\sqrt{at})$	$\dfrac{1}{\sqrt{s}} \exp\left(\dfrac{a}{s}\right)$
58	$\dfrac{1}{\sqrt{\pi a}} \sinh(2\sqrt{at})$	$\dfrac{1}{s\sqrt{s}} \exp\left(\dfrac{a}{s}\right)$

	$f(t)$	$\bar{f}(s) = \int_0^\infty \exp(-st)\, f(t)\, dt$
59	$\operatorname{erf}\left(\dfrac{t}{2a}\right)$	$\dfrac{1}{s}\exp(a^2 s^2)\operatorname{erfc}(as), \quad a>0$
60	$\operatorname{erfc}\left(\dfrac{a}{2\sqrt{t}}\right)$	$\dfrac{1}{s}\exp(-a\sqrt{s}), \quad a\geq 0$
61	$\sqrt{\dfrac{4t}{\pi}}\exp\left(-\dfrac{a^2}{4t}\right)-a\,\operatorname{erfc}\left(\dfrac{a}{2\sqrt{t}}\right)$	$\dfrac{1}{s\sqrt{s}}\exp(-a\sqrt{s}), \quad a\geq 0$
62	$\exp\{a(b+at)\}\operatorname{erfc}\left(a\sqrt{t}+\dfrac{b}{2\sqrt{t}}\right)$	$\dfrac{\exp(-b\sqrt{s})}{\sqrt{s}(\sqrt{s}+a)}, \quad a\geq 0$
63	$J_0\left(a\sqrt{t^2-\omega^2}\right)H(t-\omega)$	$(s^2+a^2)^{-\tfrac{1}{2}}\exp\left\{-\omega\sqrt{s^2+a^2}\right\}$
64	$\dfrac{1}{t}\left(e^{bt}-e^{at}\right)$	$\log\left(\dfrac{s-a}{s-b}\right)$
65	$\{\pi(t+a)\}^{-\tfrac{1}{2}}$	$\dfrac{1}{\sqrt{s}}\exp(as)\operatorname{erfc}\left(\sqrt{as}\right), \quad a>0$
66	$\dfrac{1}{\pi t}\sin(2a\sqrt{t})$	$\operatorname{erf}\left(\dfrac{a}{\sqrt{s}}\right)$
67	$\dfrac{1}{\sqrt{\pi t}}\exp(-2a\sqrt{t}), \quad a\geq 0$	$\dfrac{1}{\sqrt{s}}\exp\left(\dfrac{a^2}{s}\right)\operatorname{erfc}\left(\dfrac{a}{\sqrt{s}}\right)$
68	$C(t)=\dfrac{1}{\sqrt{2\pi}}\displaystyle\int_0^t \dfrac{\cos u}{\sqrt{u}}\,du$	$\dfrac{1}{2s}\left[\dfrac{1}{\sqrt{1+s^2}}+\dfrac{s}{1+s^2}\right]^{\tfrac{1}{2}}$
69	$S(t)=\dfrac{1}{\sqrt{2\pi}}\displaystyle\int_0^t \dfrac{\sin u}{\sqrt{u}}\,du$	$\dfrac{1}{2s}\left[\dfrac{1}{\sqrt{1+s^2}}-\dfrac{s}{1+s^2}\right]^{\tfrac{1}{2}}$
70	$\vartheta(t)=1+2\displaystyle\sum_{n=1}^\infty \exp(-n^2\pi t)$	$\left(\sqrt{s}\tanh\sqrt{s}\right)^{-1}$

Table B-5. Hankel Transforms

	$f(r)$	order	$\tilde{f}_n(k) = \int_0^\infty r\, J_n(kr)\, f(r)\, dr$
1	$H(a-r)$	0	$\dfrac{a}{k} J_1(ak)$
2	$\exp(-ar)$	0	$a(a^2 + k^2)^{-\frac{3}{2}}$
3	$\dfrac{1}{r}\exp(-ar)$	0	$(a^2 + k^2)^{-\frac{1}{2}}$
4	$(a^2 - r^2) H(a-r)$	0	$\dfrac{4a}{k^3} J_1(ak) - \dfrac{2a^2}{k^2} J_0(ak)$
5	$a(a^2 + r^2)^{-\frac{3}{2}}$	0	$\exp(-ak)$
6	$\dfrac{1}{r}\cos(ar)$	0	$(k^2 - a^2)^{-\frac{1}{2}} H(k-a)$
7	$\dfrac{1}{r}\sin(ar)$	0	$(a^2 - k^2)^{-\frac{1}{2}} H(a-k)$
8	$\dfrac{1}{r^2}(1 - \cos ar)$	0	$\cosh^{-1}\left(\dfrac{a}{k}\right) H(a-k)$
9	$\dfrac{1}{r} J_1(ar)$	0	$\dfrac{1}{a} H(a-k),\ \ a>0$
10	$Y_0(ar)$	0	$\left(\dfrac{2}{\pi}\right)(a^2 - k^2)^{-1}$
11	$K_0(ar)$	0	$(a^2 + k^2)^{-1}$
12	$\dfrac{\delta(r)}{r}$	0	1
13	$(x^2 + b^2)^{-\frac{1}{2}}$ $\times \exp\left\{-a(x^2 + b^2)^{\frac{1}{2}}\right\}$	0	$(k^2 + a^2)^{-\frac{1}{2}} \exp\left\{-b(k^2 + a^2)^{\frac{1}{2}}\right\}$

	$f(r)$	order	$\tilde{f}_n(k) = \int_0^\infty r\, J_n(kr)\, f(r)\, dr$
14	$\dfrac{\sin r}{r^2}$	0	$\begin{cases} \dfrac{\pi}{2}, & k<1 \\ \sin^{-1}\left(\dfrac{1}{k}\right), & k>1 \end{cases}$
15	$(r^2+a^2)^{-\frac{1}{2}}$	0	$\dfrac{1}{k}\exp(-ak)$
16	$\exp(-ar)$	1	$k(a^2+k^2)^{-3/2}$
17	$\dfrac{\sin ar}{r}$	1	$\dfrac{a\, H(k-a)}{k(k^2-a^2)^{\frac{1}{2}}}$
18	$\dfrac{1}{r}\exp(-ar)$	1	$\dfrac{1}{k}\left[1-\dfrac{a}{(k^2+a^2)^{\frac{1}{2}}}\right]$
19	$\dfrac{1}{r^2}\exp(-ar)$	1	$\dfrac{1}{k}\left[(k^2+a^2)^{\frac{1}{2}}-a\right]$
20	$r^n\, H(a-r)$	>-1	$\dfrac{1}{k}a^{n+1} J_{n+1}(ak)$
21	$r^n\exp(-ar)$, $(\mathrm{Re}\, a>0)$	>-1	$\dfrac{1}{\sqrt{\pi}}\dfrac{2^{n+1}\Gamma\left(n+\dfrac{3}{2}\right)ak^n}{(a^2+k^2)^{n+\frac{3}{2}}}$
22	$r^n\exp(-ar^2)$	>-1	$\dfrac{k^n}{(2a)^{n+1}}\exp\left(-\dfrac{k^2}{4a}\right)$
23	r^{a-1}	>-1	$\dfrac{2^a\, \Gamma\left[\dfrac{1}{2}(a+n+1)\right]}{k^{a+1}\, \Gamma\left[\dfrac{1}{2}(1-a+n)\right]}$
24	$r^n(a^2-r^2)^{m-n-1} H(a-r)$	>-1	$2^{m-n-1}\Gamma(m-n)a^m k^{n-m} J_m(ak)$

	$f(r)$	order	$\tilde{f}_n(k) = \int_0^\infty r\, J_n(kr)\, f(r)\, dr$
25	$r^m \exp(-r^2/a^2)$	> -1	$\dfrac{k^n\, a^{m+n+2}}{2^{n+1}\, \Gamma(n+1)} \Gamma\!\left(1 + \dfrac{m}{2} + \dfrac{n}{2}\right)$ $\times {}_1F_1\!\left(1 + \dfrac{m}{2} + \dfrac{n}{2};\; n+1;\; -\dfrac{1}{4} a^2 k^2\right)$
26	$\dfrac{1}{r} J_{n+1}(ar)$	> -1	$k^n\, a^{-(n+1)}\, H(a-k),\quad a > 0$
27	$r^n (a^2 - r^2)^m H(a-r),\ m > -1$	> -1	$2^m a^n\, \Gamma(m+1) \left(\dfrac{a}{k}\right)^{m+1} J_{n+m+1}(ak)$
28	$\dfrac{1}{r^2} J_n(ar)$	$> \dfrac{1}{2}$	$\begin{cases} \dfrac{1}{2n}\left(\dfrac{k}{a}\right)^n, & 0 < k \le a \\ \dfrac{1}{2n}\left(\dfrac{a}{k}\right)^n, & a < k < \infty \end{cases}$
29	$\dfrac{r^n}{(a^2 + r^2)^{m+1}},\quad a > 0$	> -1	$\left(\dfrac{k}{2}\right)^m \dfrac{a^{n-m}}{\Gamma(m+1)} K_{n-m}(ak)$
30	$\exp(-p^2 r^2)\, J_n(ar)$,	> -1	$(2p^2)^{-1} \exp\!\left(-\dfrac{a^2 + k^2}{4p^2}\right) I_n\!\left(\dfrac{ak}{2p^2}\right)$
31	$\dfrac{1}{r} \exp(-ar)$	> -1	$\dfrac{\left\{(k^2 + a^2)^{\frac{1}{2}} - a\right\}^n}{k^n (k^2 + a^2)^{\frac{1}{2}}}$
32	$\dfrac{r^n}{(r^2 + a^2)^{n+1}}$	> -1	$\left(\dfrac{k}{2}\right)^n \dfrac{K_0(ak)}{\Gamma(n+1)}$
33	$\dfrac{r^n}{(a^2 - r^2)^{n+\frac{1}{2}}}\, H(a-r)$	< 1	$\dfrac{1}{\sqrt{\pi}} \left(\dfrac{k}{2}\right)^n \Gamma\!\left(\dfrac{1}{2} - n\right)\!\left(\dfrac{\sin ak}{k}\right)$

Table B-6. Mellin Transforms

	$f(x)$	$\tilde{f}(p) = \int_0^\infty x^{p-1} f(x)\,dx$
1	$\exp(-nx)$	$n^{-p}\,\Gamma(p)$, $\quad \operatorname{Re} p > 0$
2	$\exp(-ax^2)$, $\quad a > 0$	$\dfrac{1}{2} a^{-(p/2)} \Gamma\!\left(\dfrac{p}{2}\right)$, $\quad \operatorname{Re} p > 0$
3	$\cos(ax)$	$a^{-p}\,\Gamma(p)\cos\!\left(\dfrac{\pi p}{2}\right)$, $\quad 0 < \operatorname{Re} p < 1$
4	$\sin(ax)$	$a^{-p}\,\Gamma(p)\sin\!\left(\dfrac{\pi p}{2}\right)$, $\quad 0 < \operatorname{Re} p < 1$
5	$(a+x)^{-1}$, $\quad |\arg a| < \pi$	$\pi\, a^{p-1}\,\operatorname{cosec}(\pi p)$, $\quad 0 < \operatorname{Re} p < 1$
6	$(a-x)^{-1}$	$\pi\, a^{p-1}\,\cot(\pi p)$, $\quad 0 < \operatorname{Re} p < 1$
7	$(1+x)^{-a}$, $\quad \operatorname{Re} a > 0$	$\dfrac{\Gamma(p)\,\Gamma(a-p)}{\Gamma(a)}$
8	$(1+x^a)^{-s}$	$\dfrac{\Gamma(p/a)\,\Gamma(s-p/a)}{a\,\Gamma(s)}$
9	$(a^2 + x^2)^{-1}$	$\dfrac{\pi}{2} a^{(p-2)}\,\operatorname{cosec}\!\left(\dfrac{\pi p}{2}\right)$
10	$\begin{cases} 1, & 0 \le x \le a \\ 0, & x > a \end{cases}$	$p^{-1} a^p$
11	$\operatorname{Ci}(x) = -\displaystyle\int_x^\infty \dfrac{\cos t}{t}\,dt$	$-p^{-1}\Gamma(p)\cos\!\left(\dfrac{p\pi}{2}\right)$, $\quad 0 < \operatorname{Re} p < 1$
12	$\operatorname{Si}(x) = \displaystyle\int_0^x \dfrac{\sin t}{t}\,dt$	$-p^{-1}\Gamma(p)\sin\!\left(\dfrac{\pi p}{2}\right)$, $\quad -1 < \operatorname{Re} p < 0$

	$f(x)$	$\tilde{f}(p) = \int_0^\infty x^{p-1} f(x)\,dx$
13	$\begin{cases}(1-x)^{a-1}, & 0<x<1\\ 0, & x\geq 1\end{cases}$	$\dfrac{\Gamma(a)\,\Gamma(p)}{\Gamma(a+p)}$
14	$\begin{cases}0, & 0<x\leq 1\\ (x-1)^{-a}, & x>1\end{cases}$	$\dfrac{\Gamma(a-p)\,\Gamma(1-a)}{\Gamma(1-p)}$
15	$\exp(-ax)\,H(x-b)$	$a^{-p}\,\Gamma(p,ab)$
16	$\exp(-ax)\,H(b-x)$	$a^{-p}\,\gamma(p,ab)$
17	${}_2F_1(a,b,c;-x)$	$\dfrac{\Gamma(p)\,\Gamma(a-p)\,\Gamma(b-p)\,\Gamma(c)}{\Gamma(c-p)\,\Gamma(a)\,\Gamma(b)}$
18	$x^{\frac{1}{2}}\,J_\nu(x)$	$\dfrac{2^{p-\frac{1}{2}}\,\Gamma\!\left[\frac{1}{2}\!\left(p+\nu+\frac{1}{2}\right)\right]}{\Gamma\!\left[\frac{1}{2}\!\left(\nu-p+\frac{3}{2}\right)\right]}$
19	$x^{-\nu}\,J_\nu(x)$	$\dfrac{2^{p-\nu-1}\,\Gamma\!\left(\dfrac{p}{2}\right)}{\Gamma\!\left(\nu-\dfrac{1}{2}p+1\right)}$
20	$P_n(x)\,H(1-x)$	$\dfrac{\Gamma\!\left(\dfrac{p}{2}\right)\Gamma\!\left(\dfrac{p}{2}+\dfrac{1}{2}\right)}{2\,\Gamma\!\left(\dfrac{p}{2}-\dfrac{n}{2}+\dfrac{1}{2}\right)\Gamma\!\left(\dfrac{p}{2}+\dfrac{n}{2}+1\right)}$
21	$\begin{cases}\log\!\left(\dfrac{a}{x}\right), & x<a\\ 0, & x\geq a\end{cases}$	$\dfrac{a^p}{p^2}$
22	$x^{-1}\log(1+x)$	$\pi(1-p)^{-1}\,\operatorname{cosec}(\pi p)$
23	$(e^x-1)^{-1}$	$\Gamma(p)\,\zeta(p)$

	$f(x)$	$\tilde{f}(p) = \int_0^\infty x^{p-1} f(x)\, dx$
24	$\left(e^x + e^{-x}\right)^{-1}$	$\Gamma(p)\, L(p)$
25	$\log\left\|\dfrac{1+x}{1-x}\right\|$	$\left(\dfrac{\pi}{p}\right) \tan\left(\dfrac{p\pi}{2}\right)$
26	$(1+x)^{-m}\, P_{m-1}\!\left(\dfrac{1-x}{1+x}\right)$	$\dfrac{\Gamma(p)\, \{\Gamma(m-p)\}^2}{\Gamma(1-p)\, \{\Gamma(m)\}^2}$
27	$x^a (1+x)^{-b}$	$\dfrac{\Gamma(a+p)\, \Gamma(b-a-p)}{\Gamma(b)}$
28	$x^{-2\nu}\, J_\nu(x)\, K_\nu(x)$	$\dfrac{2^{p-2\nu-2}\, \Gamma\!\left(\dfrac{p}{4}\right) \Gamma\!\left(\dfrac{p}{2}-\nu\right)}{\Gamma\!\left(1+\nu-\dfrac{p}{4}\right)}$
29	$\log(1+ax),\ \|\arg a\| < \pi$	$\dfrac{\pi}{p}\, a^{-p}\, \operatorname{cosec}(\pi p),\ \ -1 < \operatorname{Re} p < 0$
30	$x^{\nu+1}\, J_\nu(ax)$	$2^{p+\nu}\, a^{-(p+\nu+1)}\, \dfrac{\Gamma\!\left(\dfrac{p}{2}+\nu+\dfrac{1}{2}\right)}{\Gamma\!\left(\dfrac{1-p}{2}\right)}$
31	$(1+x^2)^{-(1+\alpha)}\, H(x-1)$	$\dfrac{\Gamma\!\left(\dfrac{p}{2}\right) \Gamma\!\left(\alpha+1-\dfrac{p}{2}\right)}{2\Gamma(\alpha+1)}$
32	$\cos(x^\alpha)$	$\dfrac{1}{\alpha}\, \Gamma\!\left(\dfrac{p}{\alpha}\right) \cos\!\left(\dfrac{p\pi}{2\alpha}\right)$
33	$\sin(x^\alpha)$	$\dfrac{1}{\alpha}\, \Gamma\!\left(\dfrac{p}{\alpha}\right) \sin\!\left(\dfrac{p\pi}{2\alpha}\right)$
34	$(1+ax)^{-n}$	$\dfrac{\Gamma(p)\, \Gamma(n-p)}{a^p\, \Gamma(n)},\ \ 0 < \operatorname{Re} p < n$

Table B-7. Hilbert Transforms

	$f(t)$	$\hat{f}_H(x) = \dfrac{1}{\pi} \displaystyle\oint_{-\infty}^{\infty} \dfrac{f(t)}{(t-x)} dt$
1	1	0
2	$\begin{cases} 0, & -\infty < t < a \\ 1, & a < t < b \\ 0, & b < t < \infty \end{cases}$	$\dfrac{1}{\pi} \log \left\| \dfrac{b-x}{a-x} \right\|$
3	$(t+a)^{-1}$, $\ \text{Im } a > 0$	$i(x+a)^{-1}$
4	$(t+a)^{-1}$, $\ \text{Im } a < 0$	$-i(x+a)^{-1}$
5	$\begin{cases} 0, & -\infty < t < 0 \\ (at+b)^{-1}, & 0 < t < \infty, \\ & a, b > 0 \end{cases}$	$\dfrac{1}{\pi}(ax+b)^{-1} \log \left\| \dfrac{b}{ax} \right\|, \quad ax \neq -b$
6	$\dfrac{t}{(t^2+a^2)}$, $\ \text{Re } a > 0$	$\dfrac{a}{(x^2+a^2)}$
7	$\dfrac{1}{(t^2+a^2)}$, $\ \text{Re } a > 0$	$-\dfrac{x}{a(x^2+a^2)}$
8	$\dfrac{\alpha t + \beta a}{(t^2+a^2)}$, $\ \text{Re } a > 0$	$\dfrac{\alpha a - \beta x}{(x^2+a^2)}$
9	$\exp(iat)$, $\ a > 0$	$i \exp(iax)$
10	$\cos(at)$, $\ a > 0$	$-\sin(ax)$
11	$\sin(at)$, $\ a > 0$	$\cos(ax)$
12	$\dfrac{a}{a^2+(t+b)^2}$, $\ a > 0$	$-\dfrac{(b+x)}{a^2+(b+x)^2}$
13	$H(t-a) - H(t-b)$, $\ b > a > 0$	$\dfrac{1}{\pi} \log \left\| \dfrac{x-b}{x-a} \right\|$

	$f(t)$	$\hat{f}_H(x) = \dfrac{1}{\pi} \displaystyle\oint_{-\infty}^{\infty} \dfrac{f(t)}{(t-x)}\, dt$
14	$\dfrac{1}{t} H(t-a), \quad a>0$	$\dfrac{1}{\pi x} \log\left\|\dfrac{a}{x-a}\right\|, \quad x \neq 0,\ x \neq a$
15	$\begin{cases} 0, & -\infty < t < -a \\ (a^2 - t^2)^{-\tfrac{1}{2}}, & -a < t < a \\ 0, & a < t < \infty \end{cases}$	$\begin{cases} (x^2 - a^2)^{-\tfrac{1}{2}}, & -\infty < x < -a \\ 0, & -a < x < a \\ -(x^2 - a^2)^{-\tfrac{1}{2}}, & a < x < \infty \end{cases}$
16	$\begin{cases} -(t^2 - a^2)^{-\tfrac{1}{2}}, & -\infty < t < -a \\ 0, & -a < t < a \\ (t^2 - a^2)^{-\tfrac{1}{2}}, & a < t < \infty \end{cases}$	$\begin{cases} 0, & -\infty < x < -a \\ (a^2 - x^2)^{-\tfrac{1}{2}}, & -a < x < a \\ 0, & a < x < \infty \end{cases}$
17	$\dfrac{\sin at}{t}, \quad a>0$	$\dfrac{1}{x}(\cos ax - 1)$
18	$\begin{cases} 0, & -\infty < t < 0 \\ \sin(a\sqrt{t}), & 0 < t < \infty,\ a > 0 \end{cases}$	$\begin{cases} \exp(-a\sqrt{\|x\|}), & -\infty < x < 0 \\ \cos(a\sqrt{x}), & 0 < x < \infty \end{cases}$
19	$\operatorname{sgn} t\, \sin\!\left(a\sqrt{\|t\|}\right), \quad a>0$	$\cos\!\left(a\sqrt{\|x\|}\right) + \exp\!\left(-a\sqrt{\|x\|}\right)$
20	$\dfrac{1}{t}(1 - \cos at), \quad a>0$	$\dfrac{1}{x}(\sin ax)$
21	$J_n(t) \sin(t-x), \quad n=0,1,\ldots$	$J_n(x)$
22	$\operatorname{sgn} t\, \|t\|^\nu J_\nu(a\|t\|),$ where $a>0,\ -\dfrac{1}{2} < \operatorname{Re} \nu < \dfrac{3}{2}$	$-\|x\|^\nu Y_\nu(a\|x\|)$
23	$\sin(at)\, J_1(at), \quad a>0$	$\cos(ax)\, J_1(ax)$

	$f(t)$	$\hat{f}_H(x) = \dfrac{1}{\pi} \displaystyle\oint_{-\infty}^{\infty} \dfrac{f(t)}{(t-x)}\, dt$
24	$\sin(at)\, J_n(bt)$, where $0 < b < a$, $n = 0,1,2,\ldots$	$\cos(ax)\, J_n(bx)$
25	$\cos(at)\, J_1(at)$, $a > 0$	$-\sin(ax)\, J_1(ax)$
26	$\cos(at)\, J_n(bt)$, $0 < b < a$ where $n = 0,1,2,\ldots$	$-\sin(ax)\, J_n(bx)$
27	$\exp(-at)\, I_0(at)\, H(t)$, $a > 0$	$\dfrac{1}{\pi} \exp(-ax)\, K_0(a|x|)$
28	$\exp(-a|t|)\, I_0(at)$, $a > 0$	$-\dfrac{2}{\pi} \sinh(ax)\, K_0(a|x|)$
29	$\operatorname{sgn} t\, \exp(-a|t|)\, I_0(at)$, $a > 0$	$\dfrac{2}{\pi} \cosh(ax)\, K_0(a|x|)$
30	$\exp(at)\, K_0(a|t|)$, $a > 0$	$\pi \exp(ax)\, I_0(ax)\, H(-x)$
31	$|t|^\nu\, Y_\nu(a|t|)$, $a > 0$, $-\dfrac{1}{2} < \operatorname{Re} \nu < \dfrac{3}{2}$	$|x|^\nu\, J_\nu(a|x|)\, \operatorname{sgn} x$
32	$\sinh(at)\, K_0(a|t|)$, $a > 0$	$\dfrac{\pi}{2} \exp(-a|x|)\, I_0(ax)$
33	$\cosh(at)\, K_0(a|t|)$, $a > 0$	$-\dfrac{\pi}{2} \exp(-a|x|)\, I_0(ax)\, \operatorname{sgn} x$

Table B-8. Stieltjes Transforms

	$f(t)$	$\tilde{f}(x) = \int_0^\infty \dfrac{f(t)}{(t+x)}\, dt$
1	$(a+t)^{-1}$, $\ |\arg a| < \pi$	$(a-x)^{-1} \log\left(\dfrac{a}{x}\right)$
2	$\dfrac{1}{(a^2+t^2)}$, Re $a > 0$	$(a^2+x^2)^{-1}\left[\dfrac{\pi x}{2a} - \log\left(\dfrac{x}{a}\right)\right]$
3	$\dfrac{t}{(a^2+t^2)}$, Re $a > 0$	$(a^2+x^2)^{-1}\left[\dfrac{\pi a}{2} + x\log\left(\dfrac{x}{a}\right)\right]$
4	t^ν, $-1 < \operatorname{Re}\nu < 0$	$-\pi x^\nu \operatorname{cosec}(\pi\nu)$
5	$\begin{cases} -1, & 2n < x < 2n+1 \\ +1, & 2n+1 < x < 2n+2 \\ & n = 0,1,2,3,\ldots \end{cases}$	$\log\left[\dfrac{x}{2}\left\{\Gamma\left(\dfrac{x}{2}\right)\Big/\Gamma\left(\dfrac{x+1}{2}\right)\right\}^2\right]$
6	$\dfrac{t^\nu}{(a+t)}$, $|\arg a| < \pi$, where $-1 < \operatorname{Re}\nu < 1$	$(a-x)^{-1}\pi(a^\nu - x^\nu)\operatorname{cosec}(\pi\nu)$
7	$\left(\dfrac{t^\nu - a^\nu}{t-a}\right)$, $-1 < \operatorname{Re}\nu < 1$	$\left(\dfrac{\pi}{a+x}\right)\left[x^\nu \operatorname{cosec}(\nu\pi) - a^\nu \operatorname{ctn}(\nu\pi)\right.$ $\left. + \dfrac{a^\nu}{\pi}\log\left(\dfrac{a}{x}\right)\right]$
8	$t^{\nu-1}(a+t)^{1-\mu}$, $|\arg a|<\pi$, $0 < \operatorname{Re}\nu < \operatorname{Re}\mu$	$\dfrac{\Gamma(\nu)\,\Gamma(\mu-\nu)}{\Gamma(\mu)}\left(\dfrac{x^{\nu-1}}{a^{\mu-1}}\right)$ $\times {}_2F_1\left(\mu-1, \nu, \mu; 1-\dfrac{x}{a}\right)$
9	$t^{-\rho}(a+t)^{-\sigma}$, $|\arg a| < \pi$, $-\operatorname{Re}\sigma < \operatorname{Re}\rho < 1$	$\pi\operatorname{cosec}(\rho\pi)x^{-\rho}(a-x)^{-\sigma}$ $\times I_{\left(1-\frac{x}{a}\right)}(\sigma,\rho)$

	$f(t)$	$\tilde{f}(x) = \int_0^\infty \dfrac{f(t)}{(t+x)} dt$
10	$\exp(-at)$, Re $a > 0$	$-\exp(ax)\, Ei(-ax)$
11	$\begin{cases} \exp(-at), & 0 < t < b \\ 0, & b < t < \infty \end{cases}$	$\exp(ax)\left[Ei(-ab - ax) - Ei(-ax)\right]$
12	$\begin{cases} 0, & 0 < t < b \\ \exp(-at), & b < t < \infty \\ & \text{Re } a > 0 \end{cases}$	$-\exp(-ax)\, Ei(-ab - ax)$
13	$\dfrac{1}{\sqrt{t}} \exp(-at)$, Re $a > 0$	$\dfrac{\pi}{\sqrt{x}} \exp(ax)\, erfc\left(\sqrt{ax}\right)$
14	$\sqrt{t}\, \exp(-at)$, Re $a > 0$	$\sqrt{\dfrac{\pi}{a}} - \pi\sqrt{x}\, \exp(ax)\, erfc\left(\sqrt{ax}\right)$
15	$t^{-v} \exp(-at)$, Re $a > 0$, Re $v < 1$	$\Gamma(1-v)\, x^{-v} \exp(ax)\, \Gamma(v, ax)$
16	$t^{v-1} \exp\left(-\dfrac{a}{t}\right)$, Re $a > 0$, Re $v < 1$	$\Gamma(1-v)\, x^{v-1} \exp\left(\dfrac{a}{x}\right) \Gamma\left(v, \dfrac{a}{x}\right)$
17	$\exp\left(-a\sqrt{t}\right)$, Re $a > 0$	$2\left[\cos\left(a\sqrt{x}\right) Ci\left(a\sqrt{x}\right) - \sin\left(a\sqrt{x}\right) Si\left(a\sqrt{x}\right)\right]$
18	$\dfrac{1}{\sqrt{t}} \exp\left(-a\sqrt{t}\right)$, Re $a > 0$	$-\dfrac{2}{\sqrt{x}}\left[\sin\left(a\sqrt{x}\right) Ci\left(a\sqrt{x}\right) + \cos\left(a\sqrt{x}\right) Si\left(a\sqrt{x}\right)\right]$
19	$(a+t)^{-1} \log\left(\dfrac{t}{a}\right)$, $\|\arg a\| < \pi$	$\dfrac{1}{2}(x-a)^{-1}\left[\log\left(\dfrac{x}{a}\right)\right]^2$

	$f(t)$	$\tilde{f}(x) = \int_0^\infty \dfrac{f(t)}{(t+x)}\, dt$
20	$(t-a)^{-1} \log\left(\dfrac{t}{a}\right)$, $a>0$	$\dfrac{1}{2}(x+a)^{-1}\left[\pi^2 + \left\{\log\left(\dfrac{x}{a}\right)\right\}^2\right]$
21	$\dfrac{1}{\sqrt{t}} \log(at+b)$, Re $a>0$, Re $b>0$	$\dfrac{2\pi}{\sqrt{x}} \log\left(\sqrt{ax}+\sqrt{b}\right)$
22	$t^\nu \log t$, $-1 < \operatorname{Re} \nu < 0$	$-\pi x^\nu \operatorname{cosec}(\nu\pi)\left[\log x - \pi \operatorname{ctn}(\nu\pi)\right]$
23	$\sin at$, $a>0$	$-\left[\sin(ax)\, Ci(ax) + \cos(ax)\, Si(ax)\right]$
24	$\sin\left(a\sqrt{t}\right)$, $a>0$	$\pi \exp\left(-a\sqrt{x}\right)$
25	$t^{-1} \sin\left(a\sqrt{t}\right)$, $a>0$	$\left(\dfrac{\pi}{x}\right)\left[1-\exp\left(-a\sqrt{x}\right)\right]$
26	$t^{-\alpha} \sin\left(a\sqrt{t} + \alpha\pi\right)$, where $a>0$, $-\dfrac{1}{2} < \operatorname{Re} \alpha < 1$	$\left(\dfrac{\pi}{x^\alpha}\right) \exp\left(-a\sqrt{x}\right)$
27	$\sin\left(a\sqrt{t} - \dfrac{b}{\sqrt{t}}\right)$, $a,b>0$	$\pi \exp\left[-\left(a\sqrt{x} + \dfrac{b}{\sqrt{x}}\right)\right]$
28	$\dfrac{1}{\sqrt{t}} \sin^2\left(a\sqrt{t}\right)$, $a>0$	$\left(\dfrac{\pi}{2\sqrt{x}}\right)\left[1-\exp\left(-2a\sqrt{x}\right)\right]$
29	$\cos(at)$, $a>0$	$\cos(ax)\, Ci(ax) - \sin(ax)\, Si(ax)$
30	$\dfrac{1}{\sqrt{t}} \cos\left(a\sqrt{t}\right)$, $a>0$	$\left(\dfrac{\pi}{\sqrt{x}}\right) \exp\left(-a\sqrt{x}\right)$
31	$\dfrac{1}{\sqrt{t}} \cos\left(a\sqrt{t} - \dfrac{b}{\sqrt{t}}\right)$, $a,b>0$	$\left(\dfrac{\pi}{\sqrt{x}}\right) \exp\left[-\left(a\sqrt{x} + \dfrac{b}{\sqrt{x}}\right)\right]$

	$f(t)$	$\tilde{f}(x) = \int_0^\infty \dfrac{f(t)}{(t+x)}\,dt$
32	$\dfrac{1}{\sqrt{t}}\cos(a\sqrt{t})\cos(b\sqrt{t})$, $a \geq b > 0$	$\dfrac{\pi}{\sqrt{x}}\exp(-a\sqrt{x})\cosh(b\sqrt{x})$
33	$t^{\left(\frac{\nu}{2}+k\right)} J_\nu(a\sqrt{t})$	$2(-1)^k\, x^{\left(\frac{1}{2}\nu+k\right)} K_\nu(a\sqrt{x})$
34	$\sin(a\sqrt{t}) J_0(b\sqrt{t})$, $0 < b < a$	$\pi\exp(-a\sqrt{x}) I_0(b\sqrt{x})$
35	$\dfrac{1}{\sqrt{t}}\sin(a\sqrt{t}) J_0(b\sqrt{t})$, $0 < a < b$	$\dfrac{2}{\sqrt{x}}\sinh(a\sqrt{x}) K_0(b\sqrt{x})$
36	$\cos(a\sqrt{t}) J_0(b\sqrt{t})$, $0 < a < b$	$2\cosh(a\sqrt{x}) K_0(b\sqrt{x})$
37	$\dfrac{1}{\sqrt{t}}\cos(a\sqrt{t}) J_0(b\sqrt{t})$, $0 < b < a$	$\dfrac{\pi}{\sqrt{x}}\exp(-a\sqrt{x}) I_0(b\sqrt{x})$
38	$J_\nu^2(at)$, $a > 0$	$2 I_\nu(a\sqrt{x}) K_\nu(a\sqrt{x})$
39	$t^\nu \exp(-at) I_\nu(at)$, $a > 0$, $\lvert\operatorname{Re}\nu\rvert < \dfrac{1}{2}$	$x^\nu \exp(ax) K_\nu(ax) \sec(\pi\nu)$
40	$t^{\frac{1}{2}(\mu-\nu)} J_\mu(b\sqrt{t}) J_\nu(a\sqrt{t})$, $0 < a < b$, $-1 < \operatorname{Re}\mu < (\operatorname{Re}\nu + 2)$	$2 x^{\frac{1}{2}(\mu-\nu)} K_\mu(b\sqrt{x}) I_\nu(a\sqrt{x})$

Table B-9. Finite Fourier Cosine Transforms

	$f(x)$	$\tilde{f}_c(n) = \int_0^a f(x) \cos\left(\dfrac{n\pi x}{a}\right) dx$
1	1	$\begin{cases} a, & n=0 \\ 0, & n \neq 0 \end{cases}$
2	x	$\begin{cases} \dfrac{a^2}{2}, & n=0 \\ \left(\dfrac{a}{n\pi}\right)^2 \left[(-1)^n - 1\right], & n \neq 0 \end{cases}$
3	x^2	$\begin{cases} \dfrac{1}{3}a^3, & n=0 \\ 2a\left(\dfrac{a}{n\pi}\right)^2 (-1)^n, & n=1,2,\ldots \end{cases}$
4	x^3	$\begin{cases} \dfrac{1}{4}a^4, & n=0 \\ \dfrac{3a^4 (-1)^n}{(n\pi)^2} + 6\left(\dfrac{a}{n\pi}\right)^4 \left[(-1)^n - 1\right], \\ \qquad n=1,2,3,\ldots \end{cases}$
5	$\begin{cases} 1, & 0 < x < \dfrac{a}{2} \\ -1, & \dfrac{1}{2}a < x < a \end{cases}$	$\begin{cases} 0, & n=0 \\ \left(\dfrac{2a}{n\pi}\right) \sin\left(\dfrac{n\pi}{2}\right), & n=1,2,3,\ldots \end{cases}$
6	$\left(1 - \dfrac{x}{a}\right)^2$	$\begin{cases} \dfrac{1}{3}a, & n=0 \\ \dfrac{2a}{(n\pi)^2}, & n=1,2,\ldots \end{cases}$
7	$\sin(bx)$	$\dfrac{ba^2}{(n\pi)^2 - (ab)^2}\left[(-1)^n \cos(ab) - 1\right],$ $\qquad n\pi \neq ab$

	$f(x)$	$\tilde{f}_c(n) = \int_0^a f(x) \cos\left(\dfrac{n\pi x}{a}\right) dx$
8	$\cos(bx)$	$\dfrac{(-1)^n ba^2 \sin(ab)}{(ab)^2 - (n\pi)^2}, \quad n\pi \ne ab$
9	$\sin\left(\dfrac{m\pi x}{a}\right)$, m an integer	$\begin{cases} 0, & n = m \\ \dfrac{m\pi}{\pi(n^2 - m^2)}\left[(-1)^{n+m} - 1\right], & n \ne m \end{cases}$
10	$\exp(bx)$	$(a^2 b)\left[\dfrac{(-1)^n \exp(ab) - 1}{(n\pi)^2 + (ba)^2}\right]$
11	$x^{-\frac{1}{2}}(a^2 - x^2)^{-\frac{1}{2}}$	$\left(\dfrac{\pi}{2}\right)^{3/2} \left(\dfrac{n\pi}{a}\right)^{\frac{1}{2}} \left\{J_{-1/4}\left(\dfrac{n\pi}{2}\right)\right\}^2$
12	$(a^2 - x^2)^{\nu - \frac{1}{2}}$	$\sqrt{\pi}\, 2^{\nu - 1}\, \Gamma\left(\nu + \dfrac{1}{2}\right) \left(\dfrac{a^2}{n\pi}\right)^{\nu} J_\nu(n\pi)$
13	$\sin\left\{b(a^2 - x^2)^{\frac{1}{2}}\right\}$	$\left(\dfrac{\pi a b}{2}\right)\left(b^2 + \dfrac{n^2 \pi^2}{a^2}\right)^{-\frac{1}{2}}$ $\times J_1\left[\left\{(ab)^2 + (n\pi)^2\right\}^{\frac{1}{2}}\right]$
14	$(a^2 - x^2)^{-\frac{1}{2}} \cos\left\{b(a^2 - x^2)^{\frac{1}{2}}\right\}$	$\left(\dfrac{\pi}{2}\right) J_0\left[\left\{(ab)^2 + (n\pi)^2\right\}^{\frac{1}{2}}\right]$
15	$J_0\left\{b(a^2 - x^2)^{\frac{1}{2}}\right\}$	$\left(b^2 + \dfrac{n^2 \pi^2}{a^2}\right)^{-\frac{1}{2}}$ $\times \sin\left[\left\{(ab)^2 + (n\pi)^2\right\}^{\frac{1}{2}}\right]$

Table B-10. Finite Fourier Sine Transforms

	$f(x)$	$\tilde{f}_s(n) = \int_0^a \sin\left(\dfrac{n\pi x}{a}\right) f(x)\, dx$
1	1	$\left(\dfrac{a}{n\pi}\right)\left[1-(-1)^n\right]$
2	x	$(-1)^{n+1}\left(\dfrac{a^2}{n\pi}\right)$
3	x^2	$\dfrac{a^3(-1)^{n-1}}{n\pi} - \dfrac{2a^3\left[1+(-1)^{n+1}\right]}{(n\pi)^3}$
4	x^3	$(-1)^n \dfrac{a^4}{\pi^5}\left(\dfrac{6}{n^3} - \dfrac{\pi^2}{n}\right)$
5	$\left(\dfrac{a-x}{a}\right)$	$\left(\dfrac{a}{n\pi}\right)$
6	$x(a-x)$	$2\left(\dfrac{a}{n\pi}\right)^3\left[1+(-1)^{n+1}\right]$
7	$x(a^2-x^2)$	$(-1)^{n+1}\, 6a\left(\dfrac{a}{n\pi}\right)^3$
8	$\exp(bx)$	$\dfrac{n\pi a}{(n\pi)^2 + (ab)^2}\left[1+(-1)^{n+1}\exp(ab)\right]$
9	$\cos(bx)$	$\dfrac{n\pi a}{(n\pi)^2 - (ab)^2}\left[1+(-1)^{n+1}\cos(ab)\right],$ $n\pi \neq ab$
10	$\sin(bx)$	$\dfrac{(-1)^n\, an\pi \sin(ab)}{(n\pi)^2 - (ab)^2},\quad n\pi \neq ab$
11	$\cosh(bx)$	$\dfrac{n\pi a}{\left[(n\pi)^2 + (ab)^2\right]}\left[1+(-1)^{n+1}\cosh(ab)\right]$

	$f(x)$	$\tilde{f}_s(n) = \int_0^a \sin\left(\dfrac{n\pi x}{a}\right) f(x)\, dx$
12	$\sin\left(\dfrac{m\pi x}{a}\right)$, m an integer	$\begin{cases} 0, & n \neq m \\ \dfrac{1}{2}a, & n = m \end{cases}$
13	$\cos\left(\dfrac{m\pi x}{a}\right)$, m an integer	$\begin{cases} \dfrac{na}{\pi(n^2 - m^2)}\left[1 + (-1)^{n+m+1}\right], & n \neq m \\ 0, & n = m \end{cases}$
14	x^{-1}	$\mathrm{Si}(n\pi)$
15	$x^{-\frac{1}{2}}(x^2 - a^2)^{-\frac{1}{2}}$	$\left(\dfrac{\pi}{2}\right)^{3/2} \left(\dfrac{n\pi}{a}\right)^{\frac{1}{2}} \left\{J_{1/4}\left(\dfrac{n\pi}{2}\right)\right\}^2$
16	$x(a^2 - x^2)^{\alpha - \frac{1}{2}}$	$\sqrt{\pi}\, 2^{\alpha-1} a^{\alpha+1} \Gamma\left(\alpha + \dfrac{1}{2}\right)$ $\times \left(\dfrac{n\pi}{a}\right)^{-\alpha} J_{\alpha+1}(n\pi)$
17	$(a^2 - x^2)^{-\frac{1}{2}} T_{2n+1}\left(\dfrac{x}{a}\right)$	$\left(\dfrac{\pi}{2}\right)(-1)^n J_{2n+1}(n\pi)$
18	$(ax - x^2)^{\alpha - \frac{1}{2}}$	$\sqrt{\pi}\, \Gamma\left(\alpha + \dfrac{1}{2}\right) \left(\dfrac{a^2}{n\pi}\right)^{\alpha} J_{\alpha}\left(\dfrac{n\pi}{2}\right)$

Table B-11. Finite Laplace Transforms

	$f(t)$	$\mathscr{L}_T\{f(t)\} = \bar{f}(s,T) = \int_0^T e^{-st} f(t)\, dt$
1	1	$\dfrac{1}{s}\left(1 - e^{-sT}\right)$
2	t	$\dfrac{1}{s^2} - \dfrac{1}{s} e^{-sT}\left(\dfrac{1}{s} + T\right)$
3	t^n	$\dfrac{n!}{s^{n+1}} - \dfrac{e^{-sT}}{s^{n+1}}\left[(sT)^n + n(sT)^{n-1} \right.$ $\left. + n(n-1)(sT)^{n-2} + \cdots + n!\right]$
4	$t^a, \ (a > -1)$	$\dfrac{1}{s^{a+1}} \gamma(a+1,\ sT)$
5	$\exp(-at), \ a > 0$	$(s+a)^{-1}\left[1 - \exp\{-T(s+a)\}\right]$
6	$t^n \exp(-at), \ a > 0$	$\dfrac{n!}{(s+a)^{n+1}} - \dfrac{e^{-(a+s)T}}{(s+a)^{n+1}}\left[\{(s+a)T\}^n\right.$ $+ n\{T(s+a)\}^{n-1}$ $\left. + n(n-1)\{T(s+a)\}^{n-2} + \cdots + n!\right]$
7	$H(t-a), \ a > 0$	$\dfrac{1}{s}\left[e^{-sa} - e^{-sT}\right] H(T-a)$
8	$\cos(at)$	$\dfrac{s}{(s^2+a^2)} + \dfrac{e^{-sT}}{(s^2+a^2)}$ $\times (a \sin aT - s \cos aT)$
9	$\sin(at)$	$\dfrac{a}{(s^2+a^2)} - \dfrac{e^{-sT}}{(s^2+a^2)}$ $\times (s \sin aT + a \cos aT)$

	$f(t)$	$\mathcal{L}_T\{f(t)\} = \bar{f}(s,T) = \int_0^T e^{-st} f(t)\, dt$
10	$e^{-at} \sin(bt)$	$\dfrac{b}{(s+a)^2 + b^2} - \dfrac{\exp(-sT)}{(s+a)^2 + b^2}$ $\times (s \sin bT + a \sin bT + b \cos bT)$
11	$e^{-at} \cos(bt)$	$\dfrac{s+a}{(s+a)^2 + b^2} + \dfrac{\exp(-sT)}{(s+a)^2 + b^2}$ $\times (b \sin bT - s \cos bT - a \cos bT)$
12	$\sinh(at)$	$\dfrac{a}{(s^2 - a^2)} - \dfrac{\exp(-sT)}{(s^2 - a^2)}$ $\times (a \cosh aT + s \sinh aT)$
13	$\cosh(at)$	$\dfrac{s}{(s^2 - a^2)} - \dfrac{\exp(-sT)}{(s^2 - a^2)}$ $\times (s \cosh aT + a \sinh aT)$
14	$t^{\frac{1}{2}}$	$-\dfrac{\sqrt{T}\,\exp(-sT)}{s} + \dfrac{\sqrt{\pi}}{2}\dfrac{\mathrm{erf}(\sqrt{sT})}{s^{3/2}}$
15	$t^{-\frac{1}{2}}$	$\dfrac{\pi}{s}\,\mathrm{erf}(\sqrt{sT})$
16	$\mathrm{erfc}\left(\dfrac{t}{2a}\right)$	$\dfrac{1}{s}\{1 - \exp(a^2 s^2)\,\mathrm{erfc}(as)\}$ $-\dfrac{e^{-sT}}{s}\,\mathrm{erfc}\left(\dfrac{T}{2a}\right)$ $+\dfrac{\exp(a^2 s^2)}{s}\,\mathrm{erfc}\left(\dfrac{T}{2a} + as\right)$

	$f(t)$	$\mathcal{L}_T\{f(t)\} = \bar{f}(s,T) = \int_0^T e^{-st} f(t)\, dt$
17	$\mathrm{erfc}(bt)$	$-\dfrac{e^{\frac{s^2}{4b^2}}}{s}\,\mathrm{erfc}\left(\dfrac{s}{2b}\right)$ $+\dfrac{e^{\frac{s^2}{4b^2}}}{s}\,\mathrm{erfc}\left(bT+\dfrac{s}{2b}\right)-\dfrac{e^{-sT}}{s}\,\mathrm{erf}(bT)$
18	$\mathrm{erf}(t)$	$-\dfrac{e^{\frac{s^2}{4}}}{s}\,\mathrm{erf}\left(\dfrac{s}{2}\right)+\dfrac{e^{\frac{s^2}{4}}}{s}\,\mathrm{erf}\left(T+\dfrac{s}{2}\right)$ $-\dfrac{e^{-sT}}{s}\,\mathrm{erf}(T)$
19	$\mathrm{erf}\left(\sqrt{t}\right)$	$\dfrac{\mathrm{erf}\left(\sqrt{T}\right)}{s(s+1)}-\dfrac{\exp(-sT)\,\mathrm{erf}\left(\sqrt{T}\right)}{s}$
20	$e^{bt}\,\mathrm{erf}\left(\sqrt{bT}\right)$	$\dfrac{\sqrt{b}\,\mathrm{erf}\left(\sqrt{sT}\right)}{\sqrt{s}\,(s-b)}+\dfrac{e^{-(s-b)T}\,\mathrm{erf}\left(\sqrt{bT}\right)}{(s-b)}$
21	$e^{bt}\,\mathrm{erfc}\left(\sqrt{bT}\right)$	$\dfrac{1}{(s-b)}\left\{1-\dfrac{\sqrt{b}}{\sqrt{s}}\,\mathrm{erf}\left(\sqrt{sT}\right)\right\}$ $-\dfrac{e^{-(s-b)T}\,\mathrm{erf}\left(\sqrt{bT}\right)}{(s-b)}$
22	$\exp(\alpha t^2),\quad \alpha>0$	$\sqrt{\dfrac{\pi}{\alpha}}\cdot\dfrac{1}{2i}\exp\left(-\dfrac{s^2}{4\alpha}\right)$ $\times\left\{\mathrm{erf}\left(T\sqrt{\alpha}-\dfrac{s}{2\sqrt{\alpha}}\right)i\right.$ $\left.+\mathrm{erf}\left(-\dfrac{si}{2\sqrt{\alpha}}\right)\right\}$

Table B-12. Z Transforms

	$f(n)$	$F(z) = \sum_{n=0}^{\infty} f(n) z^{-n}$
1	$\begin{cases} 1, & n=0 \\ 0, & n \neq 0 \end{cases}$	1
2	1	$\dfrac{z}{z-1}$
3	a^n	$\dfrac{z}{z-a}$
4	n	$\dfrac{z}{(z-1)^2}$
5	n^2	$\dfrac{z(z+1)}{(z-1)^3}$
6	$\dfrac{1}{n!}$	$\exp\left(\dfrac{1}{z}\right)$
7	$\cos nx$	$\dfrac{z(z-\cos x)}{z^2 - 2z \cos x + 1}$
8	$\sin nx$	$\dfrac{z \sin x}{z^2 - 2z \cos x + 1}$
9	$\exp(\pm nx)$	$\dfrac{z}{z - \exp(\pm x)}$
10	$n^k e^{nx}$	$\dfrac{\partial}{\partial x^k}\left(\dfrac{z}{z-e^x}\right)$
11	$n e^{-nx}$	$\dfrac{z \exp(-x)}{(z - e^{-x})^2}$

	$f(n)$	$F(z) = \sum_{n=0}^{\infty} f(n) z^{-n}$
12	$n^2 e^{-nx}$	$\dfrac{z(z+e^{-x})e^{-x}}{(z-e^{-x})^3}$
13	$\exp(-nx) \sin(an)$	$\dfrac{z \exp(-x) \sin a}{z^2 - 2ze^{-x} \cos a + e^{-2x}}$
14	$\exp(-nx) \cos(an)$	$\dfrac{z(z - e^{-x} \cos a)}{z^2 - 2z e^{-x} \cos a + e^{-2x}}$
15	$\sinh(nx)$	$\dfrac{z \sinh x}{z^2 - 2z \cosh x + 1}$
16	$\cosh(nx)$	$\dfrac{z(z - \cosh x)}{z^2 - 2z \cosh x + 1}$
17	$H(n-1)$	$\dfrac{1}{z-1}$
18	$H(n) - H(n-1)$	1
19	$H(n-m), \quad m = 1, 2, 3$	$\dfrac{1}{z^{m-1}(z-1)}$
20	$H(n-1) - H(t-2)$	$\dfrac{1}{z}$
21	$H(t-m) - H[t-(m+1)]$	$\dfrac{1}{z^m}$
22	$\dfrac{m(m-1) \cdots (m-n+1)}{n!}$	$\left(1 + \dfrac{1}{z}\right)^m$

Table B-13. Finite Hankel Transforms

	$f(r)$	order n	$\tilde{f}_n(k_i) = \int_0^a r\, J_n(rk_i)\, f(r)\, dr$
1	c, where c is a constant	0	$\left(\dfrac{ac}{k_i}\right) J_1(ak_i)$
2	$(a^2 - r^2)$	0	$\dfrac{4a}{k_i^3} J_1(ak_i)$
3	$(a^2 - r^2)^{-\frac{1}{2}}$	0	$k_i^{-1} \sin(ak_i)$
4	$\dfrac{J_0(\alpha r)}{J_0(\alpha a)}$	0	$-\dfrac{ak_i}{(\alpha^2 - k_i^2)} J_1(ak_i)$
5	$\dfrac{1}{r}$	1	$k_i^{-1}\{1 - J_0(ak_i)\}$
6	$r^{-1}(a^2 - r^2)^{-\frac{1}{2}}$	1	$\dfrac{(1 - \cos ak_i)}{(ak_i)}$
7	r^n	> -1	$\dfrac{a^{n+1}}{k_i} J_{n+1}(ak_i)$
8	$\dfrac{J_\nu(\alpha r)}{J_\nu(\alpha a)}$	> -1	$\dfrac{ak_i}{(\alpha^2 - k_i^2)} J_\nu'(ak_i)$
9	$r^{-n}(a^2 - r^2)^{-\frac{1}{2}}$	> -1	$\dfrac{\pi}{2}\left\{ J_{\frac{n}{2}}\!\left(\dfrac{ak_i}{2}\right)\right\}^2$
10	$r^n(a^2 - r^2)^{-(n+\frac{1}{2})}$	$< \dfrac{1}{2}$	$\dfrac{\Gamma\!\left(\dfrac{1}{2} - n\right)}{\sqrt{\pi}\, 2^n} k_i^{n-1} \sin(ak_i)$
11	$r^{n-1}(a^2 - r^2)^{n-\frac{1}{2}}$	$> -\dfrac{1}{2}$	$\dfrac{\sqrt{\pi}}{2} \Gamma\!\left(n + \dfrac{1}{2}\right)\!\left(\dfrac{2}{k_i}\right)^n \times a^{2n} J_n^2\!\left(\dfrac{ak_i}{2}\right)$

Answers and Hints to Selected Exercises

2.14 Exercises

1. (a) $\sqrt{\dfrac{\pi}{2}}\exp(-|k|)$,

    (c) $\dfrac{1}{\sqrt{2\pi}}(ik)^n$, (e) Hint: Put $e^x = y$, $F(k) = \dfrac{\Gamma(1-ik)}{\sqrt{2\pi}}$.

    (f) Hint: $f(x) = -\dfrac{1}{a}\dfrac{d}{dx}\exp\left(-\dfrac{1}{2}ax^2\right)$, (j) $(-i)^n f_n(k)$.

5. (b) Hint: Use $P_n(x) = \dfrac{1}{2^n n!}\dfrac{d^n}{dx^n}(x^2-1)^n$.

8. (a) $y(x) = e^{-x}\int_{-a}^{x} e^\alpha f(\alpha)d\alpha + e^x \int_x^a e^{-\alpha} f(\alpha)d\alpha$.

    For (b)-(c) use Exercises 3(a) and 3(b).

    (b) $y(x) = A\exp\left(-\dfrac{1}{4}x^2\right)$, where A is a constant.

    (e) $y(t) = \dfrac{1}{\sqrt{2\pi}}\int_{-\infty}^{\infty}\dfrac{F(k)\exp(ikt)\,dk}{(\omega^2 - k^2 + 2i\alpha k)}$.

9. (b) $f(x) = \dfrac{a}{\sqrt{\pi(a-b)}}\exp\left(-\dfrac{abx^2}{a-b}\right)$.

    (c) $f(x) = \sqrt{\dfrac{2}{\pi}}\left\{\dfrac{b}{a}(a-b)\right\}\dfrac{1}{(a-b)^2 + x^2}$.

    (d) $f(x) = \sqrt{\dfrac{2}{\pi}}\cdot\dfrac{ac}{b(x^2+c^2)}$, $c = b - a$.

    (e) $F(k) = \dfrac{1}{\pi}\cdot\dfrac{i\pi\Phi(k)}{\operatorname{sgn}k}$, $f(x) = -\dfrac{1}{\pi}\int_{-\infty}^{\infty}(x-t)^{-1}\phi(t)dt$.

10. $u(x,t) = \dfrac{1}{\sqrt{2\pi}}\int_{-\infty}^{\infty}\{A(k)\exp[i(kx+\omega t)] + B(k)\exp[i(kx-\omega t)]\}\,dk$,

    where $A(k) = \dfrac{1}{2}\left[F(k) + \dfrac{1}{i\omega}G(k)\right]$ and $B(k) = \dfrac{1}{2}\left[F(k) - \dfrac{1}{i\omega}G(k)\right]$.

11. Hint: $u(x,t) = 2\int_0^\infty A(k)\exp(-k^2 bt)\cos\{(x+at)k\}dk$

$\approx \sqrt{\dfrac{\pi}{bt}}\, A(0)\exp\left[-\dfrac{(x+at)^2}{4bt}\right]$ as $t \to \infty$,

where $A(k)$ is expanded in Taylor series and only the first term is retained at $k = 0$.

12. Hint: $\mathscr{F}^{-1}\{\cos(k^2 t)\} = \dfrac{1}{\sqrt{2t}}\cos\left(\dfrac{x^2}{4t} - \dfrac{\pi}{4}\right)$.

14. (a) Hint: Differentiate both sides of the integral

$\displaystyle\int_0^\infty e^{-ax}\sin kx\, dx = \dfrac{a}{k^2+a^2}$ with respect to a to obtain

$F_s(k) = \sqrt{\dfrac{a}{\pi}}\,\dfrac{2ak}{(k^2+a^2)^2}$.

(c) $\mathscr{F}_c\left\{\dfrac{1}{x}\right\}$ does not exist.

(d) $\sqrt{\dfrac{\pi}{2}}(a^2+k^2)^{-\frac{1}{2}}$. Hint: Use $K_0(ax) = \displaystyle\int_0^\infty \exp(-ax\cosh u)\,du$

and interchange the order of integration.

15. (a) Differentiate both sides of the integral $\displaystyle\int_0^\infty e^{-ax}\sin kx\,dx = \dfrac{k}{k^2+a^2}$

with respect to a to obtain the result.

(b) Integrate the above integral with respect to a from a to ∞ to obtain the answer.

(c) $\sqrt{\dfrac{\pi}{2}}(-i\,\text{sgn}\,k)$.

16. Use the definition of the Fourier cosine transform and integrate by parts.

17. Hint: Use the Parseval formula for the gate function.

19. Hint: $U_s(k,t) = F_s(k)\,G_c(k,t)$, where $G_c(k,t) = \exp(-\kappa k^2 t)$.

20. $u(x,y) = -\dfrac{2}{\pi}\displaystyle\int_0^\infty \dfrac{1}{k}\sin ak\cos kx\, e^{-ky}\,dk$.

21. (a) $f(x) = \dfrac{1}{\sqrt{x}}$,   (b) $f(x) = \exp(-ax)$,

(c) $f(x) = \dfrac{H(x-a)}{\sqrt{x^2-a^2}}$,   (d) $H(a-x)$.

22. Hint: Use the Fourier sine transform.
$$u(x,t) = \frac{1}{\sqrt{4\pi \kappa t}} \int_0^\infty f(\xi)\left[\exp\left\{-\frac{(x-\xi)^2}{4\kappa t}\right\} - \exp\left\{-\frac{(x+\xi)^2}{4\kappa t}\right\}\right]d\xi.$$

23. (a) $\dfrac{\pi a^3}{2}$, (b) $\dfrac{\pi}{b^2}(1-e^{-ab})$, (c) πa.

24. Hint: Use the Convolution Theorem for the Fourier cosine transform.

25. Hint: Use the Convolution Theorem for the Fourier cosine transform.

26. (b) $\dfrac{1}{2}(\pi - a)$, (d) $\pi\left[\dfrac{\exp(2\pi a)+1}{\exp(2\pi a)-1}\right]$.

28. Hint: $\mathscr{F}^{-1}\left\{{\cos \atop \sin}(atk^2)\right\} = \dfrac{1}{2\sqrt{at}}\left[\cos\left(\dfrac{x^2}{4at}\right) \pm \sin\left(\dfrac{x^2}{4at}\right)\right]$.

29. $u(x,z) = \dfrac{P}{2\pi\mu}\int_{-\infty}^\infty \dfrac{1}{\alpha}\exp(ikx - \alpha z)dx$, $\alpha = \sqrt{k^2 - \dfrac{\omega^2}{c_2^2}}$.

Hint: Write $(x,y) = r(\cos\theta, \sin\theta)$ along with $k = \dfrac{\omega}{c_2}\cos\phi$ and $\alpha = \dfrac{i\omega}{c_2}\sin\phi$ to obtain

$$u(x,z) = \dfrac{P}{2\pi i \mu}\int_{0-i\infty}^{\pi+i\infty}\exp\left[-\dfrac{i\omega r}{c_2}\sin(\theta+\phi)\right]d\phi.$$

30. $\phi(x,z,t) = -\dfrac{Pg}{2\pi}\int_{-\infty}^\infty \dfrac{\sin\omega t}{\omega}\exp(ikx + |k|z)dk$,

$\eta(x,t) = \dfrac{P}{2\pi}\int_{-\infty}^\infty \cos\omega t \exp(ikx)\,dk$, where $\omega^2 = g|k|$.

$\eta(x,t) \approx \dfrac{Pt}{2\sqrt{2\pi}}\dfrac{\sqrt{g}}{x^{3/2}}\cos\left(\dfrac{gt^2}{4x}\right)$ for $gt^2 \gg 4x$.

32. $\phi(x,z,t) = \dfrac{iP\exp(\varepsilon t)}{2\pi\rho}\int_{-\infty}^\infty \dfrac{(Uk-i\varepsilon)\exp(|k|z + ikx)}{(Uk-i\varepsilon)^2 - g|k|}dk$,

$\eta(x,t) = \dfrac{P\exp(\varepsilon t)}{2\pi\rho}\int_{-\infty}^\infty \dfrac{|k|\exp(ikx)\,dk}{(Uk-i\varepsilon)^2 - g|k|}$.

36. $u(x,t) = \dfrac{1}{2}[f(x-ct) + f(x+ct)] + \dfrac{1}{2c}\int_{x-ct}^{x+ct}g(\xi)d\xi$ for $x > ct$.

Similar result for $x < ct$.

37. Hint: $\mathcal{F}_s\{u_{xxxx}\} = \sqrt{\dfrac{2}{\pi}}\left[k^4 U_s(k, y) - k^3 u(0, y)\right]$,

$\mathcal{F}_s\{u_{xxyy}\} = \sqrt{\dfrac{2}{\pi}}\dfrac{\partial^2}{\partial y^2}\left[-k^2 U_s(k, y) + k\, u(0, y)\right]$.

43. (a) $\phi(t) = \left(1 - \dfrac{it}{a}\right)^{-p}$.

   (c) $\phi(t) = \exp(i\mu t - \lambda|t|)$.

   (d) $\phi(t) = (1 + \lambda^2 t^2)^{-1} \exp(i\mu t)$.

44. $f(x) = \dfrac{1 - \cos x}{\pi x^2}$.

45. $\phi(t) = \dfrac{1}{it}[\exp(ita) - 1]$.

47. $U(k, y) = F(k)\cos(k^2 y)$, $u(x, y) = \dfrac{1}{\sqrt{2\pi}}\displaystyle\int_{-\infty}^{\infty} F(k)\cos(k^2 y)\exp(ikx)\, dk$.

48. $u(x, y, t) = \dfrac{1}{2\pi}\displaystyle\int\!\!\int_{-\infty}^{\infty} F(k, l)\cos\left[c(k^2 + l^2)^{\frac{1}{2}} t\right]\exp[i(kx + ly)]\, dk\, dl$.

52. $u(x, t) = \dfrac{1}{2\pi}\displaystyle\int_{-\infty}^{\infty}\left[F(k)\cos(x\alpha) + \dfrac{G(k)}{\alpha}\sin x\alpha\right]\exp(ikx)\, dk$,

   where $\alpha = \dfrac{b + ika - k^2}{c^2}$.

53. $u(x, y) = \dfrac{2T_0}{\pi}\displaystyle\int_0^{\infty} \dfrac{\sin ak\,\cos xk\,\cosh y\alpha}{k\,\cosh\alpha}\, dk$, $\alpha = \sqrt{h + k^2}$.

54. (a) $u(x, y) = \dfrac{1}{4\pi^2}\displaystyle\int\!\!\int_{-\infty}^{\infty}\dfrac{F(k,l)\exp\{i(kx + ly)\}}{(k^4 + l^2 + 2)}\, dk\, dl$.

   (b) $u(x, y) = -\dfrac{1}{4\pi^2}\displaystyle\int\!\!\int_{-\infty}^{\infty}\dfrac{F(k,l)\exp\{i(kx + ly)\}\, dk\, dl}{(k^2 + 2l^2 - 3ik + 4)}$.

55. Hint: Seek a solution of the form $\psi = \phi_n(x, t)\sin n\pi y$ with $\psi_0(x, y) = \psi_{0n}(x)\sin n\pi y$ so that ϕ_n satisfies the equation

$\dfrac{\partial}{\partial t}\left[\dfrac{\partial^2}{\partial x^2}\phi_n - \alpha^2\phi_n\right] + \beta\dfrac{\partial\phi_n}{\partial x} = 0$, $\alpha^2 = (n\pi)^2 + \kappa^2$.

Apply the Fourier transform of $\phi_n(x, t)$ with respect to x and use $\Psi_n(k, 0) = \mathcal{F}\{\psi_{0n}(x)\}$.

$\phi_n(x, t) = \dfrac{1}{\sqrt{2\pi}}\displaystyle\int_{-\infty}^{\infty}\Psi_n(k, 0)\exp[i\{kx - \omega(k)t\}]\, dk$,

Integral Transforms and Their Applications 427

where $\omega(k) = -\beta k (k^2 + \alpha^2)^{-1}$.

Examine the case for $\psi_{0n}(x) = \dfrac{1}{a\sqrt{2}} \exp\left\{ik_0 x - \left(\dfrac{x}{a}\right)^2\right\}$.

3.10 Exercises

1. (a) $\dfrac{2}{s^2} + \dfrac{a^2}{s^2 + a^2}$, (b) $s(s+2)^{-2}$, (c) $\dfrac{s^2 - a^2}{(s^2 + a^2)^2}$,

   (e) $\dfrac{\exp(-3s)}{(s-1)}$, (g) $\dfrac{2}{s^3}\exp(-3s)$,

   (h) $(1+sa)s^{-2}\exp(-as)$, (i) $s\sqrt{\pi}(s-a)^{-3/2}$.

2. Hint: $\displaystyle\int_0^\infty t^{-n}\exp(-st)dt \geq e^{-s}\int_0^1 t^{-n}dt + \int_1^\infty t^{-n}\exp(-st)dt$,

   since $\exp(-st) \geq \exp(-s)$ for $0 \leq t \leq 1$. But $\displaystyle\int_0^1 t^{-n}dt$ does not exist.

5. Hint: Use (3.6.7).
6. Hint: Use definition 3.2.5 and result (3.6.7).
7. (a) $\dfrac{1}{(a^2 - b^2)}(\cos bt - \cos at)$, (b) $\left(\dfrac{t}{c^2} - \dfrac{\sin ct}{c^3}\right)$,

   (c) $(t-a)H(t-a)$, (d) $\exp(2t) - (t+1)\exp(t)$,

   (e) $\dfrac{1}{2}\exp(-t)\sin 2t$,

   (f) Hint: $\dfrac{1}{s^2(s+1)(s+2)} = \dfrac{1}{4s} + \dfrac{1}{2s^2} + \dfrac{1}{s+1} - \dfrac{1}{4(s+2)}$.

8. (a) $\dfrac{1}{2a}(\sin at + at\cos at)$, (b) $\dfrac{1}{2}\mathrm{erf}(2\sqrt{t})$, (c) $\displaystyle\int_0^t f(\tau)d\tau$,

   (d) $\dfrac{t}{2a}\sin at$, (e) $\displaystyle\int_0^t f(t-\tau)\sin\omega\tau\, d\tau$,

   (f) $\dfrac{1}{(a^2+b^2)}(b\sin bt - a\cos bt + a e^{at})$, (h) $\mathrm{erfc}\left(\dfrac{a}{2\sqrt{t}}\right)$.

9. (b) Hint: $\dfrac{1}{(\sqrt{s}-\sqrt{a})} = \dfrac{1}{\sqrt{s}}\left(\dfrac{a}{s-a}+1\right) + \dfrac{\sqrt{a}}{s-a}$.

(c) Hint: $\bar{f}(s)$ has simple poles at $s=0$ and at $s = \pm(2n+1)\dfrac{a\pi i}{b} = \pm s_n$. The residue at $s=0$ is $\dfrac{x}{a}$, and the residue at

$s = s_n$ is $\dfrac{\sinh\left\{(2n+1)\dfrac{\pi i x}{a}\right\} \exp\left\{(2n+1)\dfrac{\pi i a x}{b}\right\}}{\left(\dfrac{b}{2a}\right)\left\{(2n+1)\dfrac{\pi a i}{b}\right\}^2 \sinh\left\{(2n+1)\dfrac{\pi i}{2}\right\}}$.

Grouping the residues at $s = \pm s_n$ together and using $\sinh\left\{(2n+1)\dfrac{\pi i}{2}\right\} = i \sin(2n+1)\dfrac{\pi}{2} = i(-1)^n$, we obtain the result.

10. (a) Use result (3.6.7).
    (b) Use 10(a) and result (3.7.6).

11. (a) $J_0(at)$,   (b) $\dfrac{1}{t}\sinh at$,   (c) $1 + \dfrac{1}{2}\cdot\dfrac{t^2}{3!} + \dfrac{1.3}{2.4}\cdot\dfrac{t^4}{5!} + \cdots$,

    (d) $2\sum\limits_{n=0}^{\infty} \text{erfc}\left[\dfrac{(2n+1)x}{2\sqrt{t}}\right]$,   (e) $J_0(2\sqrt{t})$.

12. Hint: $\mathscr{L}^{-1}\left\{\dfrac{\bar{f}(s)}{s}\right\} = \int_0^t f(\tau)\,d\tau = g(t)$,

    $\mathscr{L}^{-1}\left\{\dfrac{\mathscr{L}\{g(t)\}}{s}\right\} = \mathscr{L}^{-1}\left\{\dfrac{\bar{f}(s)}{s^2}\right\} = \int_0^t g(t_1)\,dt_1$

    $= \int_0^t \left\{\int_0^{t_1} f(\tau)\,d\tau\right\} dt_1 = \int_0^t \int_0^{t_1} f(\tau)\,d\tau\,dt_1.$

13. $\dfrac{1}{s}\{\exp(s) - 1\}^{-1}$.

15. (b) Hint: Use Example 3.6.1(a).

17. $s(s^2+1)^{-1} \exp\left(-\dfrac{\pi s}{2}\right)$.

18. (f) Use result (3.6.7).

22. (a) $-\dfrac{1}{2s}\log(1+s^2)$,   (b) $\dfrac{1}{s}\log(1+s)$.

23. (a) Hint: Use (3.6.2) and then the shifting property (3.4.1).

    (c) $L_n(t) = \sum\limits_{r=0}^{n} \binom{n}{r} \dfrac{(-t)^r}{r!}$.

24. (a) Hint: Use the definition and then interchange the order of integration.

26. $\bar{f}(s) = s^{-2} \tanh\left(\dfrac{as}{2}\right)$, $s > 0$; Hint: $f(t+2a) = f(t)$.

27. (a) $f(0) = 1$, $f'(0) = 5$.

29. Hint: Use the identities
$$s\bar{f}(s)\bar{g}(s) = f(0)\bar{g}(s) + \{s\bar{f}(s) - f(0)\}\bar{g}(s)$$
$$= f(0)\bar{g}(s) + \mathcal{L}\{f'(t)\}\mathcal{L}\{g(t)\}.$$
$$s\bar{f}(s)\bar{g}(s) = g(0)\bar{f}(s) + \{s\bar{g}(s) - g(0)\}\bar{f}(s).$$

30. (a) $\bar{f}(s) \sim \dfrac{1}{s}\left(1 - \dfrac{2!}{s^2} + \dfrac{4!}{s^4} - \cdots\right).$

   (b) Hint: Put $t = x + 1$ and then write the binomial expansion of $(x^2 + 2x)^{\frac{1}{2}}$ for $|x| < 2$.

   $$K_0(s) \sim \dfrac{e^{-s}}{\sqrt{\pi}} \sum_{n=0}^{\infty} \dfrac{(-1)^n}{n!} \dfrac{\left\{\Gamma\left(n + \dfrac{1}{2}\right)\right\}^2}{(2s)^{n+\frac{1}{2}}} \quad \text{as } s \to \infty.$$

31. (a) $\displaystyle\sum_{n=0}^{\infty} \dfrac{(-1)^n \Gamma(n+1)}{s^{n+1}}$, Hint: $(1+t)^{-1} = \displaystyle\sum_{n=0}^{\infty}(-1)^n t^n$,

   (b) $\displaystyle\sum_{n=0}^{\infty} \dfrac{(-1)^n 2^{2n+1} \Gamma\left(n + \dfrac{3}{2}\right)}{(2n+1)! \, s^{n+\frac{3}{2}}}$, Hint: $\sin(2\sqrt{t}) = \displaystyle\sum_{n=0}^{\infty} \dfrac{(-1)^n 2^{2n+1} t^{\left(n+\frac{1}{2}\right)}}{(2n+1)!}$.

   (c) $\displaystyle\sum_{n=0}^{\infty} \dfrac{(-1)^n \Gamma(n+1)}{n \, s^{n+1}}$, Hint: $\log(1+t) = \displaystyle\sum_{n=1}^{\infty}(-1)^{n-1}\dfrac{t^n}{n}$,

   (d) $\displaystyle\sum_{n=0}^{\infty} \dfrac{(-1)^n \, a^{2n} \, \Gamma(2n+1)}{\left\{2^2 \cdot 4^2 \cdots (2n)^2\right\} s^{2n+1}}$, Hint: $J_0(at) = \displaystyle\sum_{n=0}^{\infty} \dfrac{(-1)^n (at)^{2n}}{2^2 \cdot 4^2 \cdots (2n)^2}$.

4.10 Exercises

1. (a) $\dfrac{1}{(a-b)}\left(e^{-bt} - e^{-at}\right),$ (b) $2e^{-t} - t^2 - 2t - 2,$

   (c) $\dfrac{1}{5}\left(2\cos t + \sin t + 3e^{-2t}\right),$ (d) $2\left(e^{2t} - 1\right).$

2. $x(t) = x_0 \exp(-kt).$

3. (a) $x(t) = \dfrac{1}{2}\left(e^{3t} + e^{-t}\right),\ y(t) = \dfrac{1}{2}\left(e^{3t} - e^{-t}\right).$

   (b) $x_1 = \dfrac{28}{9}e^{3t} - e^{-t} - \dfrac{t}{3} - \dfrac{1}{9},\ x_2 = \dfrac{28}{9}e^{3t} + e^{-t} - \dfrac{t}{3} - \dfrac{1}{9}.$

(c) $x = 15\cos t + 20\sin t - 10 e^{-t}$,
$y = 10\cos t + 5\sin t - 10 e^{-t}$,
$z = -25\sin t$.

(d) $x = \frac{1}{5}(7e^{-t} + 3e^{4t})$, $y = \frac{1}{5}(7e^{-t} - 2e^{4t})$.

4. $x(t) = \begin{pmatrix} x_1 \\ x_2 \end{pmatrix} = x_0 \begin{pmatrix} 3e^{-t} - 2 \\ 3 - 3e^{-t} \end{pmatrix}$.

5. $x(t) = x_0 e^t$, $\quad y(t) = (x_0 + y_0)e^{2t} - x_0 e^t$.

6. Write $s^4 + 2s^2(\ell + 2k^2) + \ell^2 = (s^2 + \alpha^2)(s^2 + \beta^2)$ so that
$\alpha^2 + \beta^2 = 2(\ell + 2k^2)$, $(\alpha\beta)^2 = \ell^2$ and $\alpha = \sqrt{k^2 + \ell} + k$, $\beta = \sqrt{k^2 + \ell} - k$.

7. $C(t) = \left(\frac{\alpha}{kV}\right)(1 - e^{-kt})$.

8. $p(t) = p_0 \exp\left(-\frac{ct}{k}\right)$

$+ Ac\omega\left(\omega^2 + \frac{c^2}{k^2}\right)^{-1}\left[\frac{c}{\omega k}\sin \omega t - \cos \omega t + \exp\left(-\frac{ct}{k}\right)\right]$.

10. $c(t) = c_0 \exp(-k_1 t)$.

11. Hint: $\frac{d}{dt}(c_1 + c_2 + c_3) = 0$ and so $c_1 + c_2 + c_3 = c_1(0)$.
$c_1(t) = c_1 e^{-k_1 t}$, $c_2(t) = \frac{k_1 c_1}{k_2 - k_1}(e^{-k_1 t} - e^{-k_2 t})$ and
$c_3(t) = c_1(0) - c_1(t) - c_2(t)$.

12. (a) $x(t) = \left(1 + \frac{1}{n^2 - \omega^2}\right)\cos \omega t - \frac{\cos nt}{n^2 - \omega^2}$.

(b) $x(t) = \frac{1}{3}(\sin t - \sin 2t)$.

(c) $\frac{2}{3}\sin t - \frac{1}{3}\sin 2t$, (d) $2e^{-1} + \frac{1}{16}e^{-4t} + \frac{3t}{4} - \frac{31}{16}$.

(e) $\frac{1}{16}(3\sin 2t + 5\sinh 2t)$.

(f) Hint: $\bar{x}(s) = \frac{1}{(s-1)^2 + 1} - \frac{(s+2)}{(s+2)^2 + 1}$.

14. $x(t) = a(\omega t - \sin \omega t)$, $y(t) = a(1 - \cos \omega t)$.

16. $\dot{x}(t) = \frac{eE}{m\omega}\sin \omega t$, $\dot{y}(t) = \frac{eE}{m\omega}(\cos \omega t - 1)$, $\dot{z} = 0$.

19. $\bar{y}(x,s) = \bar{f}(s) \dfrac{\sinh\left\{\dfrac{s}{c}(l-x)\right\}}{\sinh\left(\dfrac{sl}{c}\right)}.$

20. $V(x,t) = V_0\, erfc\left(\dfrac{x}{2\sqrt{\kappa t}}\right).$

21. $V(x,t) = V_0\left(t - \dfrac{x}{c}\right) H\left(t - \dfrac{x}{c}\right),$

   (i) $V = V_0\, H\left(t - \dfrac{x}{c}\right),$ (ii) $V = V_0 \cos\left\{\omega\left(t - \dfrac{x}{c}\right)\right\} H\left(t - \dfrac{x}{c}\right).$

22. $u(z,t) = Ut\left[(1+2\zeta^2)erfc(\zeta) - \dfrac{2\zeta}{\sqrt{\pi}}e^{-\zeta^2}\right]$ where $\zeta = \dfrac{z}{2\sqrt{vt}}.$

23. $q(z,t) = \dfrac{a}{2} e^{i\omega t}\left[e^{-\lambda_1 z} erfc\left\{\zeta - [it(2\Omega + \omega)]^{1/2}\right\}\right.$
   $\left. + e^{\lambda_1 z} erfc\left\{\zeta + [it(2\Omega + \omega)]^{1/2}\right\}\right]$
   $+ \dfrac{b}{2} e^{-i\omega t}\left[e^{-\lambda_2 z} erfc\left\{\zeta - [it(2\Omega - \omega)]^{1/2}\right\} + e^{\lambda_2 z} erfc\left\{\zeta + [it(2\Omega - \omega)]^{1/2}\right\}\right],$
   where $\lambda_{1,2} = \left\{\dfrac{i(2\Omega \pm \omega)}{v}\right\}^{1/2}.$

   $q(z,t) \sim a \exp(i\omega t - \lambda_1 z) + b \exp(-i\omega t - \lambda_2 z),$ $\delta_{1,2} = \left\{\dfrac{v}{|2\Omega \pm \omega|}\right\}^{1/2}.$

24. $\left(\dfrac{v}{2\Omega}\right)^{1/2}.$

25. (a) $\dfrac{1}{2}\left(t + \dfrac{3}{2}\sin 2t\right),$ (b) $(t - \cos t),$ (c) $a\, J_0(at),$
   (d) $3\sin t - \sqrt{2} \sin(\sqrt{2}\, t),$ (e) $\left(t^2 + \dfrac{2t}{a}\right).$

27. $1 - (1+t)\, e^{-t}.$

28. (a) $\dfrac{\pi}{2a^2}(1 - e^{-at}),$ (b) $\dfrac{\pi}{2}\, sgn\, t,$ (c) $\dfrac{\pi}{a} e^{-at},$
   (d) $2\pi\, e^{-at},$ (e) $\sqrt{\dfrac{\pi}{4t}}.$

29. Hint: Use the Laplace transform of sine and cosine functions.

31. $EI\, s^4\, \bar{y}(s) = W\, \exp(-as) + As + B,$
   where $A = EI\, y''(0)$ and $B = EI\, y'''(0).$

$EI\ y(x) = \dfrac{W}{6}(x-a)^3\ H(x-a) + \dfrac{A}{2}x^2 + \dfrac{B}{6}x^3,$

$y(\ell) = 0 = y''(\ell)$ gives $A = Wa\ \ell^{-2}(\ell-a)^2$ and
$B = -W\ell^{-3}(\ell-a)^2(\ell+2a).$

32. $EI\ s^4\ \bar{y}(s) = \dfrac{W}{s}\left[\exp\left(-\dfrac{\ell s}{2}\right) - \exp\left(-\dfrac{3\ell s}{2}\right)\right] + As + B,$

where $A = EI\ y''(0)$ and $B = EI\ y'''(0).$

$EI\ y(x) = \dfrac{W}{24}\left[\left(x-\dfrac{\ell}{2}\right)^4 H\left(x-\dfrac{\ell}{2}\right) - \left(x-\dfrac{3\ell}{2}\right)^4 H\left(x-\dfrac{3\ell}{2}\right)\right]$

$+ \dfrac{Ax^2}{2} + \dfrac{Bx^3}{6}.$

$y''(2\ell) = 0 = y'''(2\ell)$ gives $A = W\ell^2,\ B = -W\ell.$

33. $EI\ y^{(IV)}(x) = W[1 - H(x-\ell)],\ \ 0 < x < 2\ell.$

$EI\ y(x) = \dfrac{W}{8}\left[\dfrac{9}{8}(\ell x)^2 - \dfrac{19}{16}\ell x^3 + \dfrac{1}{3}\{x^4 - (x-\ell)^4\ H(x-\ell)\}\right].$

34. $EI\ s^4\ \bar{y}(s) = \dfrac{W}{s}[1 - \exp(-2\ell s)] + P\exp(-\ell s) + As + B,$

where $A = EI\ y''(0)$ and $B = EI\ y'''(0).$

$EI\ y(x) = \dfrac{W}{24}\left[x^4 - (x-2\ell)^4\ H(x-2\ell)\right]$

$+ \dfrac{P}{6}(x-\ell)^3\ H(x-\ell) + \dfrac{A}{2}x^2 + \dfrac{B}{6}x^3.$

The second term inside the square bracket in $y(x)$ does not contribute because the beam extends over $0 \le x < 2\ell$. Thus

$EI\ y(x) = \dfrac{W}{24}x^4 + \dfrac{A}{2}x^2 + \dfrac{B}{6}x^3, \qquad 0 \le x < \ell,$

$= \dfrac{W}{24}x^4 + \dfrac{P}{6}(x-\ell)^3 + \dfrac{A}{2}x^2 + \dfrac{B}{6}x^3,\ \ \ell < x \le 2\ell.$

$y''(2\ell) = 0 = y'''(2\ell)$ gives $A = \ell(2W\ell + P)$ and $B = -(A/\ell).$

$M\left(\dfrac{\ell}{2}\right) = EI\ y''\left(\dfrac{\ell}{2}\right) = \dfrac{\ell}{8}(9W\ell + 4P)$ and

$S\left(\dfrac{\ell}{2}\right) = EI\ y'''\left(\dfrac{\ell}{2}\right) = -\left(\dfrac{3}{2}W\ell + P\right).$

35. (a) $u_n = 3^n,$ (b) $u_n = n2^{n-1},$ (c) $u_n = (n+1)2^n,$
(d) $u_n = 2(3^n - 2^{n-1}),$ (e) $u_n = n2^n,$ (f) $u_n = A3^n + B2^n,$
where $A = (u_1 - 2u_0)$ and $B = (3u_0 - u_1),$ (g) $u_n = 3^n,$ (h) $u_n = ca^n.$

37. $u(t) = 1 + t + \dfrac{(t-1)^3}{3} + \cdots .$

38. $u(x,t) = \dfrac{1}{2\pi i} \displaystyle\int_{c-i\infty}^{c+i\infty} s^{-1} \exp\left[st - \dfrac{sx}{\sqrt{1+k^2 s^2}} \right] ds.$

39. Hint:
$$\mathscr{L}^{-1}\left[\dfrac{1}{\sqrt{s^2 - \alpha^2}} \exp\left\{ -\beta(s^2 - \alpha^2)^{1/2} \right\} \right] = I_0\left[\alpha(t^2 - \beta^2)^{1/2} \right] H(t-\alpha).$$

42. $u(x,t) = x + \exp\left[-\left(\dfrac{3\pi c}{a}\right)^2 t \right] \sin\left(\dfrac{3\pi x}{a}\right)$
$\qquad - a\left[\displaystyle\sum_{n=0}^{\infty} erfc\left\{ \dfrac{(2n+1)a + x}{2c\sqrt{t}} \right\} - \displaystyle\sum_{n=0}^{\infty} erfc\left\{ \dfrac{(2n+1)a - x}{2c\sqrt{t}} \right\} \right].$

5.5 Exercises

3(b). Hint: $\tilde{f}(k) = \left(\dfrac{Q}{\pi a k}\right) J_1(ak).$

6. $u(r,t) = \displaystyle\int_0^\infty k\, \tilde{f}(k)\, \cos(btk^2)\, J_0(kr)\, dk.$

9. Hint: The solution of the dual integral equations
$$\int_0^\infty k\, J_0(kr)\, A(k)\, dk = u_0, \qquad 0 \le r \le a,$$
$$\int_0^\infty k^2\, J_0(kr)\, A(k)\, dk = 0, \qquad a < r < \infty,$$
is $A(k) = \left(\dfrac{2 u_0}{\pi}\right) \dfrac{\sin(ak)}{k^2}.$

10. Hint: See Debnath, 1994, pp 103-105.

11. $u(r,z) = \dfrac{1}{\pi a} \displaystyle\int_0^\infty k^{-1}\, J_1(ak)\, J_0(kr)\, \exp(-kz)\, dk.$

13. Hint: $\mathscr{L}^{-1}\left[\dfrac{\exp\left\{ -k(s^2 + a^2)^{\frac{1}{2}} \right\}}{(s^2 + a^2)^{\frac{1}{2}}} \right] = H(t-k)\, J_0\left(a\sqrt{t^2 - k^2}\right).$

14. $u(r,z) = b \displaystyle\int_0^\infty k^{-1} \left(\dfrac{\sinh kz}{\cosh ka} \right) J_1(bk)\, J_0(kr)\, dk.$

15. Hint: $\mathcal{H}_0\left[\dfrac{H(a-r)}{\sqrt{a^2-r^2}}\right] = \dfrac{\sin ak}{k}$ and

$$\mathcal{L}^{-1}\left\{\dfrac{\exp(-\sqrt{s}\,k)}{\sqrt{s}(\sqrt{s}-a)}\right\} = \exp(-ak-a^2 t)\,\mathrm{erfc}\left\{\dfrac{k}{2\sqrt{t}} - a\sqrt{t}\right\}.$$

16. $u(r,z) = \dfrac{Qa}{2K}\displaystyle\int_0^\infty k^{-1}\exp(-|z|k)\,J_1(ak)\,J_0(kr)\,dk$.

17. Use the hint in exercise 9 with $a=1$ and $u_0 = 1$.

18. Hint: Use the joint Hankel and Laplace transform method.

20. Hint: Use the Hankel tranform.

$$u(r,z,t) = \dfrac{1}{\rho}\int_0^\infty k\exp(kz)\,J_0(kz)\left[\int_0^t\left\{\int_0^{r_0(\tau)}\alpha p(\alpha,\tau)\,J_0(k\alpha)\,d\alpha\right\}\right.$$
$$\left.\times\cos[\omega(t-\tau)]\,d\tau\right]dk,$$

where $\omega^2 = gk$.

6.8 Exercises

2. Hint: Substitute $e^{-t} = x$ and $g(-\log x) = f(x)$.

3. Hint: Similar to Example 6.2.1(d).

4. Hint: Use (6.2.12) and the scaling property of the Mellin transform.

5. Hint: Use $\mathcal{F}_c\{x^{-n}J_n(ax)\}$ and $\mathcal{F}_c\{x^{p-1}\}$ and then the Parseval relation for the Fourier cosine transform.

17. Hint: $\displaystyle\sum_{n=1}^\infty \dfrac{\cos kn}{n^2} = -\dfrac{k^2}{2}\cdot\dfrac{1}{2\pi i}\int_{c-i\infty}^{c+i\infty}\left(\dfrac{2\pi}{k}\right)^p \dfrac{\zeta(1-p)}{(p-1)(p-2)}\,dp,$

and the integrand has three simple poles at $p = 0, 1, 2$ with residues

$$-\dfrac{1}{2},\ \dfrac{\pi}{k},\ -\dfrac{\pi^2}{3k^2}.$$

18. Hint: $\displaystyle\sum_{n=1}^\infty e^{-nx} = \dfrac{1}{(1-e^{-x})}$.

7.13 Exercises

1. (a) $(a^2+z^2)^{-1}\left[\dfrac{\pi z}{2a}-\log\left(\dfrac{z}{a}\right)\right].$

   (b) $(a-z)^{-1}(a^\alpha-z^\alpha)\,\pi\,\mathrm{cosec}\,\pi\alpha.$

   (c) $-\exp(az)\,Ei(-az).$

   (d) $\Gamma(1-\alpha)\,z^{-\alpha}\exp(az)\,\Gamma(\alpha,\,az).$

   (e) $\left(\dfrac{\pi}{z}\right)\left[1-\exp\left(-a\sqrt{z}\right)\right].$

   (f) $z^{-1}(\cos z - 1).$

6. (a) $(a-z)^{-1}\left(z^{\alpha-1}-a^{\alpha-1}\right)\pi\,\mathrm{cosec}(\alpha\pi).$

   (b) $(a^2+z^2)^{-1}\left[\left(\dfrac{\pi z}{2a}\right)-\log\left(\dfrac{z}{a}\right)\right].$

   (c) $(a^2+z^2)^{-1}\left[\left(\dfrac{\pi a}{2}\right)+z\,\log\left(\dfrac{z}{a}\right)\right].$

9. Hint: Use $t=xu$ in the transform solution and then apply the convolution theorem for the Mellin transform.

14. Hint: $\dfrac{1}{\pi}\oint_{-\infty}^{\infty}\dfrac{f(t)}{t-x}dt=\lim_{\varepsilon\to 0}\dfrac{1}{\pi}\left[\int_{-\infty}^{x-\varepsilon}+\int_{x+\varepsilon}^{\infty}\right]\dfrac{f(t)}{t-x}dt$ and then put $t-x=u.$

8.6 Exercises

1. $\dfrac{1}{3}a^3$ when $n=0$, and $2\left(\dfrac{a}{n\pi}\right)^2 a(-1)^n,\ n=1,2,3,\ldots.$

3. $u(x,t)=\dfrac{2\pi\kappa}{a}\sum_{n=1}^{\infty}n\sin\left(\dfrac{n\pi x}{a}\right)\int_0^t f(\tau)\exp\left[-\kappa(t-\tau)\left(\dfrac{n\pi}{a}\right)^2\right]d\tau.$

4. Hint: $\tilde{f}_s(n) = \int_0^a f(x) \sin(\xi_n x)\, dx$

$$f(x) = \mathscr{F}_s^{-1}\{\tilde{f}_s(n)\} = \frac{2}{a} \sum_{n=0}^{\infty} \frac{(h^2 + \xi_n^2)\, \tilde{f}_s(n) \sin(x\xi_n)}{h + (h^2 + \xi_n^2)}$$

where ξ_n is the root of the equation $\xi \cot(a\xi) + h = 0$.

$$u(x,t) = \frac{2}{a} \sum_{n=1}^{\infty} \frac{\xi_n (h^2 + \xi_n^2)}{h + (h^2 + \xi_n^2)} \int_0^t f(\xi) \exp[-\kappa \xi_n (t - \xi)] \sin(x\xi_n)\, d\xi.$$

5. Hint: Use $\tilde{f}_c(n) = \int_0^a f(x) \cos(x\xi_n)\, dx$,

$$f(x) = 2 \sum_{n=1}^{\infty} \frac{(h^2 + \xi_n^2)\, \tilde{f}_c(n) \cos(x\xi_n)}{h + a(h^2 + \xi_n^2)},$$

where ξ_n is the root of the equation $\xi \tan(a\xi) = h$.

6. Hint: Apply the finite Fourier cosine transform.

8. Hint: $\tilde{W}_s(n,t) = W_0\, \phi(t)\, H(Ut - \ell) \int_0^\ell \sin\left(\frac{\pi n x}{\ell}\right) \delta(x - Ut)\, dx$

$$= W_0\, \phi(t)\, H\left(t - \frac{\ell}{U}\right) \sin\left(\frac{n\pi Ut}{\ell}\right).$$

12. Hint: $\dfrac{d\tilde{V}_s}{dt} + \kappa \left(\dfrac{n\pi}{a}\right)^2 \tilde{V}_s = 0, \quad \tilde{V}_s(n,t) = A \exp\left(-\dfrac{\kappa n^2 \pi^2 t}{a^2}\right),$

$$A = \tilde{V}_s(n,0) = \frac{4a V_0}{n^2 \pi^2} \sin\left(\frac{n\pi}{2}\right) = (-1)^r \frac{4a V_0}{(2r+1)^2 \pi^2}$$

where $n = (2r+1), \quad r = 0, 1, 2, \ldots$

$$V(x,t) = \left(\frac{8 V_0}{\pi^2}\right) \sum_{r=0}^{\infty} \frac{(-1)^r}{(2r+1)^2} \sin\left\{(2r+1)\frac{nx}{a}\right\} \exp\left\{-\frac{\kappa(2r+1)^2 \pi^2 t}{a^2}\right\}.$$

15. Hint: Replace P by $W_0\, d\xi\, d\eta$ and integrate with respect to ξ and η over the region $\alpha \leq \xi \leq \beta, \quad \gamma \leq \eta \leq \delta$.

$$u(x,y) = \frac{4W_0}{\mathcal{D}\pi^6} \sum_{m=1}^{\infty} \sum_{n=1}^{\infty} \left[\left\{ \cos\left(\frac{m\pi\alpha}{a}\right) - \cos\left(\frac{m\pi\beta}{a}\right) \right\} \right.$$
$$\left. \times \left\{ \cos\left(\frac{n\pi\gamma}{b}\right) - \cos\left(\frac{n\pi\delta}{b}\right) \right\} \frac{\sin\left(\frac{m\pi x}{a}\right)\sin\left(\frac{n\pi y}{b}\right)}{mn\,\omega_{mn}^4} \right].$$

16. Hint: $\dfrac{d^2\,\tilde{u}_s(m,n,t)}{dt^2} + \Omega_{mn}^2\,\tilde{u}_s(m,n,t) = 0,$ where $\Omega_{mn}^2 = \dfrac{\mathcal{D}\pi^4\,\omega_{mn}^4}{\rho h}.$

$$u(x,y,t) = \left(\frac{4}{ab}\right) \sum_{m=1}^{\infty} \sum_{n=1}^{\infty} \left\{ A_{mn} \cos(\Omega_{mn}t) + B_{mn} \sin(\Omega_{mn}t) \right\}$$
$$\times \sin\left(\frac{m\pi x}{a}\right)\sin\left(\frac{n\pi y}{b}\right).$$

9.6 Exercises

1. (a) $\dfrac{s}{(s^2-a^2)} - \dfrac{\exp(-sT)}{(s^2-a^2)}(s\cosh aT + a\sinh aT).$

    (d) $\dfrac{1}{s}(1-e^{-sT})\,H(T).$

10.8 Exercises

1. (a) $\dfrac{z^3 + 4z^2 + z}{(z-1)^4}$, use (10.4.13) and (10.3.14).

    (b) $\exp(a/z),$ (c) $\dfrac{z}{(z-e^a)^2},$ (d) $\left(1+\dfrac{1}{z}\right),$ (e) $\dfrac{z(z+a)}{(z-a)^2}.$

2. Hint: Put $b = ix$ in (10.4.14).

6. (a) $(3^{n+1} - 2^{n+1}),$ (d) $n\,a^{n-1},$

    (e) $(n-1)\,a^{n-2}\,H(n-1),$ (f) $(2^{n-1} - n),$

    (g) $2(-1)^n - (-2)^n,$ (h) $\dfrac{1}{6}\left[(-1)^n + 2^{n+3} - 3(1)^n\right],$

(i) Hint:
$$\frac{U(z)}{z} = \frac{2}{(z-1)} - \frac{1}{\left(z-\frac{1}{2}\right)}, \quad U(z) = \frac{2z}{(z-1)} - \frac{z}{\left(z-\frac{1}{2}\right)}$$

$$u(n) = (2 - 2^{-n}).$$

7. (a) $\frac{1}{16}\left[17(-3)^n + 4n - 1\right]$,   (c) $x_0 a^n + n a^{n-1}$,

   (d) $x_0 (1-a)^n + 1 - (1-a)^n$,   (e) $\frac{3}{5}\left[3^n - (-2)^n\right]$,   (i) $n a^n$.

9. (a) 1,   (b) 0,   (c) 1,

   (d) $f(0) = 0$, $m > 0$; $f(0) = 1$, $m = 0$.

11. (a) $(1 - a e^{ix})^{-1}$,   (b) $e \log\left(1 + \frac{1}{e}\right)$,   (c) $(2 \sinh x)^{-1}$.

12. $\frac{7}{4} - \frac{3}{4}\left(-\frac{1}{3}\right)^n$.

13. $U(z) = \frac{2z}{z^2 - 4}$, $u(n) = 2^{n-1}\left[1 + (-1)^{n+1}\right]$, $v(n) = 2^{n-1}\left[1 + (-1)^n\right] - 1$,
   where $n = 0, 1, 2, \ldots$.

14. Hint: $U(z) = \frac{1}{(1-z^{-1})^3} - \frac{3z^{-1}}{(1-z^{-1})^3} + \frac{4z^{-2}}{(1-z^{-1})^3}$.

16. (a) $u_n = \frac{3}{4}\left[1 - (-3)^{n-1}\right]$,   (b) $u_n = 2\left[\left(\frac{2}{3}\right)^{n-1} - 1\right]$,

   (c) $u_n = 5^{n/2}\left(2 \sin nx + \frac{1}{2} \cos nx\right)$, where $x = \tan^{-1}\frac{1}{2}$.

11.5 Exercises

1. (a) $\frac{a^2}{k_i^2}\left(ak_i - \frac{4}{ak_i}\right) J_1(ak_i)$,   (b) $\frac{ak_i}{(\alpha^2 - k_i^2)} J_0(a\alpha) J_1(ak_i)$.

10. $u(r,t) = \frac{2}{a^2} \sum_{i=1}^{\infty} \frac{J_0(rk_i)}{J_1^2(ak_i)} \int_0^t \tilde{Q}(k_i, \tau) \exp\left[-(t-\tau)k_i^2\right] d\tau$.

   (a) $u(r,t) = \frac{Q_0}{4k}(a^2 - r^2) - \left(\frac{2Q_0}{ak}\right) \sum_{i=1}^{\infty} \frac{J_0(rk_i)}{k_i^3 J_1(ak_i)} \exp(-t \kappa k_i^2)$.

(b) $u(r,t) = \dfrac{2\kappa Q_0}{k\,a^2} \displaystyle\sum_{i=1}^{\infty} \dfrac{J_0(rk_i)}{J_1^2(ak_i)} \int_0^t f(\tau) \exp\left[-\kappa(t-\tau)k_i^2\right] d\tau.$

14.5 Exercises

1. (a) $f_0(n) = \exp(-a), \quad n = 0,$
   $f_0(n) = \exp(-a)\left[L_n(a) - L_{n-1}(a)\right], \quad n \geq 1.$

   (b) $a^n(1+a)^{n+1}$,   (c) $A\,\delta_{mn}$,

   (d) $0, \ n > m; \ (-1)^n \binom{m}{n} m!, \ m \geq n$,   (e) 1 for $n = 0,1,2,3,\ldots$.

15.4 Exercises

1. (a) $2^{n-\frac{1}{2}}\,\Gamma\!\left(n+\dfrac{1}{2}\right),$ (b) $0; \ m = 0,1,2,\ldots,n-1,$

   (c) $\left(n+\dfrac{1}{2}\right)\delta_n.$

3. Hint: Use Feldheim's result (1938).
   $$H_n^2(x) = n!\,2^n \sum_{r=0}^{n} \binom{n}{r} \dfrac{H_{2r}(x)}{2^{2r}\,r!}.$$

Bibliography

The following bibliography is not by any means a complete one for the subject. For the most part, it consists of books and papers to which reference is made in the text. Many other selected books and papers related to material of the subject have been included so that they may serve to stimulate new interest in future study and research.

Abramowitz, M. and Stegun, I.A. (1964). *Handbook of Mathematical Functions*. NBS, Appl. Math. Series 55, Washington, D.C.

Antimirov, M.Ya., Kolyshkin, A.A., and Vaillancourt, Remi, (1993). *Applied Integral Transforms,* American Mathematical Society, Providence.

Bailey, W.N., (1936). On the product of two Legendre polynomials with different arguments, **41,** 215-220.

Bailey, W.N., (1939). On Hermite polynomials and associated Legendre functions, *Jour. Lond. Math. Soc.,* **14,** 281-286.

Bateman, H., (1944). *Partial Differential Equations of Mathematical Physics*, Dover, New York.

Benjamin, T.B., (1966). Internal waves of finite amplitude and permanent form, *J. Fluid Mech.,* **25,** 241-270.

Benjamin, T.B., (1967). Internal waves of permanent form in fluids of great depth, *J. Fluid Mech.,* **29,** 559-592.

Brown, H.K., (1944). Resolution of boundary value problems by means of the finite Fourier transformation; general vibration of a string, *J. Appl. Phys.* *XIV,* 609-618.

Butzer, P., Stens, R., and Wehrens, M., (1980). The continuous Legendre transform, its inverse transform and applications, *Internat. J. Math. and Math. Sci.,* **3,** 47-67.

Cahen, E., (1894). Sur la fonction $\zeta(s)$ de Riemann et sur des fonctions analogues, *Ann de l'Ec. Norm*, **11,** 75-164.

Campbell, G.A., and Foster, R.M., (1948). *Fourier Integrals for Practical Applications,* Van Nostrand, New York.

Carslaw, H.S., and Jaeger, J.C., (1953). *Operational Methods in Applied Mathematics,* Oxford University Press, Oxford.

Casey, S.T., and Walnut, D.F., (1994). Systems of convolution equations, deconvolution, Shannon sampling, and the wavelet and Gabor transforms, *SIAM Rev.* **35,** 537-577.

Chakraborty, A., (1980). Derivation of the solution of certain singular integral equation, *J. Indian Inst. Sci.,* **62B,** 147-157.

Chakraborty, A., (1988). Solutions of some special singular integral equations of applied mathematics, *Proc. of the Seminar on Techniques in Applied Mathematics and Applications,* (Edited by B.N. Mandal and B. Patra) 31-45.

Chakraborty, A., and Williams, W.E., (1980). A note on a singular integral equation, *J. Inst. Maths. Applics,* **26,** 321-323.

Churchill, R.V., (1954). The operational calculus of Legendre Transforms, *Jour. Math. and Physics,* **33,** 165-178.

Churchill, R.V., (1972). *Operational Mathematics,* (Third Edition), McGraw-Hill Book Company, New York.

Churchill, R.V., and Dolph, C.L., (1954). Inverse transforms of products of Legendre transforms, *Proc. Amer. Math. Soc.* **5,** 93-100.

Cinelli, G., (1965). An extension of the finite Hankel transform and applications, *Internat. J. Engng. Sci.,* **3,** 539-559.

Comninou, M. (1977). The interface crack, *ASME J. Appl. Mech.,* **44,** 631-636.

Conte, S.D., (1955). Gegenbauer transforms, *Quart. J. Math. Oxford (2),* **6,** 48-52.

Cooper, J.L.B., (1952). Heaviside and Operational Calculus, *Math. Gazette,* **36,** 5-19.

Copson, E.T., (1935). *An Introduction to the Theory of Functions of a Complex Variable,* Oxford University Press, Oxford.

Copson, E.T., (1965). *Asymptotic Expansions,* Cambridge University Press, Cambridge.

Crank, J., (1975). *The Mathematics of Diffusion,* (Second Edition). Clarendon Press, Oxford.

Crease, J., (1956). Propagation of Long Waves due to atmospheric disturbances on a rotating sea, *Proc. Royal Soc. London* A233, 556-570.

Debnath, L., (1960). On Laguerre transform, *Bull. Calcutta Math. Soc.,* **55,** 69-77.

Debnath, L., (1961). Application of Laguerre transform to the problem of oscillations of a very long and heavy chain, *Annali dell' Univ. di Ferrara, Sezione VII-Scienze Mathematiche,* **IX,** 149-151.

Debnath, L., (1962). Application of Laguerre transform to heat conduction problem, *Annali dell' Univ. di Ferrara, Sezione VII-Scienze Mathematiche,* **X,** 17-19.

Debnath, L., (1963). On Jacobi transforms, *Bull. Calcutta Math. Soc.,* **55,** 113-120.

Debnath, L., (1964). On Hermite transforms, *Mathematicki Vesnik,* **1,** (16), 285-292.

Debnath, L., (1967). Solution of partial differential equations by the Jacobi transform, *Bull. Calcutta Math. Soc.,* **59,** 155-158.

Debnath, L., (1968). Some operational properties of the Hermite transform, *Math. Vesnik,* **5,** (20), 29-36.

Debnath, L., (1969). On the Faltung Theorem of Laguerre transforms, *Studia Univ. Babes-Bolyai, Ser. Phys.,* **2,** 41-45.

Debnath, L., (1969). On transient development of axisymmetric surface waves, *Proc. Nat. Inst. Sci. India,* **35A,** 567-585.

Debnath, L., (1969). An asymptotic treatment of the transient development of surface waves, *Appl. Sci. Res.,* **21,** 24-36.

Debnath, L., (1978). Generalized calculus and its applications, *Internat. J. Math. Edu. Sci. Technol.,* **9,** 399-416.

Debnath, L., (1983). Unsteady axisymmetric capillary-gravity waves in a viscous liquid, *Indian J. Pure and Appl. Math.,* **14,** 540-553.

Debnath, L., (1989). The linear and nonlinear Cauchy-Poisson wave problems for an inviscid or viscous liquid, *Topics in Mathematical Analysis,* (Ed. T.M. Rassias) 123-155, World Scientific, Singapore.

Debnath, L., (1990). The Hilbert Transform and its Applications to Fluid Dynamics, *Progress of Mathematics,* **24,** 19-40.

Debnath, L., (1994). *Nonlinear Water Waves,* Academic Press, Boston.

Debnath, L. and Kulchar, A.G., (1972). On generation of dispersive long waves on a rotating ocean by wind stress distributions, *Jour. Phys. Soc. Japan,* **33,** 1464-1470.

Debnath, L., and Harrel, C.W. (1976). The operational calculus of associated Legendre transforms-I, *Indian J. Pure and Appl. Math.,* **7,** 278-291.

Debnath, L., and Thomas, J., (1976). On finite Laplace transformations with applications, *Z. Angew. Math. und Mech.,* **56,** 559-563.

Debnath, L., and Grum, William J., (1988). The fractional calculus and its role in the synthesis of special functions: *Internat. J. Math. Edu. Sci. & Technol,* Part I, **19,** 215-230; Part II, **19,** 347-362.

Debnath, L., and Mikusinski, P., (1990). *Introduction to Hilbert spaces with Applications,* Academic Press, Boston.

Debnath, L., and Rollins, D., (1992). The Cauchy-Poisson waves in an inviscid rotating stratified liquid, *Internat. J. Nonlinear Mech.,* **27,** 405-412.

Deeba, E.Y., and Koh, E.L., (1983). The Continuous Jacobi transform, *Internat. J. Math. and Math. Sci.,* **6,** 145-160.

Dimovski, I.H., and Kalla, S.L., (1988). Convolution for Hermite transforms, *Math. Japonica* **33,** 345-351.

Ditkin, V.A., and Prudnikov, A.P., (1965). *Integral transforms and Operational Calculus,* Pergamon Press, Oxford.

Doetsch, G., (1935). Integration von Differentialgleichungen vermittels der endlichen Fourier Transformation, *Math. Annalen,* **112,** 52-68.

Doetsch, G., (1950-1956). *Handbuch der Laplace-Transformation*, Verlag Birkhäuser, Basel, Vol.1, 1950; Vol. 2, 1955; Vol. 3, 1956.
Doetsch, G., (1970). *Introduction to the Theory and Applications of the Laplace Transformation.* Springer-Verlag, New York.
Dunn, H.S., (1967). A generalization of the Laplace transform, *Proc. Cambridge Philos. Soc.,* **63,** 155-161.
Erdélyi, A., (1953). *Higher Transcedental Functions,* Vols. 1 and 2, McGraw-Hill, New York.
Erdélyi, A., Magnus, W., Oberhettinger, F., and Tricomi, F. (1954). *Tables of Integral Transforms,* Vols. 1 and 2. McGraw-Hill, New York.
Erdélyi, A., (1956). *Aysmptotic Expansions,* Dover Publications, New York.
Eringen, A.C., (1954). The finite Sturm-Liouville transform, *Quart, J. Math. Oxford, (2),* **5,** 120-129.
Feldheim, E., (1938). Quelques nouvelles relations pour les polynômes d'Hermite, *Jour. Lond. Math. Soc.,* **13,** 22-29.
Firth, Jean M. (1992). *Discrete Transforms,* Chapman & Hall, London.
Fourier, J. (1822). *La Théorie Analytique de la Chaleur.* English Translation by A. Freeman, Dover Publications, 1955.
Gakhov, F.D., (1966). *Boundary Value Problems,* Pergamon Press, London.
Gamkrelidze, R.V. (ed), (1968). *Progress in Mathematics,* 2 vols., Engl. Trans.—Plenum Press, New York.
Gautesen, A.K., and Dunders, J., (1987a). The interfce crack in a tension field, *ASME J. Appl. Mech.,* **54,** 93-98.
Gautesen, A.K., and Dunders, J., (1987b). On the solution to a Cauchy principal value integral equation which arises in fracture mechanics, *SIAM J. Appl. Math.,* **47,** 109-116.
Glaeske, H.-J., (1981). Die Laguerre-Pinney transformation, *Aequationes Math.,* **22,** 73-85.
Glaeske, H.-J., (1983). On a convolution structure of a generalized Hermite transformation, *Serdica Bulgaricae Math. Publ.,* **9,** 223-229.
Glaeske, H.-J., (1986). On the Wiener-Laguerre transformation, *Riv. Téc. Ing. Univ. Zulia,* **9,** 27-34.
Glaeske, H.-J., (1987). Operational properties of generalized Hermite transformations, *Aequationes Math.,* **32,** 155-170.
Glaeske, H.-J., and Runst, T., (1987). The Discrete Jacobi transform of generalized functions, *Math. Nachr.,* **132,** 239-251.
Goldberg, R.R., (1961). *Fourier Transforms* (Cambridge Tract No. 52), Cambridge University Press, Cambridge.
Gradshteyn, I.S. and Ryzhik, I.M. (1980). *Table of Integrals, Series, and Products.* Academic Press, New York.
Harrington, W.J., (1967). A property of Mellin transforms, *SIAM Review,* **9,** 542-547.
Heaviside, Oliver, (1899). *Electromagnetic Theory,* 1950 reprint, Dover Publications, New York.
Hirschman, I.I., (1963). Laguerre transforms, *Duke Math. J.,* **30,** 495-510.
Howell, (1938). A definite integral for Legendre functions, *Phil. Mag.* **25,** 1113-1115.

Hulme, A., (1981). The potential of a horizontal ring of wave sources in a fluid with a free surface, *Proc. Roy. Soc. London,* **A375,** 295-305.

Jaeger, J.C., and Newstead, G., (1969). *An Introduction to the Laplace Transformation with Engineering Applications,* (Third Edition), Methuen Company Ltd., London.

Jerri, A.J. (1992). *Integral and Discrete Transforms with Applications and Error Analysis,* Marcel Dekker, New York.

Jones, D.S., (1965). Some remarks on Hilbert transforms, *J. Inst. Math. Applics.,* **1,** 226-240.

Jones, D.S., (1982). *The Theory of Generalized Functions.* Cambridge University Press, Cambridge.

Jones, D.S. and Kline, M. (1958). Asymptotic expansions of multiple integrals and the method of stationary phase. *J. Math. Phys.,* **37,** 1-28.

Joseph, R.I., and Adams, R.C., (1981). Extension of weakly nonlinear theory of solitary wave propagation, *Phys. Fluids,* **24,** 15-22.

Jury, E.I., (1964). *Theory and Application of the Z-Transform,* John Wiley & Sons, New York.

Kneitz, H., (1938). Lösung von Randwertprobemen bei systemen gewöhnlicher Differentialgle ichungen vermittels der endlichen Fourier transformation, *Math. Zeit. XLIV,* 266-291.

Kober, H., (1943a). A note on Hilbert transforms, *Quart. J. Math. Oxford.,* **14,** 49-54.

Kober, H., (1943b). A note on Hilbert transforms, *J. London Math. Soc.,* **18,** 66-71.

Kober, H., (1967). A modification of Hilbert transforms, the Weyl integral and functional equations, *J. Lond. Math. Soc.* **42,** 42-50.

Koschmieder, L., (1941). Die endliche Fouriersche Abbildung und ihr Nutzen bei Aufgaben der Wärmeleitung in Stäben und Platten, *Deutshe Math,* **5,** 521-545.

Kubota, T., Ko, D.R.S., and Dobbs, L., (1978). Weakly-nonlinear, internal gravity waves in stratified fluids of finite depth, *J. Hydronautics,* **12,** 157-165.

Lakshmanarao, S.K., (1954). Gegenbauer transforms, *Math, Student,* **22,** 161-165.

Lamb, H., (1904). On the propagation of tremors over the surface of elastic solid, *Phil. Trans. Royal Soc.,* A203, 1-42.

Laplace, P.S., (1820). *Théorie Analytique des Probabilitiés,* Courcier, Paris.

Lebedev, N.N., (1965). *Special functions and their applications.* Prentice-Hall, Englewood Cliffs, New Jersey.

Lighthill, M.J., (1958). *Introduction to Fourier Analysis and Generalized Functions.* Cambridge University Press, Cambridge.

Lukacs, E., (1960). *Characteristic Functions,* Griffin Statistical Monographs, New York.

Luke, Y.L., (1969). *The Special Functions and Their Approximations,* 2 vols., Academic Press, New York.

Marichev, O.I., (1983). *Handbook of Integral Transforms of Higher Transcendental Functions, Thoery and Algorithmic Tables*, Ellis Horwood, West Sussex, England

Maslowe, S.A., and Redekopp, L.G., (1980). Long nonlinear waves in stratified shear flows, **101**, 321-348.

McCully, J., (1960). The Laguerre transform, *SIAM Review*, **2**, 185-191.

Mellin, H., (1896). Üeber die fundamentale Wichtgkeit des Satzes von Cauchy für die Theorien der Gamma-und der hypergeometrischen funktionen, *Acta Soc. Fennicae*, **21**, 1-115.

Mellin, H., (1902). Über den Zusammenhang zwischen den linearen Differential- und Differezengleichugen, *Acta Math.*, **25**, 139-164.

Mikusinski, J., (1983). *Operational Calculus, Vol. 1*, (Second Edition), Pergamon Press, London.

Miller, K.S. and Ross, B. (1993). *An Introduction to the Fractional Calculus and Fractional Differential Equations*, John Wiley & Sons, New York.

Mohanti, N.C., (1979). Small-amplitude internal waves due to an oscillatory pressure, *Quart. Appl. Math.*, **37**, 92-97.

Muskhelishvili, N.J., (1953). *Singular Integral Equations*, Noordhoff, Groningen.

Myint-U, T. and Debnath, L. (1987). *Partial Differential Equations for Scientists and Engineers* (Third Edition), North Holland, New York.

Naylor, D., (1963). On a Mellin type integral transform, *J. Math. Mech.*, **12**, 265-274.

Newcomb, R.W., (1963). Hilbert transforms-Distributional theory, Stanford Electronics Laboratory, Technical Report No. 2250-1.

Oberhettinger, F. (1959). On a modification of Watson's lemma. *J. Res. Nat. Bur. Standards, Section B*, **63**, 15-17.

Oberhettinger, F. (1972). *Tables of Bessel Transforms*. Springer-Verlag, New York.

Oberhettinger, F. (1974). *Tables of Mellin Transforms*. Springer-Verlag, New York.

Okikiolu, G.O., (1965). A generalization of the Hilbert transform, *J. Lond. Math. Soc.*, **40**, 27-30.

Olver, F.W.J., (1974). *Asymptotics and Special Functions*. Academic Press, New York.

Ono, H., (1975). Algebraic solitary waves in stratified fluids, *J. Phys. Soc. Japan*, 1082-1091.

Pandey, J.N., (1972). On the Stieltjes transform of generalized functions, *Proc. Camb. Phil. Soc.*, **71**, 85-96.

Pandey, J.N., and Chaudhry, M.A., (1983). The Hilbert transform of generalized functions and applications, *Can. J. Math.* **XXXV**, 478-495.

Papoulis, A., (1963). *The Fourier Integral and its Applications*, McGraw-Hill, New York.

Pathak, R.S. and Debnath, L., (1987). Recent developements on the Stieltjes transform of generalized functions, *Internat. J. Math. and Math. Sci.*, **10**, 641-670.

Pennline, J.A., (1976). A more tractable solution to a singular integral equation obtained by solving a related Hilbert problem for two unknowns, *J. Res. Bur. Standards,* **80B** (3) 403-414.

Peters, A.S., (1972). Pairs of Cauchy singular integral equations, *Comm. Pure Applied Math.,* **25,** 369-402.

Proudman, J., (1953). *Dynamical Oceanography,* Dover Publications, New York.

Reif, F., (1965). *Fundamentals of Statistical and Thermal Physics,* McGraw-Hill, New York.

Reissner, E., (1949). Boundary value problems in aerodynamics of lifting surfaces in non-uniform motion, *Bull. Amer. Math. Soc.,* **55,** 825-850.

Riemann, B., (1876). Über die Anzahl der Primzahlen unter eine gegebenen Grösse, *Gesammelte Math. Werke,* 136-144.

Roettinger, I., (1947). A generalization of the finite Fourier transformation and applications, *Quart. Appl. Math.,* **5,** 298-310.

Sansone, G., (1959). *Orthogonal Functions,* Interscience Publishers, New York.

Scott, E.J., (1953). Jacobi transforms, *Quart. J. Math. Oxford,* **4,** 36-40.

Sen, A., (1963). Surface waves due to blasts on and above liquids, *J. Fluid Mech.,* **16,** 65-81.

Sheehan, J.P., and Debnath, L., (1972). On the dynamic response of an infinite Bernoulli-Euler beam, *Pure and Appl. Geophysics,* **97,** 100-110.

Sneddon, I.N., (1946). Finite Hankel transforms, *Phil. Mag.,* **37,** 17-25.

Sneddon, I.N., (1951). *Fourier Transforms,* McGraw-Hill, New York.

Sneddon, I.N., (1955). Functional Analysis, *Handbuch der Physik,* Vol. **2,** 198-348.

Sneddon, I.N., (1969). The inversion of Hankel transforms of order zero and unity, *Glasgow Math. J.,* **10,** 156-168.

Sneddon, I.N., (1972). *The Use of Integral Transforms,* McGraw-Hill Book Company, New York.

Sölingen, H., (1953). Algebraisierung der endlichen Hilbert-Transformation, *Z. Angew. Math. Mech.,* **33,** 280-289.

Sölingen, H., (1954). Zur Theorie der endlichen Hilbert-Transformation, *Math. Z,* **60,** 31-37.

Srivastava, H.M., (1971). The Weierstrass-Laguerre Transform, *Proc. Nat. Acad. Sci., U.S.A.,* **68,** 554-556.

Srivastava, K.N., (1965). On Gegenbauer transforms, *Math. Student,* **33,** 129-136.

Stadler, W., and Shreeves, R.W., (1970). The transient and steady-state response of the infinite Bernoulli-Euler beam with damping and an elastic foundation, *Quart. J. Mech. and Appl. Math.* **XXII,** 197-208.

Szegö, G., (1967). *Orthogonal polynomials.* Colloquium Publications, Vol. 23, (Third Edition), Amer. Math. Soc., Providence, Rhode Island.

Tamarkin, J.D., and Shohat, J.A., (1943). *The Problem of Moments,* Amer. Math. Soc., New York.

Titchmarsh, E.C., (1959). *Introduction to the Theory of Fourier Integrals,* (Second Edition). Oxford University Press, Oxford.

Tranter, C.J., (1966). *Integral Transforms in Mathematical Physics*, (Third Edition), Methuen and Company Ltd., London.
Tranter, C.J., (1950). Legendre Transforms, *Quart. J. Math.* **1**, (2), 1-8.
Tricomi, F.G., (1935). Transformazione di Laplace e polinomi di Laguerre, *Rend. Accad. Lincei, (vi)*, **21**, 235-242.
Tricomi, F.G., (1951). On the finite Hilbert transformation, *Quart. J. Math. Oxford*, **2**, 199-211.
Tuttle, D.F., (1958). *Network Synthesis*, Vol. 1. John Wiley and Sons, New York.
Ursell, F., (1983). Integrals with a large parameter: Hilbert transforms. *Math. Proc. Camb. Phil. Soc.*, **93**, 141-149.
Ursell, F., (1990). Integrals with a large parameter. A strong form of Watson's lemma. *Elasticity, Mathematical Methods and Applications* (Ed. G. Eason and R.W. Ogden). 391-395. Ellis Horwood Ltd. Chichester.
Watson, E.J., (1981). *Laplace Transforms and Applications*, Van Nostrand Reinhold, New York.
Watson, G.N., (1938). A note on the polynomials of Hermite and Laguerre, *Jour. Lond. Math. Soc.*, **13**, 29-32.
Watson, G.N., (1944). *A Treatise on the Theory of Bessel Functions*, (Second Edition). Cambridge University Press, London.
Wheelon, A.D., (1954). On the summation of infinite series in closed form, *J. Appl. Phys.*, **25**, 113-118.
Whitham, G.B., (1967). Variational methods and applications to water waves, *Proc. Roy. Soc.*, London, **A299**, 6-25.
Whittaker, E.T., and Watson, G.N., (1927). *A Course of Modern Analysis*, (Fourth Edition), Cambridge University Press, Cambridge.
Widder, D.V., (1941). *The Laplace Transform*, Princeton University Press, Princeton, New Jersey.
Wiener, N., (1932). *The Fourier Integral*, Cambridge University Press.
Williams, W.E., (1978). Note on a singular integral equations, *J. Inst. Math. Applics.*, **22**, 211-241.
Wintner, A., (1947). *Fourier Transforms of Probability Distributions*, (mimeographed notes: Princeton University Press).
Wong, R., (1989). *Asymptotic Approximations of Integrals*, Academic Press, Boston.
Wong, R. and Wyman, M., (1972). A generalization of Watson's lemma. *Can. J. Math.*, **24**, 185-208.
Wyman, M., (1963). The asymptotic behavior of the Hermite polynomials. *Can. J. Math.*, **15**, 332-349.
Wyman, M., (1964). The method of Laplace. *Trans. Roy. Soc. Canada*, **2**, 227-256.
Zadeh, L.A. and Desoer, C.A., (1963). *Linear System Theory*, McGraw-Hill, New York.
Zayed, A.I., (1987). A generalized inversion formula for the continuous Jacobi transform, *Internat. J. Math. and Math. Sci.*, **10**, 671-692.
Zemanian, A.H., (1969). *Generalized Integral Transformations*, John Wiley & Sons, New York.

Index

A

Abel's integral equation, 160-163
Absolutely integrable function, 6
Acoustic radiation problem, 201
Adams, 247
Addition theorem, 63
Admittance, 26
Airy, 48
Airy function, 44, 376
  graph of, 376
Airy stress function, 48
Angular frequency, 8
Associated Laguerre function, 384
Associated Legendre function, 327, 378
Associated Legendre transform, 327
Asymptotic expansions for
  Bessel functions, 202, 204
  Complementary error function, 371
  Gamma function, 369
  Laplace transform, 112
  One-sided Hilbert transform, 249
Autonomous system, 185
Axisymmetric acoustic radiation problem, 201
Axisymmetric biharmonic equation, 202, 206
Axisymmetric Cauchy-Poisson problem, 203, 209
Axisymmetric diffusion equation, 155, 200-201, 207
Axisymmetric Laplace equation, 207, 319
Axisymmetric wave equation, 201

B

Bailey, 356-357
Bandlimited function, 11, 19, 21
Bateman, 351
Benjamin, 244-245

Benjamin-Ono equation, 245-246, 262
Bessel's equation, 136, 198, 372
Bessel's functions, 87, 317, 372
  graphs of, 374-375
Bernoulli-Euler equation, 28, 180
Beta distribution, 80
Beta function, 95, 370
Bibliography, 441-448
Biharmonic equation, 44-45, 47, 279, 282
Bilateral Laplace transform, 232
Bilateral Z-transform, 299
Binomial distribution, 80
Blasius problem, 188
Boundary layer equation, 157
Boundary value problem, 123, 147, 163, 205, 291, 333
Bromwich, 3
Bromwich contour, 102-103
Brown, 265

C

Cahen, 211
Campbell, 387
Carson, 3
Cauchy, 1
Cauchy distribution, 80
Cauchy integral formula, 116
Cauchy principal value, 237
Cauchy-Poisson problem, 77, 180, 203, 209
Cauchy problem, 34, 36, 67-68, 74, 173, 208
Cauchy residue theorem, 103
Central limit theorem, 64
Chakraborty, 248
Characteristic function, 61
Chebyshev polynomial, 311, 381
  graphs of, 382
Churchill, 268, 270, 325, 333
Cominou, 248

Complementary error function, 56, 86, 370
 graph of, 371
Complementary incomplete gamma function, 369
Conte, 337
Continuity equation, 44
Convolution, 16, 92, 94, 270
Convolution integral equation, 159
Convolution theorem for
 finite Fourier cosine transform, 270
 finite Fourier sine transform, 270
 Fourier cosine transform, 52-54
 Fourier sine transform, 75
 Fourier transform, 16
 Gegenbauer transform, 342
 generalized Mellin transform, 229
 Hermite transform, 361-365
 Laguerre transform, 350-352
 Laplace transform, 92
 Legendre transform, 332
 Mellin transform, 217
 Z transform, 303
Cooper, 2
Copson, 113, 326
Coriolis parameter, 41, 174
Cosine integral, 119
Couette flow, 276, 277
Counting process, 144
Critical damping, 139
Crease, 177

D

D'Alembert's solution, 37, 174
Debnath, 29, 77-78, 81, 149, 176, 182, 192, 205, 209, 227, 283-294, 337-341, 345-353, 355-362
Deflection of beams, 163
Delta function, 11
Delay differential equation, 143
Density function, 61
Desoer, 305

Difference equation, 168
Differential equation for
 Airy function, 376
 Associated Laguerre function, 384
 Associated Legendre function, 379
 Bessel function, 372
 Chebyshev polynomial, 382, 383
 Gegenbauer function, 380
 Hermite function, 385
 Hermite-Weber function, 386
 Jacobi function, 380
 Laguerre function, 384
 Legendre function, 377
 modified Bessel function, 375
Differential-difference equation, 168
Diffusion equation, 34, 54, 66, 71, 75, 81, 148, 150-153, 155, 179, 200, 207, 272-273, 281, 324
Dimovski, 355, 362
Dirac, 2, 11
Dirac delta function, 11
Dirichlet conditions, 5, 6
Dirichlet function, 170
Dirichlet L-function, 232
Dirichlet problem, 32, 57, 65, 272, 333
Dirichlet series, 234
Dispersion relation, 39, 204, 244, 246
Distribution, 11
Distribution function, 60
Ditkin, 387
Doetsch, 3, 265, 387
Dolph, 325, 333
Domain of dependence, 37
Double Fourier cosine transform, 277
Double Fourier sine transform, 277
Double Fourier transform, 65, 70
Dual integral equation, 433
Duality, 14
Duhamel's integrals, 121
Dunders, 248
Dunn, 283

Duplication formula for gamma
   function, 369

E

Ekman layer, 189
Elasticity, 47
   Airy stress function, 48
   Biharmonic equation, 47, 202
   compatibility condition, 47
   equation of equilibrium, 48
   equation of motion, 48
   strain components, 47
   stress components, 47
Electric network, 141
Erdélyi, 113, 260, 343, 347,
   349, 357, 362, 387
Eringen, 337
Error function, 55, 86, 370
   graph of, 371
   properties of, 370
Euler constant, 368
Even periodic extension, 268
Existence conditions, 87
Expected value, 61-62
Exponential integral, 119
Exponential order, 87

F

Feldheim, 356
Feynman, 3
Fibonacci sequence, 309
Final value theorems of
   Laplace transform, 110-111
   Z transform, 305
Finite Fourier sine and cosine
   transforms, 265-266
   applications, 272-277, 278-280
   examples, 266-267
   inverses, 265-266
   multiple, 277
   properties of, 267-272
   tables of, 413-416
Finite Hankel transform, 317, 323
   applications, 319-323
   examples, 318
   inverse, 318
   properties of, 319
   table of, 422
Finite Hilbert transform, 248
Finite Laplace transform, 283
   applications, 290-293
   examples, 286-287
   inverse, 283
   properties of, 288-290
   table of, 417-419

Finite transforms
   Fourier cosine, 266
   Fourier sine , 265
   Gegenbauer, 341
   Hankel, 317
   Hilbert, 248
   Jacobi, 337
   Laplace, 283
Fokker-Planck equation, 76
Forced oscillations, 129
Foster, 387
Fourier, 1
Fourier-Bessel series, 317
Fourier cosine and sine transforms, 50
   applications, 54-57
   convolution, 52-54
   examples, 50-52
   inverses, 50
   Parseval's relations, 53-54
   properties of, 52-54
   tables of, 391-394
Fourier integral formula, 6
   cosine, 7
   sine, 7
Fourier integral theorem, 7
Fourier series, 6
Fourier solution, 67
Fourier transform, 8
   applications, 25-29, 29-32, 32-50,
      60-72,
   convolution, 16-18
   examples, 9-14
   inverse, 8
   multiple, 64
   properties of, 14-25
   table of, 387-390

Fractional derivatives, 116, 148, 149, 162
Fractional differential equations, 145
Fractional integral, 113, 163
Fredholm integral equation, 29
Free oscillations, 129
Fresnel integrals, 371
    graphs of, 372

G

Gakhov, 241, 248
Gamma distribution, 80
Gamma function, 85, 367
    graph of, 369
    properties of, 85, 369
Gate function, 10, 19
Gautesen, 248
Gegenbauer polynomials, 341
Gegenbauer transform, 342
    applications, 344
    convolution theorem of, 342-344
    inverse, 342
Generalized convolution, 361
Generalized Couette flow, 277
Generalized finite Mellin transform, 231
Generalized function, 11
Generalized heat equation, 340, 344
Generalized Mellin transform, 229
    convolution of, 230
    inverse, 229
Generalized Stieltjes transform, 259
    properties of, 260-261
Generating function of
    associated Legendre function, 378
    Bessel function, 373
    Chebyshev polynomials, 381, 382
    Gegenbauer polynomials, 380
    Hermite polynomials, 385
    Jacobi polynomials, 379
    Laguerre functions, 384
    Legendre functions, 377
Gradshteyn, 223
Green's function, 26, 35, 69
Group velocity, 39, 244
Grum, 227

H

Hankel integral formula, 194
Hankel transform, 4, 194
    applications of, 198-205
    examples, 195
    inverse, 194
    Parseval's relation for, 196
    properties of, 195-198
    table of, 400-402
Harmonic oscillator, 129-135,
Heat conduction equation, 148, 155, 319, 322
Heaviside, 1
Heaviside distortionless cable, 154
Heaviside expansion theorem, 105-106
Heaviside shifting theorem, 88
Heaviside unit step function, 10, 327
Helmholtz equation, 70, 77, 201
Hermite polynomials, 355, 385
Hermite transform, 355
    convolution of, 361-365
    examples, 356-358
    inverse, 355
    properties of, 358-365
Hermite-Weber function, 386
    graph of, 386
Hilbert transform, 238-242
    applications, 243-248
    examples, 238-239
    inverse, 238
    properties of, 239-241
    table of, 406-408
Howell, 346, 348
Hulme, 263
Hypergeometric function, 346, 402

J

Jacobi polynomials, 337, 379
Jacobi theta function, 24
Jacobi transform, 337
    applications of, 340-341

examples, 338-339
inverse of, 338
properties of, 339-340
Jones, 9
Joseph, 247

I

Impulse function, 25, 295
Impulse response function, 143
Impulse train, 296
Incomplete gamma function, 369
Indicial equation, 146
Initial value problems, 81, 124, 198, 206
Initial value theorems for
    Laplace transform, 109
    Z transform, 304
Inner product, 17, 18
Input function, 26
Input signal, 143, 295
Integral equations, 29-32, 159, 221-222, 257
    convolution, 29-32
    dual, 433
    Fredholm, 29
    singular, 248
    Volterra, 159
Integral transform, 3, 4
Inverse operator, 4
Isothermal curves, 33

K

Kalla, 355, 362
Kelvin ideal cable, 153
Kernel, 4, 84
Kinetic energy, 39
Klein-Gordon equation, 74, 175
Kneitz, 265
Kober, 241
Korteweg-de Vries (KdV) equation, 43, 246-247
Koschmieder, 265
Kubota, 247

L

Lagrange, 1
Laguerre polynomials, 120, 346, 383
Laguerre transform, 345
    applications, 352-353
    convolution of, 351-352
    examples, 346-347
    inverse, 345
    properties of, 348-352
Lakshmanarao, 337
Lamb, 77
Laplace, 1
Laplace distribution, 80
Laplace equation, 32-34, 56-57, 65, 199, 208-210, 219, 333
Laplace transform, 4, 83, 320
    applications, 124-184
    bilateral, 232
    Convolution Theorem, 92
    examples, 84-87
    existence theorem for, 87
    inverse, 84, 98-109
    of periodic functions, 89-90
    properties of, 88-91
    table of, 395-399
Legendre polynomials, 377
    graphs of, 378
Legendre transforms, 325
    applications, 333-334
    convolution of, 332
    definition of, 325
    examples, 326-328
    inverse, 325
    properties of, 328-333
Leibnitz, 1
Lévy-Cramér theorem, 64
Lighthill, 9
Linear operator, 4
Linear system, 295
Lommel integrals, 374
Lukacs, 64

M

Macfarlane, 227
Marichev, 387

Maslowe, 247
McCully, 345, 351
Mellin, 211
Mellin transform, 4, 211-212
   applications, 218-222, 227-229
   Convolution Theorem, 217
   examples, 212-214
   generalized, 229-232
   inverse, 211-212
   properties of, 214-218
   table of, 403-405
Mikusinski, 240
Miller, 146
Modified Bessel functions, 74, 121, 235, 375
Mohanti, 205
Moment problem, 257
Moments, 62, 293
Multiple Fourier transform, 64
   inverse, 65
Muskhelishvili, 241, 248

N

Natural frequency, 129, 139
Navier-Stokes equations, 44
Naylor, 229, 232
Neumann condition, 208
Neumann problem, 33, 208, 273
Neumann solution, 372
Newcomb, 241
Nonlinear internal waves, 243
Norm, 18, 240

O

Oberhettinger, 387
Odd periodic extension, 268
Okikiolu, 241
One-sided Hilbert transform, 248
   asymptotic expansion of, 249
Ono, 244
Orthogonal relations for
   associated Laguerre function, 385
   associated Legendre function, 379
   Bessel functions, 374
   Chebyshev polynomials, 382
   Gegenbauer polynomials, 341, 381
   Hermite polynomials, 355, 385
   Jacobi polynomials, 380
   Laguerre polynomials, 345
   Legendre polynomials, 325, 378
Output function, 26, 143, 295
Output signal, 143, 295

P

Parabolic cylinder function, 113
Parseval's relation for
   Fourier cosine transform, 53
   Fourier sine transform, 54
   Fourier transform, 17, 18
   Hankel transform, 196
   Hilbert transform, 241
   Mellin transform, 218
   Stieltjes transform, 252
Partial fractions, 99
Pennline, 248
Periodic functions, 89-90, 268
Periodic solution, 310
Peters, 248
Poisson, 1
Poisson equation, 69
Poisson integral formula, 32, 35, 334
Poisson's summation formula, 23, 24
Power spectrum, 18
Principle of superposition, 35, 70, 295
Probability density function, 61
Proudman, 174
Prudnikov, 387

R

Rayleigh, 2
Rayleigh layer, 158
Rayleigh problem, 157-158
Rectified sine wave, 90
Recurrence relations for
   associated Legendre functions, 378
   Bessel functions, 373
   Chebyshev polynomials, 382

Gegenbauer polynomials, 380
Hermite polynomials, 385
Jacobi polynomials, 380
Laguerre functions, 384
Legendre functions, 377
Redekopp, 247
Reif, 76
Renewal equation, 144
Renewal function, 145
Renewal process, 144
Residue theorem, 103
Resonance, 131
Resonant frequency, 131, 135
Reynolds number, 45
Riemann, 211
Riemann-Lebesgue lemma, 15
Riemann-Liouville fractional
    integral, 113
Riemann zeta function, 184, 212, 227
Rodrigues formula for
    associated Laguerre
      function, 384
    associated Legendre
      function, 378
    Chebyshev polynomials, 382, 383
    Gegenbauer polynomials, 380
    Hermite functions, 385
    Jacobi polynomials, 379
    Laguerre function, 383
    Legendre function, 377
Roettinger, 265
Ross, 146
Rossby wave problem, 82
Ryzhik, 223

S

Sampled function, 296, 298
Sampling integral representation, 20
Sansone, 332
Scaling property of
    finite Laplace transform, 288
    Fourier transform, 14
    Hankel transform, 195

Mellin transform, 214
Schrödinger equation, 38-39
Scott, 337
Self-reciprocal function, 9
Sen, 210
Sequence of ordinary functions, 12
Shallow water equations, 41
Shifting property of
    finite Laplace transform, 288
    Fourier transform, 14
    Laplace transform, 88
    Mellin transform, 214
Shohat, 257
Shreeves, 29
Signal, 19, 143, 242, 295
Signal function, 8, 21, 295
Signum function, 13
Simple electric circuit, 26, 137, 292
Simple harmonic oscillator, 129-136
Skin friction, 159
Slowing down of neutrons, 40
Sneddon, 205, 221, 317
Soliton solution, 246
Sommerfeld radiation condition, 43
Sound waves, 42
Special functions, 367-386
    Airy functions, 376
    associated Laguerre functions, 384
    associated Legendre functions, 378
    Bessel function, 87, 317, 372
    Beta function, 95, 370
    Chebyshev polnomials, 311, 381
    complementary error function, 56, 370
    Dirac delta function, 11
    Error function, 55, 86, 370
    Gamma function, 85, 367
    Gegenbauer polynomials, 341, 380
    Heaviside function, 10, 88, 327
    Hermite polynomials, 355, 385
    Hermite-Weber functions, 385
    Jacobi polynomials, 337, 379
    Laguerre polynomials, 345, 383
    Legendre polynomials, 326, 377
    modified Bessel function, 375
Square wave function, 89

Stadler, 29
Staircase function, 118
Stationary phase approximation, 182, 204
Steady state solution, 124, 158
Stieltjes transform, 250-251
    applications, 257-259
    examples, 251-252
    inverse, 251, 254-256
    inversion theorem, 251, 254-257
    properties of, 252-254
    table of, 409-412
Stokes problem, 157
Stokes-Ekman problem, 189
Stream function, 45

T

Tamarkin, 257
Tauberian theorems, 109, 294
Tautochroneous motion, 161
Telegraph equation, 74, 153
Titchmarsh, 241
Transfer function, 20, 143
Transform
    bilateral Laplace transform, 232
    finite Fourier cosine, 266
    finite Fourier sine, 265
    finite Hankel, 317
    finite Laplace, 283
    Fourier, 8
    Fourier cosine, 50
    Fourier sine, 50
    Gegenbauer, 341
    Hankel, 4, 194
    Hermite, 355
    Hilbert, 237, 248
    Laguerre, 345
    Laplace, 4, 83
    Legendre, 325
    Mellin, 4, 211
    Stieltjes, 250-251
    Weyl, 222
    Z, 298
Transient solution, 124
Translation property of
    Fourier transform, 14

Laplace transform, 88
Mellin transform, 214
Z transform, 301
Tranter, 221, 335-336
Triangular wave function, 120
Tricomi, 248
Tuttle, 242

U

Uniqueness theorem, 4
Unitary, 18
Unsteady Couette flow, 276-277
Unsteady viscous flow, 320, 324
Ursell, 113, 249-250, 263

V

Velocity potential, 180
Vibration problems
    forced, 281
    free, 198, 275, 278, 321
Volterra integral equation, 159
Vorticity transport equation, 45
Vorticity vector, 45

W

Water wave equations, 41, 174, 180, 203, 207
Watson, 113, 194
Watson's lemma, 112
Wave equation, 36, 40, 42, 67, 152, 157, 173, 198
Weber's solution, 372
Weyl fractional derivative, 224
Weyl fractional integral, 222
Weyl transform, 222
Wheelon, 183
Whitham, 247
Whitham equation, 247
Whittaker, 3
Widder, 251, 255-256
Williams, 248
Wong, 113, 235, 262
Wyman, 113

Z

Z transforms, 298
   applications, 308-313
   bilateral, 299
   convolution, 303-304
   examples, 299-301
   inverse, 299, 306
   properties of, 301-306
   table of, 420-421
Zadeh, 305